METHODS IN MOLECULAR BIOLOGY

Series Editor
John M. Walker
School of Life and Medical Sciences
University of Hertfordshire
Hatfield, Hertfordshire, AL10 9AB, UK

For further volumes:
http://www.springer.com/series/7651

In Vitro Mutagenesis

Methods and Protocols

Edited by

Andrew Reeves

Coskata, Inc., Warrenville, IL, USA

Editor
Andrew Reeves
Coskata, Inc.
Warrenville, IL, USA

ISSN 1064-3745   ISSN 1940-6029   (electronic)
Methods in Molecular Biology
ISBN 978-1-4939-8211-0   ISBN 978-1-4939-6472-7   (eBook)
DOI 10.1007/978-1-4939-6472-7

This Humana Press imprint is published by Springer Nature
The registered company is Springer Science+Business Media LLC
The registered company address is: 233 Spring Street, New York, NY 10013, U.S.A.

Preface

In vitro mutagenesis remains an essential tool for molecular biologists, biochemists, and metabolic engineers in performing both basic and applied research into gene and protein function. It is a cross-disciplinary research and development approach that has broad applications in many important fields. With the advent of ever more sophisticated investigative and analytical methods such as next-generation sequencing, emerging, powerful gene-editing techniques, and comprehensive data analysis software programs to process large data sets, the molecular biology field is witnessing a magnitude change in the understanding of gene and protein structural and functional relationships at the organismal level. With these powerful tools at researchers' disposal, it is now conceivable to design an in vitro mutagenesis strategy in a genetically manipulable organism, carry out the mutagenesis experiments in an efficient manner, generate, retrieve, and analyze the sequence data and impacted proteins and phenotypes in weeks or months instead of years, all with far greater insight into the understanding of the cellular and systems biology of the respective organism(s) under study. The molecular biologist's "toolkit" is in a sense now complete for designing, performing, and analyzing all manner of in vitro mutagenesis experiments with broad applications in basic research and for any conceivable commercial purpose—be it drug design, microbial cell factories development, gene therapy, or generation of model organisms, among many other applications.

In this volume on *In Vitro Mutagenesis: Methods and Protocols*, an important aim was to provide the beginner practitioner in the field and the more experienced molecular biologists, biochemists, and metabolic engineers alike a wide variety of updated, novel approaches to many powerful classical methods of performing in vitro mutagenesis such as transposon (Tn) mutagenesis, site-directed and random mutagenesis. Additionally, an important emphasis was placed on emerging, yet powerful, gene- and genome-editing and bioinformatics methods now being developed and implemented into experimental reality.

The volume is divided into seven sections: The first two sections describe detailed, novel methods for gene and genome editing of a broad cross section of the living world (e.g., mammalian, plant, viral, bacterial, and protistan systems using CRISPR/Cas9, TALEN, and Group II intron technologies). These editing technologies are enabling the facile genetic manipulation of a wide range of cellular systems in the study of fundamental metabolic processes. Importantly, they are rigorously being employed for the prospect and promise of attaining major scientific breakthroughs in human physiology and medicine. The next two sections describe (1) a variety of practical bioinformatics approaches for identifying mutagenesis targets in silico for the rational design of mutagenesis experiments and (2) a set of diverse, detailed in vitro Tn mutagenesis protocols for model microorganisms as well as for use in alternative, previously recalcitrant organisms (e.g., archaea) including genomic sequencing methods to rapidly and completely identify all the Tn insertion sites in a mutant library. The last three sections cover a wide variety of novel site-directed and random mutagenesis approaches aimed at gaining a better understanding of protein-protein and protein-cofactor interactions at the structural and functional levels along with their concomitant effects on an organism's cellular metabolism. Provided in these sections are specialized mutagenic PCR methods (e.g., is-epPCR, epPCR using heavy water, and single primer PCR, et al.) and mutagenesis and cloning methods (e.g., MUPAC, SliP-SliCE, et al.). Several of these chapters also describe state-of-the-art bioanalytical techniques and methods used by experts in the field to allow a more thorough understanding of how the specific mutation impacted the experimental outcome. The rationale being that the beginner practitioner would be able to view the experimental process from conception and design to completion through follow-up data analysis.

In keeping with the theme of the Methods in Molecular Biology series, each chapter contains an extensive Notes section which provides and elaborates on specific experimental details, tips, and tricks and thus it is hoped will allow a more rapid, successful implementation of the method by practitioners in the field whatever their experience level.

I would like to thank the authors, who are all well-known experts in their respective fields, for their contributions and for allowing me to put together what I believe is a timely, practical, and comprehensive manual on in vitro mutagenesis methods and protocols that will be embraced by a broad range of practitioners in the fields of molecular biology, biochemistry, biochemical and metabolic engineering, biophysics, among other disciplines. Moreover, this volume on in vitro mutagenesis was specifically compiled with an emphasis on providing a highly accessible manual for current and future researchers—from the beginner practitioner to the advanced investigator—who routinely perform in vitro mutagenesis experiments on all types of cells at research institutes, academia, and industrial and government laboratories.

Lastly, I would like to thank all my past and current colleagues who have inspired me to keep learning and to pursue knowledge through hard work, persistence, and investigative scientific endeavors.

Warrenville, IL, USA   *Andrew Reeves*

Contents

Contributors

TOMÁS APARICIO • *Systems and Synthetic Biology Program, Centro Nacional de Biotecnología (CNB-CSIC), Madrid, Spain*

JOSHUA T. ATKINSON • *Systems, Synthetic, and Physical Biology Graduate Program, Rice University, Houston, TX, USA*

GANG BAO • *Department of Bioengineering, Rice University, Houston, TX, USA*

KAUSHIK BISWAS • *Division of Molecular Medicine, Bose Institute, Kolkata, India*

KEVIN BROWN • *Department of Molecular Microbiology, Washington University School of Medicine, St. Louis, MO, USA*

WILLIAM H. CERNOTA • *Fermalogic, Inc., Chicago, IL, USA*

KIAN PIAW CHAI • *Faculty of Biosciences and Medical Engineering, Universiti Teknologi Malaysia, Johor, Malaysia*

CHI-PING CHAN • *School of Biomedical Sciences, The University of Hong Kong, Pokfulam, Hong Kong*

VICTOR L. DAVIDSON • *Burnett School of Biomedical Sciences, College of Medicine, University of Central Florida, Orlando, FL, USA*

TIMOTHY H. DAVIS • *Department of Bioengineering, Rice University, Houston, TX, USA*

HARSHAHARDHAN DESHMUKH • *Department of Bioengineering, Rice University, Houston, TX, USA*

BRIAN A. DOW • *Burnett School of Biomedical Sciences, College of Medicine, University of Central Florida, Orlando, FL, USA*

MARTE SINGSÅS DRAGSET • *Centre of Molecular Inflammation Research, Department of Cancer Research and Molecular Medicine, Norwegian University of Science and Technology, Trondheim, Norway*

HONG FENG • *The Key Laboratory for Bio-resources and Eco-Environment of Ministry of Education, The Sichuan Key Laboratory of Molecular Biology and Biotechnology, College of Life Sciences, Sichuan University, Chengdu, Sichuan, People's Republic of China*

DEREK J. FISHER • *Department of Microbiology, Southern Illinois University, Carbondale, IL, USA*

KIAN MAU GOH • *Faculty of Biosciences and Medical Engineering, Universiti Teknologi Malaysia, Johor, Malaysia*

TUHIN KUMAR GUHA • *Department of Microbiology, 401 University of Manitoba, Winnipeg, MB, Canada*

TOSHIHARU HASE • *Institute for Protein Research, Osaka University, Osaka, Japan*

NURIT HASPEL • *Department of Computer Science, University of Massachusetts Boston, Boston, MA, USA*

GEORG HAUSNER • *Department of Microbiology, University of Manitoba, Winnipeg, MB, Canada*

ZHENYU HE • *Center for Stem Cell Biology and Regenerative Medicine, Department of Basic Medical Sciences, School of Medicine, Tsinghua University, Beijing, China*

YANCHAO HUANG • *Marine Science & Technology Institute, Department of Environmental Science and Engineering, Yangzhou University, Yangzhou, Jiangsu, China*

DANWEI HUANGFU • *Developmental Biology Program, Sloan Kettering Institute, Memorial Sloan Kettering Cancer Center, New York, NY, USA*

ROSLI MD ILLIAS • *Faculty of Chemical and Energy Engineering, Universiti Teknologi Malaysia, Johor, Malaysia*

FILIP JAGODZINSKI • *Department of Computer Science, Western Washington University, Bellingham, WA, USA*

XUEQIU JIAN • *Center for Human Genetics, Institute of Molecular Medicine, The University of Texas Health Science Center at Houston, Houston, TX, USA*

DONG-YAN JIN • *School of Biomedical Sciences, The University of Hong Kong, Pokfulam, Hong Kong*

ALICIA M. JONES • *Biosciences Department, Rice University, Houston, TX, USA*

KEHKOOI KEE • *Center for Stem Cell Biology and Regenerative Medicine, Department of Basic Medical Sciences, School of Medicine, Tsinghua University, Beijing, China*

CHARLOTTE E. KEY • *Department of Microbiology, Southern Illinois University, Carbondale, IL, USA*

SAIJA KILJUNEN • *Helsinki University Central Hospital Laboratory Diagnostics, Helsinki, Finland; Department of Bacteriology and Immunology, Immunobiology, Research Programs Unit, University of Helsinki, Helsinki, Finland*

JU YAEN KIM • *Institute for Protein Research, Osaka University, Osaka, Japan*

MISAKI KINOSHITA • *Institute for Protein Research, Osaka University, Osaka, Japan*

KIN-HANG KOK • *Department of Microbiology, The University of Hong Kong, Pokfulam, Hong Kong*

YILIN LE • *Biofuels Institute, School of Environment, Jiangsu University, Zhenjiang, Jiangsu, People's Republic of China*

CIARAN M. LEE • *Department of Bioengineering, Rice University, Houston, TX, USA*

YOUNG-HO LEE • *Institute for Protein Research, Osaka University, Osaka, Japan*

JINSONG LI • *Group of Epigenetic Reprogramming, State Key Laboratory of Cell Biology, CAS Center for Excellence in Molecular Cell Science, Institute of Biochemistry and Cell Biology, Shanghai Institutes for Biological Sciences, Chinese Academy of Sciences, Shanghai, China; Shanghai Key Laboratory of Molecular Andrology, Institute of Biochemistry and Cell Biology, Shanghai Institutes for Biological Sciences, Chinese Academy of Sciences, Shanghai, China*

KOK JUN LIEW • *Faculty of Biosciences and Medical Engineering, Universiti Teknologi Malaysia, Johor, Malaysia*

YUXI LIN • *Institute for Protein Research, Osaka University, Osaka, Japan*

XIAOMING LIU • *Human Genetics Center, School of Public Health, The University of Texas Health Science Center at Houston, Houston, TX, USA*

SHAOJUN LONG • *Department of Molecular Microbiology, Washington University School of Medicine, St. Louis, MO, USA*

VÍCTOR DE LORENZO • *Systems and Synthetic Biology Program, Centro Nacional de Biotecnología (CNB-CSIC), Madrid, Spain*

KESEN MA • *Biofuels Institute, School of Environment, Jiangsu University, Zhenjiang, Jiangsu, People's Republic of China*

BARUN MAHATA • *Division of Molecular Medicine, Bose Institute, Kolkata, India*

GAURAV MAJUMDAR • *MRC/NHLS/UCT Molecular Mycobacteriology Research Unit, DST/NRF Centre of Excellence for Biomedical TB Research, Department of Pathology, Faculty of Health Sciences, University of Cape Town, Cape Town, South Africa*

Lenka Malinovská • *Central European Institute of Technology, Masaryk University, Brno, Czech Republic*

Natalia Markova • *Institute for Protein Research, Osaka University, Osaka, Japan*

Esteban Martínez-García • *Systems and Synthetic Biology Program, Centro Nacional de Biotecnología (CNB-CSIC), Madrid, Spain*

Rendani Mbau • *MRC/NHLS/UCT Molecular Mycobacteriology Research Unit, DST/NRF Centre of Excellence for Biomedical TB Research, Department of Pathology, Faculty of Health Sciences, University of Cape Town, Cape Town, South Africa*

D. Scott Merrell • *Department of Microbiology and Immunology, Uniformed Services University of the Health Sciences, Bethesda, MD, USA*

Troy C. Messina • *Physics Program, Berea College, Berea, KY, USA*

Toshifumi Minamoto • *Graduate School of Human Development and Environment, Kobe University, Kobe, Japan*

Ken Motohashi • *Department of Bioresource and Environmental Sciences, Faculty of Life Sciences, Kyoto Sangyo University, Kyoto, Japan*

Jana Mrázková • *Department of Biochemistry, Faculty of Science, Masaryk University, Brno, Czech Republic; National Centre for Biomolecular Research, Faculty of Science, Masaryk University, Brno, Czech Republic*

Raju Mukherjee • *MRC/NHLS/UCT Molecular Mycobacteriology Research Unit, DST/NRF Centre of Excellence for Biomedical TB Research, Department of Pathology, Faculty of Health Sciences, University of Cape Town, Cape Town, South Africa*

Hiroshi Murakami • *Department of Applied Chemistry, Graduate School of Engineering, Nagoya University, Nagoya, Japan*

Pablo I. Nikel • *Systems and Synthetic Biology Program, Centro Nacional de Biotecnología (CNB-CSIC), Madrid, Spain*

Egon A. Ozer • *Department of Medicine, Northwestern University Feinberg School of Medicine, Chicago, IL, USA*

Maria I. Pajunen • *Department of Bacteriology and Immunology, Immunobiology, Research Programs Unit, University of Helsinki, Helsinki, Finland*

Andrew Reeves • *Coskata, Inc., Warrenville, IL, USA*

Takuya Sakamoto • *Department of Mathematical and Life Sciences, Graduate School of Science, Hiroshima University, Higashi-Hiroshima, Hiroshima, Japan*

Tetsushi Sakuma • *Department of Mathematical and Life Sciences, Graduate School of Science, Hiroshima University, Higashi-Hiroshima, Hiroshima, Japan*

Harri Savilahti • *Division of Genetics and Physiology, Department of Biology, University of Turku, Turku, Finland*

Esha Sehanobish • *Burnett School of Biomedical Sciences, College of Medicine, University of Central Florida, Orlando, FL, USA*

Chong Sha • *Biofuels Institute, School of Environment, Jiangsu University, Zhenjiang, Jiangsu, People's Republic of China*

Weilan Shao • *Biofuels Institute, School of Environment, Jiangsu University, Zhenjiang, Jiangsu, People's Republic of China*

Bang Shen • *State Key Laboratory of Agricultural Microbiology, College of Veterinary Medicine, Huazhong Agricultural University, Wuhan, Hubei, People's Republic of China*

L. David Sibley • *Department of Molecular Microbiology, Washington University School of Medicine, St. Louis, MO, USA*

Jonathan J. Silberg • *Biosciences Department, Rice University, Houston, TX, USA*

VINAYAK SINGH • *MRC/NHLS/UCT Molecular Mycobacteriology Research Unit, DST/ NRF Centre of Excellence for Biomedical TB Research, Department of Pathology, Faculty of Health Sciences, University of Cape Town, Cape Town, South Africa*

CHEW-LI SOH • *Developmental Biology Program, Sloan Kettering Institute, Memorial Sloan Kettering Cancer Center, New York, NY, USA*

NAOHIRO TANIGUCHI • *PeptiDream Inc., Meguro-ku, Tokyo, Japan*

HAI-YAN WANG • *The Key Laboratory for Bio-resources and Eco-Environment of Ministry of Education, The Sichuan Key Laboratory of Molecular Biology and Biotechnology, College of Life Sciences, Sichuan University, Chengdu, Sichuan, People's Republic of China*

HONGCHENG WANG • *Biofuels Institute, School of Environment, Jiangsu University, Zhenjiang, Jiangsu, People's Republic of China*

NENGDING WANG • *Department of Microbiology-Immunology, Northwestern University Feinberg School of Medicine, Chicago, IL, USA*

DIGBY F. WARNER • *MRC/NHLS/UCT Molecular Mycobacteriology Research Unit, DST/ NRF Centre of Excellence for Biomedical TB Research, Department of Pathology, Faculty of Health Sciences, University of Cape Town, Cape Town, South Africa; Institute of Infectious Disease and Molecular Medicine, University of Cape Town, Cape Town, South Africa*

J. MARK WEBER • *Fermalogic, Inc., Chicago, IL, USA*

PENGCHENG WEI • *Key Laboratory of Rice Genetics Breeding of Anhui Province, Rice Research Institute, Anhui Academy of Agricultural Sciences, Hefei, China*

ROY K. WESLEY • *Fermalogic, Inc., Chicago, IL, USA*

PHILIP D. WEYMAN • *J. Craig Venter Institute, Synthetic Biology and Bioenergy Group, La Jolla, CA, USA*

JEANNETTE M. WHITMIRE • *Department of Microbiology and Immunology, Uniformed Services University of the Health Sciences, Bethesda, MD, USA*

MICHAELA WIMMEROVÁ • *Department of Biochemistry, Faculty of Science, Masaryk University, Brno, Czech Republic; National Centre for Biomolecular Research, Faculty of Science, Masaryk University, Brno, Czech Republic; Central European Institute of Technology, Masaryk University, Brno, Czech Republic*

YONGZHEN XIA • *State Key Laboratory of Microbial Technology, Shandong University, Jinan, People's Republic of China*

RONGFANG XU • *Key Laboratory of Rice Genetics Breeding of Anhui Province, Rice Research Institute, Anhui Academy of Agricultural Sciences, Hefei, China*

LUYING XUN • *School of Molecular Biosciences, Washington State University, Pullman, WA, USA*

TAKASHI YAMAMOTO • *Department of Mathematical and Life Sciences, Graduate School of Science, Hiroshima University, Higashi-Hiroshima, Hiroshima, Japan*

JIANBO YANG • *Key Laboratory of Rice Genetics Breeding of Anhui Province, Rice Research Institute, Anhui Academy of Agricultural Sciences, Hefei, China*

ISAAC T. YONEMOTO • *J. Craig Venter Institute, Synthetic Biology and Bioenergy Group, La Jolla, CA, USA*

KIT-SAN YUEN • *School of Biomedical Sciences, The University of Hong Kong, Pokfulam, Hong Kong*

LIKUI ZHANG • *Marine Science & Technology Institute, Department of Environmental Science and Engineering, Yangzhou University, Yangzhou, Jiangsu, People's Republic of China*

HONG-YAN ZHAO • *The Key Laboratory for Bio-resources and Eco-Environment of Ministry of Education, The Sichuan Key Laboratory of Molecular Biology and Biotechnology, College of Life Sciences, Sichuan University, Chengdu, Sichuan, People's Republic of China*

CUIQING ZHONG • *Group of Epigenetic Reprogramming, State Key Laboratory of Cell Biology, CAS Center for Excellence in Molecular Cell Science, Institute of Biochemistry and Cell Biology, Shanghai Institutes for Biological Sciences, Chinese Academy of Sciences, Shanghai, China; Shanghai Key Laboratory of Molecular Andrology, Institute of Biochemistry and Cell Biology, Shanghai Institutes for Biological Sciences, Chinese Academy of Sciences, Shanghai, China*

HAIBAO ZHU • *Department of Bioengineering, Rice University, Houston, TX, USA*

Section I

Gene and Genome-Editing Methods Part I

Chapter 1

Design and Validation of CRISPR/Cas9 Systems for Targeted Gene Modification in Induced Pluripotent Stem Cells

Ciaran M. Lee, Haibao Zhu, Timothy H. Davis, Harshahardhan Deshmukh, and Gang Bao

Abstract

The CRISPR/Cas9 system is a powerful tool for precision genome editing. The ability to accurately modify genomic DNA in situ with single nucleotide precision opens up new possibilities for not only basic research but also biotechnology applications and clinical translation. In this chapter, we outline the procedures for design, screening, and validation of CRISPR/Cas9 systems for targeted modification of coding sequences in the human genome and how to perform genome editing in induced pluripotent stem cells with high efficiency and specificity.

Key words CRISPR, Genome editing, Targeted gene knockout

1 Introduction

With the recent development of engineered nucleases such as zinc-finger nucleases (ZFNs) [1, 2], transcription activator-like (Tal) effector nucleases (TALENs) [3, 4], and clustered regularly interspaced short palindromic repeats (CRISPRs) and CRISPR-associated (Cas) proteins [5–7], we now have powerful tools for the precise induction of DNA double-stranded breaks (DSBs) in desired locations, resulting in efficient gene modification in cells and animals at preselected genomic loci. This allows targeted gene insertion or correction through the homology-directed repair (HDR) pathway with a designer DNA donor template, or gene disruption and deletion when the nuclease-induced DSBs are repaired by the nonhomologous end joining (NHEJ) pathway. In particular, the simplicity of the CRISPR/Cas9 systems has led to widespread adoption of the technology in a diverse range of biological and biomedical applications.

Andrew Reeves (ed.), *In Vitro Mutagenesis: Methods and Protocols*, Methods in Molecular Biology, vol. 1498, DOI 10.1007/978-1-4939-6472-7_1, © Springer Science+Business Media New York 2017

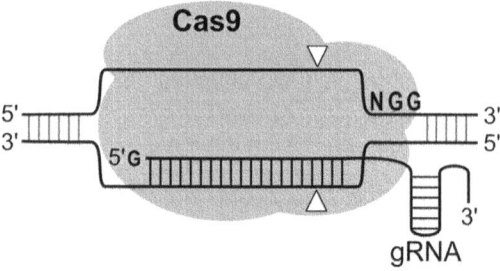

Fig. 1 A schematic of the *Spy* CRISPR/Cas9 system. The Cas9 protein recognizes the short PAM sequence NGG. The gRNA sequence corresponds to the 20 bases immediately upstream of the PAM sequence and binds to the complementary sequence on the opposite strand. A blunt-ended DNA double-stranded break is induced between the third and fourth base proximal to the PAM site (*white triangles*)

The *Streptococcus pyogenes (Spy)* CRISPR/Cas9 system targets a short stretch of DNA via both a short guide RNA (gRNA) and the Cas9 protein which recognizes the protospacer adjacent motif (PAM) NGG (*see* Fig. 1). Although there are other engineered nucleases and Cas9 orthologs, in this chapter only the *Spy* CRISPR/Cas9 systems are considered. When designing gRNAs for modifying a particular gene locus, one needs to identify some or all 20-bp sequences adjacent to the short PAM NGG, and choose those most potent for the desired gene modification, whether a gene disruption, knockout, or knockin. However, there are some caveats such as avoiding poly T stretches which can stall the PolIII transcription of gRNAs. Although there are several in silico prediction tools for gRNA design, experimental studies of specific gRNAs show no correlation between the predicted and observed efficiency of DNA cleavage (*see* Fig. 2a). Current prediction models focus on the sequence composition of the gRNA, however there may be other influencing factors such as chromatin structure or DNA accessibility. Since currently available algorithms lack predictive capability, several candidate gRNAs per target locus should be tested. Furthermore, it has been demonstrated that the CRISPR/Cas9 system can have high levels of off-target DNA cutting which can lead to DNA damage at unintended genomic loci [8–10]. Recent experimental studies have shown that in silico off-target search tools have poor predictive power [11, 12] although they represent a quick, simple method to prescreen candidate gRNAs for potential off-target sites. Here, were present a high-throughput method for design, construction, and testing of gRNAs to facilitate this process. This screening method identifies gRNAs with high levels of gene targeting for use in primary cell types such as induced pluripotent stem cells (iPSCs) and hematopoietic stem and progenitor cells (HSPCs). Recently, Cas9 orthologs from different

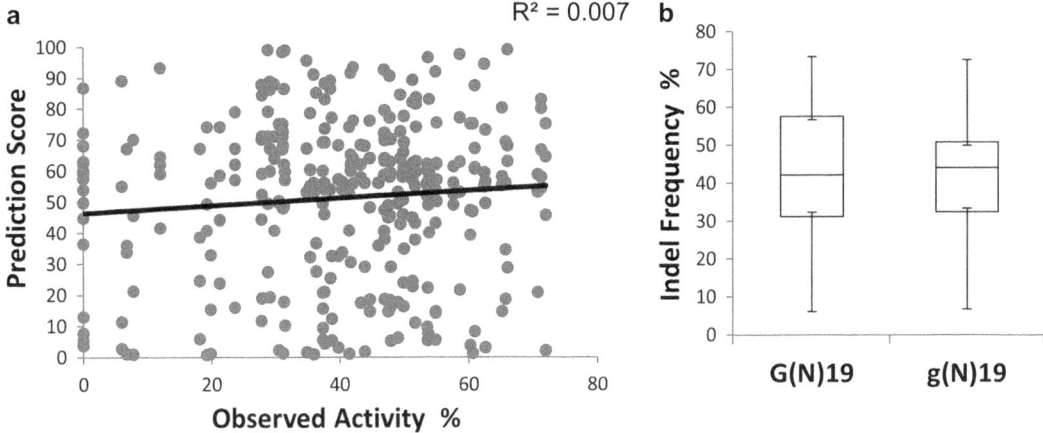

Fig. 2 Activity of *Spy* gRNAs. (**a**) The observed activity of gRNAs tested in the HEK293T cell line compared to the predicted activity score from in silico analysis. No observable correlation is apparent. Data from 78 gRNAs. (**b**) Activity of gRNAs with a matched G (GN19) and gRNAs with a mismatched G (gN19) at the 5′ end. There is no significant difference or impact when the 5′ base is a mismatch relative to the DNA target site. Data from 17 and 26 gRNAs respectively

species of bacteria with reduced off-target effects have been described [13–15]. Although this method focuses on designing *Spy* CRISPR/Cas9 systems for gene targeting, it can readily be applied to design orthologous CRISPR/Cas9 systems.

2   Materials

Unless otherwise indicated, prepare all solutions using nuclease-free water.

2.1 *Cas9, gRNA Vectors*

1. Template DNA: pX330_U6-Chimeric_BB_CBh-hSpCas9 plasmid (Addgene #42230). Store at –20 °C.

2. Oligonucleotides (10 μM concentration). Store at –20 °C.

3. Thermocycler.

4. Chemically competent *Escherichia coli*. Store at –80 °C.

5. Commercially available DNA ligation kit.

6. Sterile PCR strip tubes (0.2 mL).

7. 1.5 mL microcentrifuge tubes.

8. 6× gel loading dye (*see* **Note 1**). Store at 4 °C.

9. Table-top microcentrifuge.

10. Enzymes: *Bbs*I, alkaline phosphatase, calf intestinal (CIP). Store at –20 °C.

11. 10× T4 ligase reaction buffer. Store aliquoted at –20 °C.

12. T4 polynucleotide kinase (10 U/μL). Store at –20 °C.

13. 37 °C oven.

14. Magnetic 96-well plate.

15. Magnetic beads (*see* **Note 2**). Store at 4 °C.

16. Super optimal broth with Catabolite repression (SOC): 2% w/v tryptone, 0.5% w/v yeast extract, 10 mM NaCl, 2.5 mM KCl, 10 mM, MgCl$_2$, 10 mM MgSO$_4$, 20 mM glucose, pH 7.0.

17. Lysogeny broth (LB): tryptone 10 g, yeast extract 5 g, NaCl 10 g. Dissolve medium components in 1 L distilled H$_2$O. pH to 7.0 with 5 N NaOH. Autoclave at 121 °C, 15 psig, 20 min. Add 100 μg/mL ampicillin for selections.

18. LB + ampicillin plates: To LB liquid medium add agar to 1.5% (15 g/L) and autoclave as described above. Add ampicillin sulfate from a stock solution to 100 μg/mL. Store at 4 °C.

19. 48-deep well plates.

20. 96-deep well plates.

21. Breathable sterile membrane.

22. Buffer P1: 50 mM Tris–HCl, pH 8.0, 10 mM EDTA, 100 μg/mL RNaseA, 43 μg/mL Thymolphthalein. Store at 4 °C.

23. Buffer P2: 200 mM NaOH, 1% SDS.

24. Buffer N3: 4.2 M Guanidine-HCl, 0.9 M Potassium acetate, pH 4.8.

25. Reagent reservoirs (for multichannel pipette).

26. Biomek 3000 (*see* **Note 3**) (optional).

27. Balance.

28. Agarose.

29. 1× TAE buffer: Dilute 1:50 from a 50× TAE buffer stock composed of 242 g Tris-Base (MW = 121.1), 57.1 mL glacial acetic acid, and 100 mL 0.5 M EDTA per liter. Mix Tris base with a stir bar to dissolve in 600 mL of ddH$_2$O. Add the EDTA and acetic acid. Bring the final volume to 1 L with ddH$_2$O. Store at room temperature.

30. Glass Erlenmeyer flask.

31. Ethidium bromide (*see* **Note 4**).

32. Microwave.

33. Agarose electrophoresis apparatus.

34. UV gel box with gel imager.

2.2 In Vitro Transcription

1. GeneArt™ Precision gRNA Synthesis Kit (Thermo Fisher Scientific). Store at –20 °C.

2. Qubit™ RNA BR Assay Kit (Thermo Fisher Scientific). Store at 4 °C.

3. Qubit™ apparatus.

4. Oligonucleotides (dNTPs) (100 μM). Store at −20 °C.

2.3 Cell Culture, Transfection

1. 293T cells (ATCC).

2. 24-well plate.

3. CRISPR/Cas9 plasmids. Store at −20 °C.

4. GFP plasmid. Store at −20 °C.

5. 0.1 % Gelatin.

6. Opti-MEM I Reduced Serum Medium. Store at 4 °C.

7. 0.05 % trypsin–EDTA. Store at −20 °C.

8. 15 mL tubes.

9. Lipofectamine-2000 Transfection reagent (Thermo Fisher Scientific). Store at 4 °C.

10. Laminar flow hood with UV light source.

11. 1× phosphate buffered saline (PBS), pH 7.4, without calcium.

12. 293T cell culture media: Commercially prepared Dulbecco's Modified Eagle's Medium (DMEM) with 10% fetal bovine serum (FBS) and 2 mM L-glutamine. Store at 4 °C.

13. Flow cytometer.

2.4 iPS Feeder Layer, Transfection

1. Human iPS cells.

2. mTesR medium (commercially prepared by Stemcell Technologies). Store at 4 °C.

3. 1× Dulbecco's phosphate buffered saline (DPBS), pH 7.4, without calcium and magnesium.

4. Matrigel (BD Biosciences). Store at −80 °C.

5. 6-well tissue culture plate.

6. Gentle Cell Dissociation Reagent (commercially available from Stemcell Technologies). Store at 4 °C.

7. KnockOut™ DMEM/F-12 (Thermo Fisher Scientific). Store at 4 °C.

8. Lipofectamine CRISPRMAX™ (Thermo Fisher Scientific). Store at 4 °C.

9. In vitro transcribed guide RNA. Store at −80 °C.

10. Recombinant Cas9 protein (e.g., Thermo Fisher Scientific). Store at −20 °C.

2.5 Cleavage Detection Assay

1. QuickExtract DNA Extraction Solution (Epicentre Biotechnologies). Store at −20 °C.

2. Oligonucleotide primers (10 μM each). Store at −20 °C.

3. Platinum Taq HiFi (Life Technologies). Store at −20 °C.

4. T7 endonuclease I. Store at −20 °C.

2.6 PCR Cloning	1. NEB PCR Cloning Kit (New England Biolabs). Store at −20 °C.
	2. Chemically competent *E. coli*. Store at −80 °C.
	3. LB + 100 μg/mL ampicillin plates. Store at 4 °C.

3   Methods

3.1   gRNA Designs,
In Silico Screen

1. Identify potential gRNA target sites within the target locus. To identify target sites, search for the *Spy* Cas9 PAM sequence NGG. The 20 bases immediately upstream constitute the gRNA target site. For targeted base changes it is best to design the gRNA such that the cut site (3 bases upstream of the PAM sequence) occurs as close as possible to the base of interest. For targeted gene knockout it is best to design a gRNA to target the 5′ end of the coding region to maximize the potential to disrupt the open reading frame. The microhomology-Predictor can be used to identify gRNAs that have a higher probability of inducing a frameshift indel (www.rgenome.net/mich-calculator/).

2. To screen gRNAs for potential off-target sites, several in silico tools can be used. One such tool is COSMID which identifies sites with mismatches, insertion, or deletions [16]. Enter the gRNA sequence in the "Query Sequence" box.

3. Select the Target Genome for which off-target analysis should be performed.

4. In the Search Options enter the PAM sequence of the Cas9 ortholog and select the number of allowable mismatches and/or one base deletion or insertion.

5. COSMID can also perform PCR primer design for each off-target site identified. Primer design is available for different cleavage detection assays and is optional.

6. In addition to off-target sites with mismatches, COSMID can identify gRNAs that have multiple target sites within the genome. These gRNAs typically target repetitive elements but can appear to be highly specific in algorithms that use a repeat masker.

7. Compile a short list of candidate gRNAs that have few potential off-target sites identified.

8. The U6 promoter requires a G as a transcription start site. Therefore the first base of a gRNA must be a G. The G can be added to the 5′ end to create a 21 nt gRNA, but in our experience it is better to replace the 5′ base with G such that the gRNA is G(N)19. We have not observed any impact on gRNA activity by the presence of a mismatched G at the 5′ end of the

gRNA (*see* Fig. 2b). Design and order sense and antisense oligonucleotides to include the following sequences: 5′ caccG(N)$_{19}$ 3′,

5′aaac(N)$_{19}$C 3′.
5′-caccGNNNNNNNNNNNNNNNNNNNN-3′
3′-CNNNNNNNNNNNNNNNNNNNNcaaa-5′

9. Order oligonucleotides at a 10 µM concentration.

3.2 Cloning gRNA into pX330

1. When the oligonucleotides arrive, set up a kinase reaction as follows:

   Master Mix: (1) 10× T4 DNA ligase reaction buffer, (2) 5 U of T4 Polynucleotide Kinase (10 U/µL), (3) bring the final volume to 24.3 µL with H$_2$O, and (4) add 24.3 µL of master mix to 0.7 µL of each oligonucleotide.

2. Incubate the reaction mix at 37 °C for 30 min, 65 °C for 20 min, and then hold at 12 °C.

3. After the kinase incubation is complete add all the sense oligonucleotides to the corresponding antisense oligonucleotides and run the annealing program on a thermocycler: (1) Heat to 95 °C for 2 min; (2) Ramp cool to 25 °C over 45 min; and (3) Hold at 12 °C.

4. Digest plasmid pX330 overnight at 37 °C with *Bbs*I and dephosphorylate with CIP.

5. Gel-purify the linearized pX330 vector using a commercially available gel extraction kit. Add 0.5 M EDTA to the reaction to stop the CIP activity. Clean up the digestion mix using a commercial column-based kit. Alternatively, magnetic beads may also be used to isolate linearized vector (*see* **Note 5**).

6. Determine the concentration of vector backbone using a NanoDrop Spectrophotometer and normalize to 25 ng/µL (*see* **Note 6**).

7. Ligate the kinase and annealed double stranded oligos and the pX330 vector together using a commercially available kit and transform into chemically competent *E. coli* (*see* **Note 6**). We recommend T4 DNA ligase from New England Biolabs. A 10 min incubation at room temperature is sufficient for 100 s of colonies.

8. Set up the ligation reaction master mix as follows. For each ligation add: (1) 0.5 µL of 25 ng/µL linearized pX330, (2) 0.5 µL of T4 DNA ligase, (3) 1 µL of T4 DNA ligase buffer 10×, and (4) 7 µL of H$_2$O.

9. Aliquot 9 µL of the ligation master mix into 0.2 mL PCR tubes.

10. Add 1 µL of the annealed gRNA oligos and incubate at room temperature for 10 min.

11. Add 2.5 μL of the ligation reaction to 25 μL of chemically competent *E. coli* and incubate on ice for 30 min.

12. Heat shock the bacteria at 42 °C by placing in a water bath for 30 s.

13. Incubate on ice for 2 min and then add 175 μL of SOC medium.

14. Incubate at 37 °C for 1 hour with shaking at 250 rpm.

15. Plate 100 μL of the transformation onto LB medium plates supplemented with 100 μg/mL ampicillin and incubate the plates at 37 °C overnight (approximately 16 h).

16. After overnight incubation, pick two colonies from each plate with a sterile toothpick or pipette tip and place them individually in 3 mL of LB medium containing 100 μg/mL ampicillin in 48 deep-well plates. Seal the 48 deep-well plate with a breathable sterile membrane. Shake overnight at 250 rpm at 37 °C.

3.3 pX330 gRNA Plasmids

1. Transfer 2 mL of each culture to a 96 deep-well culture plate and pellet the cells by centrifugation at $2500 \times g$ for 10 min.

2. Invert the culture plate to discard the supernatant. Tap the inverted plate onto tissue paper to remove residual supernatant. Resuspend the cell pellets in 100 μL of buffer P1. Add 100 μL of buffer P2 and mix by gently shaking on a tabletop vortex mixer (until the cell suspension turns blue). Incubate at room temperature for 4 min. Add 100 μL of buffer N3 and mix by gently shaking on a tabletop vortex mixer until the blue color completely disappears, indicating complete neutralization of the lysis buffer.

3. Pellet the cellular debris in a centrifuge at $3700 \times g$ for 40 min. Remove 165 μL of supernatant and transfer to a 96-well PCR plate (0.2 mL). Optional: **steps 4–10** can be carried out on an automated liquid handling system such as a Biomek 3000 (*see* **Note 3**).

4. Add 15 μL of magnetic beads to each well. Mix by pipetting.

5. Add 120 μL of isopropanol to each well. Mix by pipetting. Incubate at room temperature for 5 min and transfer the plate to a magnetic plate.

6. Incubate at room temperature for 5 min. Beads will settle out from the solution. Remove the supernatant.

7. Wash the magnetic beads by adding 200 μL of 80% ethanol. Incubate at room temperature for 1 min then remove ethanol. Repeat this wash step two times for a total of three washes then allow the beads to air dry for 5 min.

8. Remove the 96-well plate from the magnet and resuspend the beads in 40 μL of H_2O.

9. Place the 96-well plate with the resuspended beads back onto the magnet and incubate at room temperature for 3 min.

10. Remove 35 μL of the eluted DNA and transfer to a 96-well plate.

11. Measure the concentration of each plasmid using a NanoDrop Spectrophotometer or similar DNA quantification apparatus.

12. Sequence plasmids using the CRISPR_Seq primer:

    5′-CGATACAAGGCTGTTAGAGAGATAATTGG-3′

3.4 In Vitro gRNA Transcription

1. Perform the in vitro transcription of gRNAs using the GeneArt Precision gRNA synthesis kit.

2. Measure the concentration of in vitro transcribed (IVT) gRNAs using a qubit fluorometer. In our experience the NanoDrop overestimates the concentration of IVT gRNAs by five to tenfold.

3.5 HEK293T Transfection

1. Approximately 24 h prior to transfection coat the 24-well plates with 0.1% gelatin and incubate at 37 °C for 30 min. After incubation, aspirate off the gelatin and proceed immediately to cell seeding.

2. Seed 80,000 cells per well with 100 μL of DMEM + 10% FBS + 2 mM L-glutamine (*see* **Note 7**).

3. Approximately 24 h after seeding cells prepare DNA plasmids in 0.2 mL tubes for transfection using Lipofectamine-2000 as follows: (1) 50 ng of GFP, (2) 1 μg of CRISPR plasmid, (3) Opti-MEM to a final volume of 25 μL.

4. Set up a Lipofectamine-2000 master mix consisting of 2 μL of Lipofectamine 2000 and 23 μL of Opti-MEM for each transfection.

5. Add 25 μL of the Lipofectamine-2000 mix to each plasmid and mix by pipetting. Incubate at room temperature for 5 min then add 50 μL of the DNA Lipofectamine-2000 mix to cells in a dropwise fashion.

6. Return the cell culture plate to the 37 °C incubator.

7. Approximately 72 h post transfection, trypsinize the cells and analyze them by flow cytometry to determine the percentage of GFP expressing cells. This acts as an estimate of the transfection efficiency.

8. Analyze the transfected cells by flow cytometry to determine the percentage of GFP-positive cells. Confirm that the transfection efficiency is >90% (*see* **Note 7**).

9. Centrifuge the cells at $1000 \times g$ and remove the supernatant. Resuspend the cell pellet in 100 μL of QuickExtract solution.

10. Run the following program on a thermocycler: 68 °C for 15 min; 95 °C for 8 min; 4 °C hold.

11. Store the cell lysate containing genomic DNA at –20 °C.

3.6 Establishing iPS Culture

1. Thaw one aliquot of Matrigel at 4 °C. This may be done over several hours or overnight.

2. Prechill a 50 mL conical tube, 10 mL serological pipet and culture plates by placing in a –20 °C freezer.

3. With chilled pipette tips, add an aliquot of the Matrigel to the appropriate volume of cold DMEM/F12 (typically ~200 μL Matrigel in 18.5 mL DMEM/F12).

4. Immediately distribute the 1.5 mL of the diluted Matrigel to each well in a prechilled 6-well tissue culture plate.

5. Store the tissue culture plate at 4 °C until needed.

6. Before plating iPS cells, place the Matrigel plate at 37 °C for at least 20 min.

7. Aspirate the Matrigel, removing as much liquid as possible without scraping the bottom of the dish.

8. Plate the iPS cells in mTesR medium. Medium change is performed every day.

3.7 Passaging iPS Cells

1. Aspirate the medium.

2. Wash the cells once with 1 mL 1× DPBS and then aspirate.

3. Add 1 mL/well (6-well plate) of Gentle Cell Dissociation Reagent and spread well over the surface.

4. Incubate the plates at 37 °C for 5 min.

5. Remove the plates from the incubator and gently wash once with 1 mL/well of 1× DPBS. Add the DPBS gently as the cells tend to detach easily.

6. Carefully aspirate the DPBS (ensure that cells do not detach) and add 2 mL of mTesR culture medium per well.

7. Use a pipette tip to physically detach the cells.

8. Observe the cells under the microscope to ensure they have detached.

9. Pipette the cells several times to fully dissociate.

10. Plate 20,000 cells in one well of a 6-well plate coated with Matrigel. Immediately distribute the cells evenly over the plate surface to avoid uneven attachment.

11. Incubate the plates at 37 °C and change the medium every day.

3.8 iPS Transfection

1. Seed 1.1×10^5 cells per well in a Matrigel-coated 24-well plate on the day before transfection.

2. Prepare the Cas9 nuclease protein and gRNA solution with Cas9 plus™ reagent as follows:

   (1) 25 µL Opti-MEM medium, (2) 500 ng recombinant Cas9 nuclease, (3) 125 ng IVT gRNA, (4) 1.5 µL Lipofectamine Cas 9 plus reagent and mix well by pipetting (Tube 1).

3. Dilute CRISPRMAX reagent in Opti-MEM I medium as follows: (1) 25 µL Opti-MEM I medium, (2) 1.5 µL Lipofectamine CRISPRMAX reagent and mix well by pipetting (Tube 2).

4. Incubate Tubes 1 and 2 separately for 5 min at room temperature.

5. Add the solution from Tube 1 to Tube 2, and mix well.

6. Incubate for 20 min at room temperature. Do not incubate for more than 30 min.

7. Aspirate the culture medium and add the mixed solution in to the well.

8. Add 100 µL Opti-MEM medium per well.

9. Incubate the plate at 37 °C for 4 h, then top off with mTesR culture medium and incubate overnight.

10. Aspirate the medium and culture the cells in fresh mTesR medium for another 3 days with daily medium changes.

11. Trypsinize the cells and centrifuge at $1000 \times g$.

12. Remove the supernatant and resuspend the cell pellet in 100 µL of QuickExtract solution.

13. Run the following program on a thermocycler: 68 °C for 15 min; 95 °C for 8 min; 4 °C hold.

14. Store the cell lysate containing genomic DNA at –20 °C.

3.9 PCR of Target Loci

1. For each target locus, a PCR product must be amplified from the genomic DNA of cells treated with CRISPR/Cas9 and from unmodified or mock treated cells for use as controls.

2. Each 50 µL PCR reaction should contain: (1) 5 µL of 10× Buffer-II, (2) 0.2 µL of Taq HiFi Polymerase, (3) Forward and reverse primers at 0.2 µM each, (4) 100 ng of genomic DNA or 1 µL of QuickExtract DNA, and (5) H_2O to 50 µL.

3. For large numbers of samples, we find it easier to break the process into the following steps.

   (1) Create a master mix consisting of Buffer-II, Taq HiFi and H_2O; (2) Aliquot 47 µL to each well; (3) Add 2 µL of the pre-mixed forward and reverse primers at 5 µM each; and (4) Add 1 µL of QuickExtract DNA.

4. After all the components are added, cover the plate with foil and use a roller to ensure that the plate is sealed tightly, especially along the edges.

5. Vortex to mix and briefly centrifuge the plate.

6. Put the plate in the thermocycler.

7. Use the following touchdown PCR thermocycler conditions: (1) 94 °C for 2 min; (2) Touchdown Cycles (10×): (1) 94 °C for 30 s; (2) 65 °C for 30 s, decreasing 0.5 °C each cycle; (3) 68 °C for 45 s. Additional Cycles (30×): (1) 94 °C for 30 s; (2) 55 °C for 30 s; (3) 68 °C for 45 s; (4) 68 °C for 10 min; (5) Hold at 4 °C.

8. Completed PCR reactions can be stored at 4 °C for several days or at –20 °C for long-term storage.

3.10 PCR Confirmation

1. Cast a 1 % agarose gel using a small-toothed comb.

2. For each sample, pipette 4 µL of ready-to-load dye (1 µL 6× loading dye and 3 µL dH$_2$O) into an empty tube.

3. Pipette 2 µL of the PCR reaction into the tube with the loading dye. Pipette up and down to mix, then load into the gel.

4. Run the gel until the bands of the ladder are well separated.

5. Unless the mock samples are heterozygous for different-sized alleles, each sample should show only a single band at the expected size. If there is no band or multiple bands, do not proceed further with the analysis and redesign the primers for the target locus.

3.11 PCR Products Purification

This protocol represents a high-throughput procedure for purifying PCR products.

1. PCR products can be purified using any method, if eluted in water. Here, we provide a high-throughput method using magnetic beads.

2. Pipette 1.8 times the PCR reaction volumes of beads into the PCR samples. Mix together by pipetting up and down ten times.

3. Incubate at room temperature for 5 min.

4. Move the 96-well reaction plate onto the magnet and let it settle for 10 min.

5. Carefully pipette and discard the liquid while not disturbing the ring of beads.

6. Pipette 200 µL of 80 % ethanol into each well and wait for 1 min.

7. Carefully pipette and discard the liquid taking care not to disturb the ring of beads.

8. Repeat **steps 6** and **7** for a total of two ethanol washes.

9. Let the plate sit for 5 min to allow any remaining ethanol to evaporate.

10. Move the plate off of the magnet.

11. Pipette the desired elution volume of water (30–40 µL recommended) into each well and pipette up and down vigorously until all the beads are back in solution.

12. Move the plate onto the magnet and let it settle for 5 min.

13. Carefully use a pipette to transfer the liquid to a new plate while not disturbing the ring of beads. It may be necessary to leave ~5 µL of elution in the original plate to avoid transferring any beads to the new plate.

14. This purified PCR product can be stored at 4 °C for several days or at –20 °C indefinitely.

15. Measure the concentration of each purified product using a NanoDrop or similar spectrophotometer or other method of choice.

3.12 CRISPR-Induced Mutations

1. For each reaction, 200 ng of purified PCR product (from CRISPR-treated or mock cell populations) is mixed with 1.8 µL of 10× NEBuffer #2 in a total volume of 18 µL.

2. Vortex and centrifuge briefly to mix.

3. Melt and reanneal the DNA by placing it in a thermocycler: (1) 10 min at 95 °C, then decreasing at 0.1 °C/s down to 25 °C.

4. Vortex and centrifuge briefly to mix.

5. Make an enzyme master mix containing 0.5 µL of T7 Endonuclease I, 0.2 µL of 10× NEBuffer #2, and 1.3 µL of sterile distilled water.

6. Add 2 µL of the master mix to each reaction, vortex and centrifuge briefly to mix, and immediately place it in a thermocycler set at 37 °C for 60 min.

7. After 60 min, immediately remove the reactions from the thermocycler, centrifuge briefly, and quench the reaction by adding 6 µL of an EDTA containing Stop solution. This can be prepared by diluting 0.5 M EDTA 1:2 with 6× DNA loading dye.

8. Vortex and centrifuge briefly to mix.

9. Quenched reactions can be stored at 4 °C for several days or at –20 °C indefinitely.

10. Load the entire reaction into a 2 % agarose gel cast with large-toothed combs. The corresponding mock samples should be run in adjacent wells to the CRISPR-treated samples. Run the gel until the bands are well separated.

11. When imaging the gel, make sure that the exposure time is properly adjusted so that none of the bands is saturated as this will interfere with accurate quantification of the band intensities.

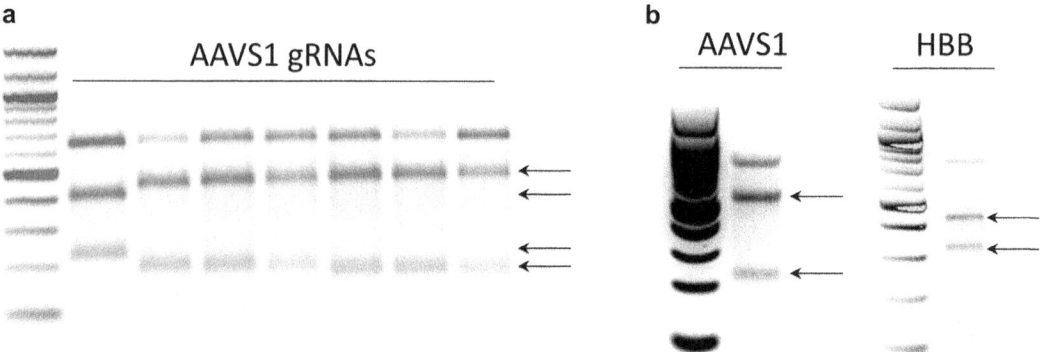

Fig. 3 T7EI data demonstrating efficient CRISPR/Cas9 gene targeting. Cut PCR products represent the percent of edited alleles in the population. Different banding patterns are created by gRNAs that cut in different locations throughout the target region amplified by PCR. (**a**) Screening of AAVS1 gRNAs in HEK293T cells. (**b**) Testing of highly active gRNAs in iPS cells

12. Ensure that each sample contains the appropriate number of bands (*see* Fig. 3a, b) (*see* **Note 8**).

13. The free software ImageJ [17] can be used to quantify the bands on the gel.

    (a) Go to the menu File → Open and select your gel image;

    (b) Use the "Rectangular" tool to select the lane containing your first sample. Only select the middle ~25–50% of the width of the lane to avoid any regions of the lane that are "smiling;"

    (c) Go to the menu → Analyze → Gels → Select First Lane (Ctrl + 1);

    (d) Use your mouse to move the rectangle to the center of your next lane. Go to the menu Analyze → Gels → Select Next Lane (Ctrl + 2). Repeat until all lanes have been selected;

    (e) Go to the menu Analyze → Gels → Plot Lanes (Ctrl + 3);

    (f) Use the "Straight" tool to draw a line under the peak of each band so that each peak has a distinct closed polygon;

    (g) Use the "Wand" tool to click on the polygon for each peak. This will calculate the area of each polygon, which represents the band intensity.

14. The percent of alleles that show evidence of NHEJ can be calculated from the band intensities according to the formulas [18]:

    (a) $f_{cut} = \dfrac{\text{Cleavage band}_1 + \text{Cleavage band}_2}{\text{Cleavage band}_1 + \text{Cleavage band}_2 + \text{Uncleaved band}}$

    (b) $\%\text{NHEJ} = 100 \times (1 - \sqrt{1 - f_{cut}})$.

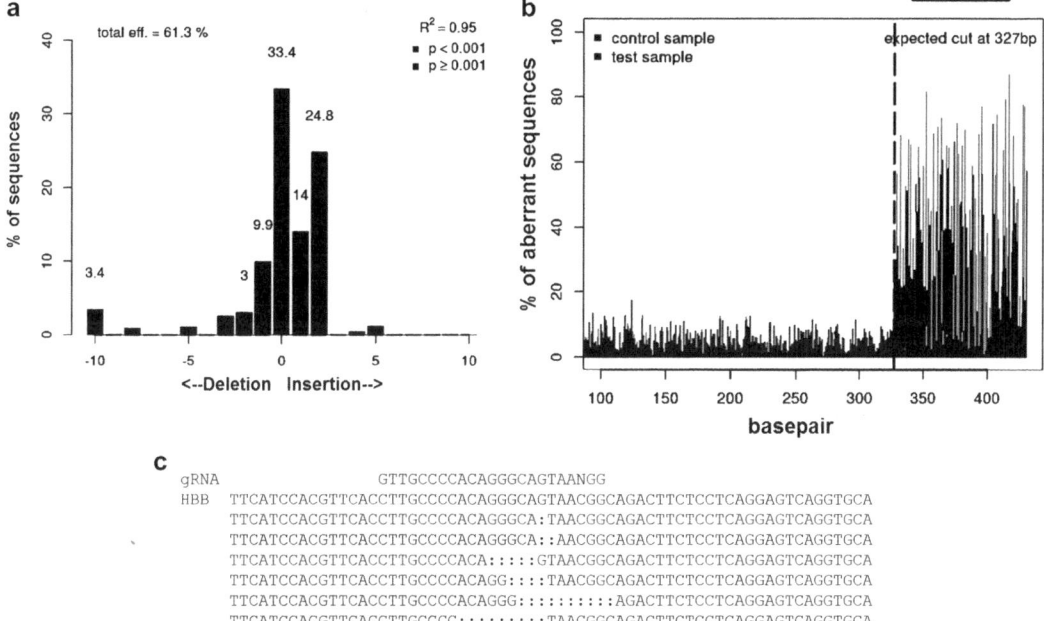

Fig. 4 Analysis of CRISPR-induced mutations. (**a**) Typical TIDE webtool output showing the indel spectrum induced by CRISPR activity. In this instance, the overall activity level was estimated to be 61.3% (Note the R^2 value). (**b**) Quality control analysis of TIDE. The *dotted line* denotes the expected CRISPR cut site. The chromatogram signal should be clean or have low aberrant signal before the cut site. The degree of aberrant signal after the cut site is proportional to the level of CRISPR activity. (**c**) Sequence analysis of PCR clones. The exact sequence of the mutant alleles can be determined by sequencing the PCR clones from treated cells

3.13 Indels Tracking by Decomposition

1. Send purified PCR product from Subheading 3.7 for Sanger sequencing using the forward or reverse primer used for PCR amplification.

2. Go to http://tide-calculator.nki.nl [19].

3. In the provided boxes enter the sample name and the 20-base gRNA target site.

4. Upload a chromatogram file (.ab1 or .scf) for both a control sample and CRISPR-treated sample.

5. Click the "update view" button and the plots will appear in two separate tabs.

6. It is important to check the quality control plot on the "decomposition" tab to ensure that the chromatogram signals are of sufficient quality for analysis (*see* Fig. 4a).

7. The R^2 value should be greater than 0.9 to ensure accurate estimation of CRISPR activity.

8. The overall cutting frequency and the frequency of each indel size is calculated revealing the spectrum of indels induced by CRISPR activity (*see* Fig. 4b).

3.14 PCR Cloning, Sequencing

1. Clone the PCR products using the NEB PCR cloning kit. Incubate the reaction for 15 min at room temperature.

2. Transform the ligation mix into competent *E. coli* cells.

3. Plate the cells onto LB agar containing 100 µg/mL ampicillin and incubate at 37 °C overnight.

4. Pick single colonies into LB medium containing 100 µg/mL ampicillin. Next day, perform a plasmid DNA miniprep procedure and have perform sequencing on target (*see* **Note 9**).

5. Align the sequence reads to the target reference sequence and analyze the sequencing reads for evidence of CRISPR-induced indels (*see* Fig. 4c).

4 Notes

1. To prevent loading dye shadow from interfering with downstream analysis of band densitometry we recommend using xylene cyanol FF. Prepare the 6× loading dye as follows: 4 g sucrose, 25 mg bromophenol blue or xylene cyanol (0.25 %), H_2O to 10 mL. Store at 4 °C.

2. Commercial magnetic beads may be used but we find the Serapure protocol by Brant Faircloth and Travis Glenn to work with similar efficiency for a fraction of the cost. Bead preparation according to the Serapure v2.2 protocol by Brant Faircloth and Travis Glenn at UCLA using the following reagents: Sera-mag SpeedBeads (Fisher # 09-981-123), PEG-8000 (Amresco 0159), 0.5 M EDTA, pH 8.0 (Amresco E177), 1.0 M Tris-HCl, pH 8.0 (Amresco E199), Tween 20 (Amresco 0777), 5 M NaCl, rare-earth magnet stand (Ambion AM10055 or NEB S1506S).

   (a) In a 50 mL conical tube using sterile stock solutions, prepare TE buffer (10 mM Tris–HCl, 1 mM EDTA, by adding 500 µL of 1 M Tris-Cl, pH 8.0, plus 100 µL of 0.5 M EDTA. Fill the conical tube to 50 mL with dH_2O.

   (b) Mix Sera-mag SpeedBeads and transfer 1 mL to a 1.5 mL microcentrifuge tube.

   (c) Place the SpeedBeads on magnet stand until the beads are drawn to the magnet.

   (d) Remove the supernatant with a P-200 or P-1000 pipetter.

   (e) Add 1 mL of TE to the beads, remove from magnet, mix, return to magnet.

   (f) Remove the supernatant with a P-200 or P-1000 pipetter.

(g) Add 1 mL of TE to the beads, remove from magnet, mix, return to magnet.

(h) Remove supernatant with a P-200 or P-1000 pipetter.

(i) Add 1 mL of TE to the beads and remove from magnet. Fully resuspend and set the microtube in a rack (i.e., not on the magnet stand).

(j) Add 9 g PEG-8000 to a new 50 mL, sterile, conical tube.

(k) Add 10 mL of 5 M NaCl (or 2.92 g) to the tube.

(l) Add 500 µL of 1 M Tris–HCl.

(m) Add 100 µL of 0.5 M EDTA.

(n) Fill the tube to ~49 mL using sterile dH_2O.

(o) Vortex for about 3–5 min until the PEG goes into solution (solution, upon sitting, should be clear).

(p) Add 27.5 µL Tween 20 and mix gently.

(q) Mix 1 mL SpeedBead plus the TE solution and transfer to a 50 mL conical tube.

(r) Fill the tube to 50 mL with dH_2O if necessary and gently mix it until the solution turns brown.

(s) Test against AMPure XP using aliquots of ladder (Fermentas GeneRuler).

(t) Wrap in tinfoil (or place in dark container) and store at 4 °C.

3. Many of the processes described herein can be automated on a liquid handling platform. Specifically, minipreps, PCR clean-ups, and sample normalization are readily adapted for such use.

4. Ethidium bromide is a commonly used for detecting DNA in gels. Ethidium bromide is a potential carcinogen. Care must be used and an isolated area for handling ethidium bromide, is suggested. Alternatively, you can use safe, but more expensive, DNA gel stains such as Gel Green or Gel Red (Biotium Inc., Hayward, CA).

5. Magnetic beads may be used to clean up the linearized pX330 vector. The fragment removed by digestion is too small to be recovered using this method. This purification method results in a higher yield of DNA compared to column-based gel extraction methods. After CIP inactivation, add 1.8× volume of beads to the pX330 digest and mix by pipetting. Proceed as outlined in **steps 8–21**, Subheading 3.7 normalizing the linearized vector to 25 ng/µL.

6. Transform 12.5 ng of the vector to determine the background level of the prepared backbone. Typically <5 colonies are observed for vector only controls compared to 50–100 colonies for gRNA transformations.

7. For the HEK293T cell line, 80,000 cells is sufficient to ensure 70–80% confluency in 24-well plates after 24 h and greater than >90% transfection efficiency. For other cell lines, the optimal cell seeding density should be determined prior to transfection by seeding cells at a range of different densities and using GFP transfection to identify the optimal conditions.

8. If the target locus is heterozygous for a SNP or small insertion or deletion then the mock sample may have extra bands present. These bands may overlap or obscure the true signal bands in samples from CRISPR-treated cells. In these cases, TIDE analysis is a better choice of assay. This chromatogram-based assay is insensitive to SNPs. If the TIDE assay returns erroneous data, sequencing of the PCR clones may reveal the true level of CRISPR activity.

9. The pMiniT cloning vector contains a toxic minigene if no insert is present. Therefore, any colonies that grow are highly likely to contain a ligated insert and no further selection is required.

Acknowledgments

This work was supported by the National Institutes of Health as an NIH Nanomedicine Development Center Award (PN2EY018244 to GB) and by the Cancer Prevention and Research Institute of Texas (RR140081 to GB).

References

1. Urnov FD, Miller JC, Lee YL, Beausejour CM, Rock JM, Augustus S, Jamieson AC, Porteus MH, Gregory PD, Holmes MC (2005) Highly efficient endogenous human gene correction using designed zinc-finger nucleases. Nature 435(7042):646–651. doi:10.1038/nature03556, PubMed PMID: WOS:000229476200043

2. Kim YG, Cha J, Chandrasegaran S (1996) Hybrid restriction enzymes: zinc finger fusions to Fok I cleavage domain. Proc Natl Acad Sci U S A 93(3):1156–1160, PubMed PMID: 8577732; PMCID: 40048

3. Christian M, Cermak T, Doyle EL, Schmidt C, Zhang F, Hummel A, Bogdanove AJ, Voytas DF (2010) Targeting DNA double-strand breaks with TAL effector nucleases. Genetics 186(2):757–761. doi:10.1534/genetics.110.120717, PubMed PMID: 20660643; PMCID: PMC2942870

4. Miller JC, Tan S, Qiao G, Barlow KA, Wang J, Xia DF, Meng X, Paschon DE, Leung E, Hinkley SJ, Dulay GP, Hua KL, Ankoudinova I, Cost GJ, Urnov FD, Zhang HS, Holmes MC, Zhang L, Gregory PD, Rebar EJ (2011) A TALE nuclease architecture for efficient genome editing. Nat Biotechnol 29(2):143–148. doi:10.1038/nbt.1755

5. Jinek M, Chylinski K, Fonfara I, Hauer M, Doudna JA, Charpentier E (2012) A programmable dual-RNA-guided DNA endonuclease in adaptive bacterial immunity. Science 337(6096):816–821. doi:10.1126/science.1225829

6. Gasiunas G, Barrangou R, Horvath P, Siksnys V (2012) Cas9-crRNA ribonucleoprotein complex mediates specific DNA cleavage for adaptive immunity in bacteria. Proc Natl Acad Sci U S A 109(39):E2579–E2586.

doi:10.1073/pnas.1208507109, PubMed PMID: 22949671; PMCID: 3465414

7. Cong L, Ran FA, Cox D, Lin S, Barretto R, Habib N, Hsu PD, Wu X, Jiang W, Marraffini LA, Zhang F (2013) Multiplex genome engineering using CRISPR/Cas systems. Science 339(6121):819–823. doi:10.1126/science.1231143, PubMed PMID: 23287718; PMCID: 3795411

8. Cradick TJ, Fine EJ, Antico CJ, Bao G (2013) CRISPR/Cas9 systems targeting beta-globin and CCR5 genes have substantial off-target activity. Nucleic Acids Res 41(20):9584–9592. doi:10.1093/nar/gkt714, PubMed PMID: 23939622; PMCID: 3814385

9. Hsu PD, Scott DA, Weinstein JA, Ran FA, Konermann S, Agarwala V, Li Y, Fine EJ, Wu X, Shalem O, Cradick TJ, Marraffini LA, Bao G, Zhang F (2013) DNA targeting specificity of RNA-guided Cas9 nucleases. Nat Biotechnol 31(9):827–832. doi:10.1038/nbt.2647, PubMed PMID: 23873081; PMCID: 3969858

10. Fu Y, Foden JA, Khayter C, Maeder ML, Reyon D, Joung JK, Sander JD (2013) High-frequency off-target mutagenesis induced by CRISPR-Cas nucleases in human cells. Nat Biotechnol 31(9):822–826. doi:10.1038/nbt.2623, PubMed PMID: 23792628; PMCID: 3773023

11. Tsai SQ, Zheng Z, Nguyen NT, Liebers M, Topkar VV, Thapar V, Wyvekens N, Khayter C, Iafrate AJ, Le LP, Aryee MJ, Joung JK (2015) GUIDE-seq enables genome-wide profiling of off-target cleavage by CRISPR-Cas nucleases. Nat Biotechnol 33(2):187–197. doi:10.1038/nbt.3117, PubMed PMID: 25513782; PMCID: 4320685

12. Lee CM, Cradick TJ, Fine EJ, Bao G (2016) Nuclease target site selection for maximizing on-target activity and minimizing off-target effects in genome editing. Mol Ther. doi:10.1038/mt.2016.1

13. Lee CM, Cradick TJ, Bao G (2016) The *Neisseria meningitidis* CRISPR-Cas9 system enables specific genome editing in mammalian cells. Mol Ther. doi:10.1038/mt.2016.8

14. Ran FA, Cong L, Yan WX, Scott DA, Gootenberg JS, Kriz AJ, Zetsche B, Shalem O, Wu X, Makarova KS, Koonin EV, Sharp PA, Zhang F (2015) *In vivo* genome editing using Staphylococcus aureus Cas9. Nature. doi:10.1038/nature14299

15. Muller M, Lee CM, Gasiunas G, Davis TH, Cradick TJ, Siksnys V, Bao G, Cathomen T, Mussolino C (2015) *Streptococcus thermophilus* CRISPR-Cas9 systems enable specific editing of the human genome. Mol Ther. doi:10.1038/mt.2015.218

16. Cradick TJ, Qiu P, Lee CM, Fine EJ, Bao G (2014) COSMID: a web-based tool for identifying and validating CRISPR/Cas off-target sites. Mol Ther Nucleic Acids 3:e214. doi:10.1038/mtna.64, PubMed PMID: 25462530; PMCID: 4272406

17. Schneider CA, Rasband WS, Eliceiri KW (2012) NIH Image to ImageJ: 25 years of image analysis. Nat Methods 9(7):671–675

18. Guschin DY, Waite AJ, Katibah GE, Miller JC, Holmes MC, Rebar EJ (2010) A rapid and general assay for monitoring endogenous gene modification. Methods Mol Biol 649:247–256. doi:10.1007/978-1-60761-753-2_15

19. Brinkman EK, Chen T, Amendola M, van Steensel B (2014) Easy quantitative assessment of genome editing by sequence trace decomposition. Nucleic Acids Res 42(22):e168. doi:10.1093/nar/gku936, PubMed PMID: 25300484; PMCID: 4267669

Chapter 2

Mutagenesis and Genome Engineering of Epstein–Barr Virus in Cultured Human Cells by CRISPR/Cas9

Kit-San Yuen, Chi-Ping Chan, Kin-Hang Kok, and Dong-Yan Jin

Abstract

The clustered regularly interspaced short palindromic repeats (CRISPR)-CRISPR associated protein 9 nuclease (Cas9) system is a powerful genome-editing tool for both chromosomal and extrachromosomal DNA. DNA viruses such as Epstein–Barr virus (EBV), which undergoes episomal replication in human cells, can be effectively edited by CRISPR/Cas9. We have demonstrated targeted editing of the EBV genome by CRISPR/Cas9 in several lines of EBV-infected cells. CRISPR/Cas9-based mutagenesis and genome engineering of EBV provides a new method for genetic analysis, which has some advantages over bacterial artificial chromosome-based recombineering. This approach might also prove useful in the cure of EBV infection. In this chapter, we use the knockout of the BART promoter as an example to detail the experimental procedures for construction of recombinant EBV in human cells.

Key words RNA-guided genome editing, Episomal viral DNA genome, Epstein–Barr virus, Genetic analysis of Epstein–Barr virus, Cure of Epstein–Barr virus infection

Abbreviations

BAC	Bacterial artificial chromosome
Cas9	CRISPR associated protein 9 nuclease
CRISPR	Clustered regularly interspaced short palindromic repeats
DSB	Double-strand break
EBV	Epstein–Barr virus
GFP	Green fluorescent protein
gRNA	Guide RNA
MOI	Multiplicity of infection
PAM	Protospacer adjacent motif
pBART	*Bam*HI-A region rightward transcript promoter
pCMV	Cytomegalovirus promoter
PCR	Polymerase chain reaction

Andrew Reeves (ed.), *In Vitro Mutagenesis: Methods and Protocols*, Methods in Molecular Biology, vol. 1498,
DOI 10.1007/978-1-4939-6472-7_2, © Springer Science+Business Media New York 2017

1   Introduction

Genetic studies are important to all areas of biology. In virology, targeted mutation of particular genetic elements on the viral genome helps to understand their function. In the early years, genomes of herpesviruses were engineered using homologous recombination in eukaryotic cells [1–4]. Subsequently, the successful cloning of herpesviral genomes into bacterial artificial chromosomes (BACs) greatly facilitated viral genome manipulation in prokaryotic cells [5–8]. However, in γ-herpesviruses including EBV and Kaposi sarcoma-associated herpesvirus, the titers of viruses recovered from BAC-transfected producer cell lines are usually very low for unknown reasons, and some transfectants even lose their ability to support lytic viral replication [7]. The low efficiency in generating high-quality γ-herpesvirus producing cell lines has become the bottleneck issue in the use of EBV BAC in the field [7]. Intensive screening of stable cells is required to obtain high-quality EBV producing cell lines and this hinders the application of EBV BAC.

In order to provide an alternative method for mutagenesis and genome engineering of EBV, we harnessed the emerging CRISPR/Cas9 technology for targeted editing of the EBV genome. CRISPR/Cas9 was originally discovered as part of the adaptive immune system in bacteria but has now been developed into a powerful method for double-strand break (DSB)-induced genome editing in eukaryotic cells [9, 10]. The EBV genome exists in infected cells as multicopy episomes [11] and this poses a unique challenge for CRISPR/Cas9 editing. Whether the multicopy EBV episome could be efficiently cleaved by CRISPR/Cas9 is one major concern and how the correctly edited version of the EBV genome might be separated from predominantly unedited viral DNA is another critical issue. In this regard, we have not only demonstrated the feasibility of CRISPR/Cas9 editing of the EBV genome but have also designed and tested different approaches to isolate the correctly edited recombinant EBV [12].

In this chapter, we detail how we harnessed CRISPR/Cas9 to edit the EBV genome in human cells. First, we describe the steps and criteria for optimal guide RNA (gRNA) design for EBV editing (see Subheading 3.1). Second, we provide the step-by-step procedures for CRISPR/Cas9 editing of the EBV genome in mammalian cells (see Subheading 3.2). The cell lines harboring edited EBV are helpful for genetic study of EBV. Third, we describe the procedures for the isolation of pure and infectious recombinant EBV created by CRISPR/Cas9 (see Subheading 3.3). The recombinant EBV can be produced and used for further infection experiments. Particularly, we supplied the protocol for insertion of the DsRed fluorescent marker to facilitate the recovery of mutant EBV. The methods described are generally applicable to the creation of both gene disruption and gene replacement in not only EBV but also other DNA viruses.

2   Materials

2.1   Molecular Biology Materials

1. PX459 vector (kindly provided by Dr. Feng Zhang, MIT, Cambridge MA).
2. Genomic DNA Purification Kit (e.g., Wizard, Promega).
3. GeneJuice (Novagen).
4. TransIT-Keratinocyte Transfection Reagent (Mirus).
5. Puromycin.
6. Filter papers and punches.
7. Agarose.
8. Agarose gel electrophoresis system.
9. RT-PCR reagents.
10. Western blotting reagents and apparatus.
11. Commercial RPMI 1640 medium (Life Technologies).
12. Goat-anti-human IgG.
13. 1× PBS.
14. pDsRed2-C1 vector (Clontech).
15. BZLF1 and gp110 expression plasmids.
16. Flow cytometer with sorting capability (e.g., BD FACSAria SORP).
17. Hi-Fidelity DNA polymerase.
18. *Dpn*I restriction enzyme.
19. Commercial DNA gel and PCR cleanup kit.
20. 1.20 μm syringe filter.
21. Amicon® Ultra-15 Centrifugal filter Ultracel®-100 kDa cutoff.

3   Methods

3.1   gRNA Design and Cloning

The steps and criteria for optimal gRNA design are described based on using the deletion of the BART promoter in the EBV genome as an example. In our design, two gRNAs (pB1 and pB2) are employed to flank the whole BART promoter (pBART) region for deletion (*see* **Note 1**). The pSpCas9(BB)-2A-Puro (PX459)-based CRISPR/Cas9 system [13], kindly provided by Dr. Feng Zhang, is accessed through Addgene (https://www.addgene.org/62988/), and is used in the following protocol.

1. The target region in the EBV genome is searched for a 19-bp sequence followed by the protospacer adjacent motif (PAM) NGG at the 3' end. Two gRNAs flanking the target region are

designed (*see* **Note 2**). For optimal gRNA binding, the GC content of the 19-bp target sequence is preferably within 40–60%. The sequences of the pBART-gRNAs are as follows:

gRNA-pB1, TAATTGCAGTGGACCCCGG AGG[PAM]

gRNA-pB2, AAGAAGCTCCTCAGCAACA TGG[PAM]

2. The target sequence is subjected to off-target analysis. Human and EBV genome sequences in the National Center for Biotechnology Information nucleotide databases are BLAST searched for matches to the 19-bp sequence together with NGG (e.g., TAATTGCAGTGGACCCCGGNGG for gRNA-B1). The stringency of the off-target analysis could be adjusted according to the purpose of the study. Only gRNAs with limited off-target hits are chosen. In our case, a sequence with a perfect match to PAM together with a match of >10 bp in the 19-bp region is avoided.

3. The 20-bp gRNA sequence with the first nucleotide being G is inserted into the PX459 vector. The addition of a G is required for optimal expression from the U6 promoter. The PX459 vector contains both gRNA and Cas9 expression cassettes.

4. The gRNA insert can be made by annealing the sense and antisense oligonucleotides. The annealed oligonucleotides are then inserted into the PX459 vector using the Zhang Lab General Cloning Protocol at the following website: (http://www.addgene.org/crispr/zhang/).

3.2 EBV Genome Editing

We have used several EBV-infected epithelial cell lines for CRISPR/Cas9 editing of the EBV genome [12]. These include the EBV-infected human embryonic kidney cell line HEK293-EBV, the human nasopharyngeal carcinoma cell lines HK1-EBV and C666-1, the human gastric adenocarcinoma cell line AGS-BX1 as well as the EBV-infected human telomerase reverse transcriptase-immortalized normal nasopharyngeal epithelial cell line NP460-EBV. CRISPR/Cas9 editing of the EBV genome is performed directly inside these cells. Cells containing the edited virus are recovered by puromycin selection. The protocol should generally be amenable to all other EBV-infected cell lines that are not too difficult to transfect, including some B lymphocytic lines.

1. Grow approximately 2×10^5 EBV-infected epithelial cells in a 6-well tissue culture plate.

2. After 24 h, transfect the cells with 1 μg of gRNA1 and 1 μg of gRNA2 with GeneJuice and TransIT transfection reagents (*see* **Note 3**).

3. Harvest the cells 72 h post-transfection. Half of the cells are collected for genomic DNA extraction. Genomic DNA is extracted using a genomic DNA purification kit.

4. Perform PCR using primers flanking the deleted region to screen for the desired CRISPR/Cas9-edited mutant virus. Successful editing will result in an additional amplicon of smaller size.

5. Transfer the remaining half of the cells into a 10-cm tissue culture dish. Twenty-four hours later, puromycin is added to select for the stable cells which contain the mutant virus (*see* **Note 4**).

6. Replace the drug-containing medium with fresh medium and cells after 48 h after puromycin treatment. Allow the cells to grow for 1 more week.

7. After 1 week, visible colonies of the selected cells should appear on the plate. Single colonies are picked by using filter paper (*see* **Note 5**).

8. Place a pre-soaked filter paper (with trypsin–EDTA) on top of the single cell colony at 37 °C for 3 min. The filter paper is then transferred with a pair of sterile forceps onto a 6-well tissue culture plate containing 2 ml of fresh culture medium. Shake the filter paper in the medium to make sure that the attached cells on the filter paper are detached into the well. The detached cells are allowed to recover and grow for 4 more days.

9. Verify the surviving clones by PCR as described in **steps 3** and **4** above. PCR products derived from both edited and unedited EBV genomes are analyzed by agarose gel electrophoresis. In general, most of the surviving clones contain a mixture of wild-type and edited forms of EBV (*see* **Note 6**). In our pBART knockout experiment conducted in HEK293-EBV cells, 3 out of 50 clones were shown to be deprived of the unedited form of EBV carrying the pBART-deleted mutant virus only [12].

10. Perform RT-PCR and Western blotting to verify the disruption of the target gene product in the puromycin-selected cell clones.

3.3 Recombinant EBV

1. To generate an EBV-infected cell line that meets all the special requirements for the creation of a recombinant EBV by CRISPR/Cas9 editing, we have established HEK293-BX1 cells by coculturing HEK293 with Akata-BX1 cells. HEK293 cells are commonly used as EBV producer cells in BAC recombineering [6, 7]. They are highly susceptible to transfection and induction of lytic replication. Establishment of HEK293-BX1 cells for the conduction of CRISPR/Cas9 editing can therefore facilitate the construction, isolation, and production of recombinant EBV. On the other hand, insertion of a selectable marker or fluorescent reporter into the EBV genome can enable drug selection and mutant tracing, which are desirable for the isolation of recombinant EBV. To this end, we provide a protocol for the addition of DsRed fluorescent marker into EBV genome during CRISPR/Cas9 editing. This greatly facilitates tracing and recovery of mutant viruses. Additionally, we also describe

an experimental approach to establish HEK293 producer cells of the mutant EBV through reinfection and sorting of DsRed+ single cells. This allows rapid and efficient isolation of the desired mutant virus created by CRISPR/Cas9 from a mixture of unedited and edited EBV genomes.

3.4 HEK293-BX1 Cells

1. Treat approximately 1×10^5 Akata-BX1 cells with 0.5% of goat-anti-human IgG in a 20-mm tissue culture dish for induction of lytic replication.

2. Wash the Akata-BX1 cells after 24 h with 1× PBS and collect by centrifugation at $200 \times g$ for 5 min.

3. Resuspend the cell pellets in 2 ml RPMI 1640. The IgG-induced Akata-BX1 cells serve as the source of EBV for the infection of HEK293 cells.

4. Grow approximately 1×10^5 HEK293 cells in a 6-well tissue culture plate 1 day before infection.

5. Wash the HEK293 cells with 1× PBS and combine with the IgG-induced Akata-BX1 cells.

6. After coculturing for 72 h, wash away the Akata-BX1 cells from the HEK293 cells with 1× PBS.

7. Observe the infection of HEK293 cells under a fluorescent microscope. Successfully infected HEK293-BX1 cells are green fluorescent protein (GFP)-positive because the BX1 virus carries a GFP fluorescent marker [12]. The efficiency of EBV infection through coculturing is usually in the range of 5–10%.

8. Trypsinize the cells and transfer them into a 10-cm tissue culture dish. Cells are then grown to confluence.

9. Single-cell sort GFP-positive cells onto a 96-well tissue culture plate by BD FACSAria SORP.

10. Recover the single cell-sorted HEK293-BX1 cells in a 96-well tissue culture plate over the next 7–10 days.

3.5 DsRed Insertion

1. The pCMV-DsRed fragment containing the cytomegalovirus promoter (pCMV) was PCR-amplified from plasmid pDsRed2-C1 using KAPA HiFi DNA polymerase. Primers with an EBV homology arm of 50 bp are used for the amplification. For the knockout of pBART, the primers are 5′-ATGGTATTG GTCGTCTTCTTCCCCTGCAGAGTAATTGCA GTGGACCCCGG<u>TGTTCTTTCCTGCGTTATCCC</u>-3′ and 5′-TAGTGCCTACGTGACTCCCTGCTCCTGCCAGTT TCCCTTCGAGGTCTCCA<u>TAAGGGATTTTGC</u>-3′. The amplification primers for pCMV-DsRed are underlined and the EBV homology arms are located at the 5′ end.

2. Treat the pCMV-DsRed fragment with 1 μl of *Dpn*I at 37 °C for 60 min to remove the pDsRed2-C1 template.

3. The pCMV-DsRed fragment is gel-purified by using a high-quality gel and PCR clean up kit.

4. One day before transfection, approximately 2×10^5 of HEK293-BX1 cells are grown in 6-well tissue culture plate.

5. Transfect cells with 0.2 μg of gRNA1, 0.2 μg of gRNA2 and 1.6 μg of pCMV-DsRed fragment.

6. Collect half of the cells 72 h post-transfection for PCR verification. Insertion of DsRed into the target region is confirmed by using primers complementary to the junction between the target region and the DsRed cassette.

3.6 HEK293 Producer Cells

1. Transfer the remaining half of the cells into 4×10-cm tissue culture dishes.

2. Next day, transfect the cells in the culture dish with 4 μg each of BZLF1 and gp110 expression plasmids to induce EBV lytic replication (*see* **Note 7**).

3. After 24 h, replace the culture medium of the transfected cells with fresh RPMI 1640. The transfected cells are grown for 5 more days for recombinant EBV production.

4. Harvest the supernatants 6 days after transfection. All supernatants are filtered through a 1.20-μm syringe filter before infection.

5. Concentrate 40 ml of supernatant into 1 ml by using an Amicon® Ultra-15 Centrifugal filter Ultracel®-100 kDa. The Centricon is spun at $1400 \times g$ for concentrating. It usually takes somewhere in the range of 30–45 min to concentrate 40 ml of supernatant to 1 ml. The concentrated virus can be stored at 4 °C for 1 week.

6. Grow approximately 2×10^5 HEK293 cells in a 6-well tissue culture plate 1 day before infection.

7. Add 1 ml of concentrated virus to 2×10^5 of HEK293 cells. The culture medium is replaced with fresh RPMI 1640 after 24 h.

8. Seventy-two hours after infection, monitor cells under a fluorescent microscope. Cells infected with wild-type EBV are GFP-positive, whereas cells infected with recombinant EBV are doubly positive for GFP and DsRed (*see* **Note 8**).

9. Sort single cells doubly positive for GFP and DsRed onto a 96-well tissue culture plate by BD FACSAria SORP.

10. Recover the sorted HEK293-ΔEBV-DsRed single cells in a 96-well tissue culture plate over the course of 7–10 days.

11. Test the recovered cells for recombinant virus production. Usually the producer cells yield 1×10^5 to 1×10^6/ml green Raji units of recombinant virus.

12. Verify the genomic pattern of the recombinant virus by PCR and deep sequencing.

13. Once verified, the recombinant viruses are now ready for use in other infection assays.

4 Notes

1. Compared to the single gRNA approach, using two gRNAs to splice out the target region enhances the editing efficiency and facilitates the subsequent screening process.

2. For abrogation of protein expression, the start codon and the first exon should preferably be removed. For other regulatory elements, the complete target region can be deleted by two gRNAs.

3. HEK293-EBV, HK1-EBV and AGS-BX1 cells are transfected with GeneJuice (Novagen) in the ratio of 1 μg of DNA to 3 μl of GeneJuice. C666-1 and NP460-EBV are transfected with TransIT-Keratinocyte Transfection Reagent (Mirus) in the ratio of 1 μg of DNA to 3 μl of TransIT Reagent.

4. PX459 vector contains a puromycin selection marker. For HEK293-EBV cells, 3 μg/ml of puromycin is used. For NP460-EBV and AGS-BX1 cells, 2 μg/ml of puromycin is used. For C666-1 and HK1-EBV cells, 0.5 μg/ml of puromycin is used.

5. Filter papers are prepared by punches. The punched filter paper is autoclaved before use.

6. EBV maintains 5–100 copies of covalently closed circular genomic DNA in latently infected cells [11]. Inside a single cell, some copies of the EBV genome may escape from CRISPR/Cas9 editing. Extensive screening is required to obtain stable cells with the pure EBV mutant.

7. BZLF1 is the key transcriptional activator mediating the switch between latent and lytic replication of EBV. gp110 is the viral glycoprotein which remarkably enhances the ability of EBV to infect human cells [7].

8. HEK293 cells are infected with a low multiplicity of infection (MOI) of EBV. Under low MOI conditions, most of the HEK293 cells will be infected with one particle of EBV, either a wild type or a recombinant.

Acknowledgments

This work was supported by Hong Kong Health and Medical Research Fund (11100602 and 12110962), S.K. Yee Medical Research Fund (2011), and Hong Kong Research Grants Council (AoE/M-06/08, HKU1/CRF/11G, C7011-15R, and T11-707/15-R).

References

1. Delecluse HJ, Hilsendegen T, Pich D, Zeidler R, Hammerschmidt W (1998) Propagation and recovery of intact, infectious Epstein-Barr virus from prokaryotic to human cells. Proc Natl Acad Sci U S A 95:8245–8250

2. Zhou FC, Zhang YJ, Deng JH, Wang XP, Pan HY, Hettler E, Gao SJ (2002) Efficient infection by a recombinant Kaposi's sarcoma-associated herpesvirus cloned in a bacterial artificial chromosome: application for genetic analysis. J Virol 76:6185–6196

3. Umene K (1999) Mechanism and application of genetic recombination in herpesviruses. Rev Med Virol 9:171–182

4. Cotter MA, Robertson ES (1999) Molecular genetic analysis of herpesviruses and their potential use as vectors for gene therapy applications. Curr Opin Mol Ther 1:633–644

5. Messerle M, Crnkovic I, Hammerschmidt W, Ziegler H, Koszinowski UH (1997) Cloning and mutagenesis of a herpesvirus genome as an infectious bacterial artificial chromosome. Proc Natl Acad Sci U S A 94:14759–14763

6. Kanda T, Yajima M, Ahsan N, Tanaka M, Takada K (2004) Production of high-titer Epstein-Barr virus recombinants derived from Akata cells by using a bacterial artificial chromosome system. J Virol 78:7004–7015

7. Feederle R, Bartlett EJ, Delecluse HJ (2010) Epstein-Barr virus genetics: talking about the BAC generation. Herpesviridae 1:6–11

8. Zhou F, Gao SJ (2011) Recent advances in cloning herpesviral genomes as infectious bacterial artificial chromosomes. Cell Cycle 10:434–440

9. Cho SW, Kim S, Kim JM, Kim JS (2013) Targeted genome engineering in human cells with the Cas9 RNA-guided endonuclease. Nat Biotechnol 31:230–232

10. Cong L, Zhang F (2015) Genome engineering using CRISPR-Cas9 system. Methods Mol Biol 1239:197–217

11. Adams A, Lindahl T (1975) Epstein-Barr virus genomes with properties of circular DNA molecules in carrier cells. Proc Natl Acad Sci U S A 72:1477–1481

12. Yuen KS, Chan CP, Wong NH, Ho CH, Ho TH, Lei T, Deng W, Tsao SW, Chen H, Kok KH et al (2015) CRISPR/Cas9-mediated genome editing of Epstein-Barr virus in human cells. J Gen Virol 96:626–636

13. Ran FA, Hsu PD, Wright J, Agarwala V, Scott DA, Zhang F (2013) Genome engineering using the CRISPR-Cas9 system. Nat Protoc 8:2281–2308

Chapter 3

Use of CRISPR/Cas Genome Editing Technology for Targeted Mutagenesis in Rice

Rongfang Xu, Pengcheng Wei, and Jianbo Yang

Abstract

Clustered Regularly Interspaced Short Palindromic Repeats (CRISPR)/CRISPR-associated protein (Cas) system is a newly emerging mutagenesis (gene-editing) tool in genetic engineering. Among the agriculturally important crops, several genes have been successfully mutated by the system, and some agronomic important traits have been rapidly generated, which indicates the potential applications in both scientific research and plant breeding. In this chapter, we describe a standard gene-editing procedure to effectively target rice genes and to make specific rice mutants using the CRISPR/Cas9 system mediated by *Agrobacterium* transformation.

Key words CRISPR/Cas9, Rice, Mutagenesis, *Agrobacterium* transformation

1 Introduction

The clustered regularly interspaced short palindromic repeats (CRISPR)/CRISPR-associated protein (Cas) technology has rapidly emerged in recent years as a powerful and robust genome-editing tool in diverse organisms [1–6]. The current CRISPR/Cas system is derived from the type II CRISPR/Cas system of *Streptococcus pyogenes* [1, 4, 7]. In this system, a synthetic single-guide RNA (sgRNA) binds directly to a 20-nt sequence followed by a 5′-NGG PAM (protospacer adjacent motif) on the target DNA to provide sequence specificity, and a Cas9 nuclease coupled with the sgRNA to induce the site-specific cleavage in the genome. The CRISPR/Cas9 tool has been shown to achieve successful genome editing in a variety of plants, including Arabidopsis, rice, tobacco, wheat, sorghum, maize, tomato, liverwort, and orange [8–19]. Unlike in animals, direct delivery of RNA into cells is technically difficult in plants. In most cases, the constructs expressing sgRNA and/or *Cas9* are co-transformed by Agrobacteria into plant cells to generate a functional CRISPR/Cas9 complex. Although the sgRNA and *Cas9* could be delivered by separate vectors, most studies have

Andrew Reeves (ed.), *In Vitro Mutagenesis: Methods and Protocols*, Methods in Molecular Biology, vol. 1498, DOI 10.1007/978-1-4939-6472-7_3, © Springer Science+Business Media New York 2017

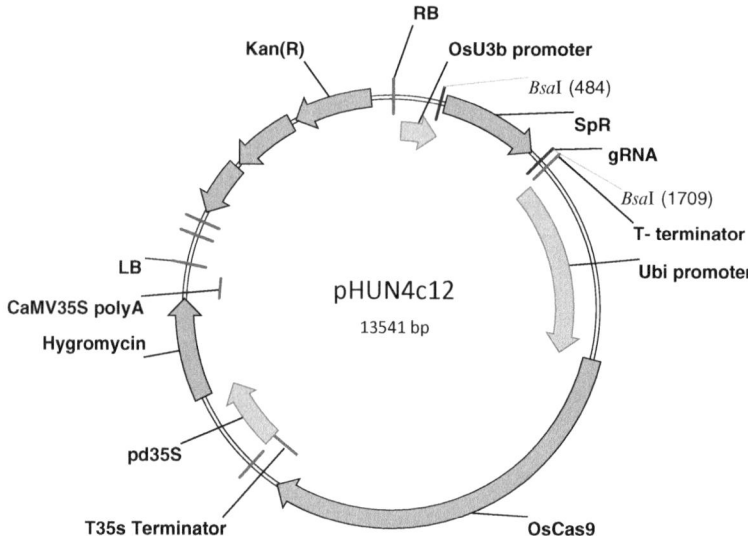

Fig. 1 Structure of the pHUN4c12 vector. The promoter, gene, and transgenic elements are labeled. *LB* left border of T-DNA, *RB* right border of T-DNA, *SpR* spectinomycin resistance gene, *gRNA* scaffold sequence of the sgRNA

used the all-in-one constructs which combine the two expression cassettes in one binary vector to increase the efficiency of mutagenesis [8, 10, 11, 20, 21].

Here, we use a typical all-in-one CRISPR/Cas9 vector, pHUN4c12 (*see* Fig. 1), and describe a standard procedure to construct and identify targeted mutants in rice.

2   Materials

All chemicals used should be of reagent grade. All solutions should be prepared with ultrapure water.

2.1   Chemicals

1. dNTPs.
2. Appropriate oligonucleotides or PCR primers.
3. Plasmid mini extraction kit.
4. DNA Gel extraction kit.
5. Isopropanol.
6. 70 and 100% ethanol.

2.2   Biological Reagents

1. pHUN4c12 vector was modified from pHUN411 vector [22], provided by Dr. Yang's lab, Anhui Academy of Agricultural Science, China.
2. *E. coli* strain XL-Blue competent cells.
3. *Agrobacterium* strain EHA105-pSOUP competent cells [23].

4. Rice seeds from Nipponbare.

5. T4 Polynucleotide Kinase (PNK).

6. T4 quick ligase kit.

7. Restriction endonuclease *Bsa*I.

8. Proofreading DNA polymerase Phusion.

9. TA cloning vector pEASY-T simple (Transgene Biotech).

10. Sequencing primer "OsU3 seq" (5′-CAGTCAGGGA CCATAGCACAGG-3′ (commercially synthesized)).

2.3 Buffers and Solutions

1. 50% NaClO solution (with final available chlorine of 2%).

2. N6 majors solution (working concentration): 463 mg/L $(NH_4)_2SO_4$, 2.83 g/L KNO_3, 125 mg/L $CaCl_2$, 90 mg/L $MgSO_4$, 400 mg/L KH_2PO_4.

3. MS iron salts (working concentration): 41.3 mg/L $Na_2EDTA \cdot 2H_2O$, 27.8 mg/L $FeSO_4 \cdot 7H_2O$.

4. B5 minors solution (working concentration): 0.75 mg/L KI, 3 mg/L H_3BO_3, 13.2 mg/L $MnSO_4 \cdot 4H_2O$, 2 mg/L $ZnSO_4 \cdot 7H_2O$, 0.25 mg/L $Na_2MoO_4 \cdot 2H_2O$, 0.025 mg/L $CuSO_4 \cdot 5H_2O$, 0.025 mg/L $CoCl_2 \cdot 6H_2O$.

5. B5 vitamins solution (working concentration): 100 mg/L myo-inositol, 1 mg/L nicotinic acid, 1 mg/L pyridoxine HCl, 10 mg/L thiamine HCl.

6. NB Medium: working solutions of N6 majors, MS iron salts, B5 minors, B5 vitamins, 500 mg/L proline, 500 mg/L glutamine, 300 mg/L casein enzymatic hydrolysate, 30 g/L sucrose, 2 mg/L 2,4-D, 3 g/L Phytagel.

7. Suspension medium: working concentrations of N6 majors, B5 minors, MS iron salts, B5 vitamins, 500 mg/L proline, 500 mg/L casein enzymatic hydrolysate, 2 mg/L 2,4-D, 20 g/L sucrose, 10 g/L glucose, 100 μmol/L acetosyringone, pH 5.20.

8. NBREC Medium: NB Medium supplemented with 250 mg/L carbenicillin.

9. NBHYG Medium: NBREC Medium supplemented with 50 mg/L hygromycin.

10. NBNB Medium: working concentrations of N6 majors, MS iron salts, B5 minors, B5 vitamins, 500 mg/L proline, 1 g/L casein enzymatic hydrolysate, 30 g/L sucrose, 30 g/L sorbitol, 500 mg/L MES, 2.5 mg/L $CuSO_4$, 1.5 mg/L NAA, 1 mg/L 6-BA, 125 mg/L carbenicillin, 2.5 g/L Phytagel, 25 mg/L hygromycin.

11. 1/2 MS-HYG Medium: 1/2 MS Medium supplemented with 25 mg/L hygromycin.

12. Genomic DNA extraction buffer: 100 mM Tris-HCl (pH 8.0), 50 mM EDTA (pH 8.0), 1 M NaCl, 1.25 % SDS (w/v). Add 1 % 2-mercaptoethanol before use.

13. 5 M potassium acetate (pH 5.2).

14. 0.1× TE buffer: 1 mM Tris–HCl (pH 8.0); 0.1 mM EDTA (pH 8.0).

3 Methods

The rice genome edited by the CRISPR-Cas system consists of three steps: (1) vector construction; (2) genetic transformation; (3) molecular identification.

3.1 Vector Construction

1. The 20-bp sequence of the targeted region ahead of the PAM site was used to synthesis oligonucleotides (*see* **Note 1**). The adapters were added onto the oligonucleotide sense and antisense sequences to link with plasmid pHUN4c12 vector (*see* **Note 2**).

2. Dissolve the oligonucleotides in ddH$_2$O as 100 μM solutions (*see* **Note 3**).

3. Mix the resuspended oligonucleotides and phosphorylate using T4 PNK at 37 °C for 1 h (*see* **Note 4**).

4. Add 50 mM NaCl (final concentration) to phosphorylated oligonucleotides (*see* **Note 5**).

5. Heat at 95 °C for 5 s and slowly cool down to room temperature (*see* **Note 6**).

6. Dilute the annealed oligonucleotides tenfold.

7. Digest pHUN4c12 plasmid with *Bsa*I at 37 °C for 2 h (*see* **Note 7**), then terminate the digestion at 65 °C for 20 min (*see* **Note 8**).

8. Ligate digested vector and the phosphorylated oligonucleotide pair using a T4 quick ligase Kit at 25 °C for 5 min (*see* **Note 9**).

9. Transform 2 μL of the ligation mixture into *E. coli* competent cells. Target clones are positively selected by kanamycin and negatively selected by spectinomycin.

10. Sanger-sequence to verify the insertion of the oligonucleotide (*see* **Note 10**).

11. Transform the sequenced-confirmed recombinant vector into *Agrobacterium* strain EHA105-pSOUP for rice plant transformation (*see* **Note 11**).

3.2 Plant Transformation

1. Surface-sterilize the dehulled rice seeds using a 50 % NaClO solution and plate on NB medium for callus induction in a 30 °C incubator in the dark for about 2 weeks.

2. Preculture well grown rice calli on the NB medium for another 3–5 days (*see* **Note 12**).

3. Suspend the *Agrobacterium* strain EHA105-pSOUP carrying the recombinant vector in suspension medium and infect rice calli for 20 min (*see* **Note 13**).

4. Plate the infected calli on the NBREC medium and incubate at 23 °C in the dark for 5 days.

5. Plate the above calli on NBHYG medium and incubated at 30 °C in the dark for ~3–4 weeks (*see* **Note 14**).

6. Transform the callus growths to NBNB medium for ~3–4 weeks to regenerate shoots (*see* **Note 15**).

7. Transfer the regenerated shoots from NBNB medium to the 1/2 MS-HYG to generate the transgenic plants (*see* **Note 16**).

3.3 Molecular Identification

1. A total of 20 mg of leaf tissue are used for each transgenic line and transferred into a 96-well plate with grinding beads (*see* **Note 17**).

2. Homogenize the samples and then add 300 µL of DNA extraction buffer to each well (*see* **Note 18**).

3. Add 150 µL of 5 M potassium acetate to each well and mix well. Centrifuge the plate at $3200 \times g$ for 15 min.

4. Transfer the supernatants to a new plate and precipitate the genomic DNA by adding 0.8 volume of isopropanol (*see* **Note 19**).

5. Wash the DNA pellets with 70% ethanol twice and air-dry on benchtop.

6. Dissolve each sample DNA with 50 µL of 0.1× TE to be used as a template for site-specific PCR (*see* **Note 20**).

7. Use PCR primers specifically designed to amplify the genomic region containing the CRISPR target.

8. Sequence the PCR product to detect and confirm the targeted mutation in the genome of the transgenic plants.

4 Notes

1. The CRISPR targeted sites are evaluated and selected with the help of CRISPR-P Web-tool [24].

2. The structure of the oligonucleotide pairs are:
Oligo I: 5′-GGCANNNNNNNNNNNNNNNNNNNN-3′
Oligo II: 3′-NNNNNNNNNNNNNNNNNNNNCAAA-5′
    NN… sequence in Oligo I represents the 20-nt sequence of the target (sense oligo); NN… sequence in Oligo II represents the complementary sequence of the target (antisense oligo).

3. The synthetic oligonucleotides are dissolved in pure water for at least 10 min at room temperature. Before adding water, the tube containing the oligonucleotides should be centrifuged for 5 s to pellet the DNA.

4. The oligonucleotide pairs are phosphorylated in the following reaction system:

Oligo I	1 μL
OligoII	1 μL
10× T4 ligase buffer	5 μL
T4 PNK	1 μL
ddH$_2$O	42 μL
Total	50 μL

5. 2.5 μL of a 1 M stock solution of NaCl is added to 50 μL of phosphorylated oligonucleotides.

6. It will take more than 2 h to cool down the mixture from 95 °C to about 25 °C (room temperature). Alternatively, it could be annealed using a PCR thermocycler under the following cycling conditions: initial denaturation at 95 °C for 5 min; annealing by decreasing the temperature from 95 to 85 °C (ramp rate at –2 °C/s) and then from 85 to 25 °C (–0.1 °C/s); and a final hold at 4 °C. Normally, it will take 40–50 min using a Life ProFlex PCR machine. The annealed oligonucleotides can be kept at 4 °C for more than a week.

7. The pHUN4c12 plasmid is digested in following reaction:

Plasmid	6 μg
BsaI	1 μL
10× digestion Buffer	4 μL
ddH$_2$O	29 μL
Total	40 μL

8. There is no need to perform the agarose gel electrophoresis and gel extraction to isolate the fragment of the linear vector.

9. The ligation mixture set up is as follows:

BsaI digested pHUN4c12	1 μL
Diluted annealed oligos	2 μL
2× T4 Quick ligase buffer	5 μL
T4 Quick ligase	0.5 μL
ddH$_2$O	1.5 μL
Total	10 μL

10. The transformed *E. coli* cells are incubated on LB plates supplemented with 50 mg/L kanamycin overnight. Then, at least 48 clones are streaked onto an LB agar plate containing 50 mg/L kanamycin and an LB plate containing 50 mg/L spectinomycin. The clones resistent to kanamycin but not to spectinomycin are selected. About 4–10 positive clones out of 48 clones are usually identified and the selected clones normally contain the target recombinant vector. The insertion of the oligonucleotide could be Sanger-sequenced using the "OsU3 seq" primer located upstream of the insertion site.

11. The pSOUP vector is introduced into the *Agrobacterium* EHA105 strain to make the EHA105-pSOUP strain following the previously described protocol [23].

12. The spherical, rough, yellowish calli are selected for transformation. The preculture of calli should not be more than 5 days old. The initial amount of calli vary depending on the species of rice. For Nipponbare, 350–500 incubated calli normally generate more than 50 independent transgenic events.

13. The agrobacteria are active after overnight incubation on plates. The agrobacteria are then collected and suspended at a density of 5×10^7 to 1.25×10^8 cfu (OD_{660} between 0.1 and 0.25). After incubation, the calli are dried on filter paper for 5–10 min until no liquid is observed on the plate surface.

14. The calli are checked every 5 days. Any contaminated calli should be removed.

15. The resistant calli are incubated at 30 °C under illumination with a light intensity of 6000 lx and a photoperiod of 16 h day/8 h night. The resistant calli that emerge from each callus are regarded as one individual event. Two to three yellowish, spherical resistance calli nodules from one resistant event are used for regeneration.

16. The shoots are incubated at 30 °C for at least 3 weeks to generate enough roots. Only one shoot from each resistant event is transferred to the rooting medium.

17. A grinding bead is put into each well of the 96-well plate before the sampling occurs. After addition of the leaf tissue, the plate is kept at –80 °C overnight.

18. The samples are homogenized twice by using a Geno 2000 homogenizer for 10 min each time.

19. The genomic DNA is precipitated at –20 °C for 2 h in isopropanol and then centrifuged at $12,000 \times g$ for 15 min at 4 °C.

20. The DNA is dissolved in 0.1× TE buffer at 4 °C overnight.

References

1. Cong L, Ran FA, Cox D, Lin S, Barretto R, Habib N, Hsu PD, Wu X, Jiang W, Marraffini LA, Zhang F (2013) Multiplex genome engineering using CRISPR/Cas systems. Science 339:819–823

2. Gratz SJ, Cummings AM, Nguyen JN, Hamm DC, Donohue LK, Harrison MM, Wildonger J, O'Connor-Giles KM (2013) Genome engineering of Drosophila with the CRISPR RNA-guided Cas9 nuclease. Genetics 194:1029–1035

3. Li D, Qiu Z, Shao Y, Chen Y, Guan Y, Liu M, Li Y, Gao N, Wang L, Lu X, Zhao Y, Liu M (2013) Heritable gene targeting in the mouse and rat using a CRISPR-Cas system. Nat Biotechnol 31(8):681–683

4. Mali P, Yang L, Esvelt KM, Aach J, Guell M, DiCarlo JE, Norville JE, Church GM (2013) RNA-guided human genome engineering via Cas9. Science 339:823–826

5. Wang H, Yang H, Shivalila CS, Dawlaty MM, Cheng AW, Zhang F, Jaenisch R (2013) One-step generation of mice carrying mutations in multiple genes by CRISPR/Cas-mediated genome engineering. Cell 153:910–918

6. Xiao A, Wang Z, Hu Y, Wu Y, Luo Z, Yang Z, Zu Y, Li W, Huang P, Tong X, Zhu Z, Lin S, Zhang B (2013) Chromosomal deletions and inversions mediated by TALENs and CRISPR/Cas in zebrafish. Nucleic Acids Res 41(14):e141. doi:10.1093/nar/gkt464

7. Jinek M, Chylinski K, Fonfara I, Hauer M, Doudna JA, Charpentier E (2012) A programmable dual-RNA-guided DNA endonuclease in adaptive bacterial immunity. Science 337:816–821

8. Jiang W, Zhou H, Bi H, Fromm M, Yang B, Weeks DP (2013) Demonstration of CRISPR/Cas9/sgRNA-mediated targeted gene modification in Arabidopsis, tobacco, sorghum and rice. Nucleic Acids Res 41(20):e188

9. Li J-F, Norville JE, Aach J, McCormack M, Zhang D, Bush J, Church GM, Sheen J (2013) Multiplex and homologous recombination-mediated genome editing in *Arabidopsis* and *Nicotiana benthamiana* using guide RNA and Cas9. Nat Biotechnol 31:688–691

10. Mao Y, Zhang H, Xu N, Zhang B, Gou F, Zhu J-K (2013) Application of the CRISPR-Cas system for efficient genome engineering in plants. Mol Plant 6:2008–2011

11. Miao J, Guo D, Zhang J, Huang Q, Qin G, Zhang X, Wan J, Gu H, Qu L-J (2013) Targeted mutagenesis in rice using CRISPR-Cas system. Cell Res 23:1233–1236

12. Nekrasov V, Staskawicz B, Weigel D, Jones JDG, Kamoun S (2013) Targeted mutagenesis in the model plant Nicotiana benthamiana using Cas9 RNA-guided endonuclease. Nat Biotechnol 31:691–693

13. Shan Q, Wang Y, Li J, Zhang Y, Chen K, Liang Z, Zhang K, Liu J, Xi JJ, Qiu J-L, Gao C (2013) Targeted genome modification of crop plants using a CRISPR-Cas system. Nat Biotechnol 31:686–688

14. Upadhyay SK, Kumar J, Alok A, Tuli R (2013) RNA-guided genome editing for target gene mutations in wheat. G3 3:2233–2238

15. Voytas DF (2013) Plant genome engineering with sequence-specific nucleases. Annu Rev Plant Biol 64:327–350

16. Brooks C, Nekrasov V, Lippman ZB, Van Eck J (2014) Efficient gene editing in tomato in the first generation using the clustered regularly interspaced short palindromic repeats/CRISPR-Associated9 system. Plant Physiol 166:1292–1297

17. Feng Z, Mao Y, Xu N, Zhang B, Wei P, Yang DL, Wang Z, Zhang Z, Zheng R, Yang L, Zeng L, Liu X, Zhu JK (2014) Multigeneration analysis reveals the inheritance, specificity, and patterns of CRISPR/Cas-induced gene modifications in *Arabidopsis*. Proc Natl Acad Sci 111:4632–4637

18. Sugano SS, Shirakawa M, Takagi J, Matsuda Y, Shimada T, Hara-Nishimura I, Kohchi T (2014) CRISPR/Cas9-mediated targeted mutagenesis in the liverwort *Marchantia polymorpha* L. Plant Cell Physiol 55:475–481

19. Gao J, Wang G, Ma S, Xie X, Wu X, Zhang X, Wu Y, Zhao P, Xia Q (2015) CRISPR/Cas9-mediated targeted mutagenesis in *Nicotiana tabacum*. Plant Mol Biol 87:99–110

20. Mikami M, Toki S, Endo M (2015) Comparison of CRISPR/Cas9 expression constructs for efficient targeted mutagenesis in rice. Plant Mol Biol 88:561–572

21. Xu R, Li H, Qin R, Wang L, Li L, Wei P, Yang J (2014) Gene targeting using the *Agrobacterium tumefaciens*-mediated CRISPR-Cas system in rice. Rice 7:1–4

22. Xing H-L, Dong L, Wang Z-P, Zhang H-P, Han C-H, Liu B, Wang X-C, Che Q-J (2014) A CRISPR/Cas9 toolkit for multiplex genome editing in plants. BMC Plant Biol 14:327. doi:10.1186/s12870-014-0327-y

23. Chandrasekaran S, Franklin M (2003) PSoup: a system for streaming queries over streaming data. VLDB J 12:140–156

24. Lei Y, Lu L, Liu H-Y, Li S, Xing F, Chen L-L (2014) CRISPR-P: a web tool for synthetic single-guide RNA design of CRISPR-system in plants. Mol Plant 7:1494–1496

Chapter 4

All-in-One CRISPR-Cas9/FokI-dCas9 Vector-Mediated Multiplex Genome Engineering in Cultured Cells

Tetsushi Sakuma, Takuya Sakamoto, and Takashi Yamamoto

Abstract

CRISPR-Cas9 enables highly convenient multiplex genome engineering in cultured cells, because it utilizes generic Cas9 nuclease and an easily customizable single-guide RNA (sgRNA) for site-specific DNA double-strand break induction. We previously established a multiplex CRISPR-Cas9 assembly system for constructing an all-in-one vector simultaneously expressing multiple sgRNAs and Cas9 nuclease or other Cas9 variants including FokI-dCas9, which supersedes the wild-type Cas9 with regard to high specificity. In this chapter, we describe a streamlined protocol to design and construct multiplex CRISPR-Cas9 or FokI-dCas9 vectors, to introduce them into cultured cells by lipofection or electroporation, to enrich the genomically edited cells with a transient puromycin selection, to validate the mutation efficiency by Surveyor nuclease assay, and to perform off-target analyses. We show that our protocol enables highly efficient multiplex genome engineering even in hard-to-transfect HepG2 cells.

Key words CRISPR-Cas9, FokI-dCas9, sgRNA, Multiplex genome engineering, All-in-one vector, Lipofection, Electroporation, Puromycin selection, Surveyor nuclease assay, Off-target analysis

1 Introduction

Currently, multiplex genome engineering in cultured cells has become a straightforward strategy by a recent innovative technology called clustered regularly interspaced short palindromic repeats (CRISPR)-CRISPR-associated protein 9 (Cas9) [1]. Multiplex genome engineering can be effected by multiple DNA double-strand breaks (DSBs) introduced by a generic Cas9 nuclease and multiple gene-specific single guide RNAs (sgRNAs) [2]. To realize highly efficient multiplex DSB induction, it is important to optimize the vector system and delivery method. Regarding the vector system, we previously reported a sophisticated strategy to construct an all-in-one vector harboring all the components required in multiplex genome engineering (i.e., a Cas9 nuclease-expressing cassette and multiple sgRNA-expressing cassettes) [3]. Furthermore, we have expanded the system by replacing the wild-type Cas9 nuclease with

Andrew Reeves (ed.), *In Vitro Mutagenesis: Methods and Protocols*, Methods in Molecular Biology, vol. 1498,
DOI 10.1007/978-1-4939-6472-7_4, © Springer Science+Business Media New York 2017

a Cas9 nickase, catalytically inactive dCas9, and FokI-dCas9, which is a chimeric fusion of the FokI nuclease domain and dCas9 [3, 4]. The principle of assembling multiple sgRNA cassettes relies on the Golden Gate method [5], which is generally used for the assembly of DNA-binding repeats of transcription activator-like effector nuclease (TALEN) [6–9]. Our multiplex CRISPR-Cas9 vector system can be applied not only for multiple targeted mutagenesis and chromosomal deletion in cultured cells [3], but also for those in mice [4] or a particular gene knockin strategy named the PITCh (precise integration into target chromosome) system [10–13].

FokI-dCas9 has been developed to maximally reduce the risk of off-target mutations, by transforming a monomeric Cas9 nuclease into a dimeric FokI-based nuclease [14, 15], as if it is an engineered site-specific nuclease such as zinc-finger nuclease (ZFN) or TALEN. It requires two sgRNA target sites with limited range of spacer lengths (around 16 bp) to efficiently induce a DSB [14, 15]. FokI-dCas9 has so far been applied both in cultured cells [14, 15] and in mice [4, 16]. Since multiplex genome engineering requires multiplex DSB induction, the risk of off-target mutation is much higher than in single gene targeting. Therefore, the all-in-one FokI-dCas9 vector is thought to be quite useful as a reliable multiple targeting vehicle in both cells and animals.

In this chapter, we comprehensively describe procedures for sgRNA design, vector construction, transfection, drug selection, and mutation analyses, along with the actual data targeting the multiple loci in human *adenomatous polyposis coli* (*APC*) gene using HepG2 cells, which is known to be a cell line that is difficult to transfect.

2 Materials

2.1 CRISPR Vectors

1. Multiplex CRISPR/Cas9 Assembly System Kit (Addgene) (*see* **Note 1**).

2. Multiplex CRISPR dCas9/FokI-dCas9 Accessory Pack (Addgene) (*see* **Note 2**).

3. Oligonucleotides for sgRNA template (*see* Table 1 for examples).

4. Luria–Bertani broth: tryptone 10 g, yeast extract 5 g, NaCl 10 g. pH to 7.2 with NaOH. For plating cells, add agar to 1.5–2.0%. For selections, add 100 μg/ml ampicillin or spectinomycin.

5. Plasmid miniprep and midiprep kits (*see* **Note 3**).

6. Thermal cycler.

7. QiaQuick Ligation Kit (*see* **Note 4**).

Table 1
Gene-specific oligonucleotides used in this study

Oligonucleotides for sgRNA template

Locus	Sense oligonucleotide (5′ → 3′)	Antisense oligonucleotide (5′ → 3′)
off-17_L	caccggcaatctgggctgcagtgg	aaacccactgcagcccagattgcc
off-17_R	caccgtgtcagccattcatacctc	aaacgaggtatgaatggctgacac
off-16_L	caccgctacttaatggttgctgggc	aaacgcccagcaaccattaagtagc
off-16_R	caccggccgaaactcaatttcccc	aaacggggaaattgagtttcggcc

Primers for mutation analyses

Locus	Forward primer (5′ → 3′)	Reverse primer (5′ → 3′)
On-target (off-17)	AAGTTTGGAGAGAGAACGCGGAATTG	GTGTATGGCAGCAGAGAGCTTCTTCTAAGTG
On-target (off-16)	CCAGGCAGACATCCCAAATAGGTG	GTTGGCTCATCTGTCTACCTGGAGATG
Off-target #1	GCAGCTACCCAGAGAAGTGAAGTTTTTCC	AGGTCAGCTATGCTGGGCTGGAG
Off-target #2	AGGCTGTGTAACTTGGAGCAAGAGAAATC	TGGTTGTGTAGGTAGCAGAGCTGGGAATC
Off-target #3	AGTGGCTGGCCCTGGAAAAGAC	CCTCCGCCCTGGCCTAAC
Off-target #4	GAAGTTCACTAAGGAGACAGCAAGGTCTGA	GAGAGGAGGAGGGACAGCATTACATATC
Off-target #5	GGACAGTGGCTCACACACCTTCTTG	CCCCTCCCTACCTCATCTCCTATTGTC
Off-target #6	GATCAAACATGAAGCTGAAGTGTAAAGACCAG	GAGGCTCTTGCCCTGCCAACTTAG
Off-target #7	GACGAGCCGATGTGCGGTCAG	GCTTCTGGCGCTGCTGGGTTTCTG

8. T4 DNA Ligase and reaction buffer.

9. *Bpi*I restriction enzyme.

10. *Eco*31I restriction enzyme.

11. 5-bromo-4-chloro-3-indolyl β-D-galactopyranoside (X-gal)/isopropyl β-D-1-thiogalactopyranoside (IPTG) solution.

12. Chemically competent bacterial cells such as XL1-Blue or XL10-Gold.

13. Standard *Taq* DNA polymerase.

14. Primers for colony PCR screening of correctly assembled clones in **step 4** in Subheading 3.4.

    CRISPR-step2-F: 5′-GCCTTTTGCTGGCCTTTTGCTC-3′

    CRISPR-step2-R: 5′-CGGGCCATTTACCGTAAGTTATGTAACG-3′

2.2 Cell Culture and Transfection

1. Cultured cells (HepG2 cells are used in this protocol).

2. CO_2 incubator.

3. Dulbecco's Modified Eagle's Medium (DMEM) supplemented with 10 % fetal bovine serum (FBS), 1× penicillin–streptomycin (PS), and 1× nonessential amino acids (NEAA) (*see* **Note 5**).

4. Opti-MEM (Optimum-Modified Eagle's Medium).

5. Trypsin–EDTA.

6. Puromycin.

7. Puromycin resistance gene (PuroR)-expressing vector (e.g., pPur Clontech).

8. 100-mm cell culture dishes.

9. 100-mm collagen I-coated cell culture dishes.

10. 6-well and 12-well collagen I-coated cell culture plates.

11. 0.22-μm filter and syringe.

12. Lipofectamine-3000 for lipofection.

13. NEPA21 electroporator.

14. Electroporation cuvettes for electroporation.

2.3 Mutation Analyses

1. DNeasy Blood and Tissue Kit (Qiagen).

2. Standard *Taq* DNA polymerase.

3. Primers for on- and off-target cleavage assay (*see* Table 1).

4. DNA gel and PCR clean up kit.

5. Surveyor Mutation Detection Kit (Transgenomic, Inc.).

6. Tris–HCl (pH 8.5).

7. KCl.

8. $MgCl_2$.

3   Methods

3.1   sgRNA Design

In this protocol, we describe design guidelines for *Streptococcus pyogenes* Cas9 and FokI-dCas9, whose protospacer adjacent motif (PAM) sequence is 5′-NGG-3′ (*see* **Note 6**).

1. For Cas9 nuclease, identify a 5′-N_{21}GG-3′ sequence to design a sgRNA. Alternatively, use a web-based software such as CRISPR design tool (http://crispr.mit.edu/) [17] and CRISPRdirect (https://crispr.dbcls.jp/) [18].

2. For FokI-dCas9, identify a 5′-CCN_{21}-3′ and 5′-N_{21}GG-3′ sequence separated by a defined spacer of length (~16 bp) (*see* **Note 7**). Alternatively, ZiFiT Targeter (http://zifit.partners.org/ZiFiT/) [14] can be used. The actual examples of sgRNA design are shown in Fig. 1a.

3.2   All-in-One Vectors

1. In this protocol, we use all-in-one CRISPR vectors constructed from the plasmid kits described below, which are available from Addgene. Although we recommend using the all-in-one vector system especially in cells with low transfection efficiency such as HepG2 cells, co-transfection of multiple CRISPR plasmids can also be used for multiple gene targeting.

2. Extract the plasmids contained in the Multiplex CRISPR/Cas9 Assembly System Kit (for both Cas9 nuclease and FokI-dCas9)

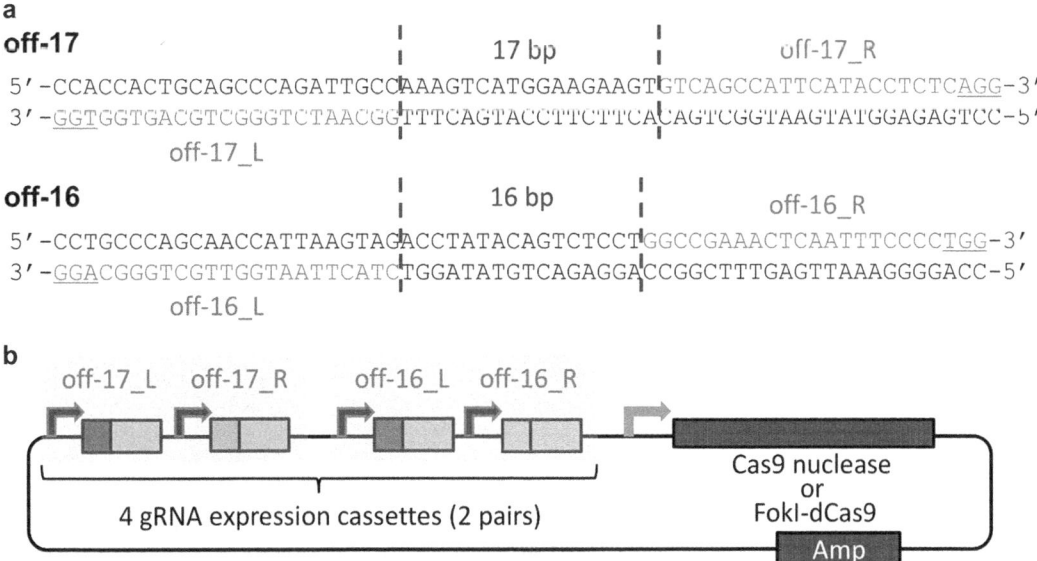

Fig. 1 Schematic illustration of target sequence (**a**) and constructed vectors (**b**) for multiplex genome editing at the human *APC* locus. *Red letters* in the sequence panels indicate sgRNA target sites. PAM sequence is *underlined*. sgRNAs are driven by U6 promoters. Cas9 nuclease and FokI-dCas9 are driven by a chicken β-actin short (CBh) promoter

and Multiplex CRISPR dCas9/FokI-dCas9 Accessory Pack (for FokI-dCas9 only) using a plasmid miniprep kit.

3. Design the oligonucleotides for each sgRNA template according to the previously reported protocol [19.]
Anneal and insert the oligonucleotides into the following vectors using the Quick Ligation Kit, T4 DNA ligase reaction buffer, *BpiI* and with the assistance of the following website (*see* **Note 8**) (https://www.addgene.org/crispr/yamamoto/multiplex-crispr-kit/#protocols-and-resources).

4. Assemble the oligonucleotide-inserted plasmids using the Golden Gate method with the Quick Ligation Kit, T4 DNA ligase reaction buffer, *Eco*31I and with the assistance of the following website (*see* **Note 9**) (https://www.addgene.org/crispr/yamamoto/multiplex-crispr-kit/#protocols-and-resources).

5. Extract the constructed all-in-one vector plasmids using a plasmid miniprep or midiprep kit. Adjust the concentration to 200 ng/μl (for lipofection) or 500 ng/μl (for electroporation). A schematic illustration of the constructed vector harboring four sgRNA cassettes and a Cas9 nuclease or FokI-dCas9 cassette is shown in Fig. 1b.

3.3 Cell Culturing

We describe the culturing method of HepG2 cells in this protocol. In our experience, programmable nuclease-expressed cells sometimes show slower proliferation than usual. To avoid this issue, a collagen I-coated dish or plate should be used and conditioned medium should be prepared before the transfection when using sensitive cells (*see* **Note 10**).

1. HepG2 cells are maintained in DMEM supplemented with 10% FBS, 1× PS, and 1× NEAA at 37 °C with 5% CO_2. Any standard cell culture plates or dishes can be used for normal passage culture, but a 100-mm collagen I-coated dish should be used just before transfection to achieve high viability and growth ability after transfection.

2. Generation of conditioned medium: Collect the culture supernatant from dishes containing cultured HepG2 cells for 48 h. Filter the supernatant using a 0.22-μm filter and syringe. Mix the equal volumes of filtered supernatant and fresh medium. The conditioned medium can be stored at 4 °C for up to 2 days.

3.4 Lipofection

Transfection can be performed by either lipofection or electroporation. Lipofection only requires a chemical reagent and does not require any specific equipment, but electroporation generally results in higher transfection efficiency. However, high mutation efficiency can be achieved with our protocol even using the lipofection method, because transfected cells are enriched with transient puromycin selection described later in this protocol. The flowchart of the cell culture experiments is shown in Fig. 2.

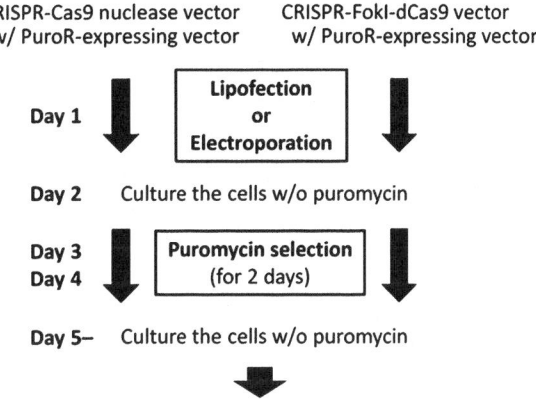

Fig. 2 General workflow of our protocol. After the mutation analysis, single cell cloning should be performed to isolate the gene knockout clones

1. The day before transfection (day 0), seed 2×10^5 cells in a 6-well collagen I-coated plate.

2. At day 1, replace the medium in the plate with 1 ml of Opti-MEM.

3. Mix 245 μl of Opti-MEM and 5 μl of Lipofectamine-3000 reagent in a microcentrifuge tube.

4. Mix 240 μl of Opti-MEM, 5 μl of P3000 reagent supplied with the Lipofectamine-3000 kit, 1.7 μg of CRISPR vector (3.4 μl), and 0.8 μg of PuroR-expressing vector (1.6 μl) in another microtube. Incubate the tubes for 5 min at room temperature.

5. Mix the two solutions and incubate for 30 min at room temperature.

6. Add the solution to the cultured cells and incubate the cells in a CO_2 incubator.

7. The day after transfection, replace the medium with fresh conditioned medium as described in **step 2**, Subheading 3.3.

3.5 Electroporation   Electroporation should be performed with specific electroporators that have been highly successful in DNA transfers using CRISPR-Cas9. As examples, Nucleofector (Lonza, Basel, Switzerland) [20–22], Neon (Thermo Fisher Scientific) [23–25], and NEPA21 are frequently used in the genome-editing field. The detailed procedures using NEPA21 are described in this protocol. The NEPA21 electroporator has been proven to work well for genome editing in various cultured cells and animal embryos, including HepG2 cells (this study), human induced pluripotent stem cells [26, 27], Jurkat cells [28], porcine fibroblast and myoblast cells [29], mouse [30] and rat [30, 31] embryos.

1. Prepare HepG2 cells with conditioned medium in a collagen I-coated 100-mm dish with 70–90 % confluency.

2. Add 700 μl of conditioned medium in a new collagen I-coated 12-well plate. Pre-incubate the plate in a CO_2 incubator.

3. Collect the cells from the 100-mm dish and adjust the cell number to 1.25×10^7 cells/ml with Opti-MEM.

4. Mix 80 μl of the cells and DNA samples, 6.7 μg of CRISPR vector (13.4 μl), and 3.3 μg of PuroR-expressing vector (6.6 μl), in a microcentrifuge tube (1×10^6 cells/100 μl).

5. Transfer the total amount (100 μl) of cells and DNA mixture to a 2 mm-gap cuvette and perform electroporation using the following parameters: Poring pulse: 150 V, 5 ms pulse, 50 ms gap, 2 pulse, 10 % decay (+ pulse orientation). Transfer pulse: 20 V, 50 ms pulse, 50 ms gap, 5 pulse, 40 % decay (± pulse orientation) (*see* **Note 11**).

6. Add 200 μl of pre-warmed HepG2 conditioned medium prepared in **step 1** in Subheading 3.5 to the cuvette, pipette the cells, and transfer the total amount of cells to the 12-well plate prepared in **step 1** in Subheading 3.5. Incubate the cells in a CO_2 incubator.

7. The day after transfection, replace the medium with fresh conditioned medium.

3.6 Puromycin Selection

When introducing an error-prone end-joining-mediated insertion or deletion without targeting vectors, stable drug selection described elsewhere [12] cannot be applied. Instead, transient selection using puromycin should be performed to enrich the transfected cells.

1. 24 h post-transfection, add puromycin to the cells (final concentration of 2 μg/ml).

2. 48 h post-transfection, replace the medium with fresh conditioned medium containing puromycin (final concentration of 2 μg/ml).

3. 72 h post-transfection, replace the medium with fresh conditioned medium not containing puromycin. A subsequent medium change should be performed every few days using conditioned medium until the cells show stable growth.

3.7 On-Target Mutation Analysis

Even if the final goal is the isolation of knockout cell clones, to know the mutation efficiency from the analysis of a bulk population of cells is quite important. There are many strategies evaluating mutation efficiencies such as heteroduplex mobility assay [32, 33], restriction fragment length polymorphism analysis [33, 34], high-resolution melt-curve analysis [35], and Surveyor nuclease assay [36]. In this protocol, we describe the procedure for Surveyor nuclease assay with actual examples (*see* Fig. 3).

1. Extract the genomic DNA from the cells using a DNAeasy Blood and Tissue Kit.

2. Perform PCR amplification around the target site using standard *Taq* polymerase. Examples of the PCR components and thermal cycling programs using LA *Taq* are shown below.

Component	Amount (μl)	Final concentration
10× LA PCR buffer	5	1×
10 μM primer mixture (F + R)	5	1 μM
2.5 mM dNTP mixture	7.5	0.375 mM
25 mM MgCl₂	5	2.5 mM
LA *Taq*	0.5	
100 ng/μl genomic DNA	2	4 ng/μl
ddH₂O	25	
Total	50	

Cycle number	Denature	Anneal	Extend
1	94 °C, 2 min		
2–34	94 °C, 30 s	67.5 °C, 30 s	72 °C, 20 s

3. Run the aliquot (~1 μl) of PCR products on a 3% agarose gel and check the amplification by ethidium bromide staining.

4. Purify the rest of the PCR products using a Gel and PCR clean up system kit (*see* **Note 12**).

5. Prepare the DNA solution for heteroduplex formation with the following composition in a PCR tube (*see* **Note 13**).

Component	Amount (μl)	Final concentration
400 ng of Purified PCR product	x	50 ng/μl
10× hybridization buffer[a]	0.8	1×
ddH₂O	$7.2 - x$	
Total	8	

[a]100 mM Tris–HCl (pH 8.5), 750 mM KCl, 15 mM MgCl₂

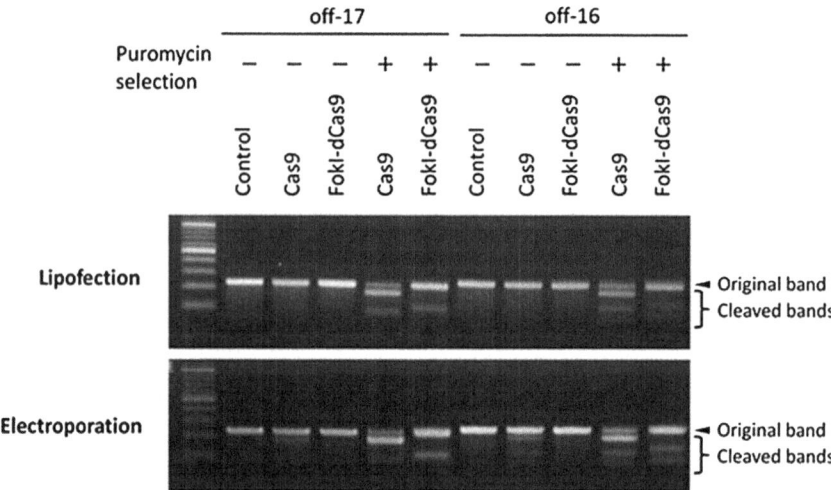

Fig. 3 Gel images of on-target mutation analysis using the Surveyor assay. Mutation frequencies dramatically increased with a transient puromycin selection in both lipofected and electroporated cells. 100 bp DNA ladder was used as a DNA size marker

6. Perform a denaturation and reannealing reaction using a thermal cycler as follows.

   (1) Denature at 95 °C for 5 min followed by (2) reannealing (slope) from 95 to 25 °C over 90 min and (3) hold at 25 °C indefinitely.

7. Add 0.4 μl of Enhancer S solution and mix well by pipetting (*see* **Note 12**).

8. Add 0.4 μl of Nuclease S solution and mix well by pipetting (*see* **Note 12**).

9. Set the tube on a thermal cycler and incubate at 42 °C for 30 min (*see* **Note 12**).

10. Run the total amount of the product on a 3 % agarose gel and stain the gel with ethidium bromide (*see* **Note 14**).

3.8 Off-Target Mutation Analysis

Although it is difficult to completely scan the potential off-target sites at a genome-wide scale, at least some high-risk sites should be checked by the Surveyor assay and/or DNA sequencing. We recommend using COSMID web tool (https://crispr.bme.gatech.edu/) [37], because COSMID can search not only mismatched sites but also the sites containing insertions or deletions [38], and produce a ranked list with scores indicating the level of risk for off-target mutations.

1. Access the COSMID web tool, choose the target genome, paste the query sequence, and set search options as follows, PAM suffix: NGG, indels: 3, Del: 2, Ins: 2.

2. Press submit button and get the results.

Table 2
Summary of potential off-target sites analyzed in this study

Locus	sgRNA	Score	Sequence[a]	Mismatch	Chromosome position
Off-target #1	off-17_L	0.38	GGCTAACTGGGCTGCAGTGGGGG—hit GGCAATCTGGGCTGCAGTGGNGG—query	2	ChrX:152286279–152286301
Off-target #2	off-16_R	0.45	CGCCGAACCTCAATTTCCCCTGC—hit GGCCGAAACTCAATTTCCCCNGG—query	2	Chr19:45320586–45320608
Off-target #3	off-17_L	0.48	AGGAAGCTGGGCTGCAGTGGAGG—hit GGCAATCTGGGCTGCAGTGGNGG—query	3	Chr14:105149531–105149553
Off-target #4	off-17_L	0.52	TGCAGCCTGGGCTGCAGTGGAGG—hit GGCAATCTGGGCTGCAGTGGNGG—query	3	Chr2:239405289–239405311
Off-target #5	off-17_L	0.52	TGCAGGCTGGGCTGCAGTGGGGG—hit GGCAATCTGGGCTGCAGTGGNGG—query	3	Chr12:31875148–31875170
Off-target #6	off-17_L	0.65	GCCCATCTGGGCTGCAGTGGGGG—hit GGCAATCTGGGCTGCAGTGGNGG—query	3	Chr5:65803876–65803898
Off-target #7	off-16_L	0.73	CTGCTTCAAGGTTGCTGGGCTGG—hit CTACTTAATGGTTGCTGGGCNGG—query	3	Chr10:119677084–119677106

[a]*Bold letters* indicate mismatched bases

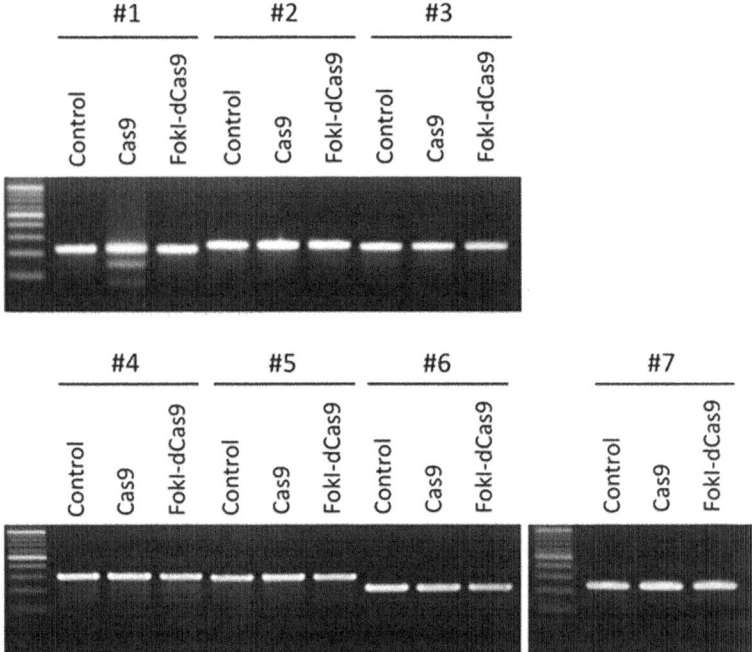

Fig. 4 Gel images of off-target mutation analysis using the Surveyor assay. Cleaved DNA fragments were detected only at #1 site in the cells transfected with CRISPR-Cas9 nuclease vector, suggesting that the scores calculated by COSMID were reliable and that the specificity of FokI-dCas9 was higher than with Cas9 nuclease, as reported previously. 100 bp DNA ladder was used as a DNA size marker

3. Repeat the search for all the sgRNA target sites used for multiplex targeting, and extract the high-risk sites according to the scores calculated. Smaller scores indicate the sites with higher risk of mutation. An actual example is shown in Table 2.

4. Perform mutation analysis for the high-risk sites by Surveyor assay as described above. The actual example is shown in Fig. 4.

4 Notes

1. The Multiplex CRISPR/Cas9 Assembly System Kit (Addgene kit #1000000055) enables the construction of an all-in-one vector expressing 2–7 sgRNAs and Cas9 nuclease or Cas9 nickase. It contains 12 ampicillin-resistant plasmids (pX330A-1x [2–7]) and pX330A_D10A-1x [2–7]) and 6 spectinomycin-resistant plasmids (pX330S [2–7]).

2. The Multiplex CRISPR dCas9/FokI-dCas9 Accessory Pack (Addgene Kit # 1000000062) enables the construction of an all-in-one vector expressing 2–7 sgRNAs and dCas9 or FokI-dCas9 along with pX330S (2–7), contained in the Multiplex CRISPR/Cas9 Assembly System Kit. The accessory pack

contains 12 ampicillin-resistant plasmids (pX330A_dCas9-1x [2–7] and pX330A_FokI-1x [2–7]).

3. The success rate of sgRNA cassette assembly with the Golden Gate method is greatly affected by the quality of the plasmid DNA. We routinely use the GenElute HP Plasmid Miniprep and Midiprep Kits (Sigma-Aldrich, St. Louis, MO).

4. The Quick Ligation Kit contains Quick Ligase and 2× reaction buffer. In our protocol, only the Quick Ligase is used and the separately sold 10× T4 DNA Ligase Reaction Buffer is used instead of the original 2× buffer.

5. The same medium can be used in other standard cell lines such as HEK293T cells, HeLa cells, and HCT116 cells, but it is important to use the optimal medium for the cells of interest other than the above cell lines.

6. It is important to check the potential off-target sites by in silico prediction using COSMID software, described in Subheading 3.8.

7. The target site for FokI-dCas9 can easily be searched using Serial Cloner software (http://serialbasics.free.fr/Serial_Cloner.html). For example, the target site with a 16-bp offset can be found by searching "5'-CCN$_{58}$GG-3'" in the sequence window of the Serial Cloner.

8. The detailed procedures can be found on Addgene's website and a list of the vectors for the different sgRNAs is provided below:

(https://www.addgene.org/crispr/yamamoto/multiplex-crispr-kit/#protocols-and-resources).

Number of target loci	Vector for sgRNA #1	Vector for sgRNA #2	Vector for sgRNA #3	Vector for sgRNA #4	Vector for sgRNA #5	Vector for sgRNA #6	Vector for sgRNA #7
For all-in-one CRISPR-Cas9 nuclease vectors							
2	pX330A-1x2	pX330S-2					
3	pX330A-1x3	pX330S-2	pX330S-3				
4	pX330A-1x4	pX330S-2	pX330S-3	pX330S-4			
5	pX330A-1x5	pX330S-2	pX330S-3	pX330S-4	pX330S-5		
6	pX330A-1x6	pX330S-2	pX330S-3	pX330S-4	pX330S-5	pX330S-6	
7	pX330A-1x7	pX330S-2	pX330S-3	pX330S-4	pX330S-5	pX330S-6	pX330S-7
For all-in-one CRISPR-FokI-dCas9 vectors							
1	pX330A_FokI-1x2	pX330S-2					
2	pX330A_FokI-1x4	pX330S-2	pX330S-3	pX330S-4			
3	pX330A_FokI-1x6	pX330S-2	pX330S-3	pX330S-4	pX330S-5	pX330S-6	

9. The detailed protocol for performing the Golden Gate method with the Quick Ligation Kit, T4 DNA ligase reaction buffer is described in the protocol available from Addgene's website: (https://www.addgene.org/crispr/yamamoto/multiplex-crispr-kit/#protocols-and-resources).

10. Neither collagen I-coated dishes and plates nor conditioned medium are necessary for robust cells such as HEK293T and HeLa cells.

11. The optimal parameters of poring pulse and transfer pulse are not identical among various cell types, and in fact, they are variable among laboratories even using the same cell line. It is important to check the transfection efficiency and cell viability using a GFP-expressing vector when starting to use a new cell line.

12. According to the instructions of the Surveyor Mutation Detection Kit, the PCR products can be used directly for the digestion assay without purification. However, we recommend purifying the products before proceeding to the digestion assay, because the unpurified products often cannot be cut clearly depending on the composition of PCR buffer.

13. Setting the optimal buffer composition is important not only for the Surveyor nuclease digestion but also for denaturation and reannealing. Note that the 10× hybridization buffer is not supplied with the kit.

14. In the Surveyor nuclease assay, a strong smear or extra bands often appear due to inappropriate reaction conditions such as buffer composition, reaction time, and the amounts of enzymes used. In most cases, the results are improved when such conditions are used as written in this protocol. If the results are not improved, even with optimization of the reaction conditions, there might be another reason such as the presence of SNPs. In these cases, it is useful for checking the mutation rate by RFLP analysis or DNA sequencing.

Acknowledgments

This work was supported by the Uehara Memorial Foundation (to T.S.), the Japan Society for the Promotion of Science (16K18478 to T.S. and 26290070 to T.Y.), and the Research Program on Hepatitis from Japan Agency for Medical Research and Development (AMED) to T.Y. The authors wish to express their thanks to Prof. Kazuaki Chayama, Department of Gastroenterology and Metabolism, Applied Life Science, Institute of Biomedical and Health Science, Hiroshima University, for his helpful suggestions concerning HepG2 cell culture.

References

1. Sakuma T, Woltjen K (2014) Nuclease-mediated genome editing: at the front-line of functional genomics technology. Dev Growth Differ 56:2–13

2. Sakuma T, Yamamoto T (2015) CRISPR/Cas9: the leading edge of genome editing technology. In: Yamamoto T (ed) Targeted genome editing using site-specific nucleases: ZFNs, TALENs, and the CRISPR/Cas9 system. Springer, Japan, pp 25–41

3. Sakuma T, Nishikawa A, Kume S, Chayama K, Yamamoto T (2014) Multiplex genome engineering in human cells using all-in-one CRISPR/Cas9 vector system. Sci Rep 4:5400

4. Nakagawa Y, Sakuma T, Sakamoto T, Ohmuraya M, Nakagata N, Yamamoto T (2015) Production of knockout mice by DNA microinjection of various CRISPR/Cas9 vectors into freeze-thawed fertilized oocytes. BMC Biotechnol 15:33

5. Engler C, Kandzia R, Marillonnet S (2008) A one pot, one step, precision cloning method with high throughput capability. PLoS One 3:e3647

6. Cermak T, Doyle EL, Christian M, Wang L, Zhang Y, Schmidt C, Baller JA, Somia NV, Bogdanove AJ, Voytas DF (2011) Efficient design and assembly of custom TALEN and other TAL effector-based constructs for DNA targeting. Nucleic Acids Res 39:e82

7. Sakuma T, Hosoi S, Woltjen K, Suzuki K, Kashiwagi K, Wada H, Ochiai H, Miyamoto T, Kawai N, Sasakura Y, Matsuura S, Okada Y, Kawahara A, Hayashi S, Yamamoto T (2013) Efficient TALEN construction and evaluation methods for human cell and animal applications. Genes Cells 18:315–326

8. Sakuma T, Ochiai H, Kaneko T, Mashimo T, Tokumasu D, Sakane Y, Suzuki K, Miyamoto T, Sakamoto N, Matsuura S, Yamamoto T (2013) Repeating pattern of non-RVD variations in DNA-binding modules enhances TALEN activity. Sci Rep 3:3379

9. Sakuma T, Yamamoto T (2016) Engineering customized TALENs using the platinum gate TALEN kit. Methods Mol Biol 1338:61–70

10. Nakade S, Tsubota T, Sakane Y, Kume S, Sakamoto N, Obara M, Daimon T, Sezutsu H, Yamamoto T, Sakuma T, Suzuki KT (2014) Microhomology-mediated end-joining-dependent integration of donor DNA in cells and animals using TALENs and CRISPR/Cas9. Nat Commun 5:5560

11. Hisano Y, Sakuma T, Nakade S, Ohga R, Ota S, Okamoto H, Yamamoto T, Kawahara A (2015) Precise in-frame integration of exogenous DNA mediated by CRISPR/Cas9 system in zebrafish. Sci Rep 5:8841

12. Sakuma T, Nakade S, Sakane Y, Suzuki KT, Yamamoto T (2016) MMEJ-assisted gene knock-in using TALENs and CRISPR-Cas9 with the PITCh systems. Nat Protoc 11:118–133

13. Sakuma T (2015) Front-line of genome editing technology for animal cell engineering. BMC Proc 9(Suppl 9):O1

14. Tsai SQ, Wyvekens N, Khayter C, Foden JA, Thapar V, Reyon D, Goodwin MJ, Aryee MJ, Joung JK (2014) Dimeric CRISPR RNA-guided FokI nucleases for highly specific genome editing. Nat Biotechnol 32:569–576

15. Guilinger JP, Thompson DB, Liu DR (2014) Fusion of catalytically inactive Cas9 to FokI nuclease improves the specificity of genome modification. Nat Biotechnol 32:577–582

16. Hara S, Tamano M, Yamashita S, Kato T, Saito T, Sakuma T, Yamamoto T, Inui M, Takada S (2015) Generation of mutant mice via the CRISPR/Cas9 system using FokI-dCas9. Sci Rep 5:11221

17. Hsu PD, Scott DA, Weinstein JA, Ran FA, Konermann S, Agarwala V, Li Y, Fine EJ, Wu X, Shalem O, Cradick TJ, Marraffini LA, Bao G, Zhang F (2013) DNA targeting specificity of RNA-guided Cas9 nucleases. Nat Biotechnol 31:827–832

18. Naito Y, Hino K, Bono H, Ui-Tei K (2015) CRISPRdirect: software for designing CRISPR/Cas guide RNA with reduced off-target sites. Bioinformatics 31:1120–1123

19. Ran FA, Hsu PD, Wright J, Agarwala V, Scott DA, Zhang F (2013) Genome engineering using the CRISPR-Cas9 system. Nat Protoc 8:2281–2308

20. Ran FA, Hsu PD, Lin CY, Gootenberg JS, Konermann S, Trevino AE, Scott DA, Inoue A, Matoba S, Zhang Y, Zhang F (2013) Double nicking by RNA-guided CRISPR Cas9 for enhanced genome editing specificity. Cell 154:1380–1389

21. Fu Y, Foden JA, Khayter C, Maeder ML, Reyon D, Joung JK, Sander JD (2013) High-frequency off-target mutagenesis induced by CRISPR-Cas nucleases in human cells. Nat Biotechnol 31:822–826

22. Smith C, Gore A, Yan W, Abalde-Atristain L, Li Z, He C, Wang Y, Brodsky RA, Zhang K, Cheng L, Ye Z (2014) Whole-genome sequencing analysis reveals high specificity of CRISPR/Cas9 and TALEN-based genome editing in human iPSCs. Cell Stem Cell 15:12–13

23. Feng Y, Sassi S, Shen JK, Yang X, Gao Y, Osaka E, Zhang J, Yang S, Yang C, Mankin HJ, Hornicek FJ, Duan Z (2015) Targeting CDK11 in osteosarcoma cells using the CRISPR-Cas9 system. J Orthop Res 33:199–207

24. Ebina H, Kanemura Y, Misawa N, Sakuma T, Kobayashi T, Yamamoto T, Koyanagi Y (2015) A high excision potential of TALENs for integrated DNA of HIV-based lentiviral vector. PLoS One 10:e0120047

25. Liang X, Potter J, Kumar S, Zou Y, Quintanilla R, Sridharan M, Carte J, Chen W, Roark N, Ranganathan S, Ravinder N, Chesnut JD (2015) Rapid and highly efficient mammalian cell engineering via Cas9 protein transfection. J Biotechnol 208:44–53

26. Li HL, Gee P, Ishida K, Hotta A (2016) Efficient genomic correction methods in human iPS cells using CRISPR-Cas9 system. Methods. doi:10.1016/j.ymeth.2015.10.015

27. Oceguera-Yanez F, Kim SI, Matsumoto T, Tan GW, Long X, Hatani T, Kondo T, Ikeya M, Yoshida Y, Inoue H, Woltjen K (2016) Engineering the AAVS1 locus for consistent and scalable transgene expression in human iPSCs and their differentiated derivatives. Methods. doi:10.1016/j.ymeth.2015.12.012

28. Matsubara Y, Chiba T, Kashimada K, Morio T, Takada S, Mizutani S, Asahara H (2014) Transcription activator-like effector nuclease-mediated transduction of exogenous gene into IL2RG locus. Sci Rep 4:5043

29. Rao S, Fujimura T, Matsunari H, Sakuma T, Nakano K, Watanabe M, Asano Y, Kitagawa E, Yamamoto T, Nagashima H (2016) Efficient modification of the myostatin gene in porcine somatic cells and generation of knockout piglets. Mol Reprod Dev. doi:10.1002/mrd.22591

30. Kaneko T, Mashimo T (2015) Simple genome editing of rodent intact embryos by electroporation. PLoS One 10:e0142755

31. Kaneko T, Sakuma T, Yamamoto T, Mashimo T (2014) Simple knockout by electroporation of engineered endonucleases into intact rat embryos. Sci Rep 4:6382

32. Ota S, Hisano Y, Muraki M, Hoshijima K, Dahlem TJ, Grunwald DJ, Okada Y, Kawahara A (2013) Efficient identification of TALEN-mediated genome modifications using heteroduplex mobility assays. Genes Cells 18:450–458

33. Nakagawa Y, Yamamoto T, Suzuki K, Araki K, Takeda N, Ohmuraya M, Sakuma T (2014) Screening methods to identify TALEN-mediated knockout mice. Exp Anim 63:79–84

34. Suzuki KT, Isoyama Y, Kashiwagi K, Sakuma T, Ochiai H, Sakamoto N, Furuno N, Kashiwagi A, Yamamoto T (2013) High efficiency TALENs enable F0 functional analysis by targeted gene disruption in Xenopus laevis embryos. Biol Open 2:448–452

35. Dahlem TJ, Hoshijima K, Jurynec MJ, Gunther D, Starker CG, Locke AS, Weis AM, Voytas DF, Grunwald DJ (2012) Simple methods for generating and detecting locus-specific mutations induced with TALENs in the zebrafish genome. PLoS Genet 8:e1002861

36. Guschin DY, Waite AJ, Katibah GE, Miller JC, Holmes MC, Rebar EJ (2010) A rapid and general assay for monitoring endogenous gene modification. Methods Mol Biol 649:247–256

37. Cradick TJ, Qiu P, Lee CM, Fine EJ, Bao G (2014) COSMID: a web-based tool for identifying and validating CRISPR/Cas off-target sites. Mol Ther Nucleic Acids 3:e214

38. Lin Y, Cradick TJ, Brown MT, Deshmukh H, Ranjan P, Sarode N, Wile BM, Vertino PM, Stewart FJ, Bao G (2014) CRISPR/Cas9 systems have off-target activity with insertions or deletions between target DNA and guide RNA sequences. Nucleic Acids Res 42:7473–7485

Chapter 5

CRISPR/Cas9-Mediated Mutagenesis of Human Pluripotent Stem Cells in Defined Xeno-Free E8 Medium

Chew-Li Soh and Danwei Huangfu

Abstract

The recent advent of engineered nucleases including the CRISPR/Cas9 system has greatly facilitated genome manipulation in human pluripotent stem cells (hPSCs). In addition to facilitating hPSC-based disease studies, the application of genome engineering in hPSCs has also opened up new avenues for cell replacement therapy. To improve consistency and reproducibility of hPSC-based studies, and to meet the safety and regulatory requirements for clinical translation, it is necessary to use a defined, xeno-free cell culture system. This chapter describes protocols for CRISPR/Cas9 genome editing in an inducible Cas9 hPSC-based system, using cells cultured in chemically defined, xeno-free E8 Medium on a recombinant human vitronectin substrate. We detail procedures for the design and transfection of CRISPR guide RNAs, colony selection, and the expansion and validation of clonal mutant lines, all within this fully defined culture condition. These methods may be applied to a wide range of genome-engineering applications in hPSCs, including those that utilize different types of site-specific nucleases such as zinc finger nucleases (ZFNs) and TALENs, and form a closer step towards clinical utility of these cells.

Key words CRISPR/Cas9, hPSCs, sgRNA, Vitronectin, Xeno-free E8 medium

1 Introduction

Human pluripotent stem cells (hPSCs) possess the capacity to both self-renew and give rise to all derivatives of the three embryonic germ layers, prompting rigorous investigations toward directing and utilizing their differentiation capabilities for regenerative medicine. Interrogating gene function in either self-renewing or differentiating hPSCs also provides a valuable platform towards further understanding human development and disease mechanisms. However, to fully realize this potential, we require more efficient and robust methods for genetic manipulation in human cells. While classical gene targeting via homologous recombination has proven to be a potent tool to dissect gene function in mouse embryonic stem cells (mESCs) [1], this process has been hampered by inherent differences in growth characteristics between mouse and human PSCs

Andrew Reeves (ed.), *In Vitro Mutagenesis: Methods and Protocols*, Methods in Molecular Biology, vol. 1498,
DOI 10.1007/978-1-4939-6472-7_5, © Springer Science+Business Media New York 2017

and by a lack of methods that are permissive for high efficiency genetic manipulation in human cells [2–4].

The recent advent of programmable site-specific nucleases including zinc finger nucleases (ZFNs), transcription activator-like effector nucleases (TALENs) and the clustered regularly interspaced short palindromic repeats (CRISPR)/ CRISPR-associated (Cas) systems have significantly facilitated genome engineering at the molecular level in a diverse range of cultured cells including hPSCs, reviewed in [5]. These chimeric nucleases enable a broad range of genetic modifications by inducing double-stranded breaks at precise locations, triggering the endogenous DNA repair machinery to activate either error-prone nonhomologous end joining (NHEJ) or homology-directed repair (HDR). Both mechanisms may be exploited for genetic manipulation by either introducing indels (via NHEJ) to create frameshift mutations that nullify gene alleles, or single nucleotide substitutions (via HDR) to recapitulate human disease mutations.

The rapid progress of genome targeting using engineered nucleases has opened up various avenues for hPSC-based disease studies through strategies such as knocking out a gene or recreating a patient mutation. However, for genome edited hPSCs to be useful for downstream cellular transplantation and disease treatment, we require a completely chemically defined xeno-free culture system for implementation in a clinical setting. Conventional methods of hPSC culture and gene targeting rely on their maintenance and expansion on a feeder layer of fibroblast cells in medium containing serum or serum replacement that incorporates animal products such as albumin. Feeders, serum (or serum replacement), and albumin have all demonstrated considerable batch variability, often requiring extensive quality control of new preparations, and uncertainty over their complex undefined components impedes the strict requirements involved in clinical utilization. For genome manipulation, feeder cells also reduce the transfection efficiency of hPSCs by sequestering transfection reagents, require considerable time and effort in preparation, and can also become depleted when drug selection is employed, necessitating further re-supplementation. Conditions for hPSCs are now beginning to transpire from feeder-dependent cells in animal-product containing media to culture on defined extracellular matrix substrates in xeno-free fully defined media [6–9]. Notably, Chen et al. [10] recently reported the development of a simplified hPSC culture medium, E8, which comprises eight completely defined components, for adherent culture of hPSCs on the defined matrix protein, vitronectin. At present, there are few studies that describe genome-editing procedures on hPSCs cultured in this fully defined culture condition [11].

We have previously described an efficient platform for genome editing in hPSCs called iCRISPR, using a combination of two gene-editing tools—the TALEN and CRISPR/Cas9 system [12]. Using this approach we have successfully created many hPSC lines with biallelic (homozygous or compound heterozygous) or heterozygous loss-of-function mutations in important developmental genes, within the context of culture conditions that depend on feeder cells and contain serum replacement in culture media as described in detail in [12]. Here we describe an optimized protocol for CRISPR/Cas9-mediated mutagenesis of hPSCs, using our iCRISPR platform, in the defined xeno-free E8 medium on vitronectin substrate. We focus on the generation of gene knockout hPSC lines through the creation of indel frameshift mutations, although our techniques and culture methods are also applicable to the generation of cell lines with precise nucleotide alterations and reporter transgene knockins [13]. These may also be adapted for general applications of TALEN- or CRISPR-mediated gene editing and are not limited to the iCRISPR system. This new, simplified, efficient protocol for genetic modification serves as a useful platform to study disease mechanisms and facilitates transfer of basic research towards clinical application.

2 Materials

2.1 Cell Culture

1. Commercially prepared E8 Medium Kit (Life Technologies).
2. Truncated recombinant human vitronectin (VTN-N).
3. Dulbecco's Phosphate-Buffered Saline (DPBS) without calcium or magnesium.
4. Penicillin–streptomycin solution (1%).
5. 0.5 M EDTA, pH 8.0.
6. Rho-associated protein kinase Y-27632 (ROCK) inhibitor.
7. Dimethyl sulfoxide (DMSO).

2.2 CRISPR Design

1. Herculase II Fusion DNA Polymerase.
2. MEGAshortscript T7 Transcription Kit.
3. MEGAclear Transcription Clean-Up Kit.

2.3 Transfection

1. Doxycycline hyclate.
2. TrypLE Select Enzyme 1×, no phenol red.
3. Opti-MEM I Reduced Serum Medium.
4. Lipofectamine RNAiMAX Reagent.
5. CO_2 incubator.
6. Beckman Coulter Vi-CELL XR or an equivalent automated cell viability analyser.

2.4 Molecular Biology Reagents

1. T7 Endonuclease I.
2. DNeasy Blood and Tissue Kit (Qiagen).
3. 10× NEB buffer 2.
4. 1× TAE buffer: 40 mM Tris acetate, 1 mM EDTA.
5. UltraPure agarose.
6. 1 % ethidium bromide solution.
7. 6× gel loading dye.
8. Gel documentation system (such as Gel Doc XR+ System, Bio-Rad).
9. 10× Lysis Buffer: 0.45 % NP-40, 0.45 % Tween 20 in 10 mM Tris base, pH 8.0. Store at 4 °C.
10. IGEPAL CA-630 (Non-ionic detergent, Sigma).
11. Tween 20.
12. Tris base.
13. Proteinase K, recombinant, PCR grade.
14. Adhesive PCR plate seals.
15. Zero Blunt TOPO PCR Cloning Kit (Life Technologies).

3 Methods

3.1 iCas9 hPSCs

A number of laboratories have now successfully used the CRISPR/Cas9 system for genome editing in hPSCs as reviewed in [5]. However, the delivery of Cas9 and a single chimeric guide RNA (sgRNA) remains a step that can limit gene editing efficiency and ease of use. We believe that inducible expression of Cas9 (iCas9) from the *AAVS1* safe harbor locus proffers a robust approach to genetic modification via the CRISPR/Cas9 system. Because this system no longer entails co-transfection of Cas9, the high transfection efficiency of the smaller sgRNA leads to significantly improved rates of genome editing in the target locus. Our laboratory has previously documented generation of iCas9 hPSCs by TALEN-mediated targeting into the *AAVS1* locus [12]. Two donor plasmids were co-electroporated with the *AAVS1*-targeting TALEN constructs. One donor plasmid (Puro-Cas9) comprised a doxycycline-inducible Cas9 expression cassette selectable with puromycin, the other (Neo-M2rtTA) contained a constitutive reverse tetracycline transactivator selectable with G418 (Geneticin). Once clonal iCas9 lines are established, Cas9 expression is induced by doxycycline treatment (*see* Fig. 1). Because the iCRISPR platform entails only the transfection of a short sgRNA oligonucleotide to perform gene editing, the system is efficient, cost-effective, convenient and has reduced technical demand, particularly if one desires to model complex human traits

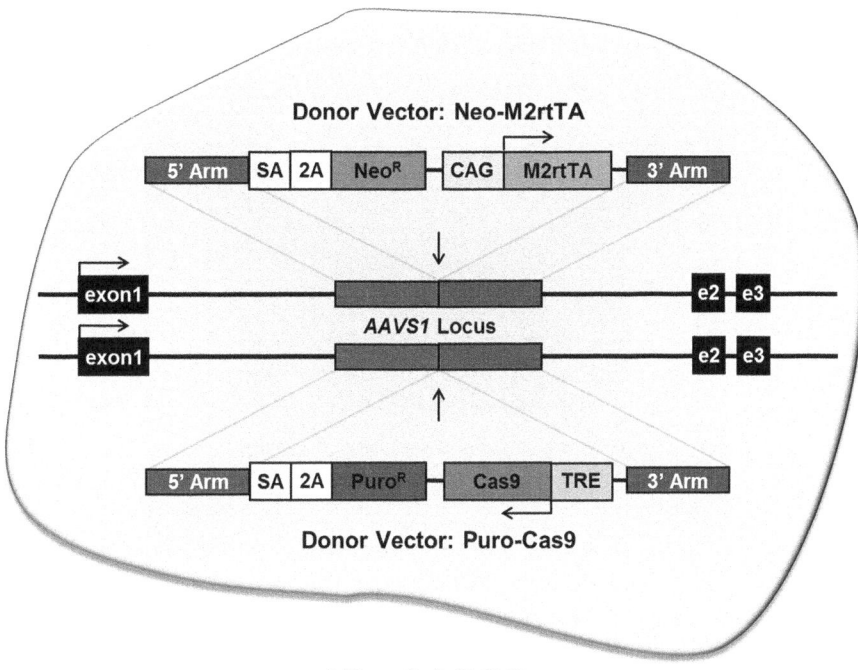

iCas9 hPSCs

Fig. 1 Generation of iCas9 hPSCs through TALEN-mediated gene targeting at the *AAVS1* locus. To establish the iCRISPR system, the Puro-Cas9 and Neo-M2rtTA donor vectors were co-electroporated into hPSCs and the clonal lines were established through drug selection. Expression of Cas9 is induced following treatment of cells with doxycycline. *SA* splice acceptor, *2A* self-cleaving 2A peptide, *NeoR* neomycin resistance gene, *CAG* constitutive synthetic promoter, *M2rtTA* reverse tetracycline-controlled transactivator sequence, *PuroR* puromycin resistance gene, *TRE* tetracycline response element

caused by multiple gene mutations. One can also envision that an integrated Cas9 line can be easily extended for CRISPR-mediated screens [14], where only the smaller and easier to infect sgRNA library would need to be introduced into the cell. For detailed information on TALEN design and donor vector construction please refer to Gonzalez et al. [12] and Zhu et al. [13]. Alternatively, the iCas9 lines can now be procured in three hESC (HUES8, HUES9, MEL1) and one hiPSC (BJ iPSC) backgrounds via the MSKCC Antibody and Bioresource Core Facility, New York, NY.

Below we detail routine hPSC culturing and passaging techniques in E8 Medium on vitronectin extracellular matrix substrate. Our protocol is based on experience working with hPSCs on the H1 (WiCell Research Institute), HUES8 (Harvard University) and MEL1 (StemCore, Stem Cells Ltd, University of Queensland) background. Other hPSC lines may have inherent differences in characteristics that may require adjustment or optimization of the protocol.

Table 1
Suggested VTN-N coating volumes for different sized culture vessels

Culture vessel	Volume of diluted VTN-N (5 μg/mL) solution
100 mm dish	5 mL
6-well plate	1 mL per well
12-well plate	500 μL per well
24-well plate	250 μL per well
96-well plate	50 μL per well

We routinely culture our hPSCs in E8 Medium on vitronectin-coated culture vessels, employing a truncated recombinant human form of vitronectin (VTN-N) that has been demonstrated to maintain pluripotency and normal growth characteristics in multiple human PSC lines [10]. A final VTN-N coating concentration of 0.5 μg/cm² is recommended, although we find cells can survive in a concentration range similar to this and generally use a working concentration of 5 μg/mL in different-sized culture vessels (*see* Table 1) (*see* **Note 1**). We routinely store VTN-N in 60 μL aliquots at −80 °C.

3.2 hPSC Cultures in E8

1. Thaw VTN-N at room temperature and coat the surface at 5 μg/mL. For example, when coating an entire 6-well plate, thaw 60 μL of stock VTN-N and dilute into 6 mL of Dulbecco's Phosphate-Buffered Saline (DPBS) without calcium or magnesium. Gently mix by pipetting the solution up and down.

2. Add 1 mL of diluted VTN-N solution to each well of a 6-well plate.

3. Incubate the coated plates at room temperature for at least 1 h (*see* **Note 2**). To ensure that VTN-N has not evaporated from wells prior to use, we recommend that coated vessels are not left out at room temperature for more than 4 h.

3.3 hPSC Maintenance in E8

1. To prepare complete E8 Medium, thaw the frozen E8 Supplement at 4 °C overnight. Transfer the entire contents of the E8 Supplement (~10 mL) into the bottle of E8 Basal Medium (500 mL). One can also add 5 mL Penicillin-streptomycin solution (1 % v/v final concentration) to the E8 Basal Medium.

2. When hPSC colonies have reached ~80 % confluence (*see* Fig. 2), aspirate expended E8 media from the vessel.

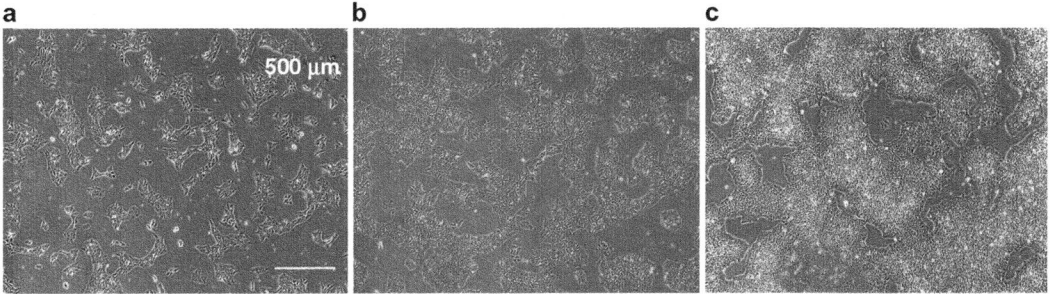

Fig. 2 Expected hPSC morphology following passage in E8 Medium on vitronectin-coated culture vessels using EDTA dissociation. *Left* to *right*: (**a**) hPSCs in culture 24 h after a 1:6 ratio passage. (**b**) hPSCs after 48 h begin to form tightly packed colonies. (**c**) hPSCs that are ready for passage are routinely around 80 % confluent

Table 2
Suggested reagent volumes for culturing in E8 conditions in different sized culture vessels

Culture vessel	DPBS for rinse	0.5 mM EDTA in DPBS	Complete E8 medium
100 mm dish	10 mL	5 mL	10 mL
6-well plate	2 mL per well	1 mL per well	2 mL per well
12-well plate	1 mL per well	500 µL per well	1 mL per well
24-well plate	500 µL per well	250 µL per well	500 µL per well
96-well plate	100 µL per well	50 µL per well	100 µL per well

3. Rinse once with DPBS without calcium and magnesium. Refer to Table 2 for suggested rinse volumes.

4. To dissociate the hPSCs, add 0.5 mM EDTA in DPBS (pre-warmed to room temperature) to the vessel (*see* **Note 3**). Refer to Table 2 for suggested dissociation volumes. Let cells sit in EDTA at room temperature for 4–5 min, at which point cells should start separating and colony edges will begin to detach. EDTA partially dissociates the cells enabling the hPSCs to adhere as small aggregates. We find that adherence as small colonies is more beneficial to cell survival than individualized hPSCs, as previously reported [9].

5. Before colonies have detached, aspirate the EDTA solution from the cells. With gentle pipetting, use complete E8 Medium to wash and disperse the colonies. Perform quickly as media components rapidly neutralize the EDTA and colonies will start to reattach to the vessel.

6. Collect the desired proportion of cells and plate onto a new pre-coated VTN-N vessel in complete E8 Medium with 10 µM Y-27632 ROCK inhibitor. For recommended plating volume, refer to Table 2. Gently sway the vessel to ensure

Single-stranded DNA sgRNA template (120-nt)

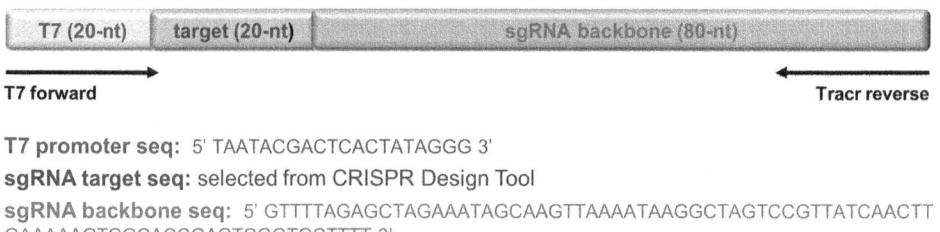

T7 promoter seq: 5' TAATACGACTCACTATAGGG 3'

sgRNA target seq: selected from CRISPR Design Tool

sgRNA backbone seq: 5' GTTTTAGAGCTAGAAATAGCAAGTTAAAATAAGGCTAGTCCGTTATCAACTT
GAAAAAGTGGCACCGAGTCGGTGCTTTT 3'

Fig. 3 Sequence structure for the 120-nt single-stranded sgRNA template used for sgRNA production. The template includes a 20-nt T7 promoter sequence, the 20-nt variable sgRNA guide sequence that specifies target recognition, and the 80-nt invariant sgRNA backbone sequence. The oligonucleotide is PCR-amplified and then the RNA is transcribed in vitro and purified in preparation for transfection

even distribution of cells, and incubate at 37 °C, in a 5 % CO_2 incubator overnight.

7. Change complete E8 Medium daily (without Y-27632 ROCK inhibitor). Depending on the growth characteristics of the particular cell line and their downstream purpose, we customarily passage cells at a ratio of 1:6–1:12, every 4–5 days (*see* Fig. 2). We have found that cells can be directly adapted from live feeder-dependent hPSC culture to VTN-N adherent E8 culture. However, cells frozen in feeder-dependent conditions may need to be thawed first onto feeders before adaptation to feeder-independent culture, to improve survival rate.

3.4 sgRNA Production

To generate gene knockout lines using our iCRISPR platform, sgRNAs are transfected into the inducible iCas9 hPSC line. A 120-nt synthetic single-stranded DNA containing the T7 promoter sequence followed by the variable 20-nt sgRNA target sequence and the remaining constant sgRNA sequence is used as the PCR template, and in vitro transcription is performed to generate sgRNAs for transfection (*see* Fig. 3) [13–15]. We have found this method more efficient and cost-effective in comparison to the transfection of an sgRNA-expressing plasmid. Cas9 endonuclease induces a double-stranded break, which can trigger DNA repair via NHEJ, leading to random indel mutations. This then allows the creation of heterozygous and biallelic (homozygous or compound heterozygous) gene mutations through codon reading frameshifts. To ensure that gene function is disrupted, we recommend designing an sgRNA such that it targets sequences immediately upstream of or within an essential functional domain of the corresponding protein. If there are multiple splice variants of the gene of interest, target a functional region that is at least relevant to the isoform in question. If a gene is not well annotated, it will be more prudent to design an sgRNA close to the initiation start codon. For the

Table 3
Primers for PCR amplification of the 120-nt sgRNA oligo

Primer	Sequence (5′–3′)
T7 forward	TAATACGACTCACTATAGGG
Tracr reverse	AAAAGCACCGACTCGGTGCC

design of the 20-nt variable sgRNA seed sequence, we routinely employ the guidelines set out by the CRISPR Design Tool (http://crispr.mit.edu) developed by the Zhang Lab at MIT [16]. The software predicts the most efficient sgRNAs based on high targeting specificity and a minimal number of off-target sites (*see* **Note 4**). We customarily design three sgRNAs per gene, and go on to generate clonal lines from at least two of these CRISPRs that can effectively induce indel mutations at the target locus, as first determined by a T7E1 Assay.

1. For each CRISPR, design a 120-nt oligo that includes the T7 promoter sequence, the variable 20-nt sgRNA recognition sequence (selected from the CRISPR Design Tool) and the constant chimeric sgRNA sequence (*see* Fig. 3). Dilute the oligo to a 100 μM stock solution, then prepare a 2.5 nM working solution. The 120-nt oligonucleotide is used as a template for PCR amplification using the T7 forward and Tracr reverse primers (*see* Tables 3, 4, and 5). Resolve 5 μL of each PCR reaction by electrophoresis on a 1% ethidium bromide-stained agarose gel to check for the appropriate amplification of the template.

2. For in vitro transcription of the 120-nt sgRNA PCR product to yield the final sgRNA product, we employ the MEGAshortscript T7 Transcription Kit, using the reagents indicated in Table 6.

3. The reaction is incubated at 37 °C for 4 h to overnight. One microliter of the TURBO DNase supplied with the kit is added to the reaction to remove the DNA template, and the reaction mix is further incubated at 37 °C for 15 min. To purify the RNA transcript, we normally use the MEGAclear Transcription Clean-Up Kit. Typically, the RNA is eluted in 100 μL RNAse-free water, yielding ~50–100 μg total RNA. When possible, the RNA concentration was adjusted to 600 or 300 ng/μL and stored at −80 °C until ready for use.

3.5 sgRNA Transfection

hPSCs are typically transfected as cultures in 24-well plates. We suggest using at least two different cell densities (usually 1:4 or 1:2 ratio) to achieve an optimal balance between cell survival and indel mutation rate, which may need to be adjusted as both CRISPR toxicity and cell line can affect the rate of survival following transfection. We generally perform each transfection in duplicate: one

Table 4
PCR reaction mix to amplify the 120-nt sgRNA oligo (50 μL)

Component	Quantity (in μL)
Distilled water (dH₂O)	35.5
5× Herculase II reaction buffer	10
Herculase II fusion DNA polymerase	0.5
dNTP mix (25 mM of each dNTP)	0.5
T7 forward (10 μM)	1.25
Tracr reverse (10 μM)	1.25
120-nt sgRNA oligo template (2.5 nM)	1

Table 5
PCR cycling parameters to amplify the 120-nt sgRNA oligo

Cycle number	Denature	Anneal	Extend
1	95 °C, 2 min		
2-31	95 °C, 20 s	60 °C, 20 s	72 °C, 1 min
32			72 °C, 3 min

well is used to collect genomic DNA for the T7E1 Assay, which provides an estimate of the rate of mutagenesis, and the other well is used for cell replating at clonal density to enable gene knockout lines to be established (*see* Fig. 4).

1. (Day-1) One day prior to transfection, when cells are ~50–60% confluent, treat iCas9 cells with 2 μg/mL doxycycline and 10 μM Y-27632 ROCK inhibitor. This will activate transcription of Cas9 endonuclease before sgRNA transfection.

2. (Day 0) Depending on the number of transfections to be performed, coat 24-well tissue culture plates with VTN-N (*see* Table 1) at room temperature for at least 1 h.

3. Dissociate doxycycline- and ROCK inhibitor-treated iCas9 cells with 1× TrypLE Select. It is important that cells are dissociated to individualized cells as opposed to EDTA-treated small aggregates for transfection. This ensures access of cells to the small sgRNA oligonucleotides and associated transfection reagents. Aspirate pre-coated VTN-N. Resuspend cells in complete E8 Medium (with 2 μg/mL doxycycline and 10 μM Y-27632 ROCK inhibitor) such that they achieve a surface area ratio of 1:4 or 1:2 and perform cell plating in duplicates

Table 6
In vitro transcription reaction mix (20 μL)

Component	Quantity (in μL)
T7 10× reaction buffer	2
T7 ATP solution	2
T7 CTP solution	2
T7 GTP solution	2
T7 UTP solution	2
T7 enzyme mix	2
PCR-amplified sgRNA DNA template	8

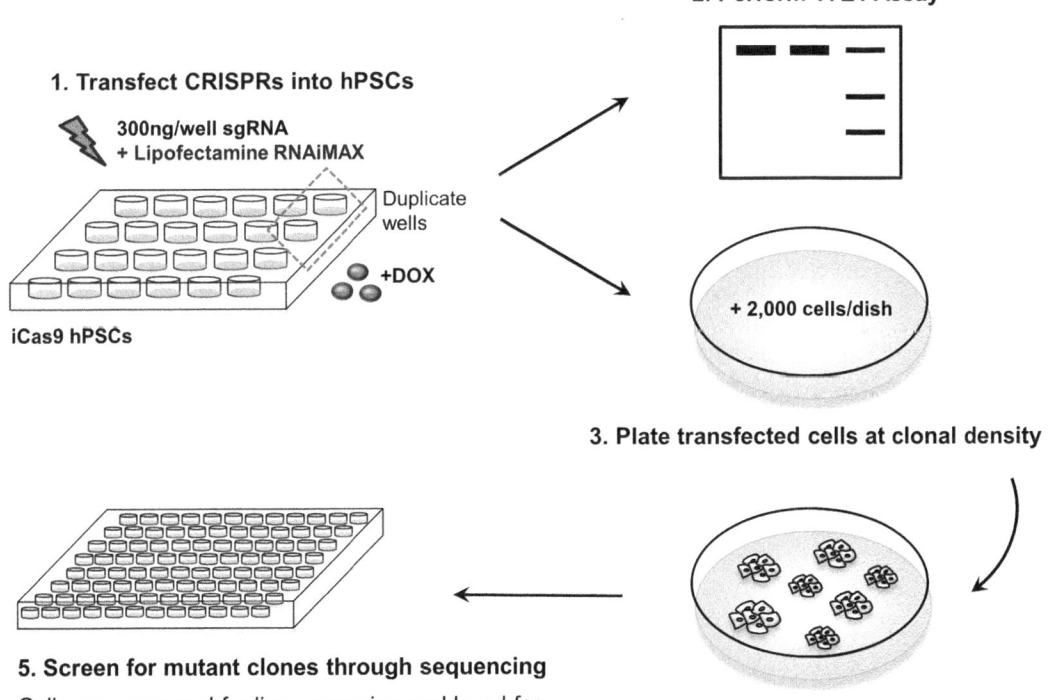

Fig. 4 Procedure for generating gene knockout mutant lines in iCas9 hPSCs in a feeder-free system. sgRNAs targeting specific loci are transfected into iCas9 hPSCs in duplicate and Cas9 expression is induced with doxycycline (DOX). Cells in one well are used to measure the rate of indel mutagenesis by the T7E1 assay. Cells in the other wells are used for replating at clonal density. Once colonies have grown, they are picked for sequence determination and expansion to establish clonal mutant lines

for each different sgRNA and density. For each dilution, plate an additional well to serve as an untransfected negative control for the ensuing T7E1 Assay.

4. For each pair of duplicate transfections, set up in a sterile 1.5 mL microcentrifuge tube: Mixture 1:50 μL Opti-MEM medium with 3 μL Lipofectamine RNAiMAX Reagent; Mixture 2: 50 μL Opti-MEM Medium with 600 ng sgRNA (typically 1–2 μL); Combine both mixtures (~100 μL total) and incubate at room temperature for 5 min.

5. Add 50 μL of each sgRNA-lipid complex dropwise to newly dissociated cells in duplicate. Each well will receive ~300 ng of sgRNA.

6. Incubate transfected cells at 37 °C in a 5% CO_2 incubator overnight.

7. (Day 1) Change complete E8 Medium with 2 μg/mL doxycycline and 10 μM Y-27632 ROCK inhibitor.

8. (Days 2–4) Change complete E8 Medium. Depending on transfection survival rates and confluency, proceed to cell harvest for genomic DNA collection or cell deposition at clonal density.

3.6 T7 Endonuclease I Assay

Cas9 cleavage triggers endogenous mechanisms of DNA repair that can eventually lead to mutations through NHEJ. These mutations include deletions, insertions, as well as deletions combined with insertions, and may occur in one allele or both. Because genome editing with engineered Cas9 nucleases will not target all intended loci with similar efficiencies, the mutation hit rate at a particular locus should be assessed prior to the selection and sequencing of isolated clones. The genomic DNA from targeted cells is first amplified by PCR. The PCR products are denatured and hybridized to allow heteroduplex formation between wild-type and CRISPR/Cas9-mutated DNA. We use the enzyme T7E1 to recognize and cleave mismatched DNA. The resulting cleaved PCR products are visualized by gel electrophoresis and band intensity quantified to provide an evaluation of mutation rate (*see* Fig. 5). Because T7E1 does not detect single base indels effectively, the T7E1 mutation rate is generally an underestimate of the CRISPR mutation rate at a specific locus [17] (*see* **Note 5**).

1. Three to four days after sgRNA transfection, cells should be 50–80% confluent depending on cell density and CRISPR toxicity. From one well of each duplicate, harvest cells with TrypLE Select and proceed to genomic DNA isolation using the DNeasy Blood and Tissue Kit. Also, extract genomic DNA from the untransfected well, which will serve as a control in the T7E1 Assay.

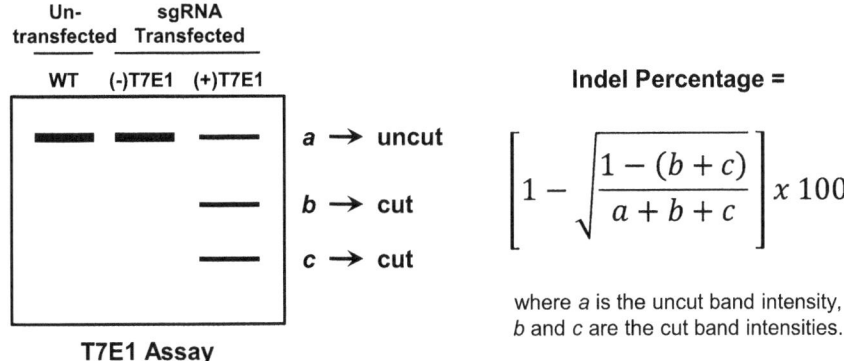

Fig. 5 T7E1 assay to measure the rate of indel mutagenesis. CRISPR-transfected cells are lysed and the genomic DNA is extracted and PCR-amplified around the target locus. PCR products are denatured and hybridized to enable heteroduplex formation between wild-type DNA and mutated DNA. T7E1 recognizes indels generated by nonhomologous end joining (NHEJ) induced by a CRISPR/Cas9-mediated double-stranded break and cuts the reannealed mismatched DNA. Cleaved products can be visualized by gel electrophoresis and the band intensity is quantified for determination of percentage indel mutagenesis

Table 7
PCR reaction mix to amplify genomic region for T7E1 Assay (25 μL)

Component	Quantity (in μL)
Distilled water (dH$_2$O)	To 25
5× Herculase II reaction buffer	5
Herculase II fusion DNA polymerase	0.25
dNTP mix (25 mM of each dNTP)	0.25
Forward primer (10 μM)	0.625
Reverse primer (10 μM)	0.625
sgRNA-transfected genomic DNA	100–200 ng

2. Perform PCR using primer pairs designed to amplify a genomic region of about 500–1000 bp flanking the CRISPR target site. We recommend using high-fidelity Herculase II Fusion DNA Polymerase, with suggested reaction mixtures and cycling conditions detailed in Tables 7 and 8. If possible, perform a gradient PCR for annealing temperature, to find the optimal temperature for each new primer pair that will enable efficient and specific amplification of targets. We routinely find an ideal range to be between 55 and 65 °C.

3. To perform the T7E1 Assay, PCR-amplified DNA is first denatured and hybridized (*see* Tables 9 and 10), then it is digested with T7E1 enzyme (*see* Table 11). DNA denaturation and hybridization is performed in a thermocycler.

Table 8
PCR cycling parameters to amplify genomic region for T7E1 Assay

Cycle number	Denature	Anneal	Extend
1	95 °C, 2 min		
2–36	95 °C, 20 s	55–65 °C, 20 s	72 °C, 1 min
37			72 °C, 3 min

Table 9
Reaction mix for DNA denaturation and hybridization (20 μL)

Component	Quantity (μL)
Unpurified PCR product	8
NEBuffer 2 10×	2
Distilled water (dH$_2$O)	10

Table 10
Thermocycling parameters for gradual denaturation and hybridization

Temperature (°C)	Duration (min)	Thermocycler conditions
95	10	
85	1	Ramp to 85 °C at 2 °C/s
75	1	Ramp to 75 °C at 0.3 °C/s
65	1	Ramp to 65 °C at 0.3 °C/s
55	1	Ramp to 55 °C at 0.3 °C/s
45	1	Ramp to 45 °C at 0.3 °C/s
35	1	Ramp to 35 °C at 0.3 °C/s
25	1	Ramp to 25 °C at 0.3 °C/s
4	Hold	

Table 11
Reaction mix for T7E1 digestion

Component	Quantity (μL)
Denatured and hybridized PCR product	10
T7E1 (10 U/μL)	0.1 (1 U)

4. Incubate reaction at 37 °C for 30 min, then proceed to the T7E1 mutagenesis quantification. The remaining 10 μL of denatured and hybridized PCR product can then be used as an undigested control during quantification.

5. To quantify the T7E1 Assay, resolve T7E1-digested hybridized samples alongside T7E1-undigested samples and the untransfected PCR-amplified sample by gel electrophoresis on an ethidium bromide-stained, 2.5% 1× TAE agarose gel (*see* Fig. 5). We recommend using Orange G gel loading dye as opposed to Bromophenol blue to prevent obstruction of the resolved bands by the gel tracking dye. Illuminate the resolved gel under UV light and image with a Gel Doc XR+ System or equivalent system. Image Lab Software (Bio-Rad) or ImageJ Software (NIH) is used to determine the relative band intensities of cut and uncut DNA to calculate the frequency of indel mutagenesis, using the formula: Indel Percentage $= (1 - (\sqrt{(1 - (b + c))/(a + b + c)}) \times 100$, where a is the intensity of the undigested PCR product, and b and c are the intensities of the T7E1-cleaved product (*see* Fig. 5). The expected size of each cleavage product can be estimated by the position of the forward and reverse primers and the intended position of Cas9 cleavage. Cas9 (from *S. pyogenes*) is known to make a blunt cut between the 17th and 18th bases in the 20-nt crRNA target sequence, three bases 5′ of the NGG PAM recognition motif [18].

3.7 Replating Transfected Cells

After determining the rate of indel mutagenesis by the T7E1 assay, identify the best transfection condition for each CRISPR. Typically, three to four days after sgRNA transfection, cells are harvested from the corresponding duplicate well for plating at clonal density to allow isolation of individually mutated clones. To accomplish this, cells are harvested by dissociation into single cells with TrypLE Select and plated at 2,000 cells per 100 mm tissue culture dish for colony growth.

1. For each CRISPR, coat 2×100 mm tissue culture dishes with VTN-N (*see* Table 1) at room temperature for at least 1 h.

2. Three to four days after sgRNA transfection, cells should be 50–80% confluent depending on cell density and CRISPR toxicity. From the remaining well (which was not harvested for the T7E1 Assay) of each duplicate, rinse once with 500 μL DPBS without calcium and magnesium.

3. To dissociate the hPSCs, add 250 μL TrypLE Select to each well. Incubate the cells at 37 °C for 4–5 min, at which point cells should start to detach. Gently pipette the colony up and down and collect in a 15 mL Falcon tube.

4. Rinse each well with 2 mL DPBS and collect the cells into the 15 mL Falcon tube.

5. Centrifuge at $200 \times g$ for 4 min at room temperature and aspirate the supernatant.

6. Resuspend in an appropriate volume of complete E8 Medium to perform a cell count. We recommend resuspending cells in 5 mL E8 Medium if using the Beckman Coulter Vi-CELL XR or an equivalent automated cell viability analyzer. If the cell count is to be performed manually via a haemocytometer, resuspend the cells in a smaller volume of E8 Medium. Determine the volume required for 2,000 cells.

7. Aspirate VTN-N coating from 100 mm dishes and add 10 mL E8 Medium with 10 μM Y-27632 ROCK inhibitor.

8. Add 2,000 cells per dish for each different CRISPR. Perform in duplicate. Gently swirl the dish and incubate cells at 37 °C, in a 5% CO_2 incubator overnight.

9. To allow the cells to settle, do not change media the following day. Two days following single-cell deposition, change with E8 medium without ROCK inhibitor. Thereafter, change the medium daily until colonies are ready for picking.

3.8 Colony Picking   We usually allow single-cell colonies to grow for 10–12 days, at which point they should reach ~2 mm in diameter, but should not have merged with neighboring colonies (*see* Fig. 6). Each clone is picked by mechanical disaggregation using a handheld pipette with filter tip and deposited into single wells of a 96-well plate. When cells have reached confluence, they are passaged onto a new 96-well plate and the remaining cells spun down for genomic DNA extraction and sequence analysis. Alternatively, colonies could be picked and deposited into duplicate 96-well plates, with one plate for maintenance and expansion, and the other for DNA preparation. The number of colonies picked per sgRNA transfected will depend on the rate of indel mutagenesis. We routinely pick 48 clones for

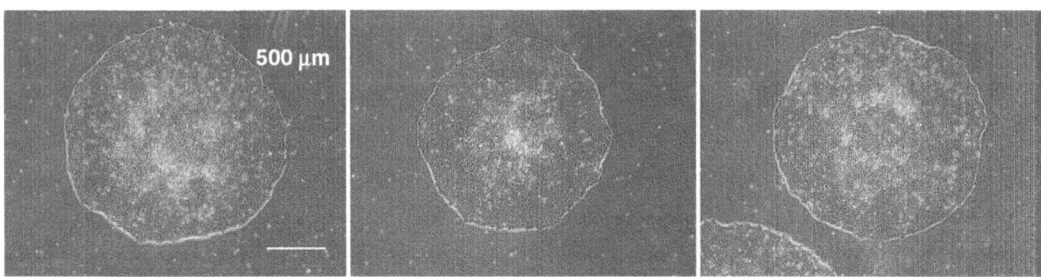

Fig. 6 Expected morphology of colonies cultured for 10–12 days following clonal density deposition in E8 medium on vitronectin-coated culture vessels. The colonies may vary slightly in size, but should not have merged with neighboring colonies

indel efficiencies >20%, and about 96 clones for indel efficiencies between 5 and 20%. It takes ~60–90 min to pick 96 clones (*see* **Note 6**).

1. For the method involving direct picking and deposition of a colony into a single well, coat the relevant number of 96-well tissue culture plates with VTN-N (*see* Table 1) at room temperature for at least 1 h.

2. Aspirate the VTN-N coating from 96-well plates, and then add 100 μL/well E8 medium with 10 μM Y-27632 ROCK inhibitor.

3. Change each 100 mm dish with fresh E8 medium before colony picking.

4. Using a manual p200 pipette set at 25 μL with a filter tip, mechanically disaggregate a well-separated colony into a crude grid structure. Dislodge colony pieces and transfer small digested clumps into one well of a new 96-well plate. To avoid cross contamination, ensure all dislodged pieces are removed before dissection of the next clone with a new filter tip. The smaller the colony pieces disaggregated, the better the survival rate (*see* **Note 7**).

5. Incubate 96-well colony plates at 37 °C, in a 5% CO_2 incubator overnight.

6. The following day, change 96-well plates with E8 Media without ROCK inhibitor (100 μL/well) (*see* **Note 8**). Thereafter, change media daily until colonies are ready for DNA extraction.

3.9 PCR-Based Screening

When cells have reached confluence, colony plates are prepared for DNA extraction and concurrently passaged into new 96-well plates. We routinely screen for mutant clones through a simplified process of crude genomic DNA extraction without phenol/chloroform purification and precipitation, followed by direct PCR amplification of the region of interest and Sanger sequencing of the unpurified PCR product. This method permits a large number of clones to be screened rapidly with the aid of multichannel pipettes, and provides an accurate and efficient method to assay for mutant clonal lines. PCR amplification is performed using the same forward and reverse primers and PCR cycling parameters as for the T7E1 assay. Sanger sequencing is performed using a forward primer internal to the one utilized in the T7E1 assay to increase amplification specificity.

1. Coat an equivalent number of 96-well tissue culture plates for passaging of colonies with VTN-N (*see* Table 1) at room temperature for at least 1 h.

2. Aspirate VTN-N coating from 96-well plates, and add 100 μL/well E8 Medium with 10 μM Y-27632 ROCK inhibitor.

3. On 96-well plates with colonies, rinse each well once with 100 μL DPBS without calcium and magnesium.

4. To dissociate the colonies, add 50 μL of 0.5 mM EDTA in DPBS (pre-warmed to room temperature) to each well. Let cells sit in EDTA at room temperature for 4–5 min, at which point cells should start separating and colony edges will begin to detach. It is important at this step to use EDTA instead of TrypLE Select as the passaging of colonies is sensitive to single-cell disaggregation, and colonies survive better when passaged in small aggregates.

5. Before colonies have detached, aspirate the EDTA solution from the cells.

6. Add 100 μL/well E8 medium with 10 μM Y-27632 ROCK inhibitor and gently pipette up and down to disperse the colonies. Passage cells onto new VTN-N coated plates. For the first passage in which genomic DNA is concurrently extracted, we customarily perform a split ratio of 1:4. For subsequent passages we routinely passage colonies at a ratio of 1:6.

7. Gently sway the newly passaged 96-well plate to ensure even distribution of cells, and incubate at 37 °C, in a 5% CO_2 incubator overnight. The following day, change 96-well plates with E8 medium without ROCK inhibitor (100 μL/well) (*see* **Note 8**). Thereafter, change medium daily until colonies are ready for passaging.

8. To the plate with the remaining cells, centrifuge at $1000 \times g$ for 4 min at room temperature and aspirate the supernatant in preparation for cell lysis.

9. Add 35 μL 1× Lysis Buffer/well. 1× Lysis Buffer is prepared from a 10× Lysis Buffer stock solution in 10 mM Tris pH 8.0 with 100 μg/mL Proteinase K.

10. Seal each plate with adhesive qPCR film or equivalent to prevent evaporation of the lysis buffer and incubate the plates at 55 °C for 2–4 h.

11. To inactivate the Proteinase K, incubate the plates at 85 °C for 20 min.

12. Store the genomic DNA plates at 4 °C until they are ready for PCR amplification.

13. Perform 25 μL PCR reactions with the same primer set as used in the T7E1 assay, using 1 μL of cell lysate for each reaction.

14. Resolve a sampling of reactions by electrophoresis on a 1% agarose gel to ensure that the PCR amplification is specific enough to proceed with sequencing.

15. We customarily submit samples for Sanger sequencing using 1 μL of unpurified PCR product, with an internal forward

primer designed to minimize problems caused by sequencing of nonspecific PCR bands.

3.10 Validation of Mutant Lines

1. Following sequence analyses of colonies, we select for clones with biallelic frameshift indel mutations (either homozygous or compound heterozygous) for knockout studies. Depending on the genetic phenotype to be determined, we also identify colonies with heterozygous mutations, with several wild-type clones generated from the same transfection experiment selected for downstream application as control lines. Mutant and wild type clones are expanded and cryopreserved until ready for experimental analysis.

2. 96-well colony plates are routinely passaged until completion of the sequence analysis. For cell line expansion, desired clones are dissociated with 0.5 mM EDTA and passaged onto VTN-N-coated 24-well plates. When cells have reached confluence, they are passaged onto VTN-N-coated 6-well plates for cryo-preservation. E8 Freezing medium is prepared using 90% complete E8 medium and 10% DMSO. We typically freeze three vials per one well of a 6-well plate.

3. We recommend resequencing of desired clones following colony expansion to ensure that no cross-contamination of clonal lines has occurred. In the case of heterozygous and compound heterozygous mutant lines, uneven overlapping peaks on chromatogram files may make it difficult to interpret sequence traces with complete accuracy. To determine the sequence of each allele in these clones, we recommend performing PCR amplification, followed by TOPO TA cloning and sequencing of individual alleles. We employ the same primer set used for the T7E1 assay for PCR amplification and clone the PCR product using the Zero Blunt TOPO PCR Cloning Kit. We typically select 8-10 individually transformed bacterial colonies for determination of monoallelic sequence reads. Where an appropriate antibody and assay exists for the protein in question, mutant lines may also be validated by Western blotting or flow-cytometric analysis.

4. The entire process to establish a clonal mutant line from CRISPR design to mutant sequencing and colony expansion takes a trained individual ~2 months, although this will inevitably vary accordingly to how many gene loci are to be targeted. We have found higher CRISPR targeting efficiencies when sgRNA transfection and clonal deposition are performed in feeder-free VTN-N/E8 conditions compared to irradiated MEF (iMEF) feeder-dependent culture conditions. We believe that this is due to increased cell survival following transfection and single-cell seeding, and the absence of iMEFs that may sequester transfection reagents. Although we discuss here only

the generation of loss-of-function mutations through indel formation in our feeder-free conditions, these methods can easily be translated to other forms of genome targeting such as the generation of precise nucleotide alterations with a homology-directed-repair template or with methods that do not utilize the iCRISPR system.

4   Notes

1. Although an optimal working concentration of vitronectin is cell-line dependent, we find that hPSCs tend not to be sensitive to vitronectin concentration and we routinely employ a working concentration of 5 μg/mL. The stock concentration of VTN-N (Life Technologies) is 500 μg/mL, so we dilute the VTN-N stock 1:100 in DPBS for general tissue culture coating.

2. VTN-N coated vessels that are in excess can be sealed with plastic wrap and stored at 4 °C for up to a week without drying. Before use, pre-warm the vessel to room temperature for at least 30 min.

3. To prevent degradation of the E8 medium components, do not warm the medium at 37 °C. Instead, warm the medium at room temperature until it is no longer cool to the touch prior to usage.

4. Recent studies have identified only a low incidence and minimal effects of off-target mutations from whole genome sequence analysis in individual hPSC mutant clones derived from CRISPR/Cas9 gene editing [19, 20]. However, off-target mutations could still lead to difficult-to-interpret results and impede functional studies. To minimize the effect of confounding phenotypes introduced by mutations at off-target sites, we suggest generating independent mutant lines using at least two different sgRNAs targeting different sequences within the same gene. If the CRISPR mutation is specific, we expect to observe a similar phenotype in cell lines generated by these same loci-targeting sgRNAs.

5. Although the percentage cut determined by the T7E1 assay may be an underestimate of the real sgRNA mutation rate, we find excellent correlation between the T7E1 assay and the number of mutant lines identified by sequencing. This emphasizes the importance of performing the T7E1 assay in parallel with clonal deposition of transfected cells and prior to mechanical picking of isolated colonies. Alternatively, the indel mutation rate can be determined through a restriction fragment length polymorphism (RFLP) assay if a CRISPR is designed such that the Cas9 cleavage site (3 bp upstream of the

PAM recognition motif) is in close proximity to a restriction enzyme cutting site (~5 bp). Since the assay directly measures the loss of a restriction site, it may be more sensitive at detecting the indel mutagenesis rate compared to the T7E1 assay.

6. It is feasible for a trained individual to mechanically pick 384 colonies (4×96-well plates) in one sitting, which should take ~4–5 h under a dissecting microscope.

7. Ensure that only well-separated colonies are picked for cell line establishment. While the majority of mutant cell lines that we have generated are homogeneous with regard to the mutated allele(s), it is possible that a given mutant cell line may have arisen from more than one cell, and this is the stage at which the cross-contamination is most likely to occur. After a clonal line is established, we highly recommend confirming the presence of the mutated allele(s) through TOPO TA cloning. Where contamination is suspected, cell lines should be subcloned.

8. We do not recommend treating cells with the ROCK inhibitor for an extended period of time as the effects of prolonged usage have not been thoroughly tested. We have found instances where it affects general cell morphology when treatment is applied continually. Where possible, replenish the medium without the ROCK inhibitor the next day following cell passage.

Acknowledgments

Our work was funded in part by NIH (R01DK096239), NYSTEM (C029156), and March of Dimes Foundation (Basil O'Connor Starter Scholar Research Award Grant 5-FY12-82).

References

1. Capecchi MR (2005) Gene targeting in mice: functional analysis of the mammalian genome for the twenty-first century. Nat Rev Genet 6(6):507–512. doi:10.1038/nrg1619

2. Zwaka TP, Thomson JA (2003) Homologous recombination in human embryonic stem cells. Nat Biotechnol 21(3):319–321

3. Costa M, Dottori M, Sourris K, Jamshidi P, Hatzistavrou T, Davis R, Azzola L, Jackson S, Lim SM, Pera M, Elefanty AG, Stanley EG (2007) A method for genetic modification of human embryonic stem cells using electroporation. Nat Protoc 2(4):792–796

4. Braam SR, Denning C, van den Brink S, Kats P, Hochstenbach R, Passier R, Mummery CL (2008) Improved genetic manipulation of human embryonic stem cells. Nat Methods 5(5):389–392. doi:10.1038/nmeth.1200

5. Hotta A, Yamanaka S (2015) From genomics to gene therapy: induced pluripotent stem cells meet genome editing. Annu Rev Genet 49:47–70. doi:10.1146/annurev-genet-112414-054926

6. Xu RH, Peck RM, Li DS, Feng X, Ludwig T, Thomson JA (2005) Basic FGF and suppression of BMP signaling sustain undifferentiated proliferation of human ES cells. Nat Methods 2(3):185–190. doi:10.1038/nmeth744

7. Ludwig TE, Bergendahl V, Levenstein ME, Yu J, Probasco MD, Thomson JA (2006) Feeder-independent culture of human embryonic stem cells. Nat Methods 3(8):637–646. doi:10.1038/nmeth902

8. Ludwig TE, Levenstein ME, Jones JM, Berggren WT, Mitchen ER, Frane JL, Crandall LJ, Daigh CA, Conard KR, Piekarczyk MS, Llanas RA, Thomson JA (2006) Derivation of human embryonic stem cells in defined conditions. Nat Biotechnol 24(2):185–187. doi:10.1038/nbt1177

9. Beers J, Gulbranson DR, George N, Siniscalchi LI, Jones J, Thomson JA, Chen G (2012) Passaging and colony expansion of human pluripotent stem cells by enzyme-free dissociation in chemically defined culture conditions. Nat Protoc 7(11):2029–2040. doi:10.1038/nprot.2012.130

10. Chen G, Gulbranson DR, Hou Z, Bolin JM, Ruotti V, Probasco MD, Smuga-Otto K, Howden SE, Diol NR, Propson NE, Wagner R, Lee GO, Antosiewicz-Bourget J, Teng JM, Thomson JA (2011) Chemically defined conditions for human iPSC derivation and culture. Nat Methods 8(5):424–429. doi:10.1038/nmeth.1593

11. Huang X, Wang Y, Yan W, Smith C, Ye Z, Wang J, Gao Y, Mendelsohn L, Cheng L (2015) Production of gene-corrected adult beta globin protein in human erythrocytes differentiated from patient iPSCs after genome editing of the sickle point mutation. Stem Cells (Dayton, OH) 33(5):1470–1479. doi:10.1002/stem.1969

12. Gonzalez F, Zhu Z, Shi ZD, Lelli K, Verma N, Li QV, Huangfu D (2014) An iCRISPR platform for rapid, multiplexable, and inducible genome editing in human pluripotent stem cells. Cell Stem Cell 15(2):215–226. doi:10.1016/j.stem.2014.05.018

13. Zhu Z, Gonzalez F, Huangfu D (2014) The iCRISPR platform for rapid genome editing in human pluripotent stem cells. Methods Enzymol 546:215–250. doi:10.1016/b978-0-12-801185-0.00011-8

14. Shalem O, Sanjana NE, Hartenian E, Shi X, Scott DA, Mikkelsen TS, Heckl D, Ebert BL, Root DE, Doench JG, Zhang F (2014) Genome-scale CRISPR-Cas9 knockout screening in human cells. Science (New York, NY) 343(6166):84–87. doi:10.1126/science.1247005

15. Liang X, Potter J, Kumar S, Zou Y, Quintanilla R, Sridharan M, Carte J, Chen W, Roark N, Ranganathan S, Ravinder N, Chesnut JD (2015) Rapid and highly efficient mammalian cell engineering via Cas9 protein transfection. J Biotechnol 208:44–53. doi:10.1016/j.jbiotec.2015.04.024

16. Hsu PD, Scott DA, Weinstein JA, Ran FA, Konermann S, Agarwala V, Li Y, Fine EJ, Wu X, Shalem O, Cradick TJ, Marraffini LA, Bao G, Zhang F (2013) DNA targeting specificity of RNA-guided Cas9 nucleases. Nat Biotechnol 31(9):827–832. doi:10.1038/nbt.2647

17. Vouillot L, Thelie A, Pollet N (2015) Comparison of T7E1 and surveyor mismatch cleavage assays to detect mutations triggered by engineered nucleases. G3 (Bethesda, MD) 5(3):407–415. doi:10.1534/g3.114.015834

18. Ran FA, Hsu PD, Wright J, Agarwala V, Scott DA, Zhang F (2013) Genome engineering using the CRISPR-Cas9 system. Nat Protoc 8(11):2281–2308. doi:10.1038/nprot.2013.143

19. Smith C, Gore A, Yan W, Abalde-Atristain L, Li Z, He C, Wang Y, Brodsky RA, Zhang K, Cheng L, Ye Z (2014) Whole-genome sequencing analysis reveals high specificity of CRISPR/Cas9 and TALEN-based genome editing in human iPSCs. Cell Stem Cell 15(1):12–13. doi:10.1016/j.stem.2014.06.011

20. Veres A, Gosis BS, Ding Q, Collins R, Ragavendran A, Brand H, Erdin S, Cowan CA, Talkowski ME, Musunuru K (2014) Low incidence of off-target mutations in individual CRISPR-Cas9 and TALEN targeted human stem cell clones detected by whole-genome sequencing. Cell Stem Cell 15(1):27–30. doi:10.1016/j.stem.2014.04.020

Chapter 6

Development of CRISPR/Cas9 for Efficient Genome Editing in *Toxoplasma gondii*

Bang Shen, Kevin Brown, Shaojun Long, and L. David Sibley

Abstract

Efficient and site-specific alteration of the genome is key to decoding and altering the genomic information of an organism. Over the last couple of years, the RNA-guided Cas9 nucleases derived from the prokaryotic type 2 CRISPR (clustered regularly interspaced short palindromic repeats) systems have drastically improved our ability to engineer the genomes of a variety of organisms including *Toxoplasma gondii*. In this chapter, we describe detailed protocols for using the CRISPR/Cas9 system adapted from *Streptococcus pyogenes* to perform efficient genetic manipulations in *T. gondii* such as gene disruption, gene tagging and genetic complementation. The technical details of the strategy, including CRISPR plasmid construction, target construct generation, parasite transfection and positive clone identification are also provided. These methods are easy to customize to any gene of interest (GOI) and will greatly accelerate studies on this important pathogen.

Key words Genetic transformation, CRISPR/Cas9, Gene editing, Protozoan parasites, Selectable markers

1 Introduction

1.1 *T. gondii* Genetics

The phylum Apicomplexa contains more than 10,000 estimated species [1], many of which are important pathogens in humans and animals including *Plasmodium* spp. [2], *T. gondii* [3], and *Cryptosporidium parvum* [4]. *T. gondii* has been developed as a model organism for studying the biology of apicomplexan parasites because of its ease of in vitro culturing and utility for cell biological and biochemical studies [5]. Forward genetic studies based on genetic crosses and linkage mapping [6], together with a range of reverse genetic approaches [7], have made *T. gondii* a model for genetic manipulation. The haploid genome of *T. gondii* is ~65 mb in size distributed among 14 linear chromosomes that encode ~8300 genes [8]. The *T. gondii* genome differs from closely related

Bang Shen and Kevin Brown contributed equally with all other contributors.

Andrew Reeves (ed.), *In Vitro Mutagenesis: Methods and Protocols*, Methods in Molecular Biology, vol. 1498, DOI 10.1007/978-1-4939-6472-7_6, © Springer Science+Business Media New York 2017

apicomplexans largely due to the expansion and diversification of several gene families encoding surface and secretory proteins that interact with the host [8]. Although initial studies focused on the three clonal lineages abundant in North America and Europe [9], recent studies have uncovered much greater genetic diversity in *T. gondii* strains from around the world [10]. These developments highlight the need for efficient unbiased genetic tools for analyzing the functions of genes in a diverse range of genetic backgrounds.

1.2 Genome Editing in T. gondii

Site-specific modification of a genome is extremely useful for biological research in order to decode the corresponding function of specific genes. Over the years, scientists have designed many different approaches to achieve this goal by trying to improve the efficiency and specificity of targeted gene modification. In the *T. gondii* research field, such practice dates back to 1993, when the methods for transient expression of foreign DNA into parasite cells was first reported [11]. In the same year, targeted gene knockout and replacement using homologous recombination were reported [11]. A number of independent positive-selection markers were developed to aid in this technology including chloramphenicol acetyl transferase (Cat), conferring resistance to chloramphenicol (Cm) [12, 13]; bleomycin-binding protein (Ble), conferring resistance to phleomycin (Ph) [14]; and mutant dihydrofolate reductase -thymidylate synthase (*DHFR-TS**), conferring resistance to pyrimethamine (Pyr) [15] (*see* Table 1). Although an important advance, gene disruption in *T. gondii* suffers from being fairly inefficient (in general the frequency of site-specific disruption is less than 0.1%) [16], and there are other limitations to consider in the choice of markers (*see* **Notes 1–4**). Techniques to compensate for the low efficiency of homologous targeting included adding long homologous flanking arms (i.e., up to several kb to improve efficiency) [17–19], and developing both positive and negative selection schemes based on the hypoxanthine xanthine guanosine phosphoribosyl transferase (*HXGPRT*) gene [20] (*see* Table 1).

The low efficiency of targeted gene disruption in *T. gondii* is due to the high frequency of random insertion of foreign DNA, which is mediated by nonhomologous end joining (NHEJ) [7]. Taking advantage of this feature, a mini-DHFR-TS* cassette was used to generate insertional mutants, for example, in the nonessential uracil phosphoribosyl transferase (*UPRT*) gene [15, 18], which also provides for negative selection (*see* Table 1). In an effort to improve this efficiency, two groups independently disrupted the *Ku80* gene (a critical component mediating NHEJ) to increase the proportion of homologous recombination [16, 21]. Using Δ*ku80* lines it is possible to target specific genes for disruption with increased efficiency using a selectable marker flanked by shorter homologous regions (i.e., 500 bp). Although deletion of *Ku80* significantly improved the efficiency of genome engineering, this approach is so far only available in the type 1 strain RH [16, 21] and type 2 strain PRU [22]. Loss of Ku80 is also a limitation

Table 1
Commonly used genetic markers for positive or negative selection in *T. gondii*

Selection	Drug	Final drug conc.	Selection cassette	Genetic locus	Advantages	Limitations
Positive	Pyrimethamine	3 μM	*DHFR-TS**	N/A	Universal	Potential occupational hazard
Positive	Chloramphenicol	20 μM	*Cat*	N/A	Universal	Requires multiple cycles
Positive	Mycophenoloic acid/xanthine	25 μg/ml, each	*HXGPRT*	N/A	Positive and negative	Requires Δ*hxgprt* strain
Positive	Phleomycin	5 μg/ml	*Ble*	N/A	Acts on extracellular parasites	Requires wild type *Ku80+* strain
Negative	5-fluorodeoxyuracil	10 μM	N/A	*UPRT*	Universal	Spontaneous resistance
Negative	Sinefungin	3×10^{-7} M	N/A	*SNFR1*	Universal	
Negative	6-thioxanthine	200 μg/ml	N/A	*HXGPRT*	Positive and negative	Requires *HXGPRT+* strain

for some selectable markers; for example, Δ*ku80* mutants show increased sensitivity to phleomycin, likely due to its mode of action as a DNA intercalating agent [16]. Additionally, loss of Ku80 in filamentous fungi has been linked to increased chromosomal instability [23]. Therefore, a tool that can be used in any strain for efficient genome modification is highly desired. Such a tool became available when the CRISPR (Clustered Regularly Interspaced Short Palindromic Repeat)/Cas9 genome-editing technology was adapted into *T. gondii* [24, 25]. Although genome editing always carries the risk of off-target effects, these are expected to be less common on *T. gondii* due to the smaller genome complexity (*see* **Note 5**). As described below, CRISPR/Cas9 has been useful for performing genome modifications such as site-specific gene disruption, deletion by complete gene knockout, and gene replacement in various strains of *T. gondii* with high efficiency.

1.3 CRISPR/Cas9 Genome Editing

CRISPR is a common adaptive immune mechanism found in bacteria and archaea where it acts against mobile genetic elements such as phages [26]. The CRISPR locus contains an array of interspaced short DNA sequences taken from the genomes of the invaders as a way to track the history of phage infections. Subsequently, this sequence information is used, along with Cas (CRISPR-*a*ssociated-*p*roteins), to defend secondary infections from the same phages [27]. The target interference mechanism of CRISPR/Cas systems has been engineered for genome editing in a variety of organisms [28–32]. For this purpose, the type 2 CRISPR system from *Streptococcus pyogenes* is the most commonly used because of its simplicity and ease of customization [28, 33]. This system requires only one protein, the RNA-guided endonuclease Cas9, and two small RNA molecules (crRNA and tracrRNA) to introduce a site-specific, double-strand break (DSB) in the target DNA. The crRNA molecule partially hybridizes with tracrRNA to form a complex recognized by the Cas9 protein, which is then guided (by crRNA) to the target DNA with a PAM sequence (typically NGG for Cas9) and causes DSB in the target [33].

Soon after the working mechanism of *S. pyogenes* Cas9 was discovered, its potential use for genome editing was recognized. By fusing the two RNA molecules together and expressing them as a chimeric RNA, called a single guide RNA (sgRNA), from a U6 promoter, Mali et al. [28] demonstrated that the Cas9 nuclease was able to introduce site-specific modifications in the genome of cultured human cells. The sgRNA molecule is optimized to contain a 20-nucleotide sequence (N_{20}) that matches the target (in the form of N_{20}NGG, where N_{20} is the same as in sgRNA and NGG is the PAM sequence) to specify the site of targeting. Modification of the N_{20} sequence in sgRNA allows CRISPR to target almost any gene in the genome (the only limitation being the necessity of the sgRNA to pair with an NGG sequence in the genome). After the pioneering work in human and mouse cell lines [28, 29], CRISPR/Cas9 technology has been quickly adapted to many other organisms [24, 32, 34–37]. The key point of CRISPR/Cas9-mediated genome editing is the introduction of site-specific DSB, which can then be repaired either by error prone NHEJ, leading to insertion and deletion (indel) mutations to inactivate genes, or by homologous recombination to precisely alter the locus using a donor template [28]. The CRISPR/Cas9 system was also engineered to regulate gene expression using a nuclease-deficient mutant of Cas9 in combination with transcriptional regulators [38, 39].

1.4 Parasite Genome Modification

The CRISPR/Cas9-mediated genome editing technology was first adapted to *T. gondii* [24, 25] and subsequently developed independently in *Plasmodium* spp. [36, 40, 41] and more recently in *C. parvum* [37]. In all cases, Cas9 was expressed from the promoter of a protein-coding gene (such as the surface antigen 1 (SAG1) or alpha tubulin (TUB1) in *T. gondii*) while the sgRNA was expressed from an endogenous U6 or exogenous T7 promoter. When both components (Cas9 and sgRNA) were present, the CRISPR/Cas9 system was shown to induce site-specific DSB and facilitate homologous recombination at the targeting sites when a donor template was provided. In addition, CRISPR/Cas9 also facilitates site-specific integration of foreign DNA in *T. gondii*, facilitating gene disruption by insertion of a selectable marker.

CRIPSR/Cas9 also improved the efficiency of genetic complementation based on negative selection and insertion at neutral sites such as *UPRT* and the recently identified locus involved in sinefungin resistance (*SNFR1*) [42]. Due to its high efficiency, CRISPR/Cas9 has even been used to perform gene disruption and allelic replacement without drug selection [25].

In this chapter, we provide detailed protocols for CRISPR/Cas9-mediated genome editing such as gene disruption, gene deletion, or gene modification in *T. gondii* (*see* Table 2). We also incorporate several useful improvements that are rapidly replacing

Table 2
CRISPR/Cas9 genome editing strategies in *T. gondii*

| CRISPR strategy | Parental *T. gondii* strain (DNA repair method) | | | |
| | Wild type (NHEJ and HR) | | Δ*ku80* (HR) | |
	Cas9/sgRNA target	Exogenous DNA	Cas9/sgRNA target	Exogenous DNA
Gene knockout by disruption				
Insertion/deletion (InDel)	Exonic near ATG	None	Not recommended due to lack of NHEJ	
Insertion of selection cassette[a]	Exonic near ATG	Selection cassette	Exonic at start ATG	Selection cassette with homology arms[b]
Gene knockout by deletion				
Single CRISPR deletion[c]	Exonic at middle	Selection cassette with homology arms	Exonic at middle	Selection cassette with homology arms[b]
Dual CRISPR deletion[d]	Exonic at ends	Selection cassette with homology arms	Exonic at ends	Selection cassette with homology arms[b]
Endogenous tagging				
N-terminal	Near start ATG	5′ Tagging cassette with homology arms	Near start ATG	5′ Tagging cassette with homology arms[b]
C-terminal[e]	Near stop codon	3′ Tagging cassette with homology arms	Near stop codon	3′ Tagging cassette with homology arms[b]
Targeted complementation				
Single CRISPR Insertion[f]	Neutral or negative selection locus	Complementation cassette with homology arms	Neutral or negative selection locus	Complementation cassette with homology arms[b]

[a]*See* Fig. 2a
[b]Homology arms as short as 40 bp are compatible with Δ*ku80* strains
[c]*See* Fig. 2b
[d]*See* Fig. 2c
[e]*See* Fig. 3
[f]*See* Fig. 4

conventional cloning: (1) Q5-mediated mutagenesis as a method to quickly modify a plasmid template to introduce large mutations [43], and (2) Gibson assembly of PCR amplified fragments into complete amplicons or plasmids [44]. Collectively, these tools will enable researchers to efficiently manipulate the *T. gondii* genome at their will and will accelerate the dissection of the complex biology of this important pathogen.

2 Materials

2.1 Molecular Cloning

1. The *UPRT* targeting CRISPR/Cas9 plasmid pSAG1::Cas9-U6::sgUPRT is available from Addgene (www.addgene.org).

2. The pUPRT::DHFR-D (Addgene) plasmid is used as a template for the Pyr-resistant selection cassette *DHFR-TS**.

3. Plasmid pUC19 (or any other convenient cloning vector).

4. Q5 mutagenesis kit (New England Biolabs).

5. Gibson assembly kit.

6. Primers:
   GOI-gRNA-Fw: NNNNNNNNNNNNNNNNNNNN GTTTTAGAGCTAGAAATAGC (N_{20} is your gene-specific sgRNA sequence) (*see* Subheading 3.1 for design guidelines).
   GOI-gRNA-Rv: AACTTGACATCCCCATTTAC.gRNA2-Fw-KpnI: CGAATTGGGTACCCAAGTAAGCAGAAGCACG CTG.gRNA2-Rv-XhoI:TCGACCTCGAGAATTAACCCTC ACTAAAGG.Seq-dualgRNA2rv: TGAAATCACGGACCGT GGAA. Other locus-specific primers for homologous template construction and gene disruption identification.

7. PCR reagents: high fidelity DNA polymerases such as the Q5 DNA Polymerase, Taq DNA Polymerase, dNTPs, and other common PCR reagents.

8. Competent cells such as *E. coli* DH5α.

9. Plasmid purification kit such the QIAprep spin miniprep kit.

10. Proteinase K (10–20 mg/ml, use as 100×).

11. Reagents for agarose gel electrophoresis.

12. QIAquick PCR purification kit.

13. QIAquick Gel/DNA Extraction Kit.

2.2 Parasite Transfection

1. Parasite strains: commonly used *T. gondii* strains (type 1 RH or GT1 strains, type 2 ME49 or PRU strains) are available from ATCC (www.atcc.org).

2. Filter holder and polycarbonate filter membranes (3 μm pore) (Whatman).

3. Cytomix buffer: 120 mM KCl, 0.15 mM CaCl$_2$, 10 mM K$_2$HPO$_4$/KH$_2$PO$_4$, 25 mM HEPES, 2 mM EDTA, 5 mM MgCl$_2$, pH 7.6.

4. ATP (0.2 M, 100×).

5. Glutathione (0.5 M, 100×).

6. Drug(s) for selection (*see* Table 1).

7. Tissue culture supplies such as 25 cm^2 cell culture flask (T25), 96-well plates, 24-well plates.

8. Electroporation cuvettes (such as BTX640 from BTX Harvard Apparatus) for parasite transfection. Square-wave electroporation system such as the BTX ECM-830 electroporator from Harvard Apparatus.

9. Commercially prepared Hanks' balanced salt solution (HBSS; e.g., ThermoFisher).

3 Methods

The efficiency of CRISPR/Cas9 has led to its use in a diverse range of wild-type strains and Δ*ku80* mutants in *T. gondii* [24, 42, 45, 46]. This section is dedicated to general methods that have been developed from such CRISPR/Cas9 genome-editing studies in *T. gondii*. A guide to selecting an appropriate *T. gondii* strain compatible with the CRISPR/Cas9 strategy of interest is presented in Table 2. Before attempting to alter the genome with CRISPR/Cas9, it is important to understand the DNA repair mechanisms that are present in the *T. gondii* strain of interest. Two pathways exist to repair DNA breaks in *T. gondii*: nonhomologous end-joining (NHEJ) and homologous recombination. In wild-type strains, NHEJ occurs at a much greater frequency than homologous recombination. Repair of double-strand DNA breaks is mediated by a mechanism conserved from yeast to humans and mutants lacking the Ku80 protein are defective in NHEJ and rely exclusively on homologous recombination [47]. In *T. gondii*, Δ*ku80* mutants are available in the type I RH strain [16, 21] and the type 2 PRU strains [22]. Such Δ*ku80* strains preferentially perform DNA repair by homologous recombination and are ideal for situations that require seamless recombination, like endogenous tagging, allelic replacement, or complete gene deletion (*see* Table 2). Conversely, Δ*ku80* strains are not as well suited for CRISPR/Cas9 gene disruptions based on insertion/deletion mutations or insertion of nonhomologous drug selection cassettes, due to their preferential repair by homologous recombination. For the latter two applications, wild-type strains are preferred (*see* Table 2). The differences in frequency of NHEJ versus homologous recombination also have practical consequences: in wild-type strains it generally requires regions of ~1 kb

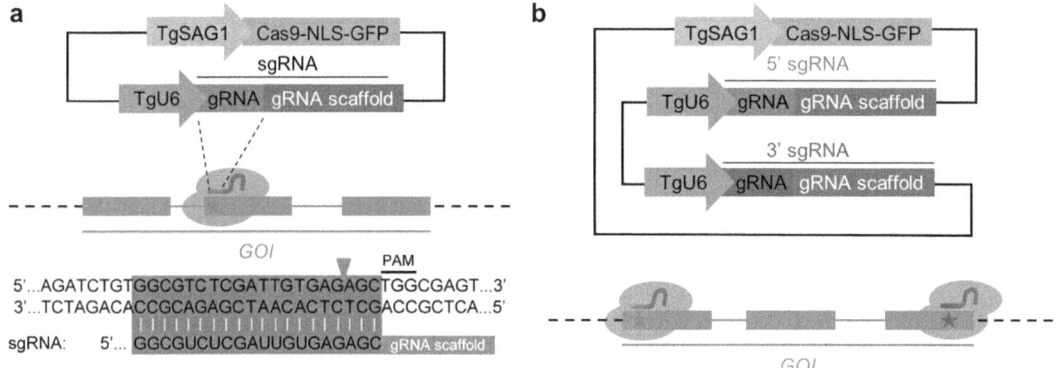

Fig. 1 Transient expression of Cas9 and guide RNA(s) from a single plasmid in *T. gondii*. (**a**) The plasmid pSag1-Cas9-U6-sgRNA is used for introducing a single Cas9-mediated double-stranded break in the gene of interest (GOI). This plasmid uses the *T. gondii* SAG1 promoter (TgSAG1) to drive expression of Cas9 protein containing a nuclear localization sequence (NLS) fused to GFP (Cas9-NLS-GFP). It also expresses a single-guide RNA (sgRNA) driven by the *T. gondii* U6 promoter (TgU6). Alignment of the sgRNA to the genome shows the site of a double-strand break (*star* and *green triangle*) directly upstream of the PAM sequence (TGG) in the GOI. (**b**) The plasmid pSAG1-Cas9-dual-U6-sgRNA is used for simultaneously introducing two Cas9-mediated double-stranded breaks in the genome. This plasmid expresses two separate sgRNAs from the TgU6 promoter to target the 5′ and 3′ regions of the GOI, respectively

or longer to favor homologous insertion, while very short regions can achieve the same outcome in Δ*ku80* strains. Such considerations are included the methods described below.

3.1 CRISPR Plasmid Construction

The general design of CRISPR single- and double-guide RNA plasmids is shown in Fig. 1. These templates form the basis for modification and for generating gene-specific CRISPR plasmids, as described below.

1. Use the web-server E-CRISP (http://www.e-crisp.org/E-CRISP/designcrispr.html) to identify gRNA sequences that target your gene of interest (GOI). The gRNA sequence is 20 bp in length and this sequence must lie immediately upstream of the PAM sequence ("NGG") in the genome.

2. Once a gRNA sequence is selected, PCR mutagenesis is used to replace the *UPRT* targeting sgRNA in the original CRISPR/Cas9 plasmid (pSAG1::Cas9-U6::sgUPRT) (*see* Fig. 1a) with your gene-specific sgRNA. We recommend using the Q5 mutagenesis kit to perform the PCR mutagenesis reaction with pSAG1::Cas9-U6::sgUPRT as template and GOI-gRNA-Fw/GOI-gRNA-Rv as primers. Recommended PCR cycling conditions are: an initial melting step at 98 °C for 1 min, followed by 25 cycles with each cycle at 98 °C for 10 s, 55 °C for 30 s, and 72 °C for 5 min, followed by a final extension at 72 °C for 2 min.

3. After PCR amplification, a small portion of the PCR product (5 μl) should be resolved on a 0.8 % agarose gel to verify successful amplification (expecting a major band with the size of 9674 bp). The rest of the PCR product is used for the kinase-ligase-digestion (KLD) reaction and in the subsequent transformation of *E. coli*.

4. Optional but recommended: the PCR product can be digested with *Dpn*I before the KLD reaction to decrease plasmid template carryover (note: the KLD reaction already includes this enzyme, so this step is not necessary). Add 1 μl *Dpn*I enzyme to 12.5 μl PCR product and incubate for 15–30 min at 37 °C. Subsequently, use the *Dpn*I-digested product (no purification procedure is required) to perform the KLD reaction as above.

5. Select clones with the correct GOI targeting sgRNA. Pick 4–6 single clones from the mutagenesis plate, individually inoculate them into 5 ml LB medium supplemented with 100 μg/ml ampicillin and grow the cultures at 37 °C for 12–16 h with vigorous shaking (>200 rpm). Isolate the plasmids from the cultures using the QIAprep spin miniprep kit and verify the miniprep products by separation on 0.8 % agarose gels by electrophoresis. To confirm the successful replacement of *UPRT*-targeting sgRNA by the designed GOI-targeting sgRNA, sequence the plasmids using primers such as M13-rev (5′-CAG GAAACAGCTATGAC) or M13R(-48) (5′-AGCGGATAACA ATTTCACACAGGA). Once the correct clones are identified, the plasmids will be amplified, purified, and used for transfection into *T. gondii*.

6. Generate a plasmid for dual CRISPR/Cas9 gene targeting (*see* Fig. 1b). In general, it is not easy to achieve targeted DSBs at two separate sites in the genome using separate sgRNA expressing plasmids that are co-transfected, likely due to the low frequency of uptake of both plasmids by a single parasite. Instead, we prefer to generate a single plasmid expressing two separate sgRNAs. Construct two separate single sgRNA plasmids, one that targets near the translation start codon (GOI 5′ sgRNA) and one that targets near the translation stop codon (GOI 3′ sgRNA) of the GOI. Once both CRISPR/Cas9 plasmids have been constructed, PCR amplify the insert containing the U6-sgRNA region from the GOI 5′ sgRNA plasmid using the primers gRNA2-Fw-KpnI (CGAATTGGGTACCCAAGTAA GCAGAAGCACGCTG) and gRNA2-Rv-XhoI (TCGACCTC GAGAATTAACCCTCACTAAAGG). Digest the U6-sgRNA amplicon with *Kpn*I and *Xho*I and gel purify the resulting fragment (678 bp). In parallel, digest the CRISPR plasmid containing the GOI 3′ sgRNA with *Kpn*I plus *Xho*I and gel purify the resulting fragment (9663 bp). Ligate the digested pU6-sgRNA PCR amplicon from the GOI 5′ sgRNA into the

prepared GOI 3′ sgRNA vector backbone to yield the dual sgRNA plasmid (*see* Fig. 1b). Repeat **step 5** to confirm the presence of the second sgRNA by sequencing the plasmids using sequencing primer, Seq-dualgRNA2rv (TGAAATCA CGGACCGTGGAA).

3.2 Marker Construction

Regardless of the specific approach, drug selection is almost always required to obtain parasites with the desired genomic changes. There are a variety of selectable markers available for use in *T. gondii*, either based on modified endogenous genes (i.e., *DHFR-TS**, *HXGPRT*), or exogenous genes that confer resistance to antibiotics (e.g., *Cat, Ble*) (summarized in Table 1). When used for selecting stable transformants on *T. gondii*, these markers must be flanked by regulatory regions containing upstream sequences that harbor promoters and downstream regions involved in regulation. Although these features are generally not precisely defined, several selectable cassettes are available that contain the appropriate regulatory regions flanking selectable makers. The most useful of these are the *DHFR-TS** minigene available in Addgene plasmid 58528 [24], and the *HXGPRT* selection cassette described in the generation of Δ*ku80* mutants [16, 24]. The following sections and the diagrams in Figs. 2, 3, and 4 rely on the use of such selection cassettes that are further modified for targeting gene deletion or editing.

Fig 2 (continued) Co-transfect the CRISPR/Cas9 plasmid and the selection cassette (preferably in the form of purified PCR products (*see* **Note 6**) into parasites and select with the appropriate drugs (*see* Table 1) (*see* **Notes 1–4**)). To screen for disruptants, assess the integration of the selection cassette into the targeting site using PCR3 as shown. Amplification from wild type parasites produces a much smaller-sized PCR3 product than the insertion mutants. (**b**) Strategy for CRISPR/Cas9-mediated gene deletion in *T. gondii*. For small genes consisting of 1–3 exons, a single sgRNA expressing CRISPR/Cas9 plasmid is efficient enough to mediate a knockout by homologous recombination. Design a gene-specific CRISPR/Cas9 plasmid (*see* Fig. 1a) to target the middle of the gene (can target an intron or exon). Separately, design a targeting construct consisting of a selection cassette with homology flanks. In wild-type cells, the homology flanks should be ~1 kb homologous to the flanking regions of the GOI. In the case of Δ*ku80* strains, short homology flanks of 40 bps are enough and can be added directly by incorporation into primers used for PCR. The targeting construct serves as a template for homologous recombination to replace the GOI with the selection cassette. To completely knock out a gene, we suggest using the fragments immediately upstream of the translation initiation site (left arm) and downstream of the stop codon (right arm) as homologous arms. To knock out the GOI, co-transfect parasites with the CRISPR/Cas9 plasmid together with the targeting construct and select with the appropriate drugs (*see* Table 1). Clone the drug-resistant pool by limiting dilution and use diagnostic PCR to confirm knockouts. The priming sites for PCR primers are indicated: PCR1 and PCR2 confirm integration of left and right homologous arms, respectively; PCR3 examines the integrity of the endogenous gene. A successful knockout clone should give positive PCR products in PCR1 and PCR2 but no product in PCR3, whereas the wild type parasites give the opposite. (**c**) Deletion of a large locus from the chromosome of *T. gondii* using a dual sgRNA CRISPR/Cas9 system. To knock out a complex locus or multiple genes/gene copies in a row, a CRISPR/Cas9 system targeting two sites in the genome can be used to increase the efficiency (*see* **Note 7**). The basic principle of this dual-targeting CRISPR/Cas9 system is the same as the single targeting one illustrated in Fig. 2b. The only difference is that the CRISPR/cas9 plasmid expresses two sgRNAs, one targeting the 5′ end and the other targeting the 3′ end of the region to be deleted. For a comparison of methods, *see* Table 2

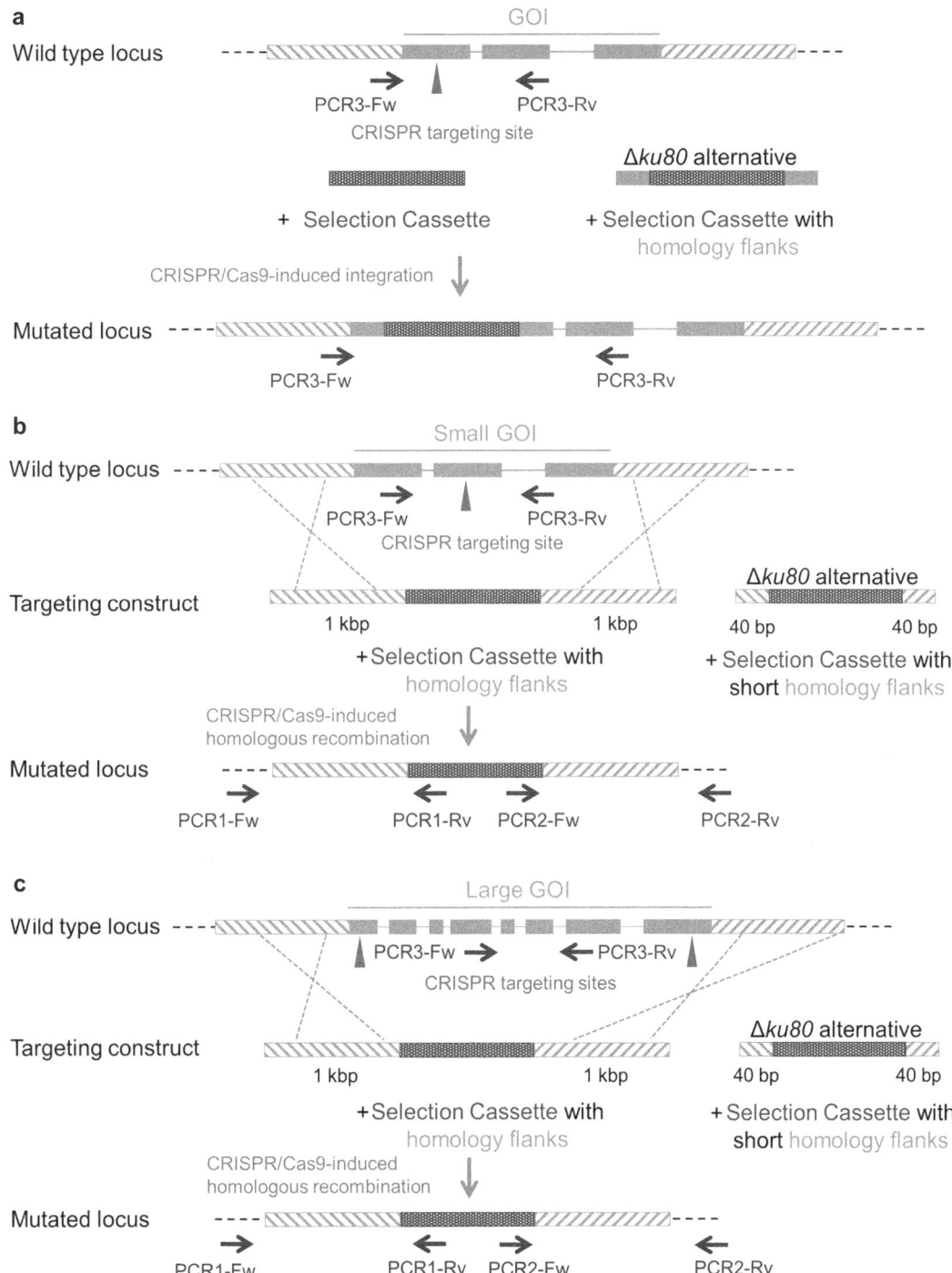

Fig 2 Strategies for CRISPR/Cas9-mediated gene disruption and deletion in *T. gondii*. (**a**) Insertion of a selection cassette to disrupt the expression of GOI using CRISPR/Cas9-mediated site-specific integration. In wild-type cells, the selection cassette can be used without homology flanks. In the case of Δ*ku80* strains, short homology flanks of 40 bps can be added directly by incorporation into the primers used for the PCR. To disrupt a GOI by insertion of a selection cassette, design a CRISPR/Cas9 plasmid (*see* Fig. 1a) to target the site of insertion. We recommend targeting the first exon close to the 5′ end since this is more likely to disrupt gene expression.

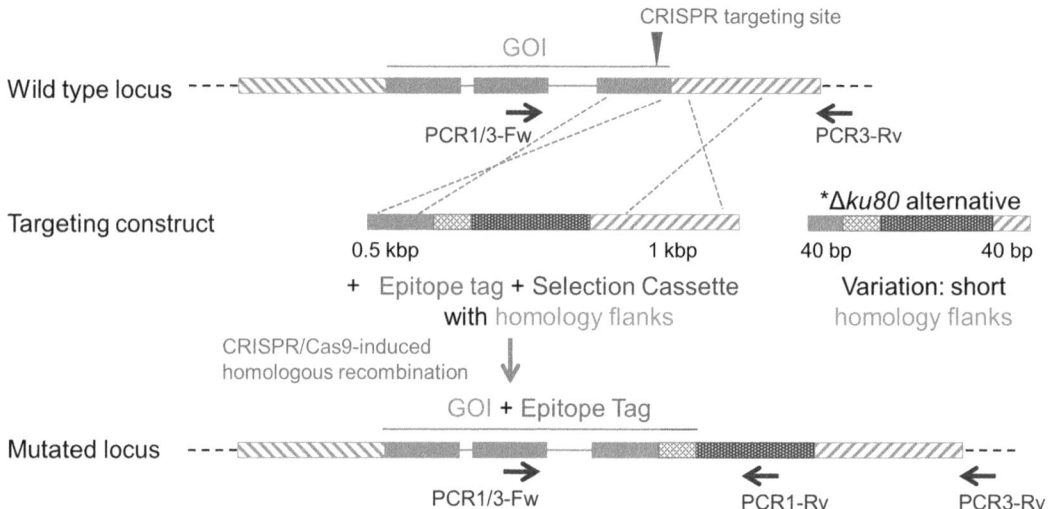

Fig. 3 Endogenous tagging with CRISPR/Cas9. Construct a CRISPR/Cas9 plasmid (*see* Fig. 1a) to induce a double-stranded break near the stop codon of the GOI. Construct a targeting construct consisting of a selection cassette surrounded by homology flanks. Two regions of homology are used: the 5′ region corresponds to the genomic sequence immediately upstream of the stop codon of the GOI. It is placed in frame with the epitope tag followed by a stop codon. The second homology region is chosen from the 3′ UTR of the GOI. Between the two regions of homology is a selection cassette allowing for drug selection using a variety of markers (*see* Table 1). In the case of Δ*ku80* strains, short homology flanks of 40 bps can be added directly by incorporation into the primers used for PCR. The targeting construct is used as a template to repair the break introduced by CRISPR/ Cas9, incorporating an epitope tag in frame at the 3′ end of the GOI. Co-transfect the CRISPR/Cas9 plasmid and the targeting construct into parasites and select with appropriate drugs (*see* Table 1). Obtain clones by limiting dilution and screen for successful complementation using the PCR 1 and 3 as shown

In general, wild-type parasites preferentially incorporate foreign DNA via NHEJ, and hence, it is useful to include regions of homology to favor homologous recombination. In wild-type strains, such regions need to be relatively large (i.e., ~1 kb). However, in Δ*ku80* strains, we have found that only ~40 bp of homology is needed for homologous recombination, which should generally lie near the DSB. Additionally, the form of DNA template used for transformation can also influence the efficiency of the desired outcome (*see* **Note 6**). Here, we summarize the general steps for preparing targeting constructs containing selection cassettes, which will be used for co-transfection with pSAG1::Cas9-U6::sgRNA(GOI) plasmid to introduce a DSB in the gene of interest.

1. When using selection cassettes that are not flanked by homology regions (*see* Fig. 2a), PCR amplify the insert or use restriction enzymes to digest the drug selection cassette from the chosen plasmid. Use a high-fidelity polymerase such as Q5 for PCR. To achieve higher success rates, the PCR prod-

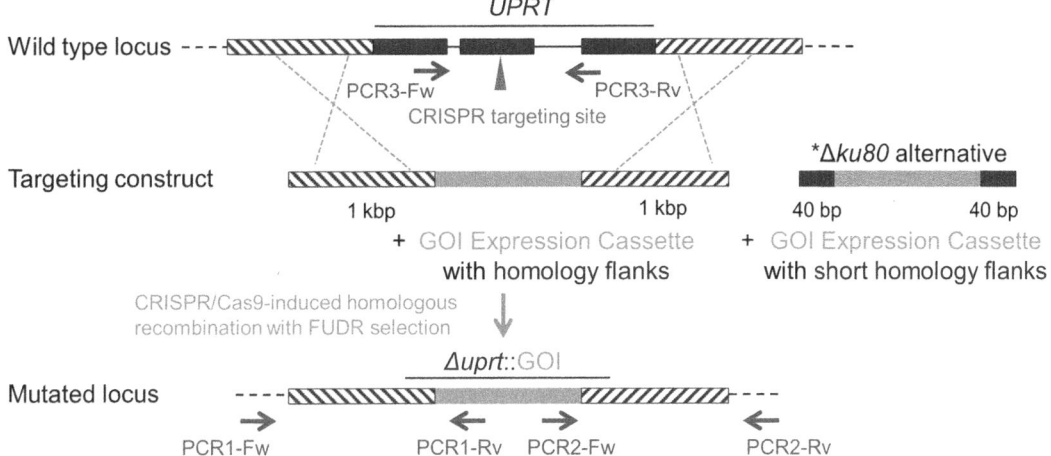

Fig. 4 Genetic complementation with CRISPR/Cas9. CRISPR/Cas9 can be used to increase the efficiency of inserting foreign DNA into the genome, for expressing transgenes or for genetic complementation (*see* **Note 5**). This is facilitated by the use of negative selectable markers that allow for selection of the GOI integration at a specific site (*see* Table 1). This is illustrated here for *UPRT*, which is neutral in tachyzoites, a negative-selectable locus suitable for genetic knockin and complementation. Design a targeting construct consisting of a minigene of the GOI (usually constructed from cDNA to reduce complexity) surrounded by flanking regions. For wild-type strains, include homology flanks of ~1 kb from the 5′ and 3′ flanking regions of the *URPT* gene. In the case of Δ*ku80* strains, short homology flanks of 40 bps can be added directly by incorporation into primers used for PCR. Co-transfect the pSAG1::Cas9-U6::sgUPRT plasmid and the targeting construct into parasites to be complemented and select with FUDR (*see* Table 1). Obtain clones by limiting dilution and screen for successful complementation using the PCR strategy shown

ucts should be purified by agarose gel electrophoresis (to get rid of nonspecific products and primers). Separate the PCR products on 0.8 % agarose gels, cut out the corresponding bands, and recover DNA from the gels using a gel extraction kit. As an estimate, ~1.5 μg is needed for each transfection.

2. For creating selection cassettes flanked by large homology regions (such as those shown in Fig. 2b, c), PCR amplify homology flanks corresponding to the GOI from genomic DNA (typically ~1 kb from upstream and ~1 kb from downstream of the target gene). The primers used for amplifying homology arms should contain additional sequences compatible for cloning into the selection plasmid by traditional restriction cloning or Gibson assembly. Design the flanking sequences so they will surround the drug selection cassette in the final construct. PCR amplify or use restriction enzymes to digest the selection cassette with its flanking regions from the plasmid backbone and gel-purify the products. As an estimate, ~1.5 μg is needed for each transfection.

3. For drug markers with short GOI homology arms (for use with Δ*ku80* strains), PCR amplify the drug selection cassette using primers that contain 40 bp of homology to the GOI. After PCR amplification, gel purify the targeting construct. As an estimate about 1.5 μg is needed for each transfection.

4. For making endogenous C-terminal tagging constructs in wild-type parasites (*see* Fig. 3), include an epitope tag sequence followed by a drug selection cassette. For making C-terminal tagging constructs, PCR amplify about 0.5 kb flanks immediately upstream and downstream of the stop codon, respectively, as homologous arms. The primers used for amplifying the homology flanks should contain additional sequences compatible for cloning into the tagging plasmid by traditional restriction enzyme cloning or Gibson assembly. Be sure that Cas9 won't also target the homology arms by introducing synonymous mutations into the primer sequences when necessary. The 5′ flank must be cloned in frame with the epitope tag. The 3′ flank should be cloned downstream of the drug selection cassette. Use restriction digestion or PCR amplification to obtain the GOI-tagging cassette from the plasmid backbone and gel purify the cassette. As an estimate, about 1.5 μg is needed for each transfection.

5. For C-terminal tagging constructs with short GOI homology arms (for use in Δ*ku80* strains), use PCR amplification to obtain the tagging construct using primers that contain the epitope tag as well as 40 bp homology flanks adjacent to the GOI stop codon (*see* Fig. 3). Be sure that Cas9 won't also target the homology arms by introducing synonymous mutations into the primer sequences when necessary. After PCR amplification, gel purify the cassette, 1.5 μg is needed for each transfection.

3.3 Parasite Transfection

1. A variety of mammalian cell lines can be used for propagating *T. gondii* using previously described protocols [7].

2. Generate parasites for transfection. Infect confluent host cells grown in T25 flasks with sufficient *T. gondii* parasites to achieve around 75% host cell lysis 2 days post-infection. Scrape the monolayers to fully capture and collect the cells and pass them through a 22 G needle 3–5 times to disrupt the host cells and release all parasites. Filter the parasites using 3.0 μm polycarbonate membranes to remove host cell debris. Centrifuge the filtered culture at $400 \times g$ for 10 min at 18 °C, resuspend the pellets in 10 ml Hanks' balanced salt solution and centrifuge again under the same conditions. Subsequently, resuspend the parasite cell pellet in Cytomix buffer to achieve a parasite density of 4×10^7 ml^{-1}.

Prepare the transfection mix in an electroporation cuvette and mix the following:

1.	Parasites resuspended in Cytomix	250 μl
2.	Targeting construct in the form of PCR or digestion product	1.5 μg
3.	GOI-specific CRISPR/Cas9 plasmid	7.5 μg
4.	ATP (0.2 M, 100×)	3 μl
5.	Glutathione (0.5 M, 100×)	3 μl
6.	Cytomix	to 300 μl

As a negative control, make the same mix but do not add the targeting construct and CRISPR/Cas9 plasmid.

3. Electroporation protocol: Mix the above components in the cuvette and electroporate the parasites with specific protocols provided by the manufacturer (e.g., for the BTX ECM-830 electroporator with 4 mm gap cuvettes, we suggest the following protocol: 1700 V, 176 μs of pulse length, two pulses with 100 ms interval). Parasites in the negative control should also be electroporated. Immediately after electroporation, transfer the parasites into T25 flasks with confluent host cells and grow them at 37 °C, 5 % CO_2.

3.4 Drug Selection and Cloning

1. At 24 h after electroporation and recovery, drugs may be added for selecting stable transformants. Alternatively, following natural egress of the electroporated parasites, collect the parasites and pass 1/5–1/3 of the culture into fresh T25 flasks with confluent cells grown in medium containing selection drugs. Repeat passages for a few times until the drug-resistant pool becomes stable (i.e., parasites in the negative control are completely killed and parasites transfected with the CRISPR/Cas9 sgRNA and targeting construct are grown well enough to be passed every ~2 days in the presence of drug selection).

2. When the drug-resistant pool shows stable growth, isolate the genomic DNA from the pool and perform a diagnostic PCR, for example, as illustrated in Fig. 2a to check whether the pool contains positive (knockout) clones. Other, more advanced methods may employ different PCR screening strategies, as discussed below.

3. If the pool is positive by diagnostic PCR, clone the pool by limiting dilution. Following natural egress of the parasites, purify the parasites as described above. Count the parasites with a hemocytometer to estimate the cell density of the filtered culture and then dilute the culture with culture medium (without drug) to achieve 2–3 (type 1 strains such as RH) or 3–5 (other types) parasites/150 μl. Subsequently, add 150 μl/well

of this diluted culture into 96-well plates with confluent host cells and let the parasites grow at 37 °C, 5 % CO_2 for 7–10 days without movement.

3.5 Identification of Positive Clones

1. After plaque development, visually check each well of the 96-well plates under an inverted-phase contrast microscope (such as a Nikon TS100) and look for wells that contain only one plaque. Transfer the parasites of positive wells into 24-well plates containing confluent host cell monolayers and grow them until they naturally egress.

2. Inoculate a small aliquot (e.g., 10 µL) from each well into a new 24-well plate to continue culturing the parasites. The rest of the lysed out culture will be used for diagnostic PCR as shown in Fig. 2a to check whether it has the expected gene disruption. Selected clones can be further examined by other means such as Western blotting. To identify epitope-tagged clones, refer to Fig. 3 for the PCR design.

3. Once positive clones are identified, they should be transferred from 24-well plates into T25 flasks and cryopreserved for future use.

3.6 Gene Disruption

In the next several sections (Subheadings 3.5–3.9) we present the protocols for specific genome editing strategies (i.e., Advanced Methods of genome editing). These advanced methods share many techniques that are universal to all (*see* Subheading 3). To avoid redundancy, the following protocols are cross-referenced to previously described methods where appropriate.

The following protocol provides step-by-step instructions to quickly knock out nonessential genes using CRISPR/Cas9 by insertional disruption (*see* Fig. 2a). We have demonstrated that the CRISPR/Cas9 system also facilitates site-specific integration of foreign DNA without homology to the GOI [24]. This feature makes gene disruption by insertional mutagenesis much easier. One can simply design a gene-specific CRISPR/Cas9 plasmid and co-transfect it with a drug selection cassette of choice (*see* Table 2) and select for disruptants that are confirmed by PCR (*see* **step 1** in Subheading 3.2).

1. Make a CRISPR/Cas9 plasmid specific for the gene of interest (GOI) (*see* Subheading 3.1).

2. Design an insertion construct to disrupt the GOI (*see* Subheading 3.2). The general design is illustrated in Fig 2a.

3. Co-transfect the CRISPR/Cas9 plasmid with the targeting construct into *T. gondii* by electroporation (*see* Subheading 3.3).

4. Select for GOI disruptants by adding drug and isolate individual clones by limiting dilution (*see* Subheading 3.4).

5. Expand individual clones in T25 flasks and save an aliquot of each to make a lysate for diagnostic PCR. Perform diagnostic PCR as shown in Fig. 2a to check for replacement of the GOI with the selection cassette (*see* Subheading 3.5). Screen ten clones initially, but be prepared to screen more, if necessary. Freeze down confirmed disruptants for future use.

3.7 Single sgRNA Deletions

The following procedures are used to delete small GOIs using a single sgRNA expressing CRISPR plasmid in combination with homologous targeting constructs (Illustrated in Fig. 2b). In general, we consider genes containing three or fewer exons and which do not exceed 5 kb to be small and well-suited for this technique. This method is especially efficient in Δ*ku80* strains but can work in wild-type strains as well.

1. Make a CRISPR/Cas9 plasmid specific for the gene of interest (GOI) (*see* Subheading 3.1).

2. Design a homologous targeting construct to disrupt the GOI (*see* **step 2** in Subheading 3.2). This step produces a plasmid containing a drug selection cassette that is flanked by long homology arms that correspond to the 5′ and 3′ UTR regions of the GOI. If working with Δ*ku80* strains, it may also be possible to perform gene deletion using short homology arms corresponding to the 5′ and 3′ UTR regions of the GOI, respectively (*see* **step 3** in Subheading 3.2).

3. Co-transfect the CRISPR/Cas9 plasmid with the homologous targeting construct into *T. gondii* by electroporation (*see* Subheading 3.3).

4. Select for GOI knockouts by drug selection and isolate individual clones by limiting dilution (*see* Subheading 3.4).

5. Expand individual clones in T25 flasks and save an aliquot of each to make lysate for diagnostic PCR as shown in Fig. 2c to check for GOI replacement with the selection marker (*see* Subheading 3.5). Screen ten clones initially, but be prepared to screen more if necessary. Freeze down confirmed GOI knockouts for future use.

3.8 Dual sgRNA Knockouts

Many genes in the *T. gondii* genome are encoded by multiple exons, which are usually distributed in a large array of genomic DNA [8]. The single gene sgRNA gene disruption strategies described in Subheadings 3.5 and 3.6 are not ideal for genes with complex exon/intron structures (*see* **Note 7**). Therefore, we developed a more efficient and reliable strategy to delete large genomic regions. This dual CRISPR strategy uses two separate sgRNAs, targeting the 5′ and 3′ ends of the GOI, that are co-expressed from the same CRISPR/Cas9 plasmid (*see* Fig. 2c). This approach has been successfully used to disrupt a ~30 kb locus in *T. gondii* containing the

ROP5 gene cluster [45] as well as a number of complex single-copy genes [46]. The procedures for the dual sgRNA strategy are as follows:

1. Generation of a dual sgRNA plasmid. Initially, two separate sgRNA plasmids are generated, targeting the 5′ and 3′ regions of the GOI. These are subsequently combined into a single plasmid by conventional cloning (*see* Fig. 1b) (*see* **step 6** in Subheading 3.1 for details).

2. Design a homologous targeting construct to disrupt the GOI (*see* **step 2** in Subheading 3.2). This step produces a plasmid containing a drug-selection cassette that is flanked by long (~1 kb) homology arms that correspond to the 5′ and 3′ UTR regions of the GOI (*see* **step 2** in Subheading 3.2). If working with Δ*ku80* strains, long homology arms are not necessary. Instead, PCR amplify a drug selection cassette with primers that contain 40 bp homology arms (*see* **step 3** in Subheading 3.2).

3. Co-transfect the dual sgRNA expressing CRISPR plasmid (*see* Fig. 1b) with the homologous replacement construct into *T. gondii* by electroporation (*see* Subheading 3.3).

4. Select for GOI knockouts by adding drug selection and isolate individual clones by limiting dilution (*see* Subheading 3.4).

5. Expand individual clones in T25 flasks and save an aliquot of each to make a lysate for diagnostic PCR.

6. Perform diagnostic PCR as diagrammed in Fig. 2c to check for GOI replacement with the selectable marker (*see* **step 1** in Subheading 3.5). Screen ten clones initially, but be prepared to screen more if necessary. Freeze down confirmed GOI knockouts for future use.

3.9 Endogenous Tagging

Recently, a CRISPR/Cas9-mediated tagging approach was used to introduce an epitope in combination with a synthesized dsDNA oligonucleotide containing a short homology region that was specific to a target gene [25]. We have modified this approach to improve the efficiency by including a resistance cassette and targeting the GOI for editing using CRISPR/Cas9 (*see* Fig. 3).

1. Make a CRISPR/Cas9 plasmid specific for the gene of interest (GOI) designed to introduce a DSB close to the translation stop codon, preferably downstream (*see* Subheading 3.1).

2. Design a homologous targeting construct to add a C-terminal epitope tag to the GOI (*see* **step 4** in Subheading 3.2). The targeting construct consists of four components: (a) a 5′ region of homology chosen from the sequence immediately upstream of the stop codon of the GOI, (b) an in-frame epitope tag with a stop codon, (c) a selection cassette, and (d) a 3′ region of

homology chosen from the 3′-UTR of the GOI (*see* **step 4** in Subheading 3.2). If working with Δ*ku80* strains, long homology arms are not necessary. Instead, PCR amplify the tagging construct using primers that contain 40 bp homology flanks adjacent to the GOI stop codon. In both cases, be sure that Cas9 won't target the homology arms by introducing synonymous mutations into the primer sequences, if necessary (*see* **step 5** in Subheading 3.2).

3. Co-transfect the CRISPR/Cas9 plasmid with the homologous targeting construct into *T. gondii* by electroporation (*see* Subheading 3.3).

4. Select for tagged parasites by drug selection and isolate individual clones by limiting dilution (*see* Subheading 3.4).

5. Expand individual clones and perform diagnostic PCR as diagrammed in Fig. 3 to check for the correct epitope-tagging of the GOI (*see* Subheading 3.5). Screen ten clones initially, but be prepared to screen more if necessary. Confirm epitope-tagging by immunofluorescence and Western blotting with appropriate commercial antibodies. Freeze down epitope-tagged parasites for future use.

3.10 Gene Complementation

Once a GOI knockout has been made, any resulting phenotypes should be complemented by adding back the deleted gene to confirm that the phenotype was due to its loss (*see* **Note 5**). When complementing a GOI, or knocking in any transgene for that matter, it is important to target the construct to a neutral locus. A neutral disruption of the locus should not affect parasite growth or the phenotypes associated with the GOI. The insertion site could either be the original endogenous locus, or a locus that is neutral but offers negative drug selection. By negative drug selection we mean that disrupting the locus renders the parasite insensitive to specific inhibitors. In *T. gondii*, three such negatively selectable markers are recommended (*see* Table 1): (1) *UPRT*, disruption of which renders the parasite insensitive to 5-fluorodeoxyribose (FUDR) [18]; (2) *SNFR1*, a putative transporter disruption of which confers sinefungin resistance [42]; and (3) *HXGPRT*, a purine salvage enzyme disruption of which confers 6-thioxanthine resistance [20]. We have successfully complemented GOIs using the *UPRT* locus [24], so we will use this locus as our example (*see* Fig. 4). Other options are provided in Table 1 (*see* **Note 7**).

1. For complementing at the *UPRT* locus, use pSAG1::Cas9-U6::sgUPRT from Addgene (www.addgene.org). For targeting a different locus, make a CRISPR/Cas9 plasmid specific for that locus (*see* Subheading 3.1).

2. Design a *UPRT*-flanked GOI complementation targeting construct using Gibson assembly. Procure plasmid pUPRT::DHFR-D

(*see* Subheading 2.1). This plasmid contains a DHFR-TS* Pyr resistance cassette that is flanked by *UPRT* 5' and 3' homology arms from UTR regions. In the next few steps, the DHFR-TS cassette will be replaced by your GOI as an expression cassette. PCR amplify the vector backbone without the DHFR-TS* cassette with pUPRT::GOI GIB F (GACAGACCGCTG ACGGAATC) and pUPRT::GOI GIB R (AGAAGCCC TGTGGACAGGTC) using Q5 polymerase and gel purify the product. Amplify and clone the GOI with primers compatible with the vector backbone. For single-exon GOIs, PCR amplify the entire endogenous locus with promoter and UTRs from genomic DNA using the primers GOI F (5'-GACCTGT CCACAGGGCTTCTN20) and GOI R (5'-GATTCCG TCAGCGGTCTGTCN20). Use Gibson assembly to clone the GOI amplicon into the pUPRT backbone to yield the targeting construct pUPRT::GOI. A similar strategy should be used for multi-exon genes, except the CDS should be cloned from cDNA to remove the introns. PCR amplify the endogenous promoter (from gDNA), CDS (from cDNA), and 3' UTR (from gDNA) with proper sequential overlaps and use a four-fragment Gibson assembly (3 GOI inserts plus the pUPRT amplicon) to yield the pUPRT::GOI targeting construct. Next, perform Q5-mediated mutagenesis on the resulting plasmid to add an epitope tag if desired. Insertions of up to 100 bp can be made in plasmids up to 20 kb using the Q5 site-directed mutagenesis kit. Use PCR amplification or restriction digestion to obtain the targeting cassette and gel purify the fragment for transfection.

3. To complement Δ*goi* mutants, co-transfect the pSAG1::Cas9-U6::sgUPRT plasmid (1.5 μg) with the targeting cassette (7.5 μg) by electroporation (*see* Subheading 3.3). This DNA ratio is inverse to knockout strategies to ensure that parasites that received the Cas9 plasmid also received the targeting construct.

4. Select for complemented Δ*goi::uprt*::GOI parasites by adding 10 μM FUDR and isolate individual clones by limiting dilution (*see* Subheading 3.4).

5. Expand individual clones in T25 flasks and save an aliquot of each to make lysate for diagnostic PCR as shown in Fig. 4 to check for GOI complementation (*see* Subheading 3.5). Screen ten clones initially, but be prepared to screen more if necessary. Confirm successful complementation by RT-PCR, IFA and Western blotting with anti-epitope tag antibodies. Freeze down confirmed complemented parasites for future use.

4   Notes

1. The *DHFR-TS** gene confers resistance to pyrimethamine (Pyr), a common therapy for human toxoplasmosis [48]. Therefore, Pyr[R] parasites pose a greater health risk to those working with them. However, *DHFR-TS** provides one of the strongest selection methods for obtaining stable transformants and thus is highly efficient. We recommend using floxed versions of *DHFR-TS** that can be removed after drug selection by addition of Cre [46] and as described in [49].

2. The Cat gene confers chloramphenicol (Cm) resistance. Chloramphenicol selection targets the apicoplast [50] and induces a delayed death phenotype, where the parasite grows normally in the first vacuole and impaired growth only occurs in the second round of infection. As a result, selection of mutants can take up to 2 weeks to complete.

3. Introduction of the *HXGPRT* gene confers mycophenolic acid (MPA) resistance in the presence of xanthine [51]. This selection is limited to strains that already have endogenous *HXGPRT* disrupted. Additionally, spontaneous resistance is sometimes encountered.

4. The Ble gene confers phleomycin (Ph) resistance. Ph is a member of the bleomycin family of antibiotics that works by intercalating into DNA. As such, this agent should not be used in *Δku80* strains due to their increased susceptibility to DNA damage [16].

5. Because the target specificity of CRISPR/Cas9 system is determined by the N_{20} sequence in sgRNA and some mismatches (especially at the 5′ end) between sgRNA and the target are allowed, off-target effects are always a concern during genome editing. Given the relatively small genome of *T. gondii*, off-target effects are not expected to be as common as in other organisms with larger genomes (i.e., mammalian). However, it does happen at low frequency, as we have shown before [24]. In this regard, phenotypic testing of a few independent clones is recommended to confirm the relationship between observed phenotypes and genetic ablation. In addition, genetic complementation is always encouraged for further confirmation. The discovery of genetic sites that can be used for negative selection greatly facilitates this approach.

6. From our experience using *Δku80* strains and the CRISPR/Cas9 system to knock out genes by homologous recombination, we often observed that how the homology template was prepared had a substantial impact on the efficiency of double-crossover homologous recombination. We found that if the

homology template was provided as a restriction digestion or as PCR products with both homologous arms immediately exposed (no extra sequence from the plasmid backbone was included), the recombination efficiency was usually the highest. Historically, linearized plasmids were used as homology templates, where the homologous arms might be flanked by extra sequences from the plasmid backbone. Using such templates, we often saw that either the recombination did not occur at all, or it only happened at one end (single crossover) but not both. However, if both homologous arms are immediately exposed in the homology template provided, the vast majority of recombination events are double crossovers resulting in replacement of the endogenous GOI with the targeting construct.

7. In the case of using single sgRNA CRISPR plasmids mediated marker insertion for gene disruption, correct insertion of a selection cassette does not guarantee gene inactivation, especially for those genes with multiple exons and functional domains. The commonly used strategy to confirm gene disruption is RT-PCR to examine the disappearance of mRNA for GOI after insertion. However, in many cases we were still able to detect low levels of transcripts after successful insertion, even for the regions downstream of the insertion site [46]. This may be due to alternative splicing, alternative start sites, or weak internal promoter activity within the inserted selection cassette. Although residual transcripts may not produce full-length wild-type proteins (due to altered reading frames), they raise questions as to whether the gene is truly inactivated. In this case, immunofluorescence assays or Western blotting with a good antibody may be helpful to clarify the results. As an alternative, the use of a dual sgRNA strategy to completely remove the CDS is recommended (*see* Subheading 3.3).

Acknowledgments

Partially supported by NIH grant AI118426, AI034036 (to LDS) and Projects 2662015PY048 and 2662015PY104 from the Fundamental Research Funds for the Central Universities (to B.S.). We thank Dr. Joshua B. Radke for a critical reading of the manuscript.

References

1. Pawlowski J, Audic S, Adl S, Bass D, Belbahri L, Berney C, Bowser SS, Cepicka I, Decelle J, Dunthorn M, Fiore-Donno AM, Gile GH, Holzmann M, Jahn R, Jirku M, Keeling PJ, Kostka M, Kudryavtsev A, Lara E, Lukes J, Mann DG, Mitchell EA, Nitsche F, Romeralo M, Saunders GW, Simpson AG, Smirnov AV, Spouge JL, Stern RF, Stoeck T, Zimmermann J, Schindel D, de Vargas C (2012) CBOL protist working group: barcoding eukaryotic richness beyond the animal, plant, and fungal kingdoms. PLoS Biol 10(11), e1001419

2. Miller LH, Ackerman HC, Su XZ, Wellems TE (2013) Malaria biology and disease pathogenesis: insights for new treatments. Nat Med 19(2):156–167

3. Dubey JP (2010) Toxoplasmosis of animals and humans. CRC Press, Boca Raton, FL

4. Checkley W, White AC Jr, Jaganath D, Arrowood MJ, Chalmers RM, Chen XM, Fayer R, Griffiths JK, Guerrant RL, Hedstrom L, Huston CD, Kotloff KL, Kang G, Mead JR, Miller M, Petri WA Jr, Priest JW, Roos DS, Striepen B, Thompson RC, Ward HD, Van Voorhis WA, Xiao L, Zhu G, Houpt ER (2015) A review of the global burden, novel diagnostics, therapeutics, and vaccine targets for cryptosporidium. Lancet Infect Dis 15(1):85–94

5. Pfefferkorn ER (1990) Cell biology of *Toxoplasma gondii*. In: Wyler DJ (ed) Modern parasite biology. W.H. Freeman, New York, NY, pp 26–50

6. Khan A, Taylor S, Su C, Mackey AJ, Boyle J, Cole RH, Glover D, Tang K, Paulsen I, Berriman M, Boothroyd JC, Pfefferkorn ER, Dubey JP, Roos DS, Ajioka JW, Wootton JC, Sibley LD (2005) Composite genome map and recombination parameters derived from three archetypal lineages of *Toxoplasma gondii*. Nucleic Acids Res 33:2980–2992

7. Roos DS, Donald RGK, Morrissette NS, Moulton AL (1994) Molecular tools for genetic dissection of the protozoan parasite *Toxoplasma gondii*. Methods Cell Biol 45:28–61

8. Lorenzi H, Khan A, Behnke MS, Namasivayam S, Swapna LS, Hadjithomas M, Karamycheva S, Pinney D, Brunk B, Ajioka JW, Ajzenberg D, Boothroyd JC, Boyle JP, Darde ML, Diaz-Miranda MA, Dubey JP, Fritz HM, Gennari SM, Gregory BD, Kim K, Saeij JP, Su C, White WH, Zhu XQ, Howe DK, Rosenthal B, Grigg ME, Parkinson J, Liu L, Kissinger JC, Roos DS, Sibley LD (2016) Local admixture of amplified and diversified secreted pathogenesis determinants shapes mosaic *Toxoplasma gondii* genomes. Nat Commun 7:10147

9. Sibley LD, Ajioka JW (2008) Population structure of *Toxoplasma gondii*: Clonal expansion driven by infrequent recombination and selective sweeps. Ann Rev Microbiol 62:329–351

10. Su CL, Khan A, Zhou P, Majumdar D, Ajzenberg D, Dardé ML, Zhu XQ, Ajioka JW, Rosenthal B, Dubey JP, Sibley LD (2012) Globally diverse *Toxoplasma gondii* isolates comprise six major clades originating from a small number of distinct ancestral lineages. Proc Natl Acad Sci U S A 109:5844–5849

11. Kim K, Soldati D, Boothroyd JC (1993) Gene replacement in Toxoplasma gondii with chloramphenicol acetyltransferase as selectable marker. Science 262(5135):911–914

12. Kim K, Boothroyd JC (1995) *Toxoplasma gondii*: Stable complementation of *sag1* (p30) mutants using *SAG1* transfection and fluorescence-activated cell sorting. Exp Parasitol 80:46–53

13. Soldati D, Boothroyd JC (1993) Transient transfection and expression in the obligate intracellular parasite *Toxoplasma gondii*. Science 260:349–352

14. Messina M, Niesman IR, Mercier C, Sibley LD (1995) Stable DNA transformation of *Toxoplasma gondii* using phleomycin selection. Gene 165:213–217

15. Donald RGK, Roos DS (1993) Stable molecular transformation of *Toxoplasma gondii*: A selectable dihydrofolate reductase-thymidylate synthase marker based on drug resistance mutations in malaria. Proc Natl Acad Sci U S A 90:11703–11707

16. Fox BA, Ristuccia JG, Gigley JP, Bzik DJ (2009) Efficient gene replacements in *Toxoplasma gondii* strains deficient for nonhomologous end joining. Eukaryot Cell 8(4):520–529

17. Donald RGK, Roos DS (1994) Homologous recombination and gene replacement at the dihydrofolate reductase-thymidylate synthase locus in *Toxoplasma gondii*. Mol Biochem Parasitol 63:243–253

18. Donald RGK, Roos DS (1995) Insertional mutagenesis and marker rescue in a protozoan parasite: Cloning the uracil phosphoribosyltransferase locus from *Toxoplasma gondii*. Proc Natl Acad Sci U S A 92:5749–5753

19. Sullivan WJ Jr, Chiang CW, Wilson CM, Naguib FN, el Kouni MH, Donald RG, Roos DS (1999) Insertional tagging of at least two loci associated with resistance to adenine arabinoside in *Toxoplasma gondii*, and cloning of the adenosine kinase locus. Mol Biochem Parasitol 103(1):1–14

20. Donald RGK, Roos DS (1998) Gene knockouts and allelic replacements in *Toxoplasma gondii*: HXGPRT as a selectable marker for hit-and-run mutagenesis. Mol Biochem Parasitol 91:295–305

21. Huynh MH, Carruthers VB (2009) Tagging of endogenous genes in a *Toxoplasma gondii* strain lacking Ku80. Eukaryot Cell 8(4):530–539

22. Fox BA, Falla A, Rommereim LM, Tomita T, Gigley JP, Mercier C, Cesbron-Delauw MF, Weiss LM, Bzik DJ (2011) Type II *Toxoplasma gondii* KU80 knockout strains enable functional analysis of genes required for cyst development and latent infection. Eukaryot Cell 10(9):1193–1206

23. Zhang J, Mao Z, Xue W, Li Y, Tang G, Wang A, Zhang Y, Wang H (2011) Ku80 gene is related to non-homologous end-joining and genome stability in Aspergillus niger. Curr Microbiol 62(4):1342–1346

24. Shen B, Brown KM, Lee TD, Sibley LD (2014) Efficient gene disruption in diverse strains of *Toxoplasma gondii* using CRISPR/CAS9. mBio 5(3):e01114-14

25. Sidik SM, Hackett CG, Tran F, Westwood NJ, Lourido S (2014) Efficient genome engineering of *Toxoplasma gondii* using CRISPR/Cas9. PLoS One 9(6), e100450

26. Barrangou R, Fremaux C, Deveau H, Richards M, Boyaval P, Moineau S, Romero DA, Horvath P (2007) CRISPR provides acquired resistance against viruses in prokaryotes. Science 315(5819):1709–1712

27. Marraffini LA (2015) CRISPR-Cas immunity in prokaryotes. Nature 526(7571):55–61

28. Mali P, Esvelt KM, Church GM (2013) Cas9 as a versatile tool for engineering biology. Nat Methods 10(10):957–963

29. Cong L, Ran FA, Cox D, Lin S, Barretto R, Habib N, Hsu PD, Wu X, Jiang W, Marraffini LA, Zhang F (2013) Multiplex genome engineering using CRISPR/Cas systems. Science 339(6121):819–823

30. Chen C, Fenk LA, de Bono M (2013) Efficient genome editing in *Caenorhabditis elegans* by CRISPR-targeting homologous recombination. Nucleic Acids Res 41(20), e193

31. Hruscha A, Krawitz P, Rechenberg A, Heinrich V, Hecht J, Haass C, Schmid B (2013) Efficient CRISPR/Cas9 genome editing with low off-target effects in zebrafish. Development 140(24):4982–4987

32. DiCarlo JE, Norville JE, Mali P, Rios X, Aach J, Church GM (2013) Genome engineering in *Saccharomyces cerevisiae* using CRISPR-Cas systems. Nucleic Acids Res 41(7):4336–4343. doi:10.1093/nar/gkt135

33. Jinek M, Chylinski K, Fonfara I, Hauer M, Doudna JA, Charpentier E (2012) A programmable dual-RNA-guided DNA endonuclease in adaptive bacterial immunity. Science 337(6096):816–821

34. Jiang W, Zhou H, Bi H, Fromm M, Yang B, Weeks DP (2013) Demonstration of CRISPR/Cas9/sgRNA-mediated targeting gene modification in Arabidopsis, tobacco, sorghum and rice. Nucleic Acids Res 41(20), e188

35. Jinek M, East A, Cheng A, Lin S, Ma E, Doudna J (2013) RNA-programmed genome editing in human cells. Elife 2, e00471

36. Wagner JC, Platt RJ, Goldfless SJ, Zhang F, Niles JC (2014) Efficient CRISPR-Cas9-mediated genome editing in *Plasmodium falciparum*. Nat Methods 11(9):915–918

37. Vinayak S, Pawlowic MC, Sateriale A, Brooks CF, Studstill CJ, Bar-Peled Y, Cipriano MJ, Striepen B (2015) Genetic modification of the diarrhoeal pathogen *Cryptosporidium parvum*. Nature 523(7561):477–480

38. Maeder ML, Linder SJ, Cascio VM, Fu Y, Ho QH, Joung JK (2013) CRISPR RNA-guided activation of endogenous human genes. Nat Methods 10(10):977–979

39. Gilbert LA, Larson MH, Morsut L, Liu Z, Brar GA, Torres SE, Stern-Ginossar N, Brandman O, Whitehead EH, Doudna JA, Lim WA, Weissman JS, Qi LS (2013) CRISPR-mediated modular RNA-guided regulation of transcription in eukaryotes. Cell 154(2): 442–451

40. Zhang C, Xiao B, Jiang YY, Zhao YH, Li ZK, Gao H, Ling Y, Wei J, Li SN, Lu MK, Su XZ, Cui HT, Yuan J (2014) Efficient Editing of Malaria Parasite Genome Using the CRISPR/Cas9 System. mBio 5(4)

41. Ghorbal M, Gorman M, Macpherson CR, Martins RM, Scherf A, Lopez-Rubio JJ (2014) Genome editing in the human malaria parasite *Plasmodium falciparum* using the CRISPR-Cas9 system. Nat Biotechnol 32(8):819–821

42. Behnke MS, Khan A, Sibley LD (2015) Genetic mapping reveals that sinefungin resistance in *Toxoplasma gondii* is controlled by a putative amino acid transporter locus that can be used as a negative selectable marker. Eukaryot Cell 14(2):140–148

43. Boube H (2013) A protocol for construction of gene targeting vectors and generation of homolgous recombinant embryonic stem cells. Methods Mol Biol 1063:337–354

44. Gibson DG, Young L, Chuang RY, Venter JC, Hutchison CA 3rd, Smith HO (2009) Enzymatic assembly of DNA molecules up to several hundred kilobases. Nat Methods 6(5):343–345

45. Behnke MS, Khan A, Lauron EJ, Jimah JR, Wang Q, Tolia NH, Sibley LD (2015) Rhoptry Proteins ROP5 and ROP18 Are Major Murine Virulence Factors in Genetically Divergent South American Strains of *Toxoplasma gondii*. PLoS Genet 11(8), e1005434

46. Long S, Wang Q, Sibley LD (2015) Analysis of non-canonical calcium dependent protein kinases in *Toxoplasma gondii* by targeting gene deletion using CRISPR/Cas9. Infect Immun 84(5):1262–1273

47. Critchlow SE, Jackson SP (1998) DNA end-joining: from yeast to man. Trends Biochem Sci 23(10):394–398

48. McCabe RE (2001) Antitoxoplasma chemo-therapy. In: Joynson DHM, Wreghitt TG (eds) Toxoplasmosis: a comprehensive clinical guide. Cambridge University Press, Cambridge, pp 319–359

49. Heaslip AT, Nishi M, Stein B, Hu K (2011) The motility of a human parasite, *Toxoplasma gondii*, is regulated by a novel lysine methyltransferase. PLoS Pathog 7(9), e1002201

50. Fichera ME, Roos DS (1997) A plastid organ-elle as a drug target in apicomplexan parasites. Nature (Lond) 390:407–409

51. Donald RGK, Carter D, Ullman B, Roos DS (1996) Insertional tagging, cloning, and expression of the *Toxoplasma gondii* hypoxanthine-xanthine-guanine phosphoribo-syltransferase gene. J Biol Chem 271(24): 14010–14019

Section II

Gene and Genome-Editing Methods Part II

Chapter 7

Generation of Stable Knockout Mammalian Cells by TALEN-Mediated Locus-Specific Gene Editing

Barun Mahata and Kaushik Biswas

Abstract

Precise and targeted genome editing using *Transcription Activator-Like Effector Endonucleases* (TALENs) has been widely used and proven to be an extremely effective and specific knockout strategy in both cultured cells and animal models. The current chapter describes a protocol for the construction and generation of TALENs using serial and hierarchical digestion and ligation steps, and using the synthesized TALEN pairs to achieve locus-specific targeted gene editing in mammalian cell lines using a modified clonal selection strategy in an easy and cost-efficient manner.

Key words Genome engineering, TALEN, Repeat variable di-residues (RVDs), Gene knockout in mammalian cells, Nonhomologous end joining (NHEJ), Homology-directed repair (HDR), Double-strand break (DSB) repair, Fok1, Restriction enzyme and ligation (REAL), T7E1 assay

1 Introduction

Precise and targeted gene editing technology-mediated genome modifications hold tremendous potential in bioengineering, biomedical research and gene therapeutics [1–4]. This easily programmable, high-end target-specific nuclease can generate double-strand breaks (DSB) in DNA inside the cells. An attempt by the cells to repair the DSBs by error-prone nonhomologous end joining (NHEJ) results in insertion/deletion of some bases (indels) generating a disrupted and/or dysfunctional protein. On the other hand, homology-directed repair (HDR) may allow incorporation of small oligo dinucleotides (ODNs)/large DNA fragments (donor vector) through homology-directed repair or nonhomologous end joining upon availability of delivered DNA within cells [5–8].

Transcription activator-like effectors (TALEs) are found in naturally occurring plant pathogens (e.g., *Xanthomonous* sp.) and they use it to subvert their host's genome regulatory networks [9]. Transcription activator-like effector nucleases are engineered DNA binding proteins generated by fusing individual TALE domains

Andrew Reeves (ed.), *In Vitro Mutagenesis: Methods and Protocols*, Methods in Molecular Biology, vol. 1498,
DOI 10.1007/978-1-4939-6472-7_7, © Springer Science+Business Media New York 2017

with the nonspecific nuclease domain of Fok1 endonucleases. The DNA binding domain of a TALEN entirely depends on modularly assembled TALE repeats. Each TALE repeat is highly conserved consisting of a 33–35-amino acid long polypeptide segment which binds to a single base. The DNA binding specificity depends on two highly variable amino acid residues at positions 12 and 13 [10, 11] known as *Repeat Variable Di*-residues (RVDs). In modularly assembled TALE repeats, total RVD composition collectively determines the fate of the TALEN targeted against the DNA sequence of interest. In most cases, RVD asparagine and isoleucine (NI) recognize A, histidine and aspartic acid (HD) recognize C, asparagine and glycine (NG) recognize T, and two asparagines recognize G, respectively [10, 11]. Finally, addition of the nuclease domain of Fok1 at the C-terminus of a paired TALE repeat allows the synthetic TAL effector nucleases to specifically generate a double-strand break at the targeted locus.

Different types of TALEN construction methods have been described [12–17]. Here, we have taken a non-PCR-based approach called the *Restriction Enzyme* digestion *and Ligation*-dependent strategy (REAL method, created by Deepak Rayon and Keith Joung [18]) to assemble the TALEN modules with some modifications.

TALEN-mediated targeted editing of endogenous genes in cultured mammalian cells has been described and its application to study gene function has widely been used [1, 2, 19]. However, it takes a rigorous clonal selection strategy and often involves screening more than 100 colonies to generate a knockout cell line as TALEN encoding plasmids do not possess any selectable marker genes. In this chapter, we describe a modified method of TALEN construction using the REAL method based on a much simpler clonal selection strategy to generate stable knockout mammalian cell lines with a higher efficiency and an increased mutational frequency.

2   Materials

2.1 TALEN Constructs, Reagents

1. ZiFit Targeter software v4.2 (http://zifit.partners.org/ZiFiT/ChoiceMenu.aspx).

2. TALEN plasmid toolkit #1000000017 (addgene).

3. Restriction enzymes and T7 Endonuclease: *Bbs*I, *Bsa*I, *Bam*HI, *Xba*I, *Kpn*I, *Bsm*BI, 10× reaction buffers, T7 Endonuclease, DNA molecular marker.

4. Ligation reagents: 2× quick ligase reaction buffer and quick ligase.

5. Competent cells and bacterial culture medium. XL1 blue recombination deficient bacterial strain (*see* **Note 1**).

6. Luria–Bertani broth: tryptone, 10 g, yeast extract, 5 g, NaCl, 5 g. Add distilled water to 1000 ml. pH to 7.0 with 5 N

NaOH. Sterilize by autoclaving. For selections, add ampicillin sulfate to 100 μg/ml. Add agarose to 1.5 % (15 g/L) for plate medium.

7. Super Optimal Broth with Catabolite Repression (SOC) Media.

8. QIAquick gel extraction kit.

9. QIAquick PCR purification kit.

10. Plasmid midi kit (Qiagen).

11. Plasmid isolation miniprep kit.

12. TALE sequencing primers: JDS2978: 5′ TTGAGGCG CTGCTGACTG 3′; JDS2980: 5′ TTAATTCAATATATT CATGAGGCAC 3′; OK-163: 5′ CGCCAGGGTTTTCCCAG TCACGAC 3′.

13. Nuclease-free agarose gel electrophoresis reagents.

2.2 TALEN Transfection

Cell lines, buffers, reagents and equipment used.

1. NIH 3T3 or Renca-v (for validation of TALEN activity) or other cell lines with a minimum 50 % transfection efficiency (*see* **Note 2**).

2. Dulbecco's Modified Eagle's Medium DMEM (commercially available from Hi-Media) with added antibiotics: contains sodium bicarbonate, Modified Eagle's Medium (MEM), sodium pyruvate, L-glutamine, 10 % fetal bovine serum (FBS), 100 U/ml penicillin, 100 μg/ml streptomycin, 50 μg/ml gentamicin sulfate.

3. Antibiotic-free DMEM medium.

4. Opti-MEM transfection compatible medium.

5. 10× trypsin–EDTA solution (Hi-Media).

6. Transfection Reagent: Lipofectamine-2000, Lipofectamine-3000, Lipofectamine-LTX with plus reagent (Invitrogen).

7. Transfection control plasmid and co-transfection plasmid: pMaxGFP or pD2EGFP-N1 (Lonza).

8. Selection antibiotic: G418 (Hi-Media).

9. DNA isolation reagent, RIPA buffer (commercially available from Pierce): 25 mM Tris–HCl, pH 7.6, 150 mM NaCl, 1 % sodium deoxycholate, 0.1 % SDS.

10. Phenol–chloroform–isoamyl alcohol.

11. 3 M sodium acetate, pH 5.2.

12. Elution buffer (EB): 10 mM Tris-Cl, pH 8.5.

13. 70 % ethanol.

14. TE buffer.

15. 6-well, 12-well, 24-well tissue culture plates.

16. SDS-PAGE electrophoresis and western-immunoblot (wet) apparatus.

17. Immunofluorescent microscope.

2.3 Genome Analysis

1. TAL Effector Nucleotide Targeter, v2.0 website: (https://tale-nt.cac.cornell.edu/node/add/talef-off-paired).

2. Sequencing analysis: e.g., Seq A6.

3. Gel documentation and imaging system.

4. DNA sequencer (e.g., ABI-3500 DNA) or outsourced service provider.

5. T7 Endonuclease with reaction buffer.

6. Amplification primers encompassing targeted locus for T7E1 assay (amplicon 500–1000 bp) and indel analysis (amplicon 100–120 bp (*see* **Note 3**).

7. Ex-Taq (Takara)/Q5 high fidelity DNA polymerase (NEB) (*see* **Note 4**).

8. Hi-resolution agarose (e.g., Metaphor, Lonza).

9. T/A cloning vector: pTZ57R/T.

10. Sequencing Primer: T7 forward or M13.

11. Sequencing reagents: 2× sequencing PCR buffer, RR mix, Hi-Di formamide.

12. Horizontal agarose gel electrophoresis apparatus.

3   Methods

3.1 Target DNA Selection

1. The DNA sequence of interest is copied and pasted in the box provided with the target base (in the third bracket) and submitted online using ZiFiT Targeter v4.2 software at: (http://zifit.partners.org/ZiFiT/ChoiceMenu.aspx) for generation of the schematic roadmap required for designing the TALEN pairs.

3.2 TALEN Pair Construction

We have used a previously published protocol for the REAL method of TALEN construction [18] for synthesis of TALEN pairs with some added modifications. The complete modified procedure is described below. Schematic representation for serial digestion and ligation is shown in Fig. 1.

For ligation of each pair of the TALE module, the left plasmid encoding TALE repeat is the amino-terminal repeat and the right one is the carboxy-terminal TALE repeat.

1. Digest 2–4 μg of the amino terminal TALE encoding plasmid with *Bam*HI and *Bsa*I and perform the following reaction at 37 °C for 12 h (see reaction mix below).

Amino terminal TALE encoding plasmid	$X\,\mu l$
10× NEB Buffer 2	3 µl
100× BSA	0.3 µl
*Bsa*I	2 µl
*Bam*HI	2 µl
Nuclease-free H$_2$O	Up to 30 µl

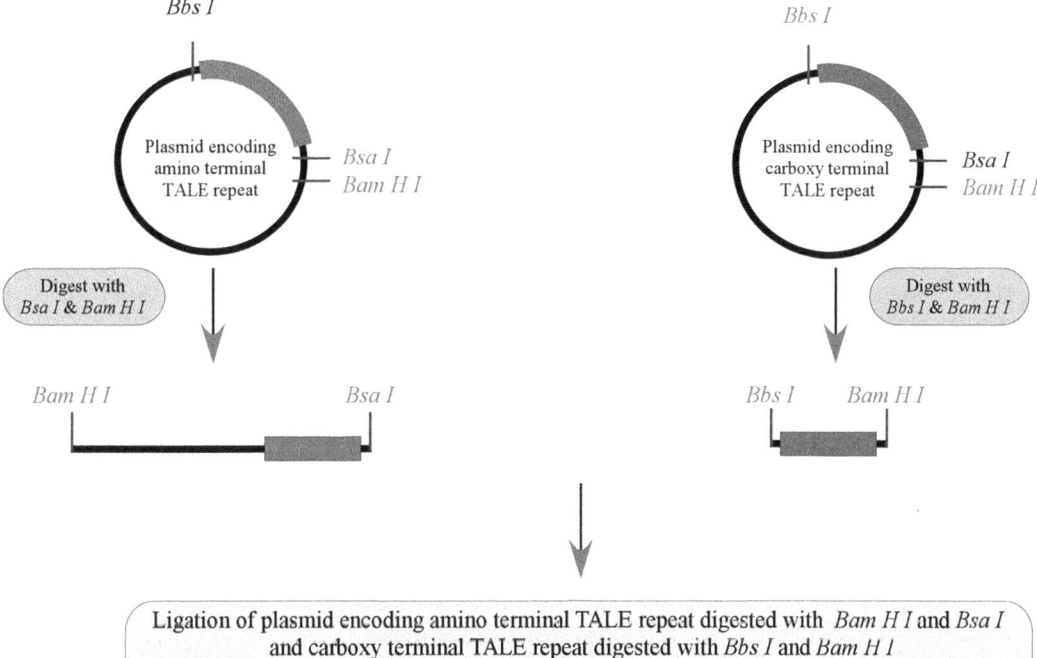

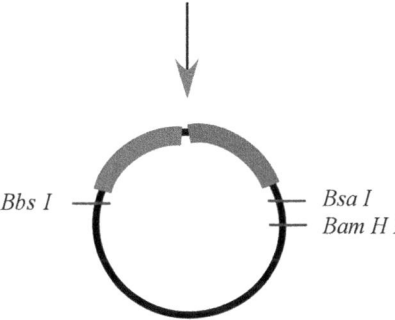

Fig. 1 Schematic illustrating the REAL method of TALEN construction. N-terminal TALE repeats containing plasmids were digested with *Bam*HI and *Bsa*I. The plasmid backbone is gel eluted whereas the C-terminal TALE repeat containing plasmids were digested with *Bbs*I and *Bam*HI and the TALE repeat was gel eluted. The plasmid backbone and TALE repeats were then ligated to generate dimers in the first step of the TALEN construction method. Using this digestion and ligation method, TALENs can be constructed in a step-wise manner

2. Run the digested product on a 1% agarose gel. Gel elute the plasmid backbone using QIAquick gel extraction column in pre-warmed 20 μl of Elution Buffer (10 mM Tris-Cl, pH 8.5) buffer. Quantify the eluted DNA.

3. Digest 15–20 μg of the carboxy-terminal TALE repeat encoding plasmid with *Bbs*I and *Bam*HI (see reaction mix below). The digestion of 15–20 μg of the carboxy terminal TALE repeat encoding plasmid is essential because only a 102 bp fragment comes out after digestion; otherwise, a very small quantity of DNA will be gel eluted. Incubate at 37 °C for 12 h.

Carboxy terminal TALE encoding plasmid	X μl
10× NEB Buffer 4	3 μl
100× BSA	0.3 μl
*Bbs*I	3 μl
*Bam*HI	3 μl
Nuclease-free H$_2$O	Up to 30 μl

4. Run the digested product on a 1% agarose gel. Gel elute the TALE repeat digested from the backbone plasmid using a QIAquick gel extraction column in pre-warmed 20 μl of EB buffer. Quantify the eluted DNA.

5. Use the online ligation calculator at (http://www.insilico.uni-duesseldorf.de/Lig_Input.html) for calculation of ligation reaction mixtures using plasmid–TALE repeat ratio of 1:3. The following protocol is then used to carry out the ligation reaction (below) in a total volume of 5 μl.

Plasmid backbone	X μl
TALE repeat	Y μl
2× Quick Ligase Buffer	2.5 μl
Quick Ligase	0.5 μl ($X + Y = 2$ μl)

6. Incubate at 25 °C for 15 min. Then place on ice for 10 min. Perform the control ligation using only plasmid backbone.

7. Thaw competent XL-1 Blue cells on ice. Add the ligation mix to 45 μl of cells, incubate 15 min on ice. Heat shock the tube at 42 °C for 60–75 s and put on ice for another 2 min. Add 500 μl of SOC medium and incubate at 37 °C for 60–90 min with moderate shaking (~200 rpm).

8. Microcentrifuge the tube at 2400 × *g* for 5 min and dissolve the pellet in 100 μl of SOC medium. Spread the cells over LB-amp agar plate. Put the plate in 37 °C and incubate for 12–15 h.

9. Normally, the original ligation results in five to tenfold more colonies than the control ligation. Inoculate two single colonies from the actual ligation plate into 5–10 ml of LB medium containing 100 μg/ml ampicillin. Incubate for 12 h in a 37 °C shaker.

10. From the overnight culture, isolate the plasmid DNA using a plasmid miniprep kit.

11. To check whether the plasmids in **step 10** have taken up the TALE repeat-encoding C-terminal TALE repeat, generate a digestion mixture (below) of each plasmid with *Xba*I and *Bam*HI. Incubate at 37 °C for 2 h.

Isolated plasmid from **step 10**	X μl (1–2 μg)
10× NEB Buffer 4	3 μl
100× BSA	0.3 μl
*Xba*1	1 μl
*Bam*H1	1 μl
Nuclease-free H₂O	Up to 30 μl

12. Run the digested product on a 2 % agarose gel and visualize with a gel documentation system. The plasmid with an N-terminal TALE repeat which was successfully ligated with a carboxy-terminal TALE repeat should yield a $[\{(M+N) \times 102\} + 33]$ bp fragment, where M and N represent TALE repeats encoded by backbone plasmid and gel-eluted carboxy terminal TALE repeats.

13. After successful completion of the first ligation step, repeat the process for the second, third and fourth steps of serial digestion and ligation (*see* Fig. 2).

3.3 Cloning Ligated TALE Arrays

Once the entire right DNA-binding TALE repeats and left DNA-binding TALE repeats have been cloned into the plasmid backbone by the REAL method mentioned above, the right and left TALE repeats then need to be cloned into the FokI nuclease-containing mammalian TALEN expression plasmid as directed by Zifit Targeter v4.2 and as described below.

1. Digest mammalian TALEN expression plasmid with *Bsm*B1 (see reaction mix below). Incubate at 37 °C.

TALEN expression plasmid (JDS70/JDS71/JDS74 or JDS 78 as per ZiFit)	X μl (2–3 μg)
10× NEB Buffer 3	3 μl
*Bsm*BI	1 μl
Nuclease-free H₂O	Up to 30 μl

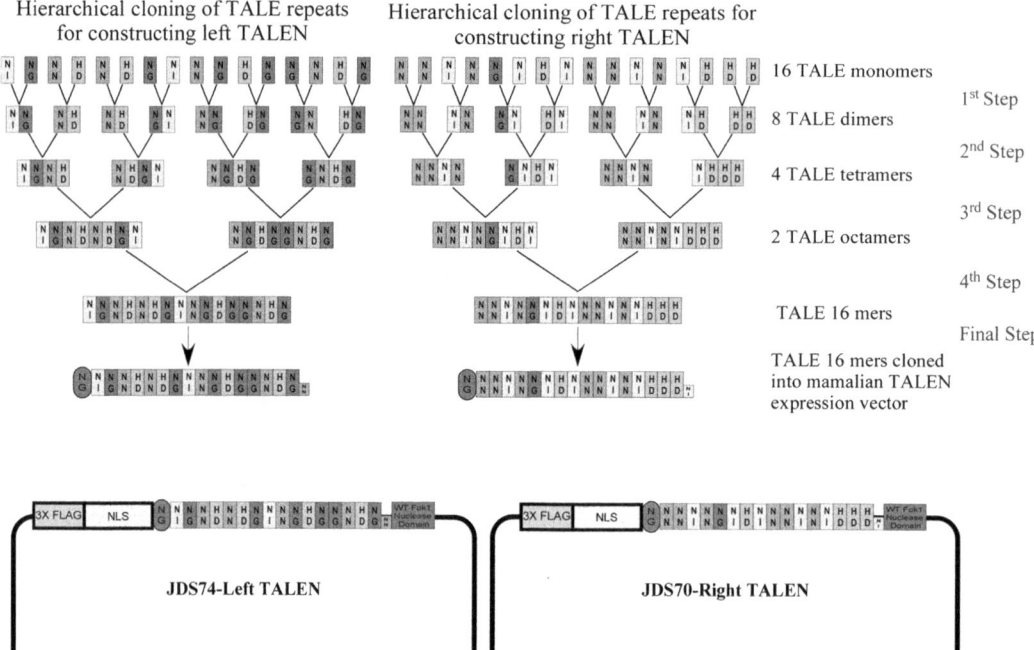

Fig. 2 Representation of the construction method of left and right TALEN using the REAL strategy. The first step involves the cloning of two TALE monomers. The second step involves the cloning of two TALE dimers. The third step involves the cloning of two TALE tetramers. The fourth step involves the cloning of two TALE octamers to generate TALE 16-mers for both the left and right TALEN. In the final step, each TALE 16-mer was then cloned into the mammalian TALEN expression vector, N-terminally tagged with FLAG and a TALEN starting TALE monomer targeting base T- and C-terminally tagged with the "0.5" TALE repeat domain with 20 amino acids and a wild type Fok1 nuclease domain. The *bottom section* of figure shows a schematic representation of the left and right TALEN expression plasmids. The *grey box* and *light green box* N-terminal to TALEN repeat modules represents the 3× FLAG and nuclear localizing (NLS) sequence, respectively. Wild type FokI nuclease domain (*red box*) is tagged at the C-terminal end of the TALEN module (Reproduced from *Sci. Rep.* 5: 9048) [4]

2. Run the digested plasmid on a 1% agarose gel and gel elute using a QIAquick gel extraction column in 20 µl of EB buffer. Quantify the DNA.

3. Digest the TALE repeat from 4th step, Fig. 2 encoding the plasmid (16-mers or as indicated by Zifit Targeter v4.2) with *Bbs*I and *Bsa*I (see reaction mix below). Incubate at 37 °C for 2–12 h.

Isolated plasmid from **step 10**, Subheading 3.2	X µl (1–2 µg)
10× NEB Buffer 4	3 µl
100× BSA	0.3 µl
BbsI	1 µl
BsaI	1 µl
Nuclease-free H$_2$O	Up to 30 µl

4. After digestion of the final TALE-encoding plasmid with *Bbs*I and *Bsa*I, the DNA fragment size will be $N \times 102$ bp, where N represents the number of TALE repeats.

5. Gel elute and quantify the TALE repeats.

6. Ligate the purified TALE array with a mammalian TALEN expression plasmid backbone and transform the ligation product into competent *E. coli* XL-1 Blue cells following **steps 4–9** in Subheading 3.2.

7. Digest the isolated plasmid with *Kpn*I and *Bam*HI (see reaction mix below). Incubate at 37 °C for 2–12 h.

Isolated plasmid from **step 7**	X μl (1–2 μg)
10× NEB Buffer 4	3 μl
100× BSA	0.3 μl
*Kpn*I	1 μl
*Bam*HI	1 μl
Nuclease-free H$_2$O	Up to 30 μl

8. Run the digested plasmid on a 1% agarose gel and visualize under a gel documentation system. Mammalian TALEN expression vector in which the final TALE repeats have been cloned successfully will show a DNA fragment size $[(N \times 102) + 605]$ bp, where N represents the number of TALE repeats.

9. Verify by sequencing the expression plasmid encoding the TALE array using the JDS 2978 and JDS 2980 sequencing primers.

3.4 Cell Culture and Transfection

TALEN-mediated DNA cleavage has been validated in several cell lines [1, 9, 12]. We have generalized and modified the protocol below.

1. Grow Renca-v or NIH 3T3 cells in DMEM high-glucose medium with 10% FBS supplemented with penicillin, streptomycin, and gentamicin. One day before transfection, plate the cells in a 12-well (1.5×10^5 cells) format in antibiotic-free DMEM.

2. In each well, transfect 1.62 μg (800 ng of left TALEN + 800 ng of right TALEN) of TALEN expression plasmid + 20 ng pEGFP-N1) of DNA using Lipofectamine-2000 in Opti-MEM medium. Use pMAX-GFP to quantify the relative transfection efficiency (*see* **Note 5**).

3. Six hours post transfection, replace the Opti-MEM containing transfection complex with complete DMEM (*see* **Note 6**).

4. 72–96 h post transfection, trypsinize the cells, count and transfer them to a 100 mm cell culture dish for G418 selection. From another well isolate the genomic DNA using a QIAquick DNA extraction kit or by RIPA–phenol–chloroform–isoamyl alcohol. Quantify the DNA.

3.5 Gene Modification, Off-Target Analysis

1. Perform a PCR reaction ($1 \times 50\ \mu l$) using the primer set encompassing the TALEN target region. Perform a control PCR from non-transfected or single TALEN-transfected cells (*see* **Note 7**).

2. Purify the PCR-amplified DNA using QIAquick PCR purification kit and quantify the DNA.

3. Use 600 ng of DNA for denaturation and renaturation. The reaction will be performed in a thermal cycler with a 5 °C decrease/30 s with an initial denaturation at 95 °C for 10 min.

4. Digest the renatured DNA using T7E1 enzyme. Perform a control digestion with each set of digestions without T7E1 (see reaction mix below). Incubate at 37 °C for 20–45 min.

Renatured PCR product	$X\mu l$ (600 ng)
10× NEB Buffer 2	2 μl
T7E1	1 μl
Nuclease-free H_2O	20 μl

5. Run the digested DNA in a 1–1.5 % agarose gel to resolve the two digested bands from the original, uncut band.

6. TALENs have a differential cutting efficiency for different targeted loci. Calculate the cutting efficiency using the formula below [20]:

$$\% \text{ gene modification} = 100 \times \left\{ 1 - \left(1 - \text{fraction cleaved}\right)^{1/2} \right\} c$$

Where the fraction cleaved was calculated from a formula $(a + b / a + b + c)$

where a and b = relative intensity of two cut bands; c = relative intensity of full length PCR fragment (*see* **Note 8**).

7. Predict the off-target site targeted by the TALEN pair using the online "TAL Effector Nucleotide Targeter, v2.0": (https://tale-nt.cac.cornell.edu/node/add/talef-off-paired). Choose 5–10 off-target sites on the basis of the low binding score and perform a repeat experiment using **steps 1–4** of this section (*see* **Note 9**).

3.6 Identification of Knockout Clones

1. After plating the cells from a 24-well plate to a 100 mm dish (**step 4**, Subheading 3.4), select the cells using G418 for 10–14 days. When each colony is barely visible to the naked eye, pick the colony using a 200 μl pipette tip and place it in a 48-well format without G418. After expansion of each colony, analyze the expression of your protein of interest by western blotting or immunofluorescence microscopy (*see* Fig. 3).

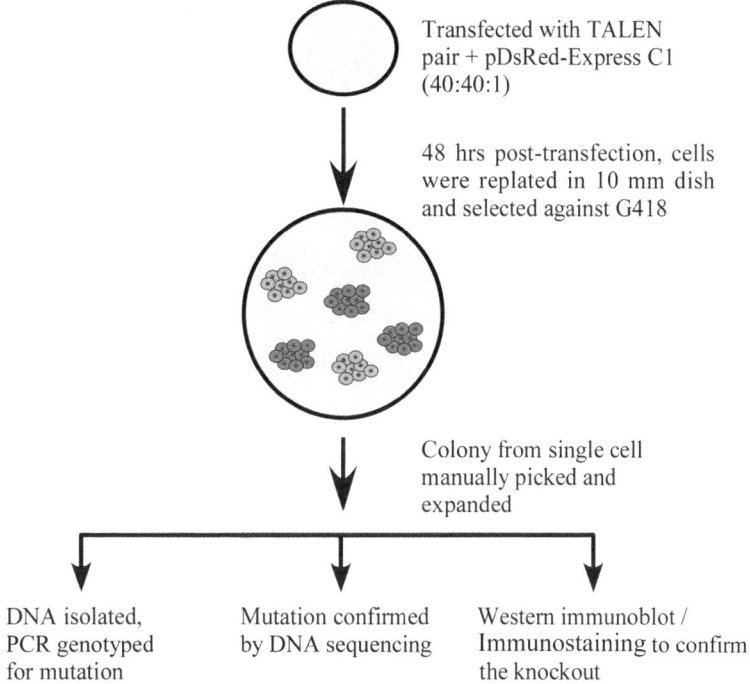

Transfected with TALEN pair + pDsRed-Express C1 (40:40:1)

48 hrs post-transfection, cells were replated in 10 mm dish and selected against G418

Colony from single cell manually picked and expanded

DNA isolated, PCR genotyped for mutation

Mutation confirmed by DNA sequencing

Western immunoblot / Immunostaining to confirm the knockout

Fig. 3 Schematic representation of the modified clonal selection strategy used for the generation of a TALEN-mediated targeted gene knockout cell line. Cells were transfected with TALEN pairs and pDsRed-Express C1 empty vector (containing a neomycin/G418 resistance cassette) at a ratio of 40:40:1 and selected against G418. Each single colony was manually picked, expanded and genomic DNA was isolated from each of the expanded colonies. PCR genotyping was performed to detect TALEN-mediated mutations, while indel mutations were confirmed by sequencing. Complete gene knockout clones were selected and expanded on the basis of expressed proteins detected by western blotting and/ or immuno-staining (Reproduced from *Sci. Rep.* 5: 9048) [4]

2. To determine indels, PCR amplify the target region using Ex-Taq DNA polymerase and clone it into a PCR cloning vector (e.g., TA-cloning vector). Verify by sequencing the cloned DNA using the T7 forward primer and determine the sequence of the indel bases.

3. PCR amplify the target region (100–120 bp, primers should be designed just beyond the target region) from the genomic DNA isolated from individual clones and run it on a 3 % agarose gel to resolve the indel mutation (*see* **Note 10**).

4   Notes

1. Competent cell quality is very important for proper transformation. Preferably use highly competent (>10^6) XL-1 blue cells. Electrocompetent cells can be purchased from different companies.

2. Transfection efficiency is one of the key factors which determines the DNA cleavage efficiency of TALEN pairs. To increase the efficiency of transfection, you can use Lipofectamine-3000, Lipofectamine-LTX or the nucleofection method.

3. For the T7E1 assay, a larger amplicon (400–800 bp) encompassing the TALEN target site is preferred. The target site should not be present exactly in the middle or at the end of the amplicon because it will generate two bands of the same size or one band very close to the original size and the other one extremely small in size after digestion with T7E1. These bands will not be resolved in an agarose gel or in the same percentage of agarose gel.

4. High-fidelity PCR enzymes are essential as normal Taq polymerase can generate random mutations which will give false results in the T7E1 assay and verified by sequencing of different clones for analyzing indel mutations generated by TALENs.

5. The transfection method other than Lipofectamine-2000 can be used. Hard-to-transfect cell lines can be transfected using Lipofectamine-3000, Lipofectamine-LTX, or nucleofection method.

6. Replacing medium 6 h post transfection is not essential. The transfection protocol using something other than Lipofectamine-2000 does not require changing media.

7. Primers designed for genotyping PCR are important. They should be specific, able to amplify 100–120 bp of DNA, and cover the TALEN targeted region. Small PCR fragments are essential to resolve in agarose for discrimination of indel mutations.

8. Digestion with T7E1 is crucial. Do not overdigest or incubate with high amounts of T7E1 enzyme. Standardize the digestion time between 20 and 45 min.

9. The online tool used for predicting off-target effects can give a huge number of off-target sites but with different binding scores. A low binding score defines the higher specificity of TALEN and vice versa. Carefully choose the off-target region with low binding scores that will have a higher chance of off-target binding in that selected region.

10. To discriminate a TALEN-mediated indel mutation in high-resolution agarose such as Metaphor agarose (Lonza), genomic PCR primers should yield single amplicons and the amplicons should be 100–120 bp in length. Metaphor agarose gels can resolve 2 bp differences in amplicon length in a 100–120 bp product. Run the gel for 3–5 h for complete resolution between 2 bp.

Acknowledgments

The authors acknowledge Nature Publishing Group (NPG) for use of methodology and figures that they have published in Scientific Reports [4] for preparation of this chapter. The work was supported by financial support from Bose Institute, Kolkata, India and the Department of Biotechnology, Government of India. Barun Mahata acknowledges the Council of Scientific and Industrial Research (CSIR), New Delhi, India for providing his Senior Research Fellowship.

References

1. Ding Q, Lee YK, Schaefer EA, Peters DT, Veres A, Kim K, Kuperwasser N, Motola DL, Meissner TB, Hendriks WT, Trevisan M, Gupta RM, Moisan A, Banks E, Friesen M, Schinzel RT, Xia F, Tang A, Xia Y, Figueroa E, Wann A, Ahfeldt T, Daheron L, Zhang F, Rubin LL, Peng LF, Chung RT, Musunuru K, Cowan CA (2013) A TALEN genome-editing system for generating human stem cell-based disease models. Cell Stem Cell 12(2):238–251

2. Xu L, Zhao P, Mariano A, Han R (2013) Targeted myostatin gene editing in multiple mammalian species directed by a single pair of TALE nucleases. Mol Ther Nucleic Acids 2:e112

3. Maier DA, Brennan AL, Jiang S, Binder-Scholl GK, Lee G, Plesa G, Zheng Z, Cotte J, Carpenito C, Wood T, Spratt SK, Ando D, Gregory P, Holmes MC, Perez EE, Riley JL, Carroll RG, June CH, Levine BL (2013) Efficient clinical scale gene modification via zinc finger nuclease-targeted disruption of the HIV co-receptor CCR5. Hum Gene Ther 24(3):245–258

4. Mahata B, Banerjee A, Kundu M, Bandyopadhyay U, Biswas K (2015) TALEN mediated targeted editing of GM2/GD2-synthase gene modulates anchorage independent growth by reducing anoikis resistance in mouse tumor cells. Sci Rep 5:9048

5. Strouse B, Bialk P, Niamat RA, Rivera-Torres N, Kmiec EB (2014) Combinatorial gene editing in mammalian cells using ssODNs and TALENs. Sci Rep 4:3791

6. Maresca M, Lin VG, Guo N, Yang Y (2013) Obligate ligation-gated recombination (ObLiGaRe): custom-designed nuclease-mediated targeted integration through nonhomologous end joining. Genome Res 23(3):539–546

7. Hockemeyer D, Wang H, Kiani S, Lai CS, Gao Q, Cassady JP, Cost GJ, Zhang L, Santiago Y, Miller JC, Zeitler B, Cherone JM, Meng X, Hinkley SJ, Rebar EJ, Gregory PD, Urnov FD, Jaenisch R (2011) Genetic engineering of human pluripotent cells using TALE nucleases. Nat Biotechnol 29(8):731–734

8. Christian M, Cermak T, Doyle EL, Schmidt C, Zhang F, Hummel A, Bogdanove AJ, Voytas DF (2010) Targeting DNA double-strand breaks with TAL effector nucleases. Genetics 186(2):757–761

9. Boch J, Bonas U (2010) Xanthomonas AvrBs3 family-type III effectors: discovery and function. Annu Rev Phytopathol 48:419–436

10. Boch J, Scholze H, Schornack S, Landgraf A, Hahn S, Kay S, Lahaye T, Nickstadt A, Bonas U (2009) Breaking the code of DNA binding specificity of TAL-type III effectors. Science 326(5959):1509–1512

11. Moscou MJ, Bogdanove AJ (2009) A simple cipher governs DNA recognition by TAL effectors. Science 326(5959):1501

12. Morbitzer R, Elsaesser J, Hausner J, Lahaye T (2011) Assembly of custom TALE-type DNA binding domains by modular cloning. Nucleic Acids Res 39(13):5790–5799

13. Zhang F, Cong L, Lodato S, Kosuri S, Church GM, Arlotta P (2011) Efficient construction of sequence-specific TAL effectors for modulating mammalian transcription. Nat Biotechnol 29(2):149–153

14. Cermak T, Doyle EL, Christian M, Wang L, Zhang Y, Schmidt C, Baller JA, Somia NV, Bogdanove AJ, Voytas DF (2011) Efficient design and assembly of custom TALEN and other TAL effector-based constructs for DNA targeting. Nucleic Acids Res 39(12):e82

15. Sander JD, Cade L, Khayter C, Reyon D, Peterson RT, Joung JK, Yeh JR (2011) Targeted gene disruption in somatic zebrafish cells using engineered TALENs. Nat Biotechnol 29(8):697–698

16. Reyon D, Tsai SQ, Khayter C, Foden JA, Sander JD, Joung JK (2012) FLASH assembly of TALENs for high-throughput genome editing. Nat Biotechnol 30(5):460–465

17. Weber E, Gruetzner R, Werner S, Engler C, Marillonnet S (2011) Assembly of designer TAL effectors by Golden Gate cloning. PLoS One 6(5):e19722

18. Reyon D, Khayter C, Regan MR, Joung JK, Sander JD (2012) Engineering designer transcription activator-like effector nucleases (TALENs) by REAL or REAL-Fast assembly. Curr Protoc Mol Biol Chapter 12:Unit 1215

19. Miller JC, Tan S, Qiao G, Barlow KA, Wang J, Xia DF, Meng X, Paschon DE, Leung E, Hinkley SJ, Dulay GP, Hua KL, Ankoudinova I, Cost GJ, Urnov FD, Zhang HS, Holmes MC, Zhang L, Gregory PD, Rebar EJ (2011) A TALE nuclease architecture for efficient genome editing. Nat Biotechnol 29(2):143–148

20. Guschin DY, Waite AJ, Katibah GE, Miller JC, Holmes MC, Rebar EJ (2010) A rapid and general assay for monitoring endogenous gene modification. Methods Mol Biol 649:247–256

Chapter 8

Efficient Generation of Gene-Modified Mice by Haploid Embryonic Stem Cell-Mediated Semi-cloned Technology

Cuiqing Zhong and Jinsong Li

Abstract

Haploid embryonic stem cells can be derived from androgenetic embryos produced by injection of sperm into enucleated oocytes or by removal of the female pronucleus from zygotes. These cells, termed AG-haESCs, can be used in place of sperm to produce the so-called semi-cloned (SC) mice. Importantly, AG-haESCs carrying *H19*-DMR and *IG*-DMR knockouts (DKO-AG-haESCs) can efficiently and stably support the generation of SC mice via intracytoplasmic AG-haESCs injection (ICAHCI), which provides a new route to obtain genetically modified mice. In this chapter, we describe the procedures for AG-haESCs culturing, enrichment of haploid cells by FACS, genomic manipulation in DKO-AG-haESCs by CRISPR/Cas9 and generation of live SC mice with gene-modified DKO-AG-haESCs.

Key words Haploid ES cells (haESCs), Fluorescence-activated cell sorting (FACS), Genetic manipulation, CRISPR-Cas9, Intracytoplasmic AG-haESCs injection (ICAHCI), Semi-cloned mice (SC mice)

1 Introduction

Mammalian haploid embryonic stem cells (haESCs) have been derived from androgenetic and parthenogenetic haploid embryos that respectively contain the genomes of sperm and oocytes from several species [1–5]. The establishment of haESCs is a remarkable progression of forward and reverse genetics due to its characteristics of being a haploid genome [5, 6]. While both parthenogenetic embryonic stem cells (PG-haESCs) and androgenetic embryonic stem cells (AG-haESCs) can be used for genetic analysis in vitro, AG-haESCs can also give rise to live SC mice via intracytoplasmic AG-haESCs injection (ICAHCI) (termed semi-cloned or SC technology), thus enabling genetic analysis in vivo [3, 7]. By adopting AG-haESC-mediated SC technology, genetically modified SC mice can be quickly and conveniently obtained via ICAHCI of gene-manipulated AG-haESCs. This extends the genetic analysis of AG-haESCs from the cellular level to the organismic level in one

Andrew Reeves (ed.), *In Vitro Mutagenesis: Methods and Protocols*, Methods in Molecular Biology, vol. 1498,
DOI 10.1007/978-1-4939-6472-7_8, © Springer Science+Business Media New York 2017

step. However, one significant application hurdle using AG-haESCs for the generation of engineered mouse models is that the efficiency of live-born SC mice is extremely low (~2% of SC embryos). Moreover, after long-term culturing of AG-haESCs especially for durable drug selection of traditional gene targeting, the birth rate of live SC mice is dramatically reduced and this greatly restricts the applications of AG-haESCs.

Most recently, we've shown that parental genomic imprints, including *H19* and *Gtl2*, cannot be stably maintained in cultured AG-haESCs, which account for aberrant development of AG-haESC-derived embryos and a low birth rate of SC mice [8]. We further demonstrate that AG-haESCs carrying deletions in *H19*-DMR and *IG*-DMR (differentially DNA methylated regions), designated DKO-AG-haESCs, can efficiently support the full-term development of SC embryos. Strikingly, the rate of SC mice born is around 20% of the transferred embryos, which is ten times higher than that of WT AG-haESCs. Intriguingly, DKO-AG-haESCs can be used to obtain SC mice with multiple genetic modifications in one step, suggesting that DKO-AG-haESCs could be used as a reproducible fertilization agent for production of live animals carrying multiple gene modifications via ICAHCI [8].

In this chapter, we describe the successful application of DKO-AG-haESC-mediated SC technology to efficiently generate SC mice carrying multiple gene-modifications. We show that haploidy of DKO-AG-haESCs can be well maintained in vitro by regular cell sorting. CRISPR/Cas9-mediated multiplex gene editing can be applied in DKO-AG-haESCs, leading to haploid cell lines carrying multiple modifications. We provide examples by the generation of DKO-AG-haESC lines carrying Tet1, 2, 3 triple-gene knockouts and Nanog-Sox2 double-reporter knockins. Finally, we show that SC mice carrying the desired genetic traits can be efficiently produced by injection of these cells into oocytes.

2   Materials

2.1   Mouse Cell Culture

1. Mouse DKO-AG-haESCs culture medium (ESC medium): For 500 ml add, 375 ml of ES DMEM (Dulbecco's Modified Eagle's Medium), 75 ml of ES FBS (fetal bovine serum), 5 ml of penicillin–streptomycin solution (100×), 5 ml of nonessential amino acids (NEAA) (100×), 5 ml of nucleosides (100×), 5 ml of L-glutamine (100×), 5 ml of 2-mercaptoethanol (100×), 100 μl of PD0325901 (5 mM, 5000×), 100 μl of CHIR99021 (15 mM, 5000×), and 50 μl of mouse leukemia inhibitory factor (LIF, 10^7 units/ml). Sterilize with a 0.22 μm filter.

2. 0.05% trypsin–EDTA.

3. Dulbecco's phosphate buffered saline (DPBS, pH 7.2).

4. Mouse embryonic fibroblast cells (MEF) medium: 860 ml of Dulbecco's Modified Eagle Medium (DMEM; commercially supplied), 120 ml of fetal bovine serum (FBS), 10 ml of penicillin–streptomycin (100×), and 10 ml of nonessential amino acids (100×). Bring volume up to 1 l. Sterilize with a 0.22 μm filter. Can scale down volume and components as necessary.

5. Commercial recovery cell-culture freezing medium (Gibco). Used for preservation.

6. 0.1 % gelatin.

7. 6-well cell culture plates.

8. 37 °C incubator.

2.2   Mouse Cell Enrichment

1. Hoechst 33342 staining buffer stock solution.

2. 37 °C water bath.

3. 40 μm cell strainer.

4. 5 ml FACS tubes.

5. Flow cytometer.

6. 15 ml tubes.

2.3   Mouse Cell Gene Modification

1. Plasmids: pX330-*mCherry*, pMD-19T.

2. *Bbs*I restriction enzyme.

3. T4 DNA ligase reaction kit.

4. Lipofectamine-2000.

5. Commercial Opti-MEM culture medium.

2.4   Semi-cloned Mice Production

1. Mice: 8-week-old B6D2F1 (C57BL/6 × DBA/2) female mice, pseudo-pregnant ICR female mice.

2. Pregnant mare serum gonadotrophin (PMSG; aka eCG).

3. Human chorionic gonadotrophin (hCG).

4. Dulbecco's phosphate buffered saline (DPBS).

5. CZB Stock medium: Into 985 ml ultrapure water add, NaCl, 4760 mg, KCl, 360 mg, $MgSO_4 \cdot 7H_2O$, 290 mg, EDTA·2Na, 40 mg, Na-lactate, 5.3 ml, D-glucose, 1 g, KH_2PO_4, 160 mg. Bring the final volume to 1 l with ultrapure H_2O.

6. H-CZB stock medium: 500 ml of CZB stock medium with 50 mg of PVA. After the PVA completely dissolves, sterilize the medium with a 0.22 μm filter.

7. Hepes-CZB buffer: H-CZB stock medium, 98.5 ml, Hepes-2Na, 520 mg, $NaHCO_3$, 42 mg, $CaCl_2 \cdot 2H_2O$, 25 mg, pyruvate 3.0 mg, glutamine (200 mM), 0.5 ml. Adjust the final volume to 100 ml. Adjust the pH to 7.4, and sterilize with a 0.22 μm filter.

8. CZB medium: CZB stock medium, 99 ml, $NaHCO_3$, 211 mg, $CaCl_2 \cdot 2H_2O$, 25 mg, pyruvate, 3.0 mg, glutamine (200 mM) 0.5 ml, BSA, 500 mg.

9. Activation medium (per ml): Ca^{2+} free CZB medium, 900 μl, $SrCl_2 \cdot 6H_2O$ (100 mM) 100 μl.

10. Hyaluronidase.

11. Mineral oil.

12. Commercially prepared potassium-supplemented simplex-optimized medium (KSOM).

13. Manipulation medium: 1 ml Hepes-CZB medium supplemented with 5 μg/ml Cytochalasin B (CB).

14. 0.05 mg/ml Demecolcine Solution (1000×).

15. 37 °C incubator.

16. Micromanipulator.

17. Piezo.

18. Injection pipet.

19. 60-mm cell culture dish.

3   Methods

3.1   Mouse DKO-AG-haESCs Culture

1. Thaw DKO-AG-haESCs.

2. Seed irradiated mouse embryonic fibroblast (MEF) feeder on gelatin-coated 6-well plate at least 1 day before ES cell thawing.

3. Remove the vial of DKO-AG-haESCs from liquid nitrogen; quickly thaw cells in a 37 °C water bath. Swirl gently.

4. Gently transfer the cell suspension to 9 ml MEF medium in a 15-ml Falcon tube. Centrifuge $160 \times g$ for 5 min. Remove and discard the supernatant.

5. Replace the MEF medium with ESC medium (2 ml per well of 6-well plate prepared as in **step 1** above). Gently resuspend the cell pellet in 1 ml ESC medium and distribute the cells into 2 wells of a 6-well cell culture plate. Incubate at 37 °C under 5 % CO_2 in air.

6. Refresh the ESC medium every day or every 2 days.

7. When the ESC colonies are large enough, maybe 2–3 days after thawing, the cells need to be split (*see* Fig. 1a).

8. Aspirate the medium from the wells.

9. Rinse with 1–2 ml DPBS buffer per well.

10. Add 400 μl of 0.05 % trypsin–EDTA to each well of a 6-well plate.

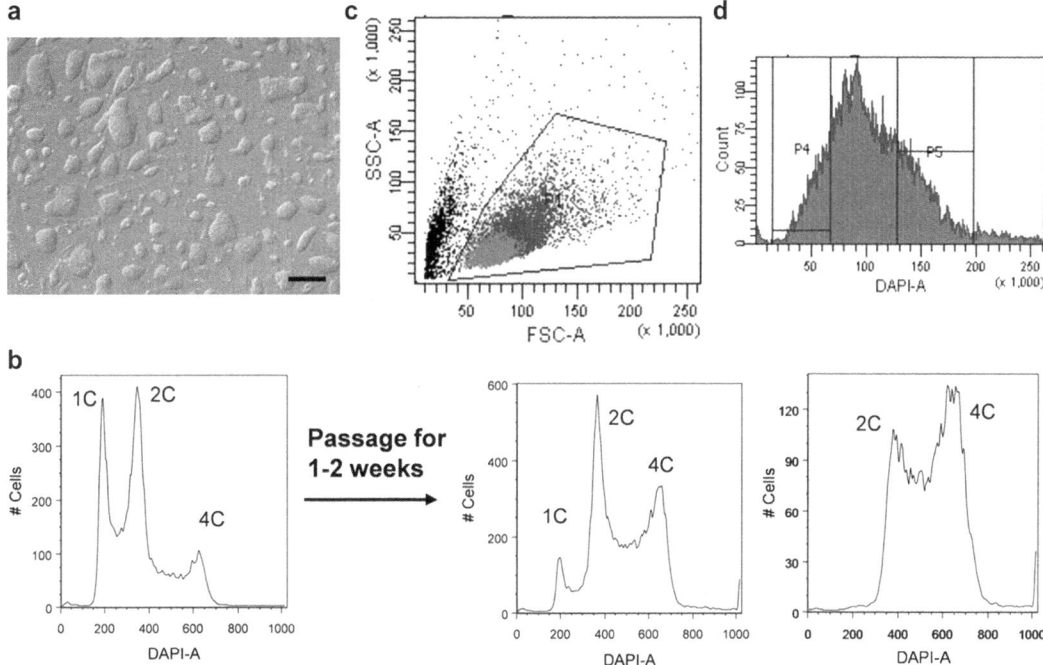

Fig. 1 Analysis and purification of haploid cells by FACS. (**a**) Morphology of established DKO-AG-haESCs. Scale bar, 200 μm. (**b**) ESCs containing high percentage of haploid cells convert into diploid cells during passaging. *Right panel* represents diploid ESCs as a negative control. (**c**) Flow analysis of DKO-AG-haESCs in BD Ariall. P1 section represents living cells, and 1C cells that will be sorted are indicated by *red dots*. (**d**) Unconventional DNA peak due to inadequate Hoechst 33342 staining. Under this condition, the 1C peak couldn't be distinguished from the 2C and 4C peaks

11. Incubate at 37 °C for about 5 min until the cells dislodge from the plate surface. To confirm colony dissociation from the plate, check cells under a microscope.

12. Add 1 ml of ESC medium per 6-wells to inactivate the trypsin. Gently pipette up and down to wash the cells and ensure that a single cell suspension is obtained.

13. Aspirate the MEF medium from fresh feeder plates, and add 2 ml ESC medium to each well of a 6-well plate. Typically, split the DKO-AG-haESCs at a ratio of 1:3–1:6. Gently shake the plate to achieve a uniform distribution of cells. Pick a small part of the cells for 1C analysis by FACS (*see* **Note 1**).

14. Replace the ESC culture medium every day and passage the cells every 2 days.

15. Aspirate the medium from the wells.

16. Add 1–2 ml DPBS per well. Then aspirate with DPBS.

17. Add 400 μl of 0.05 % trypsin–EDTA to each well of a 6-well plate.

18. Incubate at 37 °C in an incubator for about 5 min.

19. Add 1 ml of MEF medium per well of a 6-well plate to stop the trypsinization. Gently pipette up and down to wash the plate surface. Transfer the cell suspension to a 1.5-ml tube. Centrifuge at $160 \times g$ for 5 min.

20. Discard the supernatant and resuspend the cell pellet in recovery cell culture freezing medium. Generally freeze every well of cells with 200 μl of freezing medium, and 1 well of 6-well plate cells for each vial. Before cell freezing, make sure the percentage of 1C cells is high enough (approximately >15%) (*see* **Note 2**).

3.2 Purification of Haploid Cells

This procedure uses FACS to purify haploid cells.

1. Aspirate the medium from one well of the 6-well plate, and rinse with 2 ml DPBS.

2. Add 400 μl 0.05% Trypsin-EDTA per well, and incubate at 37 °C for 5 min.

3. Stop the trypsinization then collect the cell suspension into a 1.5-ml tube.

4. Centrifuge at $160 \times g$, 5 min. Remove supernatant.

5. Prepare the Hoechst 33342 staining buffer (final concentration is 15 μg/ml) by adding 6 μl Hoechst 33342 stock buffer (2.5 mg/ml) into 1 ml ESC medium. Cover with aluminum foil and avoid exposure to light.

6. Add Hoechst 33342 staining buffer to the cell pellet (about 0.5–1 ml staining buffer for cells from one well of a 6-well plate depending on the cell density). Resuspend the cell pellet (*see* **Note 3**).

7. Incubate in a 37 °C water bath for 15 min, and keep in the dark. Gently shake the tube occasionally (every 5 min) to avoid cell deposition. You may prepare a diploid ESC sample as a negative control for assisting in determining the 1C peak on the cell cytometer (*see* **Note 4**).

8. Centrifuge at $160 \times g$, 5 min and remove supernatant. After that, rinse with 1 ml DPBS once (*see* **Note 5**).

9. Centrifuge at $160 \times g$, 5 min. Add approximately 300 μl ESC medium to the cell pellet from one well of a 6-well plate, and keep all cell samples on ice.

10. Before FACS, pass each cell suspension through a 40-μm cell strainer into a 5-ml FACS tube.

11. Use a BD FACS AriaII or other cell flow cytometer for DNA content analysis. A DAPI filter can distinguish the DNA content from 1C to 4C (*see* Fig. 1b). Normally there will be three peaks (1C, 2C, and 4C). 1C represents haploid cells in the G0/G1 phase, 2C contains haploid cells in the M phase and diploid cells in the G0/G1 phase and 4C represents diploid cells in the M phase (*see* **Note 6**).

12. After this, gate the 1C peak as harvest area to purify the haploid cells. Meanwhile, put the prepared collecting tube on the right place and use ESC medium as a collection buffer.

13. Plate the harvested haploid cells onto a 6-well plate (approximately 1×10^6 cells per well). If you harvest less than 1×10^6 cells, you can plate them on a smaller plate. Change the ESC medium approximately every 1–2 days. Haploid ES cells will be adherent to the plate several hours later, and after 1–2 days small ESC colonies will appear and become larger over the next 2 days. It take 3–4 days to passage all the ESC colonies (*see* **Notes** 7 and **8**).

14. The percentage of the 1C peak needs to be analyzed every 4–6 passages by the flow cytometer, as mouse haploid ES cells tend to diploidize spontaneously during cell culture (*see* Fig. 1b). If the percentage is low, you should enrich for ES cells in the haploid state via FACS to make sure of haploid cell maintenance in vitro.

3.3 Multiplex Gene Modifications

For multiplex gene knockouts in DKO-AG-haESCs we use the Tet1, 2, 3 triple-gene knockouts as an example.

1. Synthesize sgRNA of targeted genes *Tet1*, *Tet2*, *Tet3*. Then anneal sequentially and ligate the sgRNA to a pX330-*mCherry* plasmid that was digested by *Bbs*I.

2. Set up one well of a 6-well plate with DKO-AG-haESCs, and make sure the percentage of haploid cells is not very low (1C peak >25% will be enough) and that the cells are around 60–80% confluent.

3. For the transfection sample, prepare the plasmid–Lipofectamine-2000 complex as follows: Mix Lipofectamine-2000 gently before use, and then dilute 15 µl Lipofectamine-2000 in 200 µl Opti-MEM medium. Mix and incubate for 5 min at room temperature.

4. Dilute 2 µg each pX330-mCherry plasmid of *Tet1*, *Tet2*, *Tet3* in 200 µl Opti-MEM medium (*see* **Note 9**).

5. Combine the diluted plasmids with diluted Lipofectamine-2000, and mix gently (total volume is 400 µl) and incubate for 20 min at room temperature.

6. During plasmids–Lipofectamine-2000 complexes incubation, aspirate the medium from one well of DKO-AG-haESCs, then rinse the cells with DPBS.

7. Add 400 µl 0.05% trypsin–EDTA into the dish and incubate for 5 min. Stop the trypsinization by adding 1 ml MEF medium and pipette up and down. Then transfer the cell suspension into a 1.5-ml tube.

8. Centrifuge at $160 \times g$ for 5 min. Discard the supernatant and rinse the cells with DPBS once. Resuspend the cell pellet in 400 μl plasmids-lipofectamine-2000 complexes. Incubate at 37 °C in a CO_2 incubator for 10 min.

9. Transfer the cell suspension into a 60-mm dish containing 2.5 ml Opti-MEM medium drop by drop. Mix gently by rocking the plate back and forth.

10. Incubate the cells at 37 °C in a CO_2 incubator and transfer into ESC medium after 4–6 h. About 12 h later, check the *mCherry* fluorescence protein expression under a fluorescence microscope to test the transfection efficiency. In general, the transfection efficiency is approximately 5–10 %.

11. 24–48 h after transfection, prepare the transfected cells with Hoechst 33342 staining (*see* **steps 1–10** in Subheading 3.2). Analyze both DNA content and *mCherry* protein expression by flow cytometry, and harvest the haploid cells in the G0/G1 phase expressing *mCherry* (*see* Fig. 2a).

12. Plate double-positive cells at low density into a 60-mm dish, approximately 1×10^4 cells per dish. 4–6 days after plating, when the ESC colonies become big enough, randomly pick single colonies using 2.5-μl micro-pipette tips and transfer them into 50 μl Trypsin-EDTA in one well of a 96-well plate one colony per well (*see* **Note 10**).

13. Incubate at 37 °C in a CO_2 incubator for 5 min to trypsinize the ESC colonies. Following trypsinization, pipette up and down to disperse the ESC colony into single cells.

14. Transfer half of the single ES cells into a fresh well containing MEFs in a 96-well plate for further culturing. The remaining cells are used for subsequent genotype identification.

15. For identification of *Tet1*, *Tet2*, *Tet3* triple-gene mutations in DKO-AG-haESCs (termed Tet-TKO-DAH), perform DNA sequencing of the PCR products amplified with *Tet1*, *Tet2*, *Tet3* targeted site primers (*see* Fig. 2a).

3.4 Multiplex-Reporter Knockins

We use *Nanog-Sox2* double-reporter knockins as an example.

1. Synthesize sgRNA of target genes *Sox2* and *Nanog* nearby the stop codon, then anneal and ligate the sgRNA to the pX330-*mCherry* plasmid digested with *Bbs*I.

2. For the construction of double-stranded DNA donors, amplify the sequence encoding *EGFP* and *mCherry* fluorescence protein. Then ligate them both into the vector pMD-19T.

3. Amplify the left and right arms of *Nanog* and *Sox2* from genomic DNA.

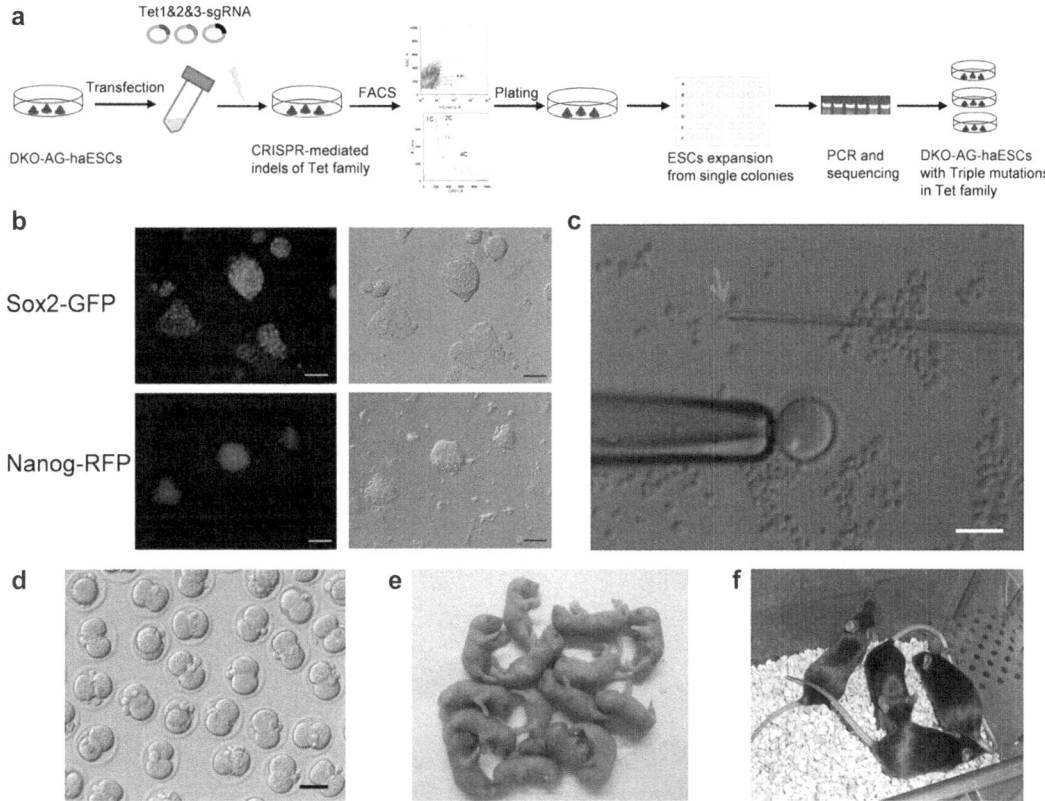

Fig. 2 Multiplex-gene modifications in DKO-AG-haESCs and generation of the corresponding gene-modified SC mice by ICAHCI. (**a**) Diagram of gene manipulation in DKO-AG-haESCs using CRISPR-Cas9 technology. Using the Tet family triple-gene knockouts as an example. (**b**) Images of DKO-AG-haESCs carrying the Sox2-GFP and Nanog-RFP double-reporter knockins. Scale bar, 100 μm. (**c**) Microscopic image of ICAHCI technology. *Arrow* indicates one smooth and small haploid cell. Scale bar, 100 μm. (**d**) Two-cell embryos generated by ICAHCI of DKO-AG-haESCs. Scale bar, 100 μm. (**e**) Newborn SC pups from ICAHCI using gene-modified DKO-AG-haESCs. (**f**) Adult SC mice

4. Insert the arms into the multiple cloning sites of pMD19T-*EGFP*/*mCherry* vectors, respectively.

5. Use one well of a 6-well plate of DKO-AG-haESCs to make the plasmid–Lipofectamine-2000 complex as follows.

6. Dilute 15 μl Lipofectamine-2000 into 200 μl Opti-MEM medium. Mix and incubate for 5 min at room temperature.

7. Dilute 3 μg pX330-mCherry plasmid of *Sox2* and 6 μg of its donor vector in 200 μl Opti-MEM medium (*see* **Note 9**).

8. Combine the diluted plasmids with diluted Lipofectamine-2000, mix gently (total volume is 400 μl) and incubate for 20 min at room temperature.

9. Perform cell transfection, FACS and cell plating as above (*see* **steps 4–11** in Subheading 3.3).

10. 4–6 days later, pick single colonies expressing the *GFP* protein into a 96-well plate. Expand the ESC colonies and subsequently identify knockin cells by PCR and sequencing analysis with primers spanning each arm (*see* **Note 11**).

11. Manipulate the *Nanog* reporter knockin in DKO-AG-haESCs with the *Sox2-GFP* reporter knockin. After completing the manipulations, the DKO-AG-haESC cell lines with *Sox2* and *Nanog* reporter knockins are established (*see* Fig. 2b) (*see* **Note 12**).

3.5 SC Mice with Multiplex Modifications

1. Superovulate 8-week-old B6D2F1 (C57BL/6×DBA2) female mice by injecting intraperitoneally 5 IU (1-ml syringe) PMSG followed by injecting 5 IU hCG 48 h later.

2. Coat one well of a 12-well plate with Tet-TKO-DAH cells (1C percentage should be as high as possible) (*see* **Notes 13** and **14**).

3. Arrest the cells at M phase by culturing them in ESC medium containing 0.05 μg/ml demecolcine for 8 h. The cell colonies will disperse due to the treatment with demecolcine.

4. Trypsinize Tet-TKO-DAH cells for 5 min, and add 1 ml ESC medium to stop the trypsinization. Pipette up and down to disperse the ESC colonies into single cells, then transfer the cell suspension to a 1.5-ml tube. Centrifuge at $160 \times g$ for 5 min.

5. Aspirate the supernatant, and add 200 μl Hepes-CZB to resuspend the cell pellet. Keep the cell suspension on ice.

6. Aliquot drops (50 μl per drop) of CZB medium, KSOM medium and activation medium covered with mineral oil in one 60-mm dish and incubate at 37 °C in a CO_2 incubator before use.

7. 14-h post-hCG injection, collect mice oviducts to a 100 μl drop of Hepes-CZB and obtain oocyte masses from the intumescentia (enlarged) part of the oviducts with a 1-ml syringe into a fresh drop of Hepes-CZB under a stereomicroscope.

8. Add 100 μl hyaluronidase to the oocyte masses and incubate at 37 °C for 5 min. Pipette up and down using a glass pipet to separate the cumulus cells from the MII oocytes. Rinse the oocytes with several drops of Hepes-CZB, and then transfer to and rinse further in CZB medium. Distribute about every 50 oocytes to a drop of CZB medium. After that, perform cell trypsinization for injection (*see* **steps 4** and **5** in Subheading 3.4).

9. Assemble a holder and an injection pipet of the piezo-drill micromanipulation system. Prepare a 100-mm dish with eight drops of Hepes-CZB with 5 μg/ml CB and a drop of Hepes-CZB covered with mineral oil.

10. Fix the oocytes using a holder and drill the zona pellucid using an injection pipet. Pick small haploid cells as donor (you can

distinguish haploid cells from diploid cells by cell size), and then break the membrane of the haploid cell with a piezo. Inject the cell nucleus into oocytes and then withdraw the injection pipet quickly and aspirate the membrane gently to seal the hole of the oocyte cytomembrane (*see* Fig. 2c) (*see* **Note 15**).

11. Reconstructed embryos are cultured in CZB medium for 1 h. Transfer them into activation medium without CB (rinse embryos several times in different drops of activation medium to remove any CB in manipulation medium before transferring into activation medium). Incubate at 37 °C in a CO_2 incubator for 5 h.

12. Following activation, rinse the embryos several times with KSOM medium and transfer them into drops of KSOM embryo culture medium. Keep at 37 °C in a CO_2 incubator for another 24 h in order to reach the 2-cell embryo stage (*see* Fig. 2d).

13. Transfer every 10–15 2-cell SC embryos into each oviduct of pesudopregnant ICR females in 0.5-day postcoitum (dpc) pesudopregnant ICR female mice.

14. Check the pregnancy state of the recipient mice. Normally, recipient mothers at 19.5 days of gestation can naturally deliver pups (*see* Fig. 2e). If not, perform cesarean section and quickly remove the pups from the uteri. After cleaning up any fluid, keep the pups in a warm environment and raise them by lactating mothers.

4   Notes

1. During the first passage after cell thawing, pick the small parts of cells to analyze the haploid percentage by flow cytometer to make sure there are enough haploid cells, because haploid cells die during cell freezing and thawing. If the percentage of haploid ES cells is high, passage several more times and freeze some vials for storage. If the haploid cell percentage is low (<10%), perform FACS immediately (*see* Fig. 1b middle panel), maybe 2–3 days after the initial passage. During the early periods of haploid ESC culturing, perform FACS or cell analysis frequently to evaluate the speed of cell diploidization.

2. It is useful to write down the 1C percentage on the cryotube.

3. Strictly control the concentration and staining volume of Hoechst 33342. If the concentration of Hoechst 33342 is too high, most haploid cells will not survive after FACS. Also make sure that the volume for cell staining is appropriate. If the volume is too small and cells are too dense, the peaks during the cell cycle will not separate smoothly (*see* Fig. 1d) because the cells are not fully stained by the Hoechst 33342 dye. In this

situation, restain with Hoechst 33342 for FACS. You can decrease the incubation time to 5–10 min. Experience is extremely important in this step.

4. When most of cells are haploid, only two peaks (1C and 2C) are observed instead of the typical three peaks. It is uncertain whether haploid cells exist (you may take them as 2C and 4C peaks by accident). In this case, you need a negative or positive control to be sure of the 1C peak position.

5. Rinse with DPBS once after Hoechst 33342 staining to remove residual dye, otherwise residual stain may be toxic to the cells.

6. Use FSC and SSC parameters to distinguish live cells from apoptotic cells and MEF cells, but pay attention to cells located in the position of low FSC and SSC, which contains mostly haploid cells. Because haploid cells are small, the demarcation line should include cells located in the lower left corner shown as a red dot (see Fig. 1c).

7. Haploid cells can be cultured and maintained in a feeder-free system. However, it will be better to plate the FACS-purified haploid cells on MEF feeder due to potential damage caused by Hoechst 33342 staining and laser irradiation.

8. Mark the dates of passaging on the plate after you seed the FACS-enriched haploid cells so you can decide the timing of the next cell sorting.

9. The volume of Lipofectamine-2000, plasmids and transfection system components may not be exactly the same as described in the protocol. Adjustments may be made based on previous experimentation.

10. The number of colonies you pick for identifying gene knockouts and knockins depend on the efficiency of the genetic modifications. Generally, the efficiency of gene knockout in DKO-AG-haESCs reaches approximately 90% of transfected cells. As for gene knockins, it is less efficient (about 10%). Moreover, the efficiency varies according to the targeted genes.

11. Sometimes you may not observe the expression of reporters systems in cells. This does not mean that there is no correct reporter knockin colonies in dish only that the signal is below the level of detection. The possible reason could be that the protein expression level isn't high enough in embryonic stem cells, but it could be expressed highly enough in differentiated cell types of interest. Therefore, to make sure that the desired cell lines are established, genotyping analysis needs to be performed.

12. It is straightforward to obtain multiple genes knockouts in a single pX330-mCherry plasmid transfection because of the high efficiency of CRISPR-Cas9-induced DNA cleavage. As for multiplex-gene knockins, one may need to manipulate the gene editing process individually. Cells with a high concentration

of haploid cells must be used for plasmid transfection to increase the yield of positive haploid cells for genetic modification.

13. By preparing haploid cells for ICAHCI, you can use cells without passaging them after FACS or by passaging them several times. Make sure the percentage of haploid cells is high enough (>30 % for the best). If the percentage is too low, it is difficult and time-consuming to pick small haploid cells. As a result, the birth efficiency of SC mice will be reduced.

14. Generally, the cell colonies should not be too big and dense before you add demecolcine to arrest the M phase cells, because this will result in a slow cell proliferation rate, leading to insufficient arrest. If the cell colonies are really big, passage them once in 1 day before cell arrest at an appropriate density.

15. Choose small and smooth haploid cells as donors for ICAHCI. This is extremely crucial for high-birth efficiency of SC mice (*see* Fig. 2c). Small cells can partially represent haploid but not diploid cells, and well-arrested M phase cells will be smooth without bulges at the edge of cells.

Acknowledgments

This study was supported by grants from the Ministry of Science and Technology of China (2014CB964803 and 2015AA020307), the National Natural Science Foundation of China (91319310, 31225017, and 31530048), the Chinese Academy of Sciences (XDA01010403 and XDB19010203), and the Shanghai Municipal Commission for Science and Technology (12JC1409600 and 13XD1404000).

References

1. Li W, Li X, Li T, Jiang MG, Wan H et al (2014) Genetic modification and screening in rat using haploid embryonic stem cells. Cell Stem Cell 14:404–414

2. Yang H, Liu Z, Ma Y, Zhong C, Yin Q, Zhou C et al (2013) Generation of haploid embryonic stem cells from Macaca fascicularis monkey parthenotes. Cell Res 23:1187–1200

3. Yang H, Shi L, Wang BA, Liang D, Zhong C, Liu W et al (2012) Generation of genetically modified mice by oocyte injection of androgenetic haploid embryonic stem cells. Cell 149:605–617

4. Leeb M, Wutz A (2011) Derivation of haploid embryonic stem cells from mouse embryos. Nature 479:131–134

5. Elling U, Taubenschmid J, Wirnsberger G, O'Malley R, Demers SP et al (2011) Forward and reverse genetics through derivation of haploid mouse embryonic stem cells. Cell Stem Cell 9:563–574

6. Leeb M, Dietmann S, Paramor M, Niwa H, Smith A (2014) Genetic exploration of the exit from self-renewal using haploid embryonic stem cells. Cell Stem Cell 14:385–393

7. Li W, Shuai L, Wan H, Dong M, Wang M et al (2012) Androgenetic haploid embryonic stem cells produce live transgenic mice. Nature 490:407–411

8. Zhong C, Yin Q, Xie Z, Bai M, Dong R et al (2015) CRISPR-Cas9-mediated genetic screening in mice with haploid embryonic stem cells carrying a guide RNA library. Cell Stem Cell 17:221–232

Chapter 9

Insertion of Group II Intron-Based Ribozyme Switches into Homing Endonuclease Genes

Tuhin Kumar Guha and Georg Hausner

Abstract

Fungal mitochondrial genomes act as "reservoirs" for homing endonucleases. These enzymes with their DNA site-specific cleavage activities are attractive tools for genome editing, targeted mutagenesis and gene therapy applications. Herein, we present strategies where homing endonuclease open reading frames (HEases ORFs) are interrupted with group II intron sequences. The ultimate goal is to achieve in vivo expression of HEases that can be regulated by manipulating the splicing efficiency of the HEase ORF-embedded group II introns. That addition of exogenous magnesium chloride ($MgCl_2$) appears to stimulate splicing of nonnative group II introns in *Escherichia coli* and the addition of cobalt chloride ($CoCl_2$) to the growth medium antagonizes the expression of HEase activity (i.e., splicing). Group II introns are potentially autocatalytic self-splicing elements and thus can be used as molecular switches that allow for temporal regulated HEase expression. This should be useful in precision genome engineering, mutagenesis, and minimizing off-target activities.

Key words Homing endonucleases, Biotechnology, Group II introns, Temporal regulation, Targeted mutagenesis, Homing endonuclease assay

1 Introduction

Homing endonucleases (HEases) are site-specific DNA cleaving enzymes that are encoded by homing endonuclease genes (HEGs) which are frequently found embedded within mobile elements [1] but sometimes HEGs can be freestanding [2]. HEases promote their own mobility and the mobility of the elements that host them by introducing site-specific, double-stranded breaks in cognate alleles that lack HEGs or intron/intein insertions, thereby stimulating the double-stranded DNA repair process which involves homologous recombination [1, 3]. The LAGLIDADG family of HEGs (LAG HEase) are frequently encoded within fungal mitochondrial group I introns [3] and these enzymes recognize long asymmetrical 12–40 bp of DNA sequence as their target sites. Due to their target site specificity, HEases have applications in DNA

Andrew Reeves (ed.), *In Vitro Mutagenesis: Methods and Protocols*, Methods in Molecular Biology, vol. 1498, DOI 10.1007/978-1-4939-6472-7_9, © Springer Science+Business Media New York 2017

sequence assembly or synthetic biology [4], as genome-editing tools by promoting gene replacements via homologous repair [5–10], as a gene targeting tool by promoting mutations generated by nonhomologous end-joining repair [7–10], or as rare-cutting enzymes that are part of cloning vectors [9]. Sometimes, procedures involving in vivo gene targeting for the temporal regulation of HEase activity might be essential in order to minimize off-target activities of the enzyme [11].

In this chapter, we describe an on/off "switch" system that provides an opportunity for the temporal control of HEase activity in *Escherichia coli*. Splicing of group II introns requires the intron RNA to fold into a splicing competent tertiary structure that requires interactions between intron and flanking exon sequences. The so-called intron binding sequences (IBS), located upstream of the intron insertion site, are needed for splicing as they interact with the corresponding exon binding sequences (EBS1 and EBS2) present within the intron [12, 13]. Group II-intron derived ribozymes are metalloenzymes [14, 15] and they require positive cations like magnesium (Mg^{2+}) for catalysis [16]. Strategies will be presented with regard to inserting group II intron sequences in vitro into expression constructs and for manipulating the in vivo splicing of these introns by stimulated splicing with the addition of Mg^{2+} or antagonizing splicing by the addition of cobaltous ion (Co^{2+}) in the form of cobalt chloride. It should be noted that this strategy of using ribozyme-based switches could be applied to other protein-based genome editing tools such as TALENS, Zinc-finger endonucleases (ZFNs), and the cas9 (CRISPR)-based systems.

2   Materials

2.1   Molecular Biology Reagents

1. All buffers use DNAse and RNAse-free sterile water.

2. Commercially synthesized HEase ORF (with group II intron) cloned into an expression vector.

3. *E. coli* competent cells such as DH5α and BL21 (λDE3).

4. Temperature controlled water bath (42 °C/65 °C).

5. Super-Optimal broth with Catabolite repression (SOC): 2% tryptone, 0.5% yeast extract, 10 mM NaCl, 2.5 mM KCl, 10 mM $MgCl_2$, 10 mM $MgSO_4$, 20 mM glucose.

6. Shaker incubator (37 °C).

7. Rotary shaker incubator (37 °C).

8. Lysogeny broth (LB), LB agar plates: 10 g/L tryptone, 5 g/L yeast extract, 10 g/L NaCl. pH to 7.0 with 5 N NaOH. Sterilize by autoclaving. Add 15 g/L agar. Pre-warm in an incubator.

9. 100 mg/mL kanamycin sulfate stock solution.

10. 60 mg/mL ampicillin sulfate stock solution.

11. 70% ethanol.

12. 95% ethanol.

13. Miniprep DNA purification kit.

14. PCR reaction mixture (total volume 50 μL) ingredients (μL/reaction): 10× Taq DNA polymerase buffer 5 μL, 50 mM $MgCl_2$ 0.5 μL, 2.5 mM dNTP 4 μL, 40 pmol each forward and reverse primer 0.5 μL each, H_2O 38.25 μL, DNA template 1 μL (~10–100 ng), and Taq DNA polymerase 0.25 μL (~2.5 U).

15. 1× Tris–EDTA (TE) buffer (DNA storage buffer): 10 mM Tris–HCl, pH 7.6, 1 mM $Na_2EDTA \cdot 2H_2O$.

16. Commercially available restriction enzymes and their respective buffers.

17. RNA purification kit.

18. Thermoscript Reverse Transcriptase Kit.

19. BigDye® Terminator sequencing system.

20. Endonuclease reaction buffer #3: 50 mM Tris–HCl, pH 8.0, 10 mM $MgCl_2$, 100 mM NaCl supplemented with 1 mM DTT.

21. Ultra-pure agarose and gel apparatus.

22. Agarose gel loading buffer (6×): 3 mL glycerol (30%), 25 mg bromophenol blue (0.25%) dH_2O to 10 mL.

23. Tris–borate EDTA buffer: 1× TBE buffer, 89 mM Tris–borate, 10 mM EDTA, pH 8.0.

24. Microcentrifuge.

2.2 Protein Work Reagents

1. Modified LB medium: 10 g/L tryptone, 5 g/L yeast extract, 5 g/L NaCl, and MilliQ water to 1 L. Add 200 μL of 5 N NaOH and autoclave. This recipe contains half the amount of NaCl.

2. Terrific broth (TB) medium (optional): tryptone 16 g, yeast extract 10 g, NaCl 5 g, adjust volume to 1 L. pH to 7.0 with 5 N NaOH and autoclave at 121 °C, 20 min, 15 psig.

3. Qiagen Nickel-NTA Superflow resin and column.

4. SDS PAGE: 30% acrylamide–bis solution (37.5:1), 10% ammonium persulfate, TEMED, 1 M Tris–HCl, pH 8.8, 0.5 M Tris–HCl pH 6.8, 10% (w/v) of sodium dodecylsulfate (SDS) stock solution in ddH_2O.

5. Cell Lysis (CL) buffer: 50 mM Tris–HCl, pH 8.0, 0.3 M NaCl.

6. Wash Buffer 1 (WB1): CL + 25 mM imidazole.

7. Wash Buffer 2 (WB2): CL + 50 mM imidazole.

8. Wash Buffer 3 (WB3): CL + 100 mM imidazole.

9. Elution Buffer 1 (EB1): CL + 250 mM imidazole.

10. Elution Buffer 2 (EB2): CL + 500 mM imidazole.

11. Dialysis Buffer: 50 mM Tris–HCl, pH 8.0, 150 mM NaCl and 1 mM DTT.

12. Protein storage buffer: 50 mM Tris–HCl, pH 8.0, 50 mM NaCl, 1 mM dithiothreitol (DTT), 30 % (w/v) glycerol.

13. 2× protein loading dye: 66 mM Tris–HCl, pH 6.8, 26.3 % (w/v) glycerol, 2.1 % SDS, 0.01 % bromophenol blue.

14. Amicon concentrator, Ultrafiltration membranes (desired molecular weight cutoff), and Amicon Ultra-4 Centrifugal filters (select for desired molecular weight cutoff).

15. High-speed centrifuge and rotors (e.g., SLA1500 and SS34 rotors).

16. Sonicator (with probe).

17. Spectrophotometer.

3   Methods

3.1 Expression Vector Design

First, one has to select a HEase sequence that is known to be functional and the sequence has to be codon-optimized for being expressed in *E. coli*. HEases are commercially available and can be engineered to intended target sequences, but one can start with "native HEases" and see if some by chance cut within a gene of interest. Suitable webserver resources to aid in codon optimization and potentially evaluate the expression of the HEase sequence in *E. coli* (or other hosts) are http://genomes.urv.es/OPTIMIZER/ [17] and http://mbs.cbrc.jp/ESPRESSO/TopPage.html [18] respectively (*see* Subheading 3.2). The choice of group II introns is obviously critical. Group II introns have a wide distribution and are found in all three domains of life [16]. They have been primarily classified based on structural details such as their RNA folding [19–21], their intron-encoded proteins (if present), and depending on how these introns fold, it tends to have implications on their exon sequences involved in generating splicing competent folds [12, 13]. The secondary structure of Group II Intron RNA can be viewed as a central wheel from which six "fingers," i.e., domains (I through VI) emerge. Domain I contains the exon-binding sequences (EBS) that ultimately interact with elements within the flanking exon sequences (referred to as intron-binding sequences=IBS). So it is important to investigate the choice of intron and be aware of the splicing requirements for the intron. Based on the current literature, the following group II introns have been well characterized and may offer good starting points: *Chaetomium thermophilum* mtDNA mS1247 nested group II intron [22, 23]; the mtDNA rI1 of *Scenedesmus obliquus* [24, 25]; and the bacterial Ll.LtrB, Ecl5, Rmint1, and B.hl1-B introns [16]. All of these introns have been well characterized with regard to their requirements for exon recognition and splicing conditions in various hosts (reviewed in ref. [16]). It is also best to choose introns that lack ORFs or remove ORFs if present and select introns that have rather "simple" exon recognition requirements. For example group II B introns require three intron–exon interactions (i.e., EBS1, 2, and 3 plus corresponding IBS 1, 2, and 3) whereas it has been shown that the rI1 group II B intron actually will splice efficiently in *E. coli* as long as the IBS1 sequence is provided in the upstream exon [24, 25]. The fewer interactions needed by the intron means less manipulation of the HEase ORF sequence is required.

1. Select a known functional HEase for the insertion of a ribozyme-based switch (*see* **Note 1**).

2. For selecting a suitable group II intron that could serve as a "switch" examine either group IIA intron or group IIB introns from the NCBI Genbank database at: (http://www.ncbi.nlm.nih.gov/genbank/), or consult a group II database (http://webapps2.ucalgary.ca/~groupii/) [26] and/or the Comparative RNA web site (http://www.rna.icmb.utexas.edu/ [27] (*see* **Note 2**).

3. Determine the intron binding sites (IBS). They are usually upstream (6–12 nucleotides) from the intron insertion site; however, depending on the type of intron IBS components present they can be downstream of the introns native insertion site [16] (*see* **Note 3**).

4. Prior to the group II intron sequence being inserted into the HEase ORF, it is necessary to match the required intron-based EBS sequences with the exon's ORF potential IBS sequences (*see* **Note 4**). This may require some manipulation of the HEases sequence and determine where the intron is inserted.

3.2 Gene Synthesis

1. A codon-optimized version of the HEG sequence should be synthesized to account for differences between the fungal mitochondrial and bacterial genetic code and codon biases (*see* **Note 5**). Note: *DO NOT* modify the selected IBS sequence(s) and the internal intron sequences.

2. Clone the ORF sequence with the embedded group II intron sequence in an expression plasmid with an inducible T7 promoter (e.g., pET28b+) for overexpression. We will refer to this HEase ORF interrupted by a group II intron construct as "ORF-switch" in the text.

3. Sequence the "ORF-switch" plasmid in order to confirm the orientation and to ensure that the ORF is in frame with the vector that provides the start codon and the N-terminal 6×-His-tag.

3.3 HEase Substrate Design

1. Construct a substrate plasmid by inserting a DNA segment that contains the target site for the HEase, such as an allele that does not contain the HEase (and/or associated intron) sequence.

2. Generate a control plasmid by inserting a DNA fragment that lacks the HEase target site such as the allele with the HEG (and/or intron) insertion (*see* **Note 6**).

3. Synthesize and clone the substrate sequence into any suitable plasmid (e.g., pUC57 vector).

4. Sequence the substrate plasmid to ensure that the insert is in place (*see* **Note 7**).

5. Transform the plasmids (substrate and control) into *E. coli* DH5α separately (*see* Subheading 3.4) and then purify the constructs from ~5 mL LB overnight cultures with any suitable plasmid purification kit.

3.4 Chemical Transformation

For transforming the "ORF-switch" construct, *E. coli* BL21 cells are recommended as they are efficient for the overexpression of heterologous proteins and for maintaining the substrate or non-substrate control constructs. *E. coli* DH5α cells can also be considered as an alternative. The chemical transformation method will be detailed below as a standard procedure. Readers must take into account which constructs are being transformed into which cell line (*see* **Note 8**).

1. Transfer 1 μL of the plasmids into vials containing 100 μL of chemically competent *E. coli* cells and mix gently. Avoid pipetting up and down.

2. Incubate the vials on ice for 5–30 min (*see* **Note 9**).

3. Heat-shock the cells for 1 min at exactly 42 °C without shaking.

4. Transfer the vials onto ice and keep for 2 min.

5. Add 300 μL of pre-warmed SOC medium at room temperature to the vials.

6. Tightly cap the tubes and shake horizontally (~200 rpm) at 37 °C for 1 h.

7. Spread 100–150 μL of the mixture onto a warm LB agar plate containing the appropriate antibiotic(s) and incubate at 37 °C till the colonies are clearly visible (usually 16–24 h).

3.5 Analyzing Clones

1. From the LB agar plate in **step 7**, Subheading 3.4, identify single colonies and inoculate several into 5 mL of LB medium and incubate at 37 °C, 250–300 rpm for 14–18 h.

2. 3 mL of the LB culture is collected for extracting plasmid DNA using a minprep plasmid kit [28]; however, one can also perform colony PCR (*see* **step 7**, Subheading 3.4) screening [29] to confirm colonies that maintain the plasmid of interest.

3. Perform restriction enzyme digestion to confirm the presence of the correct construct/plasmid. Ideally, one should use a restriction enzyme or a combination of enzymes that cut once in the vector and once in the insert.

4. Resolve and visualize restriction digests by agarose gel electrophoresis [28].

3.6 Gel Electrophoresis

1. Preparation of a 1 % agarose gel: Add 1 g ultra-pure agarose to 100 mL (volume depends on size of gel tray, adjust accordingly) of 1× TBE buffer then mix and melt agarose in a microwave oven. Once the agarose has completely dissolved, allow it to cool to about 55–60 °C and pour into an assembled gel casting tray with positioned comb. Allow the gel to solidify at room temperature and carefully remove the comb and place the gel into an electrophoresis box containing 1× TBE buffer.

2. Mix each DNA sample with the agarose gel loading buffer and load the samples into the wells of the gel. Electrophorese at 80–120 V until the tracking dye migrates to the positive electrode end of the gel. Resolved DNA fragments are sized with a DNA ladder (such as 1 kb plus DNA ladder by Invitrogen/Life technologies).

3. Stain the nucleic acids by soaking the gel in 1× TBE buffer supplemented with 0.5 µg/mL ethidium bromide (EtBr) for 30 min.

4. Expose the stained gel to ultraviolet light to visualize and document.

3.7 Cell Storage

1. Once a colony with a construct of interest has been identified, mix 0.85 mL of the culture with 0.15 mL of 50 % sterile glycerol and transfer it to a cryovial and store at –80 °C. For simplicity, the HEase expression construct containing the group II intron glycerol stock will be referred to as the "ORF-switch" stock. Other constructs such as the substrate plasmids, etc. can be preserved in the same manner.

2. As an additional backup, always store an aliquot of purified plasmid DNA at –20 °C.

3.8 In Vivo RNA Splicing Assay

Reverse Transcriptase PCR (RT-PCR) needs to be employed to examine in vivo splicing activity of the HEase ORF group II intron. Specifically, the determination of the concentration of exogenous $MgCl_2$ that has to be added to the growth medium in order to induce splicing of the group II intron. The plasmid-derived HEG transcript has to be evaluated to verify that splicing has occurred and to ensure that splicing in *E. coli* maintains the expected intron–exon junctions. This is important otherwise frameshift mutations could be introduced.

1. Inoculate the "ORF-switch" stock in 10 mL of LB medium supplemented with appropriate antibiotic (e.g., 100 µg/mL of kanamycin for constructs if cloned in pET28b+) and 0.25% w/v glucose. Also, inoculate the control (no HEase ORF) vector (e.g., pET28b+ in BL21) in another LB medium with the same concentration of antibiotic and glucose.

2. Incubate the cultures overnight in a rotary incubator at 37 °C.

3. Prepare several 50 mL LB culture flasks and supplement with 1 mM, 5 mM, 10 mM, 20 mM up to 100 mM of magnesium chloride ($MgCl_2$).

4. Inoculate the 50 mL LB culture flasks with 500 µL of the overnight cultures (*see* **step 2**, Subheading 3.8). As a negative control, inoculate a 50 mL LB culture flask with any added $MgCl_2$ with 500 µL of the overnight culture.

5. Grow the 50 mL cultures at 37 °C with agitation till the OD_{600nm} reaches 0.65 (*see* **Note 10**).

6. Pellet the bacterial cells from 10 mL aliquots from the above cultures (including the negative controls) by centrifuging for 3 min at $4000 \times g$.

7. Lyse the pelleted cells and extract the RNA using any bacterial RNA extraction kit. Make sure you set aside at least 1 µg of RNA for in vitro translation (*see* Subheading 3.9 below).

8. Treat the extracted RNA samples with 2 U of DNaseI and incubate at 37 °C for 15 min. Stop the reaction by adding 1 µL EDTA (50 mM) followed by a 10 min incubation at 65 °C.

9. Remove 2 µL from each of the reaction mixtures and perform a standard PCR reaction by using the HEase ORF-specific primers in order to confirm the complete elimination of any residual DNA from the extracted samples.

10. Run a 1% agarose gel. This PCR reaction should not yield any amplification products to indicate the removal of all DNA from the RNA sample (*see* **Note 11**).

11. Perform RT-PCR to make cDNA from the transcript of interest contained within the extracted RNA samples using a standard RT-PCR kit.

12. The cDNA obtained in **step 11** above can now be used as a template for performing standard PCR using HEase ORF-specific primers. This will now determine the splicing potential of the group II intron and show possible splicing intermediates. A successful group II splicing event should yield a single PCR product corresponding to the difference between the distance of the forward and reverse primers minus the nucleotide length of the inserted group II intron used for the study (*see* **Note 12**).

13. Gel-excise the PCR amplicon corresponding to the desired length as mentioned above using a Gel/PCR DNA extraction kit.

14. Sequence the gel extracted fragment utilizing the primers required for obtaining the amplicon in order to determine whether the correct splicing occurred (i.e., investigate the intron–exon splicing junction (*see* **Note 13**).

15. Note the concentration of MgCl$_2$ added in the LB culture flask(s) that yielded the correct splicing product.

3.9   In Vitro HEase Expression

1. To assess whether the HEase can be expressed in an "*E. coli*" environment, perform in vitro translation with the RNA extracted from the *E. coli* bacterial cells grown in LB medium which was supplemented with the predetermined MgCl$_2$ concentration that induced proper splicing of the group II intron (*see* **step 7**, Subheading 3.8).

2. For in vitro translation, one can use a commercial in vitro protein synthesis kit (e.g., PURExpress in Vitro Protein Synthesis Kit, New England Biolabs) (*see* **Note 14**).

3. After a minimum incubation of at least 3 h at 37 °C, mix 2.5 μL of the reaction mixture with 2.5 μL of the 2× protein loading dye and resolve the proteins in SDS-PAGE.

4. Analyze for the presence of the desired HEase protein by comparing the resolved proteins and scanning those with the expected molecular weight based on the protein ladder.

3.10   In Vivo HEase Overexpression

It is assumed that the overexpression conditions for the functional HEase are known (*see* Subheading 3.11, below). However, readers can reassess the overexpression conditions as follows:

1. Inoculate small flasks (50 mL of LB medium containing the appropriate antibiotic supplemented with 0.25 % w/v glucose) with 500 μL of an overnight culture of *E. coli* (which was transformed with the "ORF-switch" construct).

2. Supplement the LB medium with the predetermined concentration of MgCl$_2$. Inoculate another small flask with *E. coli* BL21 containing only the control plasmid (plasmid containing no insert).

3. Grow the cultures (with agitation) at 37 °C until the OD$_{600}$ reaches 0.65, induce the cells with 0.2 mM IPTG (low) and 1 mM IPTG (high) to the respective flasks, and shift the flasks to various temperatures. (Several trials may be required to optimize the concentration of IPTG in the range of 0.1–1 mM and temperature (15–37 °C) for proper induction (i.e., stable protein expression).

4. Incubate the flasks at the various temperatures for 6 h to overnight.

5. Pellet the cells via centrifugation at 4000×*g* for 10 min at 4 °C.

6. Discard the supernatant and resuspend the cells in 2 mL of cell lysis buffer.

7. Sonicate the cells thoroughly for short pulses (~15 s) to lyse. Keep the vials on ice during the entire period.

8. Centrifuge the cell lysate at $16,000 \times g$, 15 min at 4 °C and collect the crude protein extract in microcentrifuge tubes. Keep on ice.

9. Determine the concentration of the crude protein mixture (A_{260}/A_{280} ratio) using a spectrophotometer.

10. Analyze the samples by SDS-PAGE using about 8 μg of each protein extract plus the same amount of protein from the control sample(s).

11. Check the SDS-PAGE gel for overexpression of the protein of interest by scanning for a band in the expected size range that is absent in the control lane. One can also perform western blotting with any commercially available anti-His antibody to further confirm the presence of the His-tag on the overexpressed protein, which is required for purification in downstream steps. Once the specific parameters have been determined for protein overexpression one can proceed to the large-scale HEase overexpression procedure.

3.11 Overexpression of HEase

1. Inoculate 10 mL LB medium (supplemented with the appropriate antibiotic and 0.25 % w/v glucose) with a small amount (~10 μL) of the "ORF-switch" glycerol stock and incubate overnight at 37 °C in a rotatory incubator.

2. Inoculate 1 L of LB medium (supplemented with 100 μg/mL of kanamycin and 0.25 % w/v glucose plus the optimal amount of MgCl$_2$; *see* Subheading 3.8) with 5 mL of the overnight culture prepared in **step 1** above.

3. Grow the culture at 37 °C with agitation and induce with IPTG (*see* Subheading 3.10) when the OD$_{600}$ reaches ~0.65 and grow further at the predetermined conditions for over expression.

4. Harvest the cells by centrifugation at $4000 \times g$ for 10 min and freeze the pellet at −80 °C.

3.12 HEase Purification

1. Thaw the pellet in a warm water bath and resuspend it in 10 mL of CL buffer per 1 g wet weight of cells. Stir the suspension for 30 min at 4 °C in order to make it homogeneous.

2. Lyse the cells using a French press two times (as needed) and centrifuge the lysate at $16,000 \times g$ for 30 min at 4 °C to pellet the cell debris.

3. Add the clear lysate to 3 mL of Ni-NTA resin and incubate at 4 °C with shaking for 30–60 min.

4. Load the crude-extract onto a Ni-NTA super-flow column.

5. Carry out the following series of washings with wash (1) 30 mL of WB1 buffer; wash (2) 30 mL of WB2 buffer; and wash (3): 30 mL of WB3 buffer. Collect and save 1 mL of each wash.

6. Elute the protein in Elution buffer EB1 and if necessary EB2. Collect the eluted samples in 1.5 mL microfuge tubes as 700 μL fractions.

7. Remove excess imidazole by dialyzing in the dialysis buffer using a slide-a-lyzer dialysis cassette with a desired molecular weight (MW) cutoff.

8. Check the concentration of the protein using the absorbance (A_{280}) function of a spectrophotometer and analyze the fractions by performing SDS-PAGE (*see* **Note 15**).

9. Pool the desired fractions to a final volume of 9 mL in a protein storage buffer.

10. Concentrate the fractions using a centrifugal filter with a predetermined molecular weight cutoff and centrifuge at $4000 \times g$ at 4 °C until the sample is concentrated to a final volume of 500 μL. Store the protein in small aliquots (20 μL) at −80 °C. Check the concentration of the protein before freezing. Do not freeze-thaw the purified HEase.

3.13 Cleavage Assays

1. Combine 15 μL of substrate plasmid (25 μg/mL), 5 μL in vitro endonuclease reaction buffer supplemented with 1 mM DTT, 5 μL of HEase protein (~50 μg/mL), and 25 μL ultrapure H_2O. In addition, the linearized substrate plasmid can be tested as a substrate for endonuclease activity.

2. Set up a parallel reaction as in **step 3** (below) but with a control plasmid which contains an insert that covers the HEase/intron containing allele. This would be a negative control that should not be cleaved by the HEase.

3. Incubate the cleavage reactions at 37 °C and take 10 μL aliquots at the following time intervals: 0, 30 and 60 min. Stop the reactions by adding 2 μL of 200 mM EDTA (pH 8.0) and 1 μL of proteinase K (1 mg/mL) to each 10 μL aliquots followed by incubation for 30 min at 37 °C.

4. Resolve the cleavage reaction products on a 1 % agarose gel. Samples representing an untreated version of the substrate, ideally a restriction enzyme-linearized version, and the negative control plasmid(s) should be resolved on this gel along with a suitable molecular weight marker.

3.14 Cleavage Site Mapping

1. Treat the substrate plasmid with HEase under optimal conditions (as outlined in Subheading 3.13).

2. Resolve the cleaved substrate plasmid on a 1 % agarose gel and excise the DNA fragment from the gel with any suitable PCR product gel clean-up/extraction system.

3. Treat the linearized substrate plasmid with T4 DNA polymerase under conditions that generate blunt ends [30].

The reaction mixture contains 40 μL of HEase-treated linearized plasmid (25 μg/mL), 2 μL T4 DNA polymerase (5 U/μL), 20 μL 5× T4 DNA polymerase buffer, 20 μL dNTP mixture (0.5 mM) and the total volume is adjusted to 100 μL with sterile distilled water.

4. Incubate the reaction mixture at room temperature (~24 °C) for 20 min, place it on ice for 5 min and terminate the reaction by incubating for 10 min at 70 °C.

5. Purify the T4 DNA polymerase-treated linearized DNA (now blunt-ended) and then add 2 μL of T4 DNA ligase (1 U/μL) in the presence of 10 μL 5× ligase buffer in a total volume of 40 μL. Incubate the ligation reaction at room temperature for 2 h to generate the desired religated plasmid.

6. Dilute the ligation reaction fivefold and use 10 μL of this dilution to transform chemically competent *E. coli* DH5α cells.

7. Transformed *E. coli* cells are grown overnight at 37 °C in 5 mL of LB medium (supplemented with the appropriate antibiotics).

8. Purify the plasmid from the transformed overnight cultures with a suitable plasmid purification kit and sequence the recovered plasmid.

9. Compare the chromatogram for the obtained sequence with the sequence of the original uncleaved substrate plasmid or sequence the uncleaved substrate plasmid in parallel with the HEase cleaved/T4 DNA polymerase-treated substrate plasmid using the same primers for both types of constructs.

10. Nucleotides missing in the sequence of the HEase/T4 DNA polymerase-treated substrate plasmid when compared to the original untreated substrate sequence define the nucleotides removed by T4 DNA polymerase. This approach works for LAGLIDADG-type HEases that typically generate four nucleotide 3′ overhangs at their cleavage sites. These staggered cuts are blunt ended by the T4 DNA polymerase [1, 2, 30].

3.15 MgCl$_2$ Effects on HEase Expression

We have noticed that manipulating the exogenous [Mg^{2+}] (i.e., in the medium) stimulates group II intron splicing and thus the removal of the intron acts like a switch that can control the expression of the HEase. In order to evaluate the appropriate amount of Mg^{2+} (suggested range from 1 to 10 mM) to be added to the medium an in vivo endonuclease assay has to be established. This assay is based on a two-plasmid in vivo endonuclease assay where two compatible plasmids, a HEase "donor" plasmid ("ORF-switch" plasmid as the ORF contains a group II intron sequence) and a HEase "substrate" plasmid both need to be maintained in *E. coli* BL21 (λDE3). *See* Subheading 3.8 on evaluating the splicing potential of the group II intron. The plasmids have different (compatible) origins of replication and can be selected by antibiotic selection (kanamycin (Kan) and chloramphenicol (Cm), respectively). For example, the pET28b+ vector (ColE1 origin of replication and Kan^R) can be used for the overexpression of the HEase and a second plasmid pACYC184 (p15A origin of replication and Cm^R) can be used to provide the target site for the HEase. The "donor" plasmid hosts the HEase ORF with the group II intron at an appropriate location to facilitate suitable EBS/IBS interactions and the "substrate" plasmid has a sequence inserted that offers the HEase a cleavage target site. Successful expression and production of the HEase will lead to the loss of the substrate plasmid and the Kan^R marker. The loss of cell viability is an indicator of intron splicing which will lead to the production of a functional HEase. To ascertain the group II intron as an on/off switch for in vivo HEase expression the medium can be supplemented 10 μM of CoCl$_2$. Co^{2+} appears to negate the stimulatory effect of exogenous Mg^{2+} on intron splicing, possibly because Co^{2+} interferes with the uptake of Mg^{2+} into *E. coli* cells [31, 32]. In summary, the addition of Mg^{2+} stimulates the expression of the HEase and the addition of Co^{2+} is inhibitory to HEase expression (*see* Fig. 1).

1. Co-transform *E. coli* BL21 with two plasmids, one containing the "ORF-switch" (i.e., the HEase ORF plus the group II intron) plasmid and the other being the substrate plasmid. Make sure the plasmids are compatible (different origin of replications) and also have different antibiotic selection markers (e.g., the "ORF-switch" plasmid has the Kan^R cassette while the substrate construct has the Cm^R cassette).

2. Repeat **step 1** above to co-transform the control vector and substrate plasmid (*see* Subheading 3.4). Store the positive, co transformed clones as 50% glycerol stocks.

3. Grow seed cultures overnight at 37 °C derived from the glycerol stocks in culture tubes containing 5 mL LB medium plus the appropriate antibiotics. Add 1% glucose to the medium to prevent leaky expression (if T7 promoter containing vectors are used).

4. Inoculate 50 mL of LB broth (containing appropriate antibiotics, 1% glucose) with 500 μL of the 5 mL seed cultures. Add the predetermined concentrations of MgCl$_2$ to the medium. Label the flask as LB + Mg^{2+}. For additional 50 mL LB flasks, inoculate with the same amount of seed culture and keep all the supplements constant but do not add MgCl$_2$. Label this culture flask as LB − Mg^{2+} (negative control).

5. In order to antagonize the stimulatory effect of MgCl$_2$ on splicing of group II introns, in one LB culture flask add 10 μM of CoCl$_2$ along with the desired concentration of MgCl$_2$. Label this flask as LB + Mg + Co.

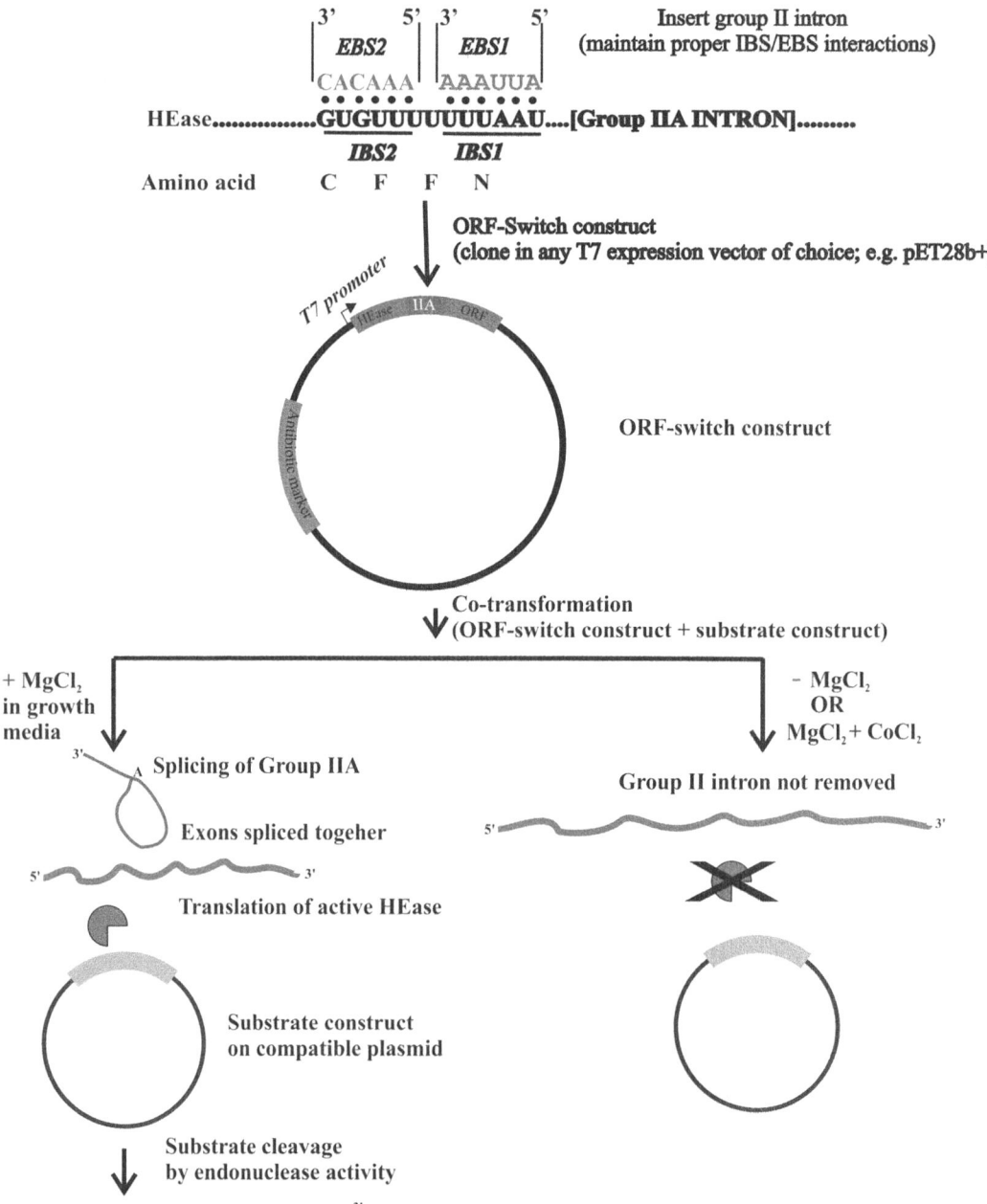

Fig. 1 Schematic overview of the considerations and implementations required for utilizing group II introns-based ribozymes as switches for the expression of HEases within *E. coli*

6. Grow the cells at 37 °C with vigorous shaking (~210 rpm) till the OD_{600} reaches 0.65.

7. Induce protein overexpression with the predetermined concentration of IPTG in LB cultures with Mg^{2+}, without Mg^{2+} and an LB culture with Mg^{2+} and Co^{2+}.

8. Incubate the flasks with vigorous shaking (210 rpm) at the predetermined (optimal) temperature for at least 4–6 h.

9. Perform a serial dilution for each of the above cultures and plate the diluted cells (out to 10^{-6}) on pre-warmed (37 °C) LB agar plates containing only the antibiotic that was selected by the substrate plasmid (e.g., if the substrate plasmid contains a chloramphenicol-resistance gene, plate the diluted cells on the LB agar chloramphenicol plates). Perform at least two biological and three technical replicates for each of the above cultures.

10. Incubate the plates at 37 °C until the colonies are clearly visible and count the number of colonies in order to get the mean cfu/mL values and standard deviations. This will establish suitable parameters for setting up conditions for the temporal expression of a HEase that could cut an intended target during a specific growth phase of the bacterium.

4 Notes

1. In order to find potential HEGs see Hafez et al. [33] as well as the LAGLIDADG homing endonuclease database [34].

2. Use group IIC intron with caution as these tend to have three IBS/EBS interactions for establishing splicing competent folds this can complicate the design of the construct. We use group IIA and IIB introns as they tend to have fewer IBS/EBS interactions.

3. Some group II introns require one IBS (IBS1), some require two IBS sequences (IBS1 and IBS2) while some group II intron categories require three sets of sequences to satisfy all IBS/EBS interactions for proper splicing [16].

4. As an example, it first has to be determined with a great degree of certainty what the group II introns EBS sequences are. From there one can proceed and scan the HEase ORF sequence for a location that provides compatible (complementary) IBS sequences, keeping in mind that with regard to RNA U can interact with A or G. If a group II intron has been selected and it requires an IBS1 sequence that is aacagg, one would scan the nucleotide sequences of the HEase ORF and try to locate a match for this sequence. If the sequence is found in the middle (or near the middle) of the ORF sequence, that should be an ideal location for inserting a group II intron. Keep in mind there might be additional IBS sequences required (IBS2, etc.). It is important to find a suitable location in the HEase ORF sequence that would maintain all the required IBS/EBS interactions with minimal modification to the HEase coding sequence.

5. Several online programs assist in codon optimization, e.g., http://www.encorbio.com/protocols/Codon.htm, http://genomes.urves/OPTIMIZER/. Several commercial suppliers

will perform codon optimization and gene synthesis such as GenScript (http://www.genescript.com/), GeneArt (Thermo Fisher Scientific), Gene Oracle (Sigma-Aldrich), etc.

6. The latter plasmid should not be cleaved by the HEase as the cleavage site is disrupted by the HEase/intron sequence. One could also obtain substrates and controls by using PCR products of alleles that lack the HEase/intron insertion and alleles that contain the HEase/intron; however, some HEases appear to prefer plasmid DNAs as substrates (i.e., supercoiled templates).

7. For evaluating the potential inserts within the pUC57 vector, use the M13 Forward primer (M13F) and the M13 Reverse primer (M13R). One must use the respective vector-specific primers to sequence the insert to confirm that the correct sequence is present.

8. *E. coli* BL21 is specifically designed for the overexpression of genes regulated by the T7 promoter. However, DO NOT use this strain for the propagation and maintenance of plasmids as this strain has leaky T7 RNA polymerase expression, which might lead to instability and eventual loss of the plasmid.

9. Sometimes incubation for 1 h in ice leads to better transformation compared to 5 min.

10. Check the OD_{600} of the cultures in order to see whether high concentrations of $MgCl_2$ are detrimental to bacterial cell growth.

11. The presence of PCR products indicates residual DNA contamination.

12. It is always possible to observe a RT-PCR product that was generated due to the presence of unspliced transcripts or some other splicing intermediates.

13. Alternative splicing is a possibility and such an event might happen if alternative EBS/IBS interactions can be established. This would also shift the intron–exon junction and thus could alter the coding sequence.

14. Although the PURExpress kit is designed for coupled transcription/translation from an expression construct containing a T7 promoter, direct translation is also possible provided purified RNA (1–5 μg) with a proper ribosome-binding site (RBS) is incubated within the in vitro translation reaction mixture.

15. The imidazole concentration in the wash buffer should be adjusted depending on the affinity of the protein for the nickel resin.

Acknowledgments

This work is supported by a Discovery grant from the Natural Sciences and Engineering Research Council of Canada (NSERC) to G.H. T.K.G. would like to acknowledge funding support from the Faculty of Science Graduate Award program (University of Manitoba) and the University of Manitoba Faculty of Graduate Studies GETS program.

References

1. Stoddard BL (2006) Homing endonuclease structure and function. Q Rev Biophys 38:49–95

2. Gimble FS (2000) Invasion of a multitude of genetic niches by mobile endonuclease genes. FEMS Microbiol Lett 185:99–107

3. Hausner G (2012) Introns, mobile elements and plasmids. In: Bullerwell CE (ed) Organelle genetics: evolution of organelle genomes and gene expression. Springer, Berlin, pp 101–131

4. Liu JK, Chen WH, Ren SX, Zhao GP, Wang J (2014) iBrick: a new standard for iterative assembly of biological parts with homing endonucleases. PLoS One 9:e110852

5. Stoddard BL, Scharenberg AM, Monnat RJ Jr (2008) Advances in engineering homing endonucleases for gene targeting: ten years after structures. In: Bertolotti R, Ozawa K (eds) Progress in gene therapy 3: autologous and cancer stem cell gene therapy. World Scientific Press, Hackensack, NJ, pp 135–167

6. Marcaid MJ, Muñoz IG, Blanco FJ, Prieto J, Montoya G (2010) Homing endonucleases: from basics to therapeutic applications. Cell Mol Life Sci 67:727–748

7. Takeuchi R, Lambert AR, Mak AN, Jacoby K, Dickson RJ, Gloor GB, Scharenberg AM, Edgell DR, Stoddard BL (2011) Tapping natural reservoirs of homing endonucleases for targeted gene modification. Proc Natl Acad Sci U S A 108:13077–13082

8. Stoddard BL (2011) Homing endonucleases: from microbial genetic invaders to reagents for targeted DNA modification. Structure 19:7–15

9. Hafez M, Hausner G (2012) Homing endonucleases: DNA scissors on a mission. Genome 55:553–569

10. Prieto J, Molina R, Montoya G (2012) Molecular scissors for in situ cellular repair. Crit Rev Biochem Mol Biol 47:207–221

11. Posey KL, Gimble FS (2002) Insertion of reversible redox switch into a rare cutting DNA endonuclease. Biochemistry 41:2184–2190

12. Olga F, Nora Z (2007) Group II introns: structure, folding and splicing mechanism. Biol Chem 388:665–678

13. Michel F, Costa M, Westhof E (2009) The ribozyme core of group II introns: a structure in want of partners (Review). Trends Biochem Sci 34:189–199

14. Donghi D, Pechlaner M, Finazzo C, Knobloch B, Sigel RKO (2013) The structural stabilization of the κ three-way junction by Mg(II) represents the first step in the folding of a group II intron. Nucleic Acids Res 41:2489–2504

15. Sigel RKO (2005) Group II intron ribozymes and metal ions—a delicate relationship. Eur J Inorg Chem 12:2281–2292

16. Lambowitz AM, Belfort M (2015) Mobile bacterial group II introns at the crux of eukaryotic evolution. Microbiol Spectr 3:MDNA3-0050-2014

17. Puigbò P, Guzmán E, Romeu A, Garcia-Vallvé S (2007) OPTIMIZER: a web server for optimizing the codon usage of DNA sequences. Nucleic Acids Res 35:W126–W131

18. Hirose S, Noguchi T (2013) ESPRESSO: a system for estimating protein expression and solubility in protein expression systems. Proteomics 13:1444–1456

19. Toor N, Hausner G, Zimmerly S (2001) Coevolution of group II intron RNA structures with their intron-encoded reverse transcriptases. RNA 8:1142–1152

20. Lambowitz AM, Zimmerly S (2004) Mobile group II introns. Annu Rev Genet 38:1–35

21. Lambowitz AM, Zimmerly S (2011) Group II introns: mobile ribozymes that invade DNA. Cold Spring Harb Perspect Biol 3:a003616

22. Hafez M, Majer A, Sethuraman J, Rudski SM, Michel F, Hausner G (2013) The mtDNA rns gene landscape in Ophiostomatales and other fungal taxa: twintrons, introns and intron encoded proteins. Fungal Genet Biol 53:71–83

23. Guha TK, Hausner G (2014) A homing endonuclease with a switch: characterization of a twintron encoded homing endonuclease. Fungal Genet Biol 65:57–68

24. Kück U, Godehardt I, Schmidt U (1990) A self-splicing group II intron in the mitochondrial large subunit rRNA (LSUrRNA) gene of the eukaryotic alga Scenedesmus obliquus. Nucleic Acids Res 18:2691–2697

25. Hollander V, Kück U (1999) Group II intron splicing in Escherichia coli: phenotypes

of cis-acting mutations resemble splicing defects observed in organelle RNA processing. Nucleic Acids Res 27:2339–2344

26. Candales MA, Duong A, Hood KS, Li T, Neufeld RA, Sun R, McNeil BA, Wu L, Jarding AM, Zimmerly S (2012) Database for bacterial group II introns. Nucleic Acids Res 40(Database issue):D187–D190

27. Cannone JJ, Subramanian S, Schnare MN, Collett JR, D'Souza LM, Du Y, Feng B, Lin N, Madabusi LV, Müller KM, Pande N, Shang Z, Yu N, Gutell RR (2002) The comparative RNA web (CRW) site: an online database of comparative sequence and structure information for ribosomal, intron, and other RNAs. BMC Bioinformatics 3:2

28. Green MR, Sambrook R (2012) Molecular cloning, a laboratory manual, 4th edn. Cold Spring Harbor Laboratory Press, Cold Spring Harbor, NY

29. Dafa'alla TH, Hobom G, Zahner H (2000) Direct colony identification by PCR-Miniprep. Mol Biol Today 1:65–66

30. Bae H, Kim KP, Song JM, Kim JH, Yang JS, Kwon ST (2009) Characterization of intein homing endonuclease encoded in the DNA polymerase gene of *Thermococcus marinus*. FEMS Microbiol Lett 297:180–188

31. Nelson DL, Kennedy EP (1971) Magnesium transport in *Escherichia coli*. Inhibition by cobaltous ion. J Biol Chem 246:3042–3049

32. Nelson DL, Kennedy EP (1972) Transport of magnesium by a repressible and a nonrepressible system in Escherichia coli. Proc Natl Acad Sci U S A 69:1091–1093

33. Hafez M, Guha TK, Shen C, Sethuraman J, Hausner G (2014) PCR-based bioprospecting for homing endonucleases in fungal mitochondrial rRNA genes. Methods Mol Biol 1123:37–53

34. Taylor GK, Petrucci LH, Lambert AR, Baxter SK, Jarjour J, Stoddard BL (2012) LAHEDES: the LAGLIDADG homing endonuclease database and engineering server. Nucleic Acids Res 40(Web Server issue):W110–W116

Chapter 10

Generating a Genome Editing Nuclease for Targeted Mutagenesis in Human Cells

Zhenyu He and Kehkooi Kee

Abstract

Gene targeting and editing is an essential tool for both basic research and clinical application such as gene therapy. Several endonucleases have been invented to fulfill these purposes, including zinc finger nucleases, TALEN, and CRISPR/Cas9. Although all of these systems can target DNA sequence with high efficiency, they also exert off-target effects and genotoxicity. The off-target effects might not hinder their usage in animal models because the correctly targeted cells can be selected for further studies. However, the off-target effects could cause mutations which may be damaging or cancerous to the patients. In this chapter, we describe a genome-editing nuclease method which relies on modifying specific amino acids on a monomeric endonuclease, I-SceI, to recognize a targeted sequence in the human genome. This nuclease is small in size and shows a much lower genotoxicity compared to other nucleases including CRISPR/Cas9.

Key words I-SceI, Gene editing, Gene targeting, Homologous directed recombination, β-thalassemia

1 Introduction

Most gene-editing technologies rely on using sequence-specific DNA endonucleases which can generate DNA double-strand breaks (DSBs) at the targeted sequences. The majority of the DSBs will be repaired by the nonhomologous end joining (NHEJ) mechanism and a minority will be repaired by homology-directed repair (HDR) in the presence of the DNA template. The most widely used targeting system now is CRISPR/Cas9 because of its ease of designing targeting sequences. It originates from bacteria and relies on two key elements: a 20-bp-guided RNA complimentary to a target DNA for detection and a Cas9 endonuclease for cutting [1]. This system overcomes the complexity of designing target sequences by modifying the sequence of RNA instead of reengineering the amino acids which function in detection in two other gene targeting-systems: zinger finger nucleases (ZFN) and TALEN. However, many researchers have reported that off-target activity was evident when these three endonucleases are applied in

Andrew Reeves (ed.), *In Vitro Mutagenesis: Methods and Protocols*, Methods in Molecular Biology, vol. 1498,
DOI 10.1007/978-1-4939-6472-7_10, © Springer Science+Business Media New York 2017

targeting specific loci in human cells. For example, two recent studies reported evidence for CRISPER/Cas9-induced genotoxicity in the human β-globin gene [2].

In our study, we developed an alternative system to reduce the off-target rate in the correction of β-globin gene mutations to treat β-thalassemia by reengineering I-SceI, a monomeric endonuclease. I-SceI was first discovered in yeast [3]. Some researchers have used it in human DNA recombination experiments and found it caused a low level of genotoxicity [4, 5] because the human genome lacks the sequences that could be recognized by I-SceI. I-SceI has an 18-bp recognition sequence and the central five nucleotides (ATAAC) act as the cutting site. This sequence is only seven nucleotides different from a DNA sequence close to the second most prevalent mutations in Asian β-thalassemia (β-IVS654) patients [5]. Based on the crystal structure of I-SceI bound to DNA, we found four nucleotides that directly interact with the I-SceI protein, corresponding to residues N152, K193, R48, and N15 [5] (*see* Fig. 1). Subsequently, we reengineered these amino acids one by one so that the modified I-SceI can recognize the β-globin sequence and induce HDR with low genotoxicity in human cells. In this protocol, we describe the procedure and provide a detailed methodology using this reengineering approach (*see* Fig. 2).

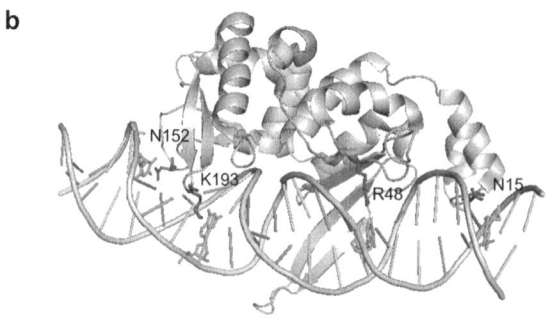

Fig. 1 Corresponding amino acids and DNA sequences for the reengineering of I-SceI to target the β-globin sequence. (**a**) Schematic diagram of the eGFP reporter with a modified I-SceI recognition sequence. Two nucleotides were modified in the first round of selection as shown in the listed sequences. (**b**) A computational model of I-SceI binding to its target DNA sequence based on crystal structure data deposited in the protein databank (pdb), highlighting the interactions of the four amino acids (N152, K193, R48, and N15) directly contacting the nucleotides

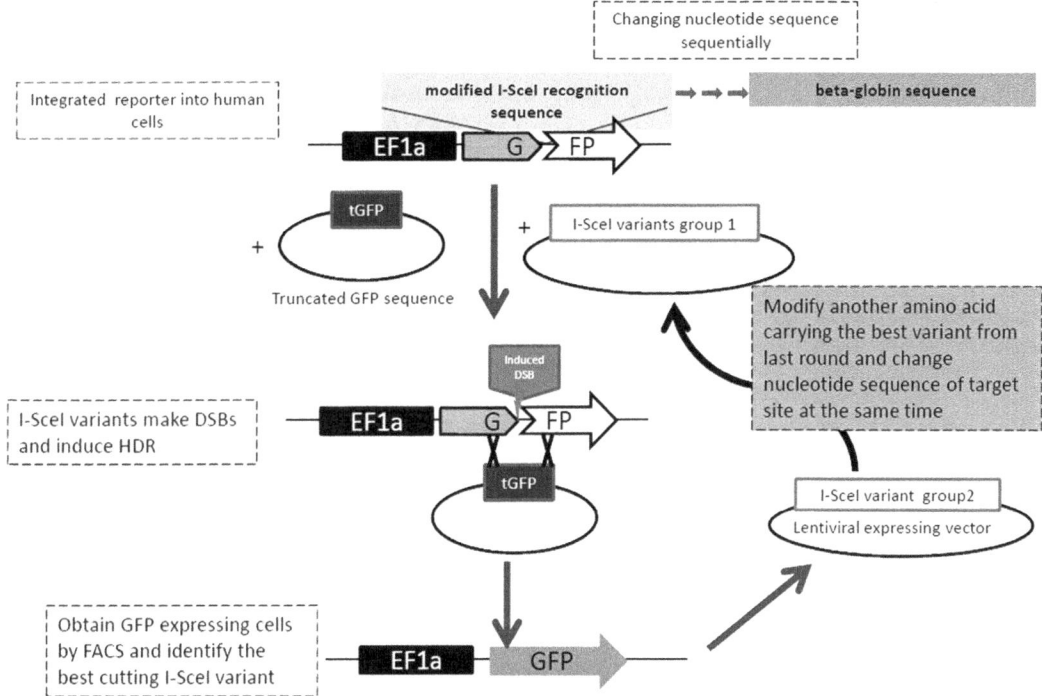

Fig. 2 Strategy and procedures for reengineering I-SceI to target the β-globin sequence

2   Materials

2.1   Plasmids and Vectors

Unless otherwise indicated, all vectors were generated in reference [5] and information found therein.

1. p2k7- EF1α-I-SceI carrying the original I-SceI coding sequence.
2. pENTY5′-TOPO-EF1α carrying the EF1α promoter sequence.
3. pENTY/D-TOPO vector (pENTY/D-TOPO Cloning Kit, Life Technologies).
4. pENTY-tGFP, a donor plasmid for homology-directed repair, carrying a truncated eGFP coding sequence.
5. p2k7-EF1α-eGFP vector carrying a full length eGFP coding sequence.
6. Virus assembly vectors: Vsvg, Δ8.9.

2.2   Oligo Primers

1. Primers F1/R1, for full length I-SceI amplification. The sequences are:
   F1: 5′-CACCATGAAAAACATCAAAAAAAACCAGG-3′.
   R1: 5′-TTACTTAAGAAAAGTTTCGGAGG-3′.
2. For I-SceI variant (N152) degenerate primers: Overlap forward (FV-N152): TAAATGGGATTACNNNAAAAACT

CTACC and overlap reverse (RV-N152): GGTAGAGTTTT
TNNNGTAATCCCATTTA (N: A, T, C, or G).

3. Forward and reverse primers for tGFP amplification:
5′ATATGCGGCCGCGTGAGCAAGGGCGAGGAG3′ and
5′ATATGGCGCGCCCTACTTGTACAGCTCGTCCA3′.

*2.3 Molecular
Biology Reagents*

Reagents required for PCR, DNA restriction analysis, DNA ligation, DNA purification, transformation, plasmid extraction, and LR integration reaction.

1. DNA polymerase: KOD-Plus-Neo.

2. High-Fidelity NotI restriction enzyme.

3. Hi-Fidelity KpnI restriction enzyme.

4. Appropriate restriction enzyme buffer (10×).

5. T4 DNA ligase and buffer.

6. DNA gel purification kit.

7. Plasmid miniprep purification kit.

8. Chemically competent *E. coli* cells.

9. One-shot *E. coli* chemically competent stbl3 cells (Invitrogen).

10. LR enzyme and buffer kit (Gateway LR Clonase Plus Enzyme Mix, Life Technologies).

11. Tris-EDTA buffer.

*2.4 Media,
Transfection Reagents*

1. Luria–Bertani (LB) medium: tryptone 10 g, yeast extract 5 g, and NaCl 10 g. Add distilled H_2O to 1 l and pH to 7.2 with 5 N NaOH. Autoclave. For agar plates add agar to 1.5 % final concentration.

2. 293FT cells cultured in 90 % Dulbecco's Modified Eagle Medium (DMEM) with 10 % fetal bovine serum (FBS), 0.1 mM MEM nonessential amino acids, 2 mM l-glutamine, 1 % penicillin–streptomycin, and 500 μg/mL geneticin, 5 % CO_2. Incubate at 37 °C.

3. HT1080 cells cultured in 90 % DMEM with 10 % FBS, 1 % penicillin–streptomycin, 5 % CO_2. Incubate at 37 °C.

4. 293FT-eGFP-I-SceI cell line and four 293 FT-eGFP-insert sequence cell lines cultured as 293FT.

5. Transfection reagent Lipofectamine (Lipofectamine-2000).

6. Transfection reagent Opti-MEM.

7. Gelatin.

2.5 Viral Reagents

1. Infection agent, polybrene (Millipore).

2. Filter: Millex-HV 0.45um filter unit (Millipore).

3. Blasticidin.

4. Geneticin.

5. Crystal violet stain.

6. Phosphate-buffered saline (PBS).

7. Fluorescence-activated cell sorter (e.g., BD Calibur).

3 Methods

3.1 I-SceI Variant Construction

N152 (asparagine) modification as an example, first round.

1. Use overlapping PCR to carry out site-direct mutagenesis (change N152 to all other 19 amino acids). Amplify the 5′-fragment and the 3′-fragment from the template vector p2k7-EF1α-I-SceI separately with the primers carrying sequence variants (T_m of 62 °C, 35 cycles). Mix primers F1 with RV-N152 and R1 with FV-N152 separately in two PCR reactions.

2. Purify the PCR products carrying 5′-fragment variants and 3′-fragment variants with a gel midi purification kit.

3. Add 5′ fragments and 3′ fragments in 1:1 ratio and anneal by denaturing at 94 °C for 5 min and cooling to 68 °C for 20 min in KOD-Plus-Neo buffer, dNTP, $MgSO_4$, KOD-Plus-Neo, and ddH_2O.

4. Add I-SceI F1 and R1 primers and amplify full-length I-SceI carrying the sequence variants at N152 (T_m of 62 °C, 35 cycles).

5. Purify the PCR product using a gel midi purification kit.

6. Mix the PCR products (4 μl), buffer (1 μl), and pENTY/D--TOPO vector (1 μl) at room temperature for 5 min.

7. Transform the mix into TOP10 chemically competent E. coli cells and spread them on an LB agar plate containing kanamycin sulfate at 50 μg/ml.

8. Pick >20 colonies and inoculate them into LB medium containing kanamycin. Sequence to confirm the clones (see **Note 1**).

9. Align the sequences to find the vectors expressing the 19 other amino acids.

10. Clone each I-SceI variant into the p2k7 vector by using the Gateway LR recombination clonase kit.

11. Mix the EF1α vector (100 ng), pENTY/D-TOPO-I-SceI-N152 variants vector (100 ng), p2K7-Bsd vector (300 ng), LR Clonase plus Enzyme mix (2 μl), and TE buffer to a total volume of 10 μl. Leave the mix at 25 °C for 16 h.

12. Transform the mix into E. coli Stbl3 competent cells, pick a single colony and confirm the correct sequence.

13. Extract each vector (p2k7-EF1α-I-SceI N152 and the 19 other variants) using a miniprep plasmid kit.

14. Repeat the above procedures to modify I-SceI-K193 (lysine) in the second round after first round selection described in the following sections, R48 (arginine) in the third round after second round selection and N15 (asparagine) in the final round (*see* **Note 2**) for a total of four mutations.

3.2 HDR Measurements

1. Insert the 18 bp I-SceI recognition sequence (IRS) into the eGFP coding sequence at position 329 by overlapping PCR (T_m of 65 °C, 35 cycles).

2. Recombine the eGFP-IRS with p5′D-TOPO-EF1α into p2k7 vector to produce p2K7-EF1α-eGFP-IRS by Gateway LR Clonase kit according to the instructions in the kit.

3. Change two nucleotide sequences of the 18-bp IRS from TAG to AAA (-7, -5 position) (*see* Fig. 1) for the N152 binding site by overlapping PCR using p2K7-EF1α-eGFP-IRS as the template to produce p2K7-EF1α-eGFP-IRS1 vector.

4. Change the nucleotide at the three position of IRS (G to A) in the second round, nucleotides at positions 6 and 8 (GGT to TGA) in the third round and finally nucleotides at positions 9 and 11 (AAT to TAA). Each round uses a modification of the vector produced from the previous round as the DNA template for overlapping PCR.

5. Produce lentiviral supernatant using these reporter vectors and transduce 293FT cells.

3.3 Lentiviral Supernatant

1. Prepare 293FT cells on a 0.1% gelatin-coated T175 so that the cells are ~90% confluent on the day of transfection.

2. For each T175 transfection, premix:
   A: 120 μl Lipofectamine + 5 ml Opti-MEM.
   B: 10 μg Vsvg + 15 μg Δ8.9 + 10 μg lentivectors carrying I-SceI-insert (total of six types from **step 1** in Subheading 3.1) in 10 ml Opti-MEM.

3. Allow A and B to sit for 5 min at room temperature. Then, mix them together vigorously by vortexing and let the mixture sit for 20 min at room temperature.

4. Pour the 15 ml A + B mixture into the 293FT cells gently, swirl a bit to distribute the mixture evenly in the flask and keep the flask in a 37 °C incubator for 6 h.

5. Replace the transfection mixture with 18 ml of 10% FBS in DMEM and l-glutamine medium and transfer the flasks to a lentiviral room (*see* **Note 3**).

6. Incubate the transfected 293FT at 37 °C for ~72 h.

7. Harvest the supernatant (~15–18 ml from each flask) in a 50 ml conical tube and centrifuge at $751 \times g$ for 5 min.

8. Filter the supernatant through a Millex-HV 0.45 μm filter unit.

9. The supernatant is now ready for infection or aliquoting to −80 °C.

3.4 Viral Infection Efficiency

1. Place HT1080 cells in a 6-well plate so that the cells are ~50% confluent on the day of infection.

2. Dilute the lentiviral supernatant 10×, 100×, 1000×, 10^4×, 10^5×, and 10^6× and add a total of 1 ml to each plate.

3. Change the lentiviral supernatant every 2 days with 14 ml HT1080 medium, 14 µl polybrene, and a 1:1000 blasticidin solution (2 µg/ml) for 10 days.

4. Stain with crystal violet to determine the number of resistant cells on each plate and confirm the appropriate concentration.

3.5 Transduction

1. Place the infectable 293FT cells in a 6-well plate with confluency of ~90%.

2. Add polybrene at a ratio of 1:1000 to the lentivirus with a final concentration of 8 µg/ml.

3. Lay 1 ml of viral supernatant with the determined concentration to cover the well and incubate for 6 h at 37 °C.

4. Add 293FT medium to 3 ml (e.g., 2 ml viral supernatant + 1 ml medium) at the end of 6 h and let it incubate overnight.

5. Aspirate off the supernatant completely and wash twice with 4 ml PBS.

6. Add the 293FT medium to the well and bring it back to the regular incubator.

7. Add blasticidin (2 µg/ml) to the medium after 1 day of regular culture incubation.

8. Generate a single-cell suspension of the resistant cells after >3 days of blasticidin selection. Calculate the cell density in the suspension. Transfer a single cell into each well of a 96-well plate (*see* **Note 4**).

9. Expand the single cell clone to a cell line carrying a specific reporter (*see* **Note 5**).

3.6 Selection of I-SceI Variants

1. Place the 293FT cells carrying p2K7-EF1α-eGFP-IRS1 in a 6-well plate so that the cells are ~50% confluent on the day of transfection

2. For each transfection, premix:
A: 4 µl Lipofectamine and 0.5 ml Opti-MEM.
B: 1 µg pENTY-tGFP vector and 1 µg p2K7-EF1α-I-SceI-N152 variants (first round: N152 and 19 variants) in 0.5 ml Opti-MEM. (The control group is to only transfect 1 µg pENTY-tGFP.)

3. Let A and B sit for 5 min at room temperature. Mix them together vigorously by vortexing and let the mixture sit for 20 min at room temperature.

4. Pour the A + B mixture gently (total volume ~1 ml) into 293FT-eGFP-I-SceI single-variant cells, swirl a bit to distribute the mixture evenly in the well. Keep the flask in the 37 °C incubator for 6 h.

5. Gently remove the Lipofectamine and add 2 ml of 293FT medium to each well from the side without disrupting the cell layer. Let the transfected 293FT cells incubate at 37 °C for 36 h.

6. Use FACS to measure the HDR level by examining the percentage of cells expressing eGFP (*see* Fig. 3) (*see* **Note 6**).

7. Pick the I-SceI that exhibits the highest HDR level in the 293FT reporter cell line as the template for the next round of selection.

8. Prepare 293FT cell lines carrying p2K7-EF1α-eGFP-IRS2 for I-SceI variants at K193 and repeat the steps described above (*see* Fig. 3) (*see* **Note 7**).

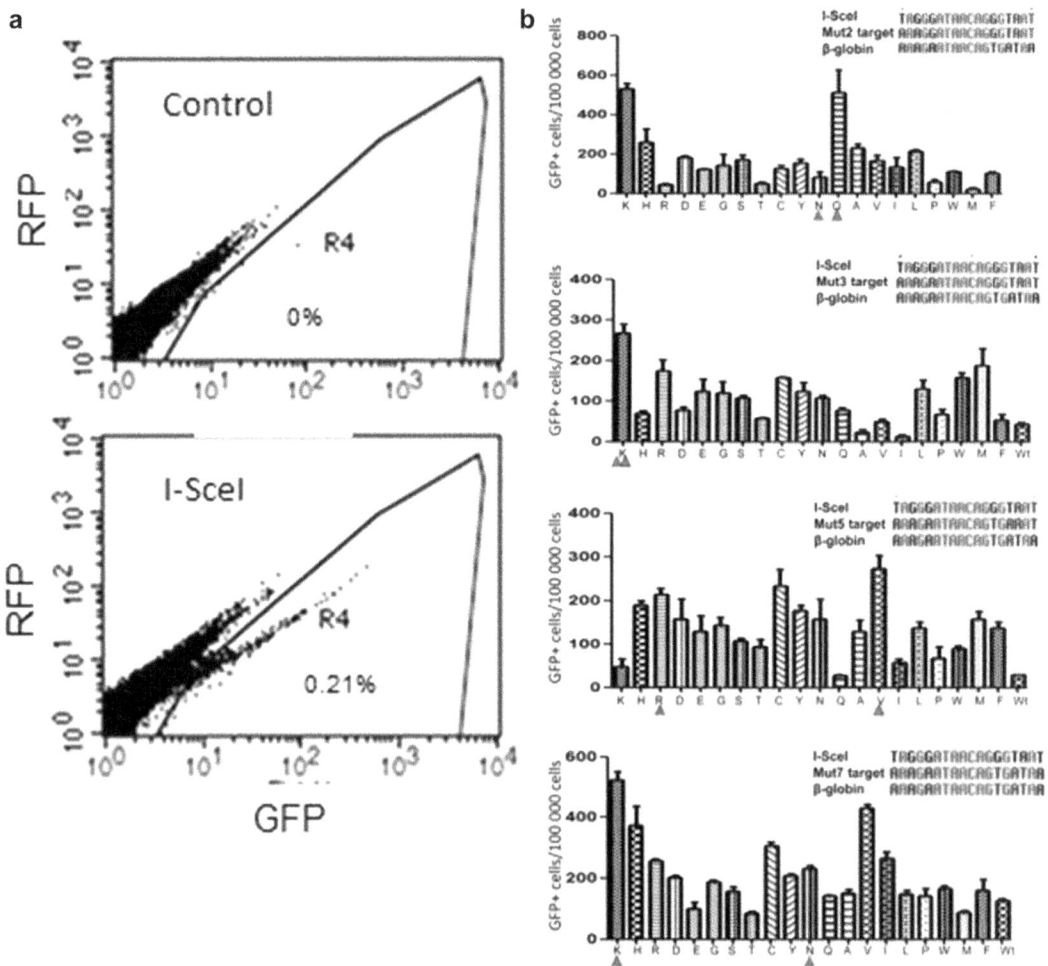

Fig. 3 Using eGFP reporter system to measure HDR in 293FT cells. (**a**) FACS analysis of the control with only donor vector and I-SceI-induced HDR using a reporter containing the I-SceI recognition sequence. (**b**) Sequential screening of I-SceI variants which induced the highest HDR at the indicated recognition sequences. Nineteen variants and the wild-type sequence were tested for their ability to induce HDR at the targeted sequence is indicated at the *upper right corner*. The *blue arrowhead* indicates the original amino acid and the *red arrowhead* indicates the selected amino acid that yields the highest HDR in each round of screening. Wild-type (wt) I-SceI was included in all screenings (N in first round was the wild-type I-SceI)

9. Repeat the same procedures for R48 and N15 with the corresponding IRS reporters.

4 Notes

1. The frequency of picking up each variant is relatively random. The degenerate primers provide more than one sequence combination for each amino acid variant. We usually pick and sequence 20–30 colonies to cover all 19 variants.

2. When we modified the I-SceI protein, the second round modification was based on the results of the first round selection. After the first round selection, N152 was replaced by Q152 and this new modified I-SceI protein was used for K193 modification. Similarly, the third round modification was based on the result of the second round and the final round modification was based on the result of the third round.

3. Handle the flask with extra caution from this point on, assuming that the infectious lentivirus is present in the supernatant after the transfection.

4. It takes about 1 week for each single-cell clone to grow up and be ready for passaging and expansion.

5. Each cell line is expanded from a 96-well plate to a 12-well, from a 12-well plate to a 6 well, from a 6-well plate to a T25 flask. When the cells are expanded to a T25 flask, some are frozen as stocks and kept at liquid nitrogen for long-term storage.

6. The control is to transfect only with the donor vector carrying a truncated eGFP and is needed to set up the gating so that there is no cell expressing eGFP without I-SceI variants.

7. I-SceI carrying original amino acids may still induce the highest HDR. For example, K193 exhibits the highest HDR among the 20 different variants, so we kept this amino acid and continue the next round of selection.

Acknowledgments

This work is funded by National Natural Science Foundation of China (81470011) and the Ministry of Science and Technology and Technology of China (2012CB966702).

References

1. Mussolino C, Cathomen T (2013) RNA guides genome engineering. Nat Biotechnol 31(3):208–209

2. Cradick TJ, Fine EJ, Antico CJ, Bao G (2013) CRISPR/Cas9 systems targeting beta-globin and CCR5 genes have substantial off-target activity. Nucleic Acids Res 41(20):9584–92

3. Moure CM, Gimble FS, Quiocho FA (2003) The crystal structure of the gene targeting homing endonuclease I-SceI reveals the origins of its target site specificity. J Mol Biol 334(4):685–95

4. Szczepek M, Brondani V, Büchel J, Serrano L, Segal DJ, Cathomen T (2007) Structure-based redesign of the dimerization interface reduces the toxicity of zinc-finger nucleases. Nat Biotechnol 25(7):786–93

5. Lin J, Chen H, Luo L, Lai Y, Xie W, Kee K (2015) Creating a monomeric endonuclease TALE-I-SceI with high specificity and low genotoxicity in human cells. Nucleic Acids Res 43(2):1112–1122. doi:10.1093/nar/gku1339

Chapter 11

Use of Group II Intron Technology for Targeted Mutagenesis in *Chlamydia trachomatis*

Charlotte E. Key and Derek J. Fisher

Abstract

Dissecting the contribution of genes to virulence in fulfillment of Molecular Koch's postulates is essential for developing prevention and treatment strategies for bacterial pathogens. This chapter will discuss the application of a targeted, intron-based insertional mutagenesis method for creating mutants in the obligate, intracellular bacterial pathogen *Chlamydia trachomatis*. The methods employed for intron targeting, mutant selection, and mutant verification will be outlined including available selection markers, gene targeting strategies, and potential pitfalls.

Key words *Chlamydia trachomatis*, Group II intron, Insertional mutagenesis, Chemical transformation, Gene inactivation

1 Introduction

Chlamydia spp. are obligate, intracellular bacteria that infect a wide range of animals [1]. The primary pathogens of concern for humans are *C. pneumoniae* and *C. trachomatis*, which cause respiratory tract infections or sexually transmitted infections and the ocular infection trachoma, respectively [2–4]. Prior to 2011, researchers lacked methods for genetically manipulating *Chlamydia*, which greatly hindered research on these important pathogens [5]. Development of a chemical transformation and plasmid-selection protocol by Wang et al. ushered in a new era of chlamydial research by opening the door to the field of chlamydial genetics [6]. In addition to the numerous shuttle vectors now available for *Chlamydia*, targeted chromosomal mutations can be created through intron-based insertional mutagenesis (discussed in this chapter) [7]. Ethyl methanesulfonate (EMS)- and *N*-ethyl *N*-nitrosourea (ENU)-based chemical mutagenesis techniques also have been employed to create random chromosomal mutations [8–10].

To assess gene function and contribution to pathogenesis, we created targeted gene insertion mutants in *C. trachomatis* utilizing

Andrew Reeves (ed.), *In Vitro Mutagenesis: Methods and Protocols*, Methods in Molecular Biology, vol. 1498, DOI 10.1007/978-1-4939-6472-7_11, © Springer Science+Business Media New York 2017

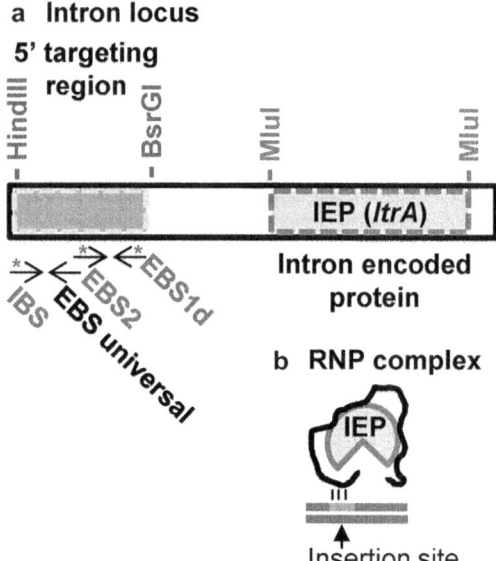

Fig. 1 Diagram of the group II intron. The wild-type GII intron has two primary "functional" regions, the 5′ region involved in targeting and the intron-encoded protein (IEP) required for intron splicing, folding, gene targeting, and insertion (**a**). The targeting mechanism uses base pairing between DNA sequences in the target gene and RNA sequences in the intron along with interactions between the IEP and DNA (**b**). Base pairing and intron insertion is mediated by the ribonucleoprotein complex (RNP, **b**) formed by the intron RNA and IEP. Target gene recognition is controlled by altering bases within the 5′ region of the intron using PCR-based site directed mutagenesis with primers IBS, EBS2, EBS universal, and EBS1d to change select bases, allowing for base pairing with the desired target gene. Base changes, indicated by *red asterisks* in **a**, are determined using a proprietary algorithm (*see* Fig. 5). New targeting regions are swapped with the template targeting region by digestion of the intron and PCR product with *Hind*III and *Bsr*GI followed by ligation of the PCR product with the digested intron (*see* Fig. 3). For mutagenesis, the wild-type IEP, *ltrA*, is expressed *in trans* and alternative IEPs, including *bla* or *aadA*, are inserted between MluI restriction sites allowing for antibiotic selection of mutants

a mobile Group II (GII) intron, Ll.LtrB IIA, from *Lactococcus lactis* [7, 11]. Group II introns are mobile DNA elements that move through an RNA-based retro-homing process with the assistance of an intron-encoded protein (IEP) that possesses RNA maturase, reverse transcriptase, DNA binding, and DNA endonuclease activity (*see* Fig. 1) [12]. Under native conditions, transcription of the intron is driven by the host-gene promoter resulting in production of the host gene mRNA carrying the intron sequence (*see* Fig. 2). The IEP is then translated followed by IEP-mediated forward splicing of the intron from the mRNA restoring the wild-type version of the host gene mRNA, thus preventing loss of host gene function. Note that while the wild-type transcript is restored, the DNA copy

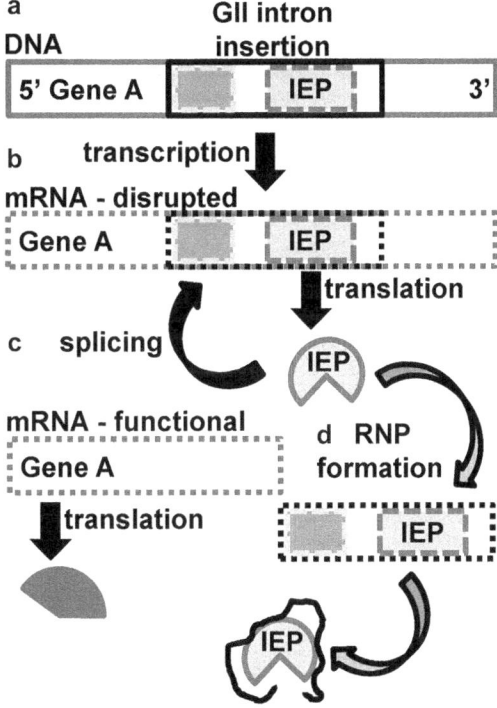

Fig. 2 Overview of intron splicing and RNP formation. Expression of the intron (**a**) is driven by the promoter controlling gene A, resulting in a mRNA containing gene A and the intron (**b**). Presence of the intron would result in production of a truncated protein A due to disruption of the gene A open reading frame by the intron sequence. Translation of the IEP open reading frame located within the gene A/intron mRNA produces LtrA (**b**). The LtrA protein will recognize the 5′ and 3′ intron ends and assist forward splicing of the intron out of the gene A mRNA transcript, restoring the wild-type transcript and allowing for production of the full length protein A (**c**). The LtrA will then assist in intron folding and formation of the catalytic RNP complex (**d**). This complex will then search for a DNA-insertion target through DNA-RNA base pairing and initiate reverse splicing. Removal of the IEP region of the intron prevents production of LtrA and consequently inhibits splicing of the intron from the gene A transcript precluding production of protein A. For mutagenesis, loss of the IEP is achieved by expressing it *in trans* from a suicide plasmid

of the intron remains stably inserted in the genome. After intron splicing, the IEP-intron ribozyme complex "searches" for DNA insertion sites via base pairing between sequences in the 5′ region of the intron and the DNA target along with interactions between the IEP and target gene. Once a target is located, the intron is reverse spliced creating a DNA–RNA hybrid, and the intron RNA is reverse-transcribed by the IEP forming a DNA–cDNA hybrid. Host cell enzymes then repair the single stranded cDNA section (*see* refs. 13, 14 for an in-depth discussion of intron function and applications). To adapt the Ll.LtrB IIA for mutagenesis, the IEP is

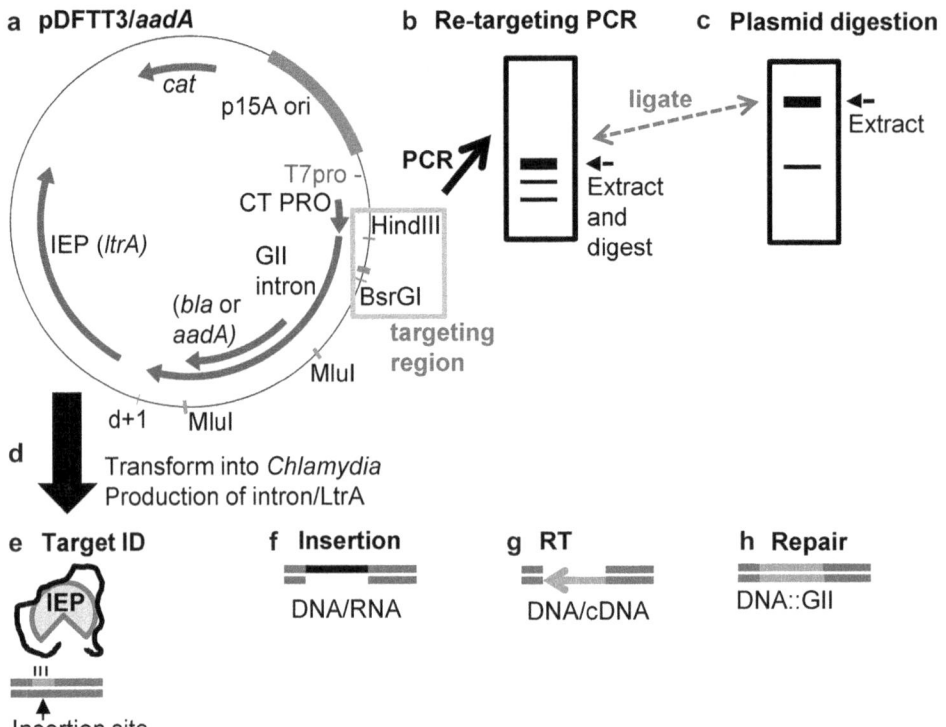

Fig. 3 Outline of GII intron mutagenesis. The intron is delivered to *C. trachomatis* on a suicide vector (**a**). Maintenance of the vector is performed using *E. coli*. The vector is selected for in *E. coli* using chloramphenicol (*cat*), and *C. trachomatis* mutants are selected using either spectinomycin (*aadA*) or ampicillin (*bla*), depending on the intron selection marker. Retargeting of the intron requires PCR resulting in three products that can be separated using DNA gel electrophoresis (**b**). The largest band (~350 bp) is extracted and the purified DNA is digested with *Hin*dIII and *Bsr*GI. The vector is also digested with the same enzymes, products are separated using DNA gel electrophoresis, and the vector backbone (~7.5 kb) is extracted (**c**). The purified, digested PCR product and vector backbone are ligated together and transformed into *E. coli*. The plasmid can then be isolated from an *E. coli* broth culture and the targeting region is sequence verified using the T7 promoter primer (T7pro). The resultant vector is transformed into *C. trachomatis* and intron and IEP expression are driven by a chlamydial promoter (CT PRO) resulting in formation of the IEP/intron RNP (**d** and **e**). The target is identified through base pairing (**e**), the intron is inserted (**f**) and reverse-transcribed (**g**) by the IEP forming a DNA–cDNA hybrid, followed by repair of the second strand by host cell enzymes (**h**). As the plasmid is lost during continuous passage of *Chlamydia*, the IEP will not be present in the bacterium and the intron cannot be spliced from gene transcripts resulting in permanent gene disruption

removed from the intron and expressed *in trans* from the intron donor plasmid (*see* Fig. 3). Loss of the donor plasmid during mutant construction removes the LtrA IEP and prohibits intron splicing, leading to a disrupted mRNA transcript that cannot be restored to the wild-type version. Expression of the LtrA IEP *in trans* also allows antibiotic resistance markers to be carried in place of the IEP by the intron for use in antibiotic-mediated selection of mutants. Targeting of the intron to the gene of interest is performed by mutating the 5′ intron sequence to allow base pairing between the RNP and DNA target.

Creation of GII-insertion mutants begins with selection of the target gene and identification of DNA insertion sites followed by "rewriting" of the 5′ intron sequence using PCR. The modified 5′ fragment is then swapped for the template 5′ intron sequence using DNA restriction digestion and ligation. The vector with the retargeted intron is then introduced into *C. trachomatis* using $CaCl_2$ chemical transformation and mutants are selected in cell culture using antibiotics (based on the intron-encoded resistance marker [15]). Clonal mutants are then isolated using a plaque assay or limiting dilution and verified through molecular approaches (PCR, Southern Blotting, and DNA sequencing) to confirm intron insertion and western blotting to confirm loss of protein production.

2 Materials

2.1 Cell Culturing

1. HeLa cells (*see* **Note 1**).

2. *Chlamydia trachomatis* L2 434/Bu (*see* **Note 2**).

3. Dulbecco's Modified Eagle Medium (DMEM) supplemented with GlutaMAX (Thermo Scientific).

4. Antibiotics: Ampicillin (100 mg/ml in distilled H_2O [dH_2O], filter sterilized), chloramphenicol (20 mg/ml in ethanol), spectinomycin (50 mg/ml in dH_2O, filter sterilized); 0.45 μm PVDF syringe top filters and 20 ml Luer lock syringes.

5. Lysogeny broth (LB) and agar plates: 10 g tryptone, 5 g yeast extract, 5 g NaCl per liter of dH_2O. Add 15 g agar per liter for plates (sterile, 100 mm × 15 mm petri dishes).

6. Fetal bovine serum (FBS).

7. 100× nonessential amino acids.

8. Cycloheximide (200 μg/ml in water, filter sterilized).

9. Tissue culture-treated 6-well plates.

10. Clinical centrifuge with swinging bucket plate adapters and 50 ml conical tube adaptors.

11. Refrigerated Centrifuge with fixed angle rotor for 50 ml conical centrifuge tubes.

12. Sterile, disposable 15 and 50 ml polypropylene conical centrifuge tubes.

13. Sterile 1.5 ml Eppendorf tubes.

14. BSL2 biological safety cabinet and disposal biohazard bags.

15. Humidified CO_2 Cell Culture Incubator.

16. Sonicator with 0.12 in. diameter probe.

2.2 Plasmids and E. coli Strains

1. Chemically competent *Escherichia coli* XL-1 Blue or DH5α for cloning and plasmid maintenance (*see* **Note 3**).

2. pDFTT3 or pDFTT3*aadA* (*see* **Note 4**), available from the corresponding author [7, 15].

2.3 Molecular Biology Reagents

1. JumpStart REDTaq ReadyMix (Sigma-Aldrich) (*see* **Note 5**).

2. DNA Primers (*see* **Note 6**).

3. FastDigest (such as Thermo Scientific) *Hin*dIII, FastDigest *Bsp*1407I (*Bsr*GI isoschizomer), and 10× restriction enzyme buffer.

4. FastAP Thermosensitive Alkaline Phosphatase.

5. T4 DNA ligase and 10× ligase buffer.

6. PCR thermal cycler.

7. TAE (50×) DNA electrophoresis running buffer: Tris–HCl, acetic acid, EDTA 242 g Tris base, 57.1 ml glacial acetic acid, 18.6 g EDTA, distilled deionized H_2O (ddH_2O) to 1 l. Store at room temperature.

8. 1.5 and 0.8 % agarose gels.

9. DNA electrophoresis unit.

10. UV transilluminator.

11. Plasmid miniprep kit.

12. Agarose gel extraction kit.

13. PCR Purification kit.

14. 37 °C degree water bath.

15. 37 °C degree incubator.

16. Sterile, DNase free 1.5 ml Eppendorf tubes.

17. 0.2 ml PCR tubes.

18. Microcentrifuge.

19. 5× $CaCl_2$—250 mM $CaCl_2$, 50 mM Tris Base pH 7.4 in ddH_2O. Filter sterilize or autoclave and store at room temperature.

20. Sucrose, phosphate buffered glutamic acid (SPG): 7.5 % w/v sucrose, 17 mM Na_2HPO_4, 3 mM NaH_2PO_4, 5 mM L-glutamic acid, pH 7.4 in ddH_2O. Filter sterilize and store at –20 °C.

21. Sterile, DNase free pipette filter tips; various capacity pipetters; pipet controller; disposable serological pipets.

22. Sterile, molecular grade water.

23. 3 mm sterile glass distillation beads.

24. Scalpel.

3 Methods

3.1 Intron Construction

1. Select the target gene from the published *C. trachomatis* L2 434/Bu genome sequence [16] and identify intron-insertion sites through TargeTron at: (http://www.sigmaaldrich.com/

life-science/functional-genomics-and-rnai/targetron.html)   or
Targetronics, LLC (http://www.targetrons.com/) (*see* **Note 7**).

2. Select two insertion sites per target gene and order primers for
   intron-retargeting (*see* **Note 8**).

3. Hydrate primers with molecular grade water and create the
   PCR primer master mix. Store primers and master mix at
   −20 °C.

   (a) Primer stock concentrations: IBS (100 μM), EBS1d
       (100 μM), EBS2 (20 μM), EBS universal primer (20 μM).
       The IBS, EBS1d, and EBS2 sequences will vary based on the
       target gene (*see* Fig. 5). The EBS universal primer sequence
       is: 5′ CGAAATTAGAAACTTGCGTTCAGTAAAC 3′ (*see*
       Note 9).

   (b) PCR primer master mix: 2 μl of each primer stock and
       12 μl of molecular grade water. Mix by vortexing.

4. Set up a 50 μl PCR reaction with the REDTaq ready mix: 23 μl
   REDTaq, 1 μl primer master mix, and 1 μl of template (use
   either 1 μl of the template supplied by Sigma-Aldrich or 1 μl of
   pDFTT3 at a concentration of 1 ng/μl). Thermal cycler conditions for REDTaq PCR: (1) Initial denaturation at 94 °C, 30 s;
   (2) 30 cycles of denaturation at 94 °C, 15 s; annealing at
   55 °C, 30 s; extension at 72 °C, 30 s; and (3) a final extension
   at 72 °C for 2 min.

5. Run the entire PCR product on a 1.5 % agarose gel and visualize the PCR products using ethidium bromide staining and
   UV transillumination. Three bands should be visible, a promi-
   nent band at 350 bp and two lighter bands at ~250 bp and
   ~100 bp (*see* Fig. 3b). Excise the 350 bp band using a scalpel
   and collect the gel slice in a 1.5 ml Eppendorf tube.

6. Purify the DNA from the gel slice using a Gel extraction kit
   and elute with 25 μl of elution buffer.

7. Digest the purified PCR product from **step 6** and 1 μg of the
   desired intron-donor plasmid (pDFTT3 or pDFTT3*aadA*) in
   30 μl reactions using *Hin*dIII and *Bsr*GI. If using the restriction enzymes and buffer supplied with the TargeTron kit, use
   3 μl of buffer and 1 μl of each enzyme. Mix the samples by
   vortexing briefly and then incubate the samples at 37 °C for
   30 min followed by 30 min at 60 °C. If using Thermo Scientific
   fast digest enzymes, use 3 μl of buffer and 1 μl of each enzyme
   and incubate the sample at 37 °C for 1 h after mixing.

8. After digestion, add 1 μl of FastAP to the plasmid digestion
   reaction and incubate at 37 °C for 10 min.

9. Run the plasmid digestion on a 0.8 % agarose gel and visualize
   the DNA with ethidium bromide staining and UV transillumination. The plasmid backbone should be present at ~7.5 kbp

along with the ~340 bp 5′ intron region removed through digestion (*see* Fig. 3c). Excise the 7.5 kbp band and extract the DNA using the Gel extraction kit. Elute the DNA with 30 μl of elution buffer.

10. Clean up the digested PCR products using a PCR Purification kit and elute the DNA in 30 μl of elution buffer.

11. Quantify the concentration of DNA for the digested, purified plasmid and PCR products using a spectrophotometer at A_{260nm} where an A_{260nm} of $1 = 50$ μg/ml of pure dsDNA.

12. Prepare a ligation reaction in a PCR tube containing a 3:1 ratio of insert to vector using the DNA concentrations calculated in **step 11**. The total reaction volume should be 10 μl and include 1 μl of T4 DNA ligase, 1 μl of 10× ligase buffer, the digested plasmid, and the digested PCR product. Mix the ligation reaction by brief vortexing and incubate the reaction overnight at 4 °C.

13. Use the entire ligation reaction to transform chemically competent *E. coli* (*see* **Note 10**).

14. Plate the entire transformation mix on LB agar plates containing chloramphenicol at 20 μg/ml and incubate at 30 °C for 2 days.

15. Select three colonies per transformant and perform two serial isolation streaks on LB agar plates containing chloramphenicol at 20 μg/ml and incubate each passage at 30 °C for 1–2 days.

16. Pick two isolates per transformation and use a single colony from each isolate to inoculate 7 ml broth cultures containing 20 μg/ml chloramphenicol. Grow the cultures with shaking at 30 °C for 20 h and purify the plasmid from each isolate using the Plasmid isolation kit. Quantify the plasmid DNA and perform DNA sequencing of the 5′ retargeted region using the T7 promoter primer (5′ TAATACGACTCACTATAGGG 3′) to confirm the presence of the desired mutations. Store the plasmid DNA at –20 °C.

3.2 DNA Transformation

1. Seed $\sim 8 \times 10^5$ HeLa cells per well in a 6-well tissue culture dish with DMEM, 10% FBS and grow until confluent (~1 day) in a cell culture incubator at 37 °C with 5% CO_2 (*see* **Note 11**).

2. Set up the chlamydial transformation reaction when the HeLa cells are confluent. Use *C. trachomatis* elementary bodies (EBs) which is the infectious form of the organism.

    (a) Thaw the *C. trachomatis* stocks on ice. 3.4×10^6 EBs will be required for each transformation reaction (*see* **Note 12**).

    (b) Calculate the volume of plasmid, EBs, and water needed for each reaction. The total reaction volume will be 67 μl and will contain 1 μg of plasmid DNA, 13.4 μl of 5× $CaCl_2$, and 3.4×10^6 EBs.

(c) Pipet the appropriate amount of EBs, plasmid DNA, sterile molecular grade water, and $CaCl_2$ into a 1.5 ml Eppendorf tube, in the order written, and mix by pipetting up and down ten times. Let the transformation reaction sit at room temperature for 30 min. Mix every 5 min by gentle flicking of the tube.

3. After 30 min, pipet the transformation reaction into a 15 ml conical tube containing 4 ml of ice cold SPG. Mix the SPG/transformation solution by pipetting up and down with a serological pipet.

4. Remove the medium from the 6 well tissue culture dish and immediately pipet 2 ml per well of the 4 ml SPG/transformation solution from **step 3**. Each 6 well plate may be used for three individual transformations.

5. Centrifuge the plate at $545 \times g$ for 1 h at room temperature in a clinical centrifuge. The start of the centrifugation step is time zero (T_0). This step also represents the first infection indicated as passage 0 (P_0).

6. Remove the transformation medium and add 2 ml of pre-warmed (37 °C) DMEM, 10% FBS and incubate the cells as indicated in **step 1** in Subheading 3.2 (used for all cell culture incubation steps).

7. At 12 h post infection, remove the medium and add 2 ml of pre-warmed medium with antibiotic per well (DMEM, 10% FBS, 5 µg/ml ampicillin [for pDFTT3-series plasmids] or 500 µg/ml spectinomycin [for pDFTT3*aadA*-series plasmids]) and return the cells to the incubator.

8. Harvest and passage of transformants:

(a) At 40–48 h post-infection (*see* **Note 13**), add sterile 3 mm glass beads to cover ~1/3 of the well and rock the plate to dislodge the cells.

(b) Transfer the medium to a 50 ml conical tube, pooling identical transformants (4 ml total, 2 ml from each well).

(c) Add 1 ml of ice cold DMEM to each well, rock the plate to wash the well, and transfer the medium to the conical tube from **step 8b**. The total volume will be ~6 ml.

(d) Add glass beads to fill the cone of the conical tube and lyse the cells via continuous vortexing on high speed for 2 min.

(e) Centrifuge the samples at $173 \times g$ for 5 min at room temperature in a clinical centrifuge to pellet unlysed cells and large debris (*see* **Note 14**).

(f) Transfer 2×3 ml of the **step 8e** supernatant to confluent monolayers of HeLa cells grown in a 6 well tissue culture plate and infect via centrifugation (*see* **step 5** in Subheading 3.2). After centrifugation, remove the medium and add 2 ml of pre-warmed infection medium per well: DMEM, 10% FBS, 5 µg/ml ampicillin (for pDFTT3-series plasmids) or 500 µg/ml spectinomycin (for pDFTT3*a*-*adA*-series plasmids), 1 µg/ml cycloheximide, and 1× NEAA. Incubate the cells for 40–48 h. This infection step constitutes P_1.

9. Repeat **steps 8a–8f** through P_4 (*see* **Note 15**). If no transformants are present at the end of P_4, the experiment is terminated (*see* **Note 16**). Transformants present at P_4 are processed for molecular analysis and isolation of clones.

10. P_4 samples are harvested using **steps 8a–8e**. After **step 8e**, transfer the supernatant to a new 50 ml conical tube and centrifuge at $13,000 \times g$ for 10 min at 4 °C in a fixed angle centrifuge rotor. Discard the supernatant using a serological pipet and add 1 ml of SPG. Suspend the pellet using a vortexer followed by brief sonication (~5 s) using a flame sterilized probe and 20% sonication amplitude. Store the sample at –80 °C in 200 µl aliquots.

11. Use one aliquot to obtain genomic chlamydial DNA for PCR analysis of the mutant locus (*see* **Note 17**). Recommended PCR reactions include: (1) an intron specific PCR reaction, (2) a target gene specific PCR reaction, and (3) a linkage reaction using target gene and intron-specific primers to assess intron orientation (*see* Fig. 4).

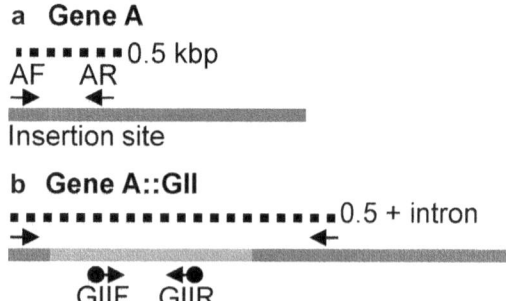

Fig. 4 PCR verification of mutants. Initial assessment of mutants is performed using PCR and primers specific to the target gene and the intron. Target gene primers (AF and AR) should flank the insertion and yield a small, but easily detectable PCR product (**a**). Mutants will produce an enlarged PCR product when assessed with the gene specific primers due to intron insertion (**b**). Intron specific primers (GIIF and GIIR) can be used to confirm presence of the intron and can be used in combination with gene specific primers to assess intron orientation. For example, GIIF and AR would yield a product for a sense insertion while GIIF and AF would yield a product for an antisense insertion. The mutant PCR product (from AF and AR in **b**) should also be sequenced to verify the exact intron-insertion location

12. Once the mutant has been verified, clones should be isolated using the plaque assay or limiting dilution (*see* **Note 18**). Clones should be re-verified using locus-specific PCR, DNA sequencing of the disrupted locus, and western blotting analysis to ensure absence of protein production.

4 Notes

1. Any cell line supporting growth of *C. trachomatis* should, in theory, work for transformation and mutagenesis. We have successfully used both HeLa cells and a mouse L2 fibroblast cell line for mutagenesis ([7, 15, 17 and unpublished results]). As the transformation protocol uses confluent cell monolayers for infection, adherent cells would be preferred over cells that grow in suspension.

2. While we see no reason that intron-based mutagenesis would not work with other *C. trachomatis* serovars and *Chlamydia* species, we have not tested the method outside of serovar L2. Note that *C. trachomatis* is a BSL2 pathogen and should be handled accordingly with appropriate laboratory safe guards and training. Consult the BMBL handbook for detailed information [18]. The appropriate institutional safety office also should be contacted regarding pathogen handling and for approval of recombinant DNA work.

3. We have used both *E. coli* DH5α and XL-1 Blue for cloning and plasmid maintenance. There appears to be a problem with stability of the plasmid in both strains which manifests itself as a plasmid that runs too high in agarose gel electrophoresis. The normal, supercoiled plasmid runs at approximately 6 kb. Occasionally, plasmid preparations will result in a plasmid that runs at >10 kb. These plasmid preparations will not work for mutagenesis or cloning procedures as they often fail to digest with *Bsr*GI and/or *Hind*III. Growing transformants at 30 °C rather than the commonly used 37 °C appears to reduce the chance of obtaining an altered plasmid.

4. Due to NIH restrictions, the *bla*-based intron in pDFTT3 should not be used with *C. trachomatis* serovars D–K as amoxicillin is still prescribed as an alternative therapy for pregnant women with chlamydial sexually transmitted infections. Use of the *aadA*-based intron in pDFTT3*aadA* is permissible for all serovars/species. Plasmid maps and sequences have been published [7, 15].

5. The Jumpstart REDTaq ReadyMix comes with the TargeTron Gene Knockout System kit available from Sigma-Aldrich (catalog number TA0100). The kit also includes the EBS universal primer, template DNA, restriction buffer, *Bsr*GI, and *Hind*III

and either three or ten accesses to the insertion prediction algorithm. The DNA polymerase may be substituted with other high-fidelity polymerases such as Phusion High Fidelity polymerase from Thermo Scientific. We also routinely use *Bsr*GI and *Hin*dIII from alternative suppliers and order the EBS universal primer from Integrated DNA Technologies.

6. The sequences for the primers required for intron retargeting (IBS, EBS2, and EBS1d) are provided by the algorithm run by either Sigma-Aldrich (http://www.sigmaaldrich.com/life-science/functional-genomics-and-rnai/targetron.html) or Targetronics, LLC (http://www.targetrons.com/).

7. To obtain targeting information, the DNA sequence of the target gene is submitted to the chosen company for analysis. We have only used Sigma-Aldrich for predictions. When selecting target genes keep in mind two primary weaknesses associated with insertional mutagenesis: (1) Intron-insertion disrupts, but does not delete, the target gene and consequently truncated protein products may still be produced from the targeted gene. Choosing an insertion site as close as possible to the 5′ end of the gene reduces the chance of producing a functional protein, and (2) Insertion mutations are prone to polar effects. Gene complementation of mutants is critical for concluding that mutant phenotypes are due to the gene of interest and not due to polar effects on genes surrounding the targeted gene.

8. The algorithm will return a list of predicted insertion sites (ranked based on predicted insertion efficiency) and the orientation of intron insertion. The intron can insert in either a sense or antisense configuration, referencing the location of the 5′ intron end in relation to the 5′ end of the gene. We choose two insertion sites per gene for vector construction. An example algorithm result list is provided in Fig. 5.

9. As an alternative to constructing the modified 5′ intron region using PCR, one can design the modified 5′ region in silico for synthetic construction as a gBlock (Integrated DNA Technologies). An example of the gBlock approach can be found at: http://www.targetrons.com/lacZ-LtrB-results.txt.

10. Electrocompetent *E. coli* can be substituted for chemically competent *E. coli*. Competent *E. coli* may be purchased or prepared (*see* ref. 19 for details).

11. Multiple 6-well plates can be seeded at lower densities concurrently with the plate destined for use at P_0 to allow for immediate passage of transformants following harvests. Cells should be seeded to reach confluency at 2-day intervals to coincide with harvesting of later passaged transformants.

12. *C. trachomatis* EB stocks from either crude preparations or density-dependent gradient preparations may be used for

a Insertion Location

	Gene 5`exon	+1 Gene 3`-exon	Score	E-value
366|367s	TAAGCAAGCTGGTAGCGAA -intron-	GATCTTCTT	10.54	0.013
196|197a	TCAGAAATTTCTTGAATAC -intron-	GTTTTTCTC	6.88	0.250
260|261a	TAGAGCATTAAACTCATCC -intron-	ATTTTTTTT	5.91	0.448

b Primers for retargeting the Ll.LtrB intron (used with EBS universal)

366|367s-IBS   AAAAAAGCTTATAATTATCCTTA**GCTGGC**A**GCGAA**GTGCGCCCA
GATAGGGTG

366|367s-EBS1d CAGAT<u>TGTAC</u>AAATGTGGTGATAACAGATAAGTC**AGCGAAGA**TAA
CTTACCTTTCTTTGT

366|367s-EBS2   TGAACGCAAGTTTCTAATTTCG**A**T**TCCAGCT**CGATAGAGGAAAGT
GTCT

196|197a-IBS   AAAAAAGCTTATAATTATCCTTA**TTTCTC**GAATAC**GTGCGCCCA
GATAGGGTG

196|197a-EBS1d CAGAT<u>TGTAC</u>AAATGTGGTGATAACAGATAAGTC**GAATACGT**TAA
CTTACCTTTCTTTGT

196|197a-EBS2   TGAACGCAAGTTTCTAATTTCG**GT TAGAAAT**CGATAGAGGAAAGT
GTCT

260|261a-IBS   AAAAAAGCTTATAATTATCCTTA**TTAAAC**TCATCC**GTGCGCCCA
GATAGGGTG

260|261a-EBS1d CAGAT<u>TGTAC</u>AAATGTGGTGATAACAGATAAGTC**TCATCCAT**TAA
CTTACCTTTCTTTGT

260|261a-EBS2   TGAACGCAAGTTTCTAATTTCG**GT TTTTAAT**CGATAGAGGAAAGT
GTCT

Fig. 5 Example algorithm result list. The target gene is entered into the algorithm and a list of insertion sites ranked by predicted insertion efficiency (Score and E-value) are provided (**a**). The insert location provided is based on the open reading frame with the A in the ATG designated as position 1. The lowercase a and s behind the insertion locations indicate the orientation of the intron relative to the gene. The +1 position in *red* indicates the base that should be present in the intron-donor vector at the d + 1 position (*see* Fig. 4) for optimal forward splicing of the intron from the donor intron transcript. This position could be altered in the donor vector using site-directed mutagenesis. The primers required for the PCR reactions used to target the intron to each insertion site are listed in (**b**). The *Hind*III restriction site in the IBS primers and the *Bsr*GI restriction site in the EBS1d primers are *underlined*. Bases in the intron that will be mutated through PCR are in *bold* (shown with *asterisks* in Fig. 1a)

transformation (*see* ref. 20 for details on maintenance of *Chlamydia* and stock preparation). We have not noticed differences in transformation efficiencies between EB stocks prepared using the two alternative methods. EB stock titers can be determined using the inclusion forming unit assay [20].

13. If the desired mutation is predicted to affect the developmental cycle [21], the harvest time can be altered. For example, a defect delaying the developmental cycle would increase the growth period required before harvesting.

14. This step is a potential stopping point. If you cannot proceed with the next infection (**step 8f**), the supernatant from **step 8e**

should be transferred to a new 50 ml conical tube and centrifuged at $13,000 \times g$ for 10 min at 4 °C. The pellet, which contains the EBs, should be suspended in 1 ml of SPG and frozen at −80 °C until cells are ready to infect.

15. It is important to monitor the cells throughout each passage. Cells will appear heavily infected at P_0 (apparent $MOI \geq 1$) with inclusions containing aberrant RBs (ampicillin treated) or inclusions containing EBs lacking wild-type "movement" (spectinomycin treated). The number of infected cells in the monolayer should decrease from P_0 to P_1 and may not be apparent or very few in number at P_2. The majority of inclusions present from P_2 through P_4 should no longer carry the drug-sensitive phenotypes discussed for P_0. If cells are heavily infected at P_3 (an apparent $MOI > 0.1$), the P_4 infection should be performed with differing amounts of supernatant to prevent superinfection and subsequent premature death of the HeLa cells. We also recommend saving some material from P_3 in case P_4 results in premature cell death.

16. We target two different insertion sites per gene and make three attempts to obtain mutants. Failure to obtain mutants could indicate that the targeted gene is essential, that polar effects inactivated an essential gene, or that intron targeting was unsuccessful. We recommend becoming proficient with chlamydial transformation and performing control transformations with each experiment to rule out mutagenesis failure due to issues with transformation.

17. We use the Qiagen DNeasy Blood and Tissue Kit for purification of chlamydial genomic DNA using the cultured cell protocol. Other methods for purification of genomic DNA from bacteria should also be suitable.

18. For information on clone isolation using plaque assay or limiting dilution *see* refs. [22] and [23], respectively.

Acknowledgments

This work was supported by NIH grants 1R21AI115238-01 and 1R15AI109566-01A1 to DJF. We thank Anna Hooppaw and Jae Claywell for reviewing our manuscript.

References

1. Horn M (2008) *Chlamydiae* as symbionts in eukaryotes. Annu Rev Microbiol 62:113–131. doi:10.1146/annurev.micro.62.081307.162818

2. Campbell LA, Rosenfeld ME (2015) Infection and atherosclerosis development. Arch Med Res 46(5):339–350. doi:10.1016/j.arcmed.2015.05.006

3. Darville T (2013) Recognition and treatment of chlamydial infections from birth to adolescence. Adv Exp Med Biol 764:109–122

4. Hu VH, Holland MJ, Burton MJ (2013) Trachoma: protective and pathogenic ocular immune responses to *Chlamydia trachomatis*. PLoS Negl Trop Dis 7(2):e2020. doi:10.1371/journal.pntd.0002020, PNTD-D-12-00752 [pii]

5. Hooppaw AJ, Fisher DJ (2015) A coming of age story: *Chlamydia* in the post-genetic era. Infect Immun. doi:10.1128/IAI.01186-15

6. Wang Y, Kahane S, Cutcliffe LT, Skilton RJ, Lambden PR, Clarke IN (2011) Development of a transformation system for *Chlamydia trachomatis*: restoration of glycogen biosynthesis by acquisition of a plasmid shuttle vector. PLoS Pathog 7(9):e1002258. doi:10.1371/journal.ppat.1002258, PPATHOGENS-D-11-00473 [pii]

7. Johnson CM, Fisher DJ (2013) Site-specific, insertional inactivation of *incA* in *Chlamydia trachomatis* using a group II intron. PLoS One 8(12):e83989. doi:10.1371/journal.pone.0083989

8. Kari L, Goheen MM, Randall LB, Taylor LD, Carlson JH, Whitmire WM, Virok D, Rajaram K, Endresz V, McClarty G, Nelson DE, Caldwell HD (2011) Generation of targeted *Chlamydia trachomatis* null mutants. Proc Natl Acad Sci U S A 108(17):7189–7193. doi:10.1073/pnas.11022291081102229108

9. Kokes M, Dunn JD, Granek JA, Nguyen BD, Barker JR, Valdivia RH, Bastidas RJ (2015) Integrating chemical mutagenesis and whole-genome sequencing as a platform for forward and reverse genetic analysis of *Chlamydia*. Cell Host Microbe 17(5):716–725. doi:10.1016/j.chom.2015.03.014S1931-3128(15)00131-6

10. Nguyen BD, Valdivia RH (2012) Virulence determinants in the obligate intracellular pathogen *Chlamydia trachomatis* revealed by forward genetic approaches. Proc Natl Acad Sci U S A 109(4):1263–1268. doi:10.1073/pnas.1117884109, 1117884109 [pii]

11. Zhong J, Karberg M, Lambowitz AM (2003) Targeted and random bacterial gene disruption using a group II intron (targetron) vector containing a retrotransposition-activated selectable marker. Nucleic Acids Res 31(6):1656–1664

12. Lambowitz AM, Zimmerly S (2011) Group II introns: mobile ribozymes that invade DNA. Cold Spring Harb Perspect Biol 8:a003616. doi:10.1101/cshperspect.a003616, a003616 [pii] cshperspect.a003616 [pii]

13. Enyeart PJ, Chirieleison SM, Dao MN, Perutka J, Quandt EM, Yao J, Whitt JT, Keatinge-Clay AT, Lambowitz AM, Ellington AD (2013) Generalized bacterial genome editing using mobile group II introns and Cre-lox. Mol Syst Biol 9:685. doi:10.1038/msb.2013.41, msb201341 [pii]

14. Enyeart PJ, Mohr G, Ellington AD, Lambowitz AM (2014) Biotechnological applications of mobile group II introns and their reverse transcriptases: gene targeting, RNA-seq, and non-coding RNA analysis. Mob DNA 5(1):2. doi:10.1186/1759-8753-5-2, 1759-8753-5-2 [pii]

15. Lowden NM, Yeruva L, Johnson CM, Bowlin AK, Fisher DJ (2015) Use of aminoglycoside 3′ adenyltransferase as a selection marker for *Chlamydia trachomatis* intron-mutagenesis and *in vivo* intron stability. BMC Res Notes 8:570. doi:10.1186/s13104-015-1542-9

16. Thomson NR, Holden MT, Carder C, Lennard N, Lockey SJ, Marsh P, Skipp P, O'Connor CD, Goodhead I, Norbertczak H, Harris B, Ormond D, Rance R, Quail MA, Parkhill J, Stephens RS (2008) *Chlamydia trachomatis*: genome sequence analysis of lymphogranuloma venereum isolates. Genome Res 18(1):161–171. doi:10.1101/gr.7020108, gr.7020108 [pii]

17. Thompson CC, Griffiths C, Nicod SS, Lowden NM, Wigneshweraraj S, Fisher DJ, McClure MO (2015) The Rsb phosphoregulatory network controls availability of the primary sigma factor in *Chlamydia trachomatis* and influences the kinetics of growth and development. PLoS Pathog 11(8):e1005125. doi:10.1371/journal.ppat.1005125

18. Chosewood LC, Wilson DE (2009) Centers for Disease Control and Prevention (U.S.), National Institutes of Health (U.S.). Biosafety in microbiological and biomedical laboratories, vol 21-1112, 5th edn. U.S. Dept. of Health and Human Services, Public Health Service, Centers for Disease Control and Prevention, National Institutes of Health, Washington, DC

19. Green MR, Sambrook J (2012) Molecular cloning: a laboratory manual, 4th edn. Cold Spring Harbor Laboratory Press, Cold Spring Harbor, NY

20. Scidmore MA (2005) Cultivation and laboratory maintenance of *Chlamydia trachomatis*. Curr Protoc Microbiol Chapter 11:Unit 11A 1. doi:10.1002/9780471729259.mc11a01s00

21. Abdelrahman YM, Belland RJ (2005) The chlamydial developmental cycle. FEMS Microbiol Rev 29(5):949–959

22. Banks J, Eddie B, Schachter J, Meyer KF (1970) Plaque formation by *Chlamydia* in L cells. Infect Immun 1(3):259–262

23. Mueller KE, Fields KA (2015) Application of beta-lactamase reporter fusions as an indicator of effector protein secretion during infections with the obligate intracellular pathogen *Chlamydia trachomatis*. PLoS One 10(8):e0135295. doi:10.1371/journal.pone.0135295

Section III

Bioinformatics Approaches for Identifying and Analyzing Mutagenesis Targets

Chapter 12

In Silico Approaches to Identify Mutagenesis Targets to Probe and Alter Protein–Cofactor and Protein–Protein Functional Relationships

Brian A. Dow, Esha Sehanobish, and Victor L. Davidson

Abstract

When performing site-directed mutagenesis experiments to study protein structure–function relationships, ideally one would know the structure of the protein under study. It is also very useful to have structures of multiple related proteins in order to determine whether or not particular amino acid residues are conserved in the structures either in the active site of an enzyme at the surface of a protein or at a putative protein–protein interface. While many protein structures are available in the Protein Data Base (PDB), a structure of the protein of interest may not be available. In the study of reversible and often transient protein–protein interactions it is rare to have a structure of the complex of the two interacting proteins. In this chapter, methods are described for comparing protein structures, generating putative structures of proteins with homology models based on the protein primary sequence, and generating docking models to predict interaction sites between proteins and cofactor–protein interactions. The rationale used to predict mutagenesis targets from these structures and models is also described.

Key words Homology model, Ligand docking, Protein Data Bank (PDB), Protein docking, Structural alignment

1 Introduction

Many enzymes are initially synthesized as inactive apoenzymes which require an exogenous cofactor to form the active holoenzyme. Cofactors are organic molecules which are often derived from vitamins, metal ions, or organometallic molecules. Some cofactors are tightly and permanently bound to the enzyme while others are dissociable. In either case, specific interactions between amino acid residues of the protein and the cofactor are critical for cofactor binding and for modulating the biological activities of the protein-bound cofactor. An ideal approach to obtain insight into exactly how the protein environment of the cofactor and specific interactions with amino acid residues influence the cofactor function is to perform site-directed mutagenesis of specific amino acid

Andrew Reeves (ed.), *In Vitro Mutagenesis: Methods and Protocols*, Methods in Molecular Biology, vol. 1498,
DOI 10.1007/978-1-4939-6472-7_12, © Springer Science+Business Media New York 2017

residues in the proximity of the cofactor. Such studies can elucidate how the identity and physical properties of particular residues influence cofactor binding and activity.

Specific protein–protein interactions are critical in biology. In order to perform their biological and metabolic functions certain proteins must associate, sometimes tightly and sometimes transiently. During energy metabolism redox proteins transfer electrons to each other. Such interprotein electron transfer requires protein–protein interaction which must be specific to prevent potentially reactive electrons from being misdirected and thus waste energy and damage the protein. Other important protein–protein interactions include binding of soluble proteins to membrane-associated protein receptors during cell signaling which regulates biological activities. Thus, an ideal approach to obtain insight into exactly how specific protein–protein interactions are stabilized is to perform site-directed mutagenesis of specific amino acid residues that are present at the protein–protein interface in order to see how the identity and physical properties of a particular residue influence the affinity of the protein–protein interaction. If the structure of the protein–protein complex is not known, then using site-directed mutagenesis to alter residues that are predicted to be important for the protein–protein interaction can provide information that validates the prediction.

In order to identify targets for site-directed mutagenesis one should know the primary sequence of the protein and the structure of the protein or protein complex. If structural information is not available it may be possible to construct models of individual proteins and protein–protein complexes in silico, in most cases using free online tools and information that is readily available in online databases. The general methods for accomplishing this are described below.

2   Materials

The methods that are described employ well-established and freely available online tools and databases. These include the Protein Data Bank (PDB), ExPASy which is the Swiss Institute for Bioinformatics Resource Portal, The National Center for Biotechnology Information (NCBI), PyMOL, and ZDOCK. Other related websites as well as several commercially available programs may also be used which provide similar functionality. As such, the only materials needed are a computer and Internet access.

3   Methods

3.1   Downloading Primary Protein Sequences

The primary sequences of known proteins and predicted sequences from genes of putative proteins may be obtained from the NCBI as well as several other databases. Sequences may be downloaded in a

FASTA format which uses the one-letter amino acid sequence codes and is used to input protein sequences in most programs that require sequence information.

3.2 Visualizing Protein Structures

1. Visualization of 3D structures of proteins requires a file which contains the 3D coordinates for each atom of the protein. Such files may be obtained from the Protein Data Base are called PDB files. Each protein structure in this data base has a PDB ID. To obtain a PDB file go to RCSB Protein Data Bank at www.rcsb.org/. If you know the PDB ID then simply enter that. If you do not, then enter the name of the protein and you may then choose from any of the structures of that protein which have been deposited. The protein structure may be visualized on this website but it is usually desirable to download the PDB file to use in another program. Simply follow the directions provided in the program for doing this, select "download Files", then "PDB File (text)."

2. Many commercial and free programs are available for visualizing and manipulating protein structures. A popular program which is often used to create figures of protein structures in publications is PyMOL (www.pymol.org). Another popular alternative is Swiss-PDB Viewer which is available from ExPASy (www.expasy.org). To illustrate this point, the FASTA sequence and two representations of the structure of the protein amicyanin are shown in Fig. 1 (*see* **Note 1**).

3.3 Constructing a Homology Model

If the structure of a protein has not yet been determined, it may be possible to construct a model of the protein structure using the primary sequence of that protein and the known structure of another protein, if there is sufficient similarity in sequence and structure of the two proteins. This section describes how this may be done using the Swiss Model program.

1. Obtain the FASTA sequence of the protein of interest as described in Subheading 3.1.

2. Go to www.swissmodel.expasy.org [1] and start "modelling." You will need to enter the FASTA sequence and then "Build model."

3. The program will perform an extensive search and provide a summary of all the results, showing the templates which have the most sequence similarity and generating homology models of the protein of interest.

4. For any model of interest, a file in PDB format may be downloaded. These models can then be visualized and manipulated using PyMOL or a similar program in the same manner that one would visualize and analyze the PBD file of a protein of known structure. An example of the structure of a homology model of a protein and the structure of the template protein from which it was derived is shown in Fig. 2.

a   **Amicyanin, PDB ID: 2OV0, FASTA sequence**

DKATIPSESPFAAAEVADGAIVVDIAKMKYETPELHVKVGDTVTWINREAMPH
NVHFVAGVLGEAALKGPMMKKEQAYSLTFTEAGTYDYHCTPHPFMRGKVVVE

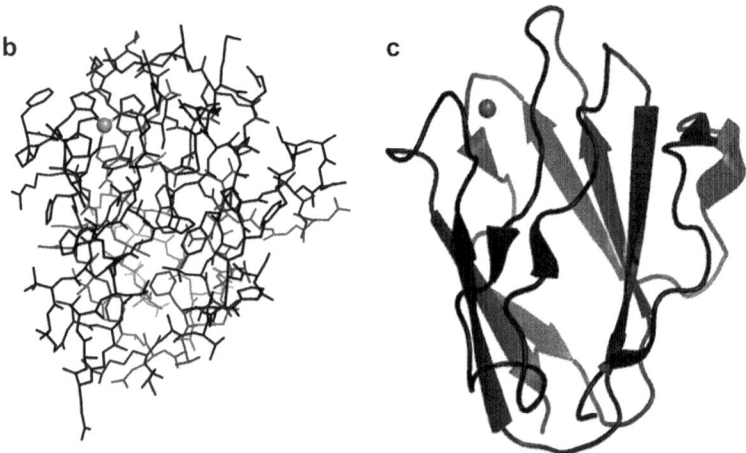

Fig. 1 FASTA sequence and structures of the protein amicyanin generated using PyMOL. (**a**) The FASTA sequence of amicyanin from *Paracoccus denitrificans* with its PDB ID. The sequence uses the standard one-letter symbols for the amino acids. (**b**) The structure is presented in "line" style which shows all covalent bonds between atoms in the structure, with the copper represented as a *sphere*. (**c**) The structure is presented in "cartoon" style which shows the secondary structure features of the proteins; *broad arrows* for beta-sheet broad ribbon for alpha helix and *lines* for unordered segments of the protein, with the copper represented as a *sphere*

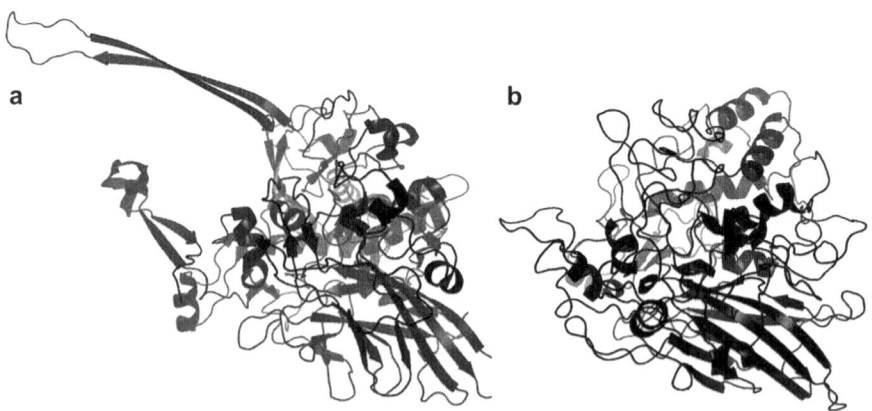

Fig. 2 Homology model of the protein GoxA constructed using the protein LodA as a template. (**a**) The structure of the lysine-ε-oxidase (LodA) from *Marinomonas mediterranea* presented in "cartoon" style which shows the secondary structure features of the proteins; *broad arrows* for beta-sheet broad ribbon for alpha helix and *lines* for unordered segments of the protein. (**b**) The homology model of the structure of the glycine oxidase (GoxA) from *M. mediterranea*, which was generated using the Swissmodel program from ExPASy. The structure of LodA (PDB ID: 2YMW) was used as a template structure for the FASTA sequence of GoxA (gene designation, Marme_1655) which was obtained from the NCBI. The sequence of GoxA exhibits 22.8 % identity and 34.6 % similarity to the sequence of LodA [8]

3.4 Comparing Structures to Identify Residues of Interest

A rational approach for identifying targets for site-directed mutagenesis is to compare the structures of similar proteins. One can then identify residues which are conserved structurally that may be important for function and residues that are not conserved that may be responsible for differences in function. The methods for overlaying and comparing the structures of two proteins using PyMOL are described. Alternative programs such as the RCSB Pairwise Structure Alignment program at www.rcsb.org/pdb/ workbench/workbench.do [2] could also be used for this purpose.

1. In PyMOL open the file of the first protein. In the same window go to "file" then "open" and open the file of the second protein.

2. Go to "action" and click on "align". This will align the proteins with respect to each another. The alignment can also be done based on certain residues if needed (*see* **Note 2**). Alignments using different methods can also be done via the command prompt. Examples of these methods include "super" and "cealign", which are preferred for aligning selections with low sequence similarity. The "super" method does a sequence-independent structure-based dynamic alignment followed by a series of refinement cycles to align individual atoms. "Cealign" aligns proteins by a sequence-independent, structural combinatorial extension alignment algorithm to best align the residues of the proteins [3].

3. To initiate the alignment process enter "super protein1, protein2" or "cealign protein1, protein2" into the PyMOL command prompt window, where "protein1" and "protein2" refer to the object names of the proteins within PyMol.

4. When the command has completed, the objects are automatically docked and visualized within the same window in PyMol. An example of the overlay of the structures of two related proteins and their active sites is shown in Fig. 3.

3.5 Constructing Protein–Protein Docking Models

There are many different software- and server-based utilities for predicting protein–protein interactions. Many of them are very similar in several aspects including the protocol for job submissions and the options for which molecular forces to take into account. However, they may differ in their specific parameters and algorithms for determining and ranking binding predictions. Below is a detailed method for submitting a job and analyzing the results for a protein–protein docking model using the ZDOCK at www. zdock.umassmed.edu [4].

1. Obtain the PDB IDs and files for the two proteins of interest.

2. On the ZDOCK site input the protein PDB IDs into the fields labeled "Input Protein 1" and "Input Protein 2." Alternatively, you may upload your own PDB files by selecting "PDB File" from the drop-down menu.

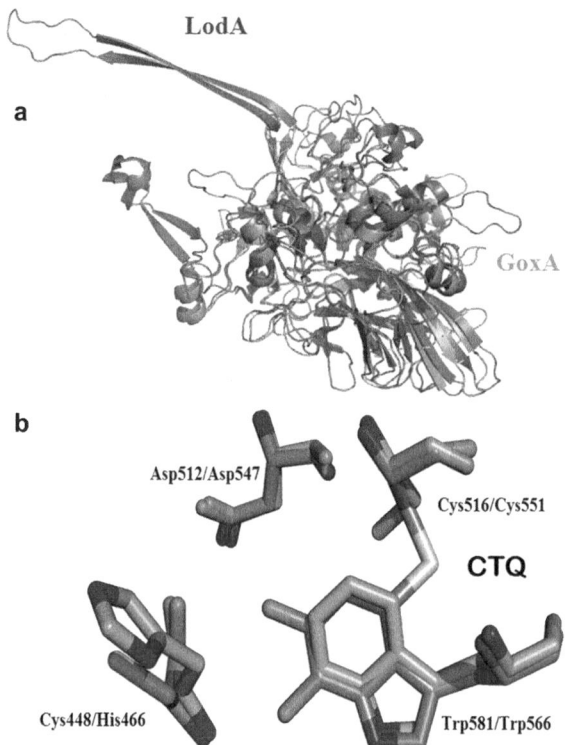

Fig. 3 Alignments of the structures of LodA and GoxA. (**a**) Alignment of the overall structure of LodA (*purple*) and homology model of GoxA (*orange*) which were shown individually in Fig. 2. (**b**) Alignment of the structures of the active sites of the enzymes which includes the cysteine tryptophylquinone (CTQ) cofactor. The amino acid resides in each protein are given with the residues for LodA on the left and the residues from GoxA on the *right*. The structures are colored by atom with oxygen *red*, nitrogen *blue*, and sulfur *yellow*. The carbon atoms and bonds to the carbon atoms are colored *purple* for LodA and *orange* for GoxA

3. After uploading one may select specific segments of the protein to use for the docking prediction. If you do not wish to select residues to use and/or exclude from the docking model, you may select the "skip residue selection" checkbox.

4. Click "submit."

5. If you did not select the "skip residue selection" checkbox, you may now select from the lists of residues which ones to exclude from the docking prediction and which ones to use as the putative binding site. Then click "ok" to confirm your selections.

6. A link to your results page, where you can visualize and download the predicted docking models will be sent to your email. On this page, you are presented the top five ranked predicted models. You may visualize any predicted model by entering its respective number in the "Get prediction number:" text box.

7. The top ten predictions as well as the entire ZDOCK output may be downloaded by clicking on the "Top Ten Predictions" link or the "ZDOCK Output" link, respectively.

8. The downloaded files are in PDB format and may be opened with any program that recognizes PDB files. A docking model predicting the interaction of two proteins is shown in Fig. 4.

3.6 Constructing Protein–Ligand Docking Models

There are many different software and server based utilities for predicting protein–ligand interactions. This would be useful in predicting protein–cofactor interactions, particularly for dissociable cofactors. It would also be useful in predicting interactions between enzyme substrates and cofactors and amino acid residues in the active site of the enzyme. Below is a detailed method for submitting a job and analyzing the results for a protein–ligand docking model using the SwissDock program at http://www.swissdock.ch/docking.

1. Obtain the PDB IDs and files for the two proteins of interest.

2. In SwissDock under the "Target Selection" heading, input the PDB ID and click "Search." A list of search results will appear.

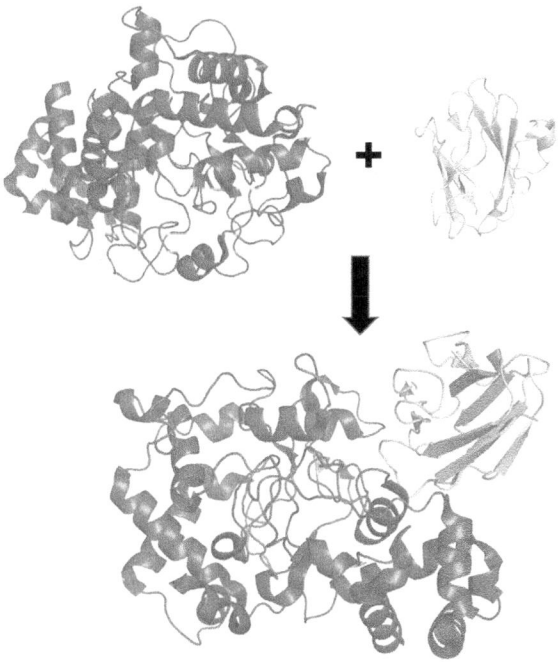

Fig. 4 Docking model of the interaction of two proteins, amicyanin and MauG. The structures of the individual proteins MauG (PDB ID: 3L4M, in *red*) and amicyanin (PDB ID: 2OVO in *purple* are shown on *top*). The best structural model of the complex of MauG and amicyanin that was predicted using the ZDOCK utility is shown on the *bottom*

Select the one you wish to use by clicking the green plus sign next to its name. Alternatively, you may upload a PDB file by selecting "Upload File."

3. Select the specific segments of the protein that you wish to use, if applicable.

4. Click "Select for Docking."

5. Under the "Ligand Selection" heading, input the ZINC ID, ligand name or category (like scaffolds or sidechains), or URL. ZINC is an online database of small molecule structures found at zinc.docking.org/ [5].

6. Click "Search" and a list of search results will appear. Select the one you wish to use by clicking the green plus sign next to its name. Alternatively, you may upload a MOL2 or CHARMM file containing the coordinates of the structure of the ligand by selecting "Upload File."

7. In the "Job Name" field, enter a title for your job.

8. From the "Docking Type" drop-down box, you may select "Accurate", "Fast", or "Very Fast." "Accurate" uses the most CPU time and "Very fast" uses the least CPU time (*see* **Note 3**).

9. The putative docking site(s) can be restricted to a local area by defining the coordinate of the origin as well as size of the local area in the "Definition of region of interest" section. The defined box should encompass the putative binding site(s). If you wish to perform a blind dock, leave these fields blank.

10. In the "Flexibility" field, you may choose the amount of flexibility of the protein that should be considered during docking. This is done by setting a distance (*see* **Note 4**).

11. Click "Start Docking" to begin your job. A results page will open, where your results will automatically display, once finished. This page will show the predicted binding sites and orientations as organized by cluster. A cluster is a distinct localized area where multiple binding orientations are predicted. Within each cluster, the results are organized first by FullFitness (kcal/mol) then by estimated ΔG (kcal/mol). FullFitness is a fitness function which first takes into account electrostatic and van der Waals interactions. Next, the solvent effect is considered, and the search space is adjusted accordingly [6, 7]. Overall, the predictions are ranked and organized by their respective free energies. Output PDB files may be downloaded from this page by clicking on the "Download your predictions" link at the bottom of the page (*see* **Note 5**).

3.7 Rationale for Selection of Amino Acid Residues on Which to Perform Site-Directed Mutagenesis

From inspection of known protein structures, homology models, comparison of structures of related proteins, and docking models it should be possible to identify specific amino acid residues which may be important for protein–cofactor and protein–protein interactions. A general approach for deciding which residues are most likely to be important is provided below.

1. *Identifying relevant electrostatic interactions.* Amino acid residues with acidic side-chains (aspartic acid and glutamic acid) and with basic side-chains (lysine, arginine, and histidine) may interact with oppositely charges segments of cofactors or substrates in an enzyme active site or oppositely charged residues at protein–protein interfaces. Conversion of such a residue to one which lacks charge (e.g., glutamic acid to glutamine) or one with opposite charge (e.g., glutamic acid to lysine) would allow one to test the hypothesis that the charge of the residue was important for function.

2. *Identifying relevant hydrogen bonding interactions.* Amino acid residues with polar side-chains that contain carboxylic, amino, amide, hydroxyl and thiol moieties may serve as hydrogen bond donors or acceptors for segments of cofactors or substrates in an enzyme active site or other polar residues at protein–protein interfaces. Conversion of such a residue to one which is unable to participate in hydrogen bonding (e.g., serine to alanine) would allow one to test the hypothesis that the ability of a residue to form a hydrogen bond was important for function.

3. *Identifying relevant hydrophobic interactions.* Amino acid residues with nonpolar side-chains may participate in hydrophobic interactions with segments of cofactors or substrates in an enzyme active site or other nonpolar residues at protein–protein interfaces. Conversion of such a residue to one which is polar or charged (e.g., valine to serine or aspartic acid) would allow one to test the hypothesis that the ability to form hydrophobic interactions was important for function (*see* **Note 6**).

4   Notes

1. Other options in addition to "line" and "cartoon" for displaying the structure are available including "ribbon" and "stick" models. One can select any of these and decide which method of visualizing the structure is most suitable.

2. One can select specific atoms. For example, if the alignment is supposed to be based on the position of the cofactor, then one can select atoms of the cofactor and then click "align." In this way, the program would be asked to align only certain amino acid residues in the proximity of the cofactor to provide information specifically on protein–cofactor interactions.

3. The selected docking time determines the number of sampled binding modes, number of minimization steps to relax the ligand, and number of putative binding modes that are sent back to the user in the results. If the ligand has less than 15 rotatable bonds and/or is likely to fit exactly into a binding pocket of your target, the fast or very fast docking types are likely to be sufficient.

4. If the distance is set to 0 Å, the protein will be considered rigid. If the distance is set to 3 Å, the side chains that are in close contact with the ligand in its reference binding mode will be co-optimized during docking. If the distance is set to 5 Å, the side chains that are farther from the ligand in its reference binding mode will also be co-optimized during docking.

5. Predicted binding modes may also be visualized in the UCSF Chimera program (www.cgl.ucsf.edu/chimera).

6. In addition to these criteria for determining which residue to target for mutagenesis, one may also wish to target residues which are structurally conserved in similar structures, as this suggests that the identity of the residue may be important for function.

References

1. Arnold K, Bordoli L, Kopp J, Schwede T (2006) The SWISS-MODEL workspace: a web-based environment for protein structure homology modelling. Bioinformatics 22:195–201

2. Ye Y, Godzik A (2003) Flexible structure alignment by chaining aligned fragment pairs allowing twists. Bioinformatics 19(Suppl 2):246–255

3. Shindyalov IN, Bourne PE (1998) Protein structure alignment by incremental combinatorial extension (CE) of the optimal path. Protein Eng 11:739–747

4. Pierce BG, Wiehe K, Hwang H, Kim BH, Vreven T, Weng Z (2014) ZDOCK server: interactive docking prediction of protein-protein complexes and symmetric multimers. Bioinformatics 30:1771–1773

5. Irwin JJ, Sterling T, Mysinger MM, Bolstad ES, Coleman RG (2012) ZINC: a free tool to discover chemistry for biology. J Chem Inf Model 52:1757–1768

6. Grosdidier A, Zoete V, Michielin O (2011) SwissDock, a protein-small molecule docking web service based on EADock DSS. Nucleic Acids Res 39:W270–277

7. Grosdidier A, Zoete V, Michielin O (2011) Fast docking using the CHARMM force field with EADock DSS. J Comput Chem 32:2149–2159

8. Campillo-Brocal JC, Chacon-Verdu MD, Lucas-Elio P, Sanchez-Amat A (2015) Distribution in microbial genomes of genes similar to lodA and goxA which encode a novel family of quinoproteins with amino acid oxidase activity. Bmc Genomics 16:231

Chapter 13

In Silico Prediction of Deleteriousness for Nonsynonymous and Splice-Altering Single Nucleotide Variants in the Human Genome

Xueqiu Jian and Xiaoming Liu

Abstract

In silico prediction methods have increasingly been valuable and popular in molecular biology, especially in human genetics, for deleteriousness prediction to filter and prioritize huge amounts of DNA variation identified by sequencing human genomes. There is a rich collection of available methods developed upon different levels/aspects of knowledge about how DNA variations affect gene expression. Given the fact that their predictions are not always consistent or even opposite of what was expected, using consensus prediction or majority vote among these methods is preferred to trusting any single one. Because querying different databases for different methods is both tedious and time-consuming for such big data sets, one database integrating predictions from multiple databases can facilitate the process. In this chapter, we describe the general steps of obtaining comprehensive predictions and annotations for large numbers of variants from dbNSFP, the first and probably the most widely used database of its kind.

Key words dbNSFP, dbscSNV, Single nucleotide variant, Nonsynonymous, Splice site, In silico, Functional prediction, Database, Protocol

1 Introduction

In silico identification and prediction of deleteriousness of a DNA variant has become a rising field of study in molecular biology in recent decades. The application of in silico methods is potentially more valuable in genetic studies of human diseases in view of the enormous number of human genomic DNA base pairs (~3 billion). As the revolutionary next generation sequencing technologies are currently widely used, many more novel genetic variants have been identified as potential targets for disease susceptibility. A typical whole exome sequencing study can identify millions of variants in coding regions, which comprise ~1 % of the entire human genome. As the cost of sequencing continues to drop, large-scale whole genome sequencing becomes a reality, by which tens of millions of variants can be identified. Among these variants, only a

Andrew Reeves (ed.), *In Vitro Mutagenesis: Methods and Protocols*, Methods in Molecular Biology, vol. 1498,
DOI 10.1007/978-1-4939-6472-7_13, © Springer Science+Business Media New York 2017

small portion is believed to be disease-related. A widely acknowledged assumption is that not only should a pathogenic variant directly change the amino acid, disrupt normal splicing, or affect any regulatory element acting in any step of gene expression, etc., but these changes must also have a deleterious effect on protein functions. However, it is impractical to experimentally validate them one by one, prompting the development of other fast and economic methods for filtering and prioritizing these variants.

Dozens of in silico methods for predicting the deleteriousness of a variant are available, among which similar or different sources of information or their combination are utilized in the prediction model. In many cases, different methods may give opposite predictions for the same variant. Unfortunately, no consensus has been reached for the choice of the "gold standard" method because each has its own strengths and weaknesses. One common solution is to use consensus prediction or majority vote among these methods rather than to trust any single one. Because querying different databases for different methods is both tedious and time-consuming for such big data sets, one database integrating predictions from multiple databases is of great value. In this chapter, we take the first and probably the most widely used database of its kind, dbNSFP, as an example, to demonstrate the general steps of using in silico methods for identifying and predicting the deleteriousness of a DNA variant.

dbNSFP is a database of human nonsynonymous single nucleotide variants (nsSNVs) and their functional predictions and annotations [1–3]. The variants collected in its version 3.0 extended to all potential nsSNVs, splice-site SNVs (ssSNVs), and SNVs within splicing consensus regions (scSNVs). It compiles 18 functional prediction scores and 6 conservation scores, as well as other related information including allele frequencies observed in different large datasets, various gene IDs from different databases, functional descriptions of genes, gene expression and gene interaction information, etc. nsSNVs and ssSNVs are incorporated into dbNSFP, and scSNVs comprise an attached database, dbscSNV [4]. Table 1 summarizes some general information about the dbNSFP v3.0 and attached dbscSNV. Please refer to the dbNSFP website (https://sites.google.com/site/jpopgen/dbNSFP) for more details and the most up-to-date information.

Generally speaking, dbNSFP has three main characteristics from the viewpoint of end-users:

1. Comprehensive collection: all potential SNVs within the defined regions and most mainstream prediction scores are included, which most research needs can be fulfilled by searching a single database.

2. Ease of use: only very basic knowledge about using command line interface is sufficient to search the database, which can be mastered quickly even by those with very little programming skills.

Table 1
Summary of dbNSFP v3.0

Variant	Definition	Database	Number	Score
nsSNV	Exonic SNVs causing amino acid substitution (missense) or protein truncation (nonsense) by changing the codon	dbNSFP	80,622,428	SIFT [5], 2×PolyPhen2 [6], LRT [7], MutationTaster [8], MutationAssessor [9], FATHMM [10], PROVEAN [11], VEST3 [12], CADD [13], DANN [14], fathmm-MKL [15], MetaSVM [16], MetaLR [16], 4×fitCons [17], GERP++ [18], 2×phyloP [19], 2×phastCons [20], SiPhy [21]
ssSNV	SNVs at the first two and last two sites of an intron	dbNSFP	2,209,599	
scSNV	−3 to +8 at the 5′ splice site and −12 to +2 at the 3′ splice site (excluding ssSNVs)	dbscSNV	15,030,459	ada [4], rf [4]

Detailed information can be found in the website

3. Stand-alone database: internet connection is not necessary as long as the database is in the local machine, which end-users can take full control of usage without worrying about issues such as internet intermittence or web-server breakdown.

2 Materials

1. The database must be downloaded before use (~76 GB for version 3.0 unzipped files), so a local machine with internet connection is required or the database can be copied from a portable hard drive.

2. The database only supports a command line search, so the command line interface of the operating system used should be available (e.g., LINUX, OS X, Windows).

3. The search program is written in Java, so the Java Runtime Environment version 1.6 or higher must be installed.

3 Methods

1. *Download the database.* Go to https://sites.google.com/site/jpopgen/dbNSFP and download the current version of a zipped dbNSFP distribution file from the download link provided by

the developer. Unzip the file to a directory and the database is ready to use. Besides the database and the Java search program, the following files should also be found in the directory: readme file for the database, an instruction file for the search program, the source code of the Java search program, the license for using the source code, and sample input files for various types of format. For instructions on downloading the attached database, dbscSNV (*see* **Note 1**).

2. *Preparation of the input file*. The search program allows for quite a few types of input format (*see* Fig. 1). The most commonly used format defining a SNV is a four-column plain text each representing its chromosome number (1–22, X, Y, and M), genomic position (hg38 (default), hg19, and hg18), reference allele, and alternative allele, respectively (chr pos ref alt). More specifically, for an nsSNV its corresponding reference amino acid and alternative amino acid can be added (chr pos ref alt refAA altAA). More generally, alleles at a position do not need to be specified (chr pos) in which case all SNVs recorded in the database at the specified position will be outputted (*see* **Note 2**). For other supported custom formats *see* **Note 3**. Alternatively, standard VCF files can be used as input (*see* **Note 4**).

3. *Run the search program*. The main command is:

   java [search program name] -i [input file name] -o [output file name]

   Take the example input file tryhg38.in as an example, in the command line type:

   java search_dbNSFP30a –i tryhg38.in -o tryhg38.out

   Available options are listed in Table 2 (*see* **Note 5**).

4. *Read and interpret the output*. Use either a text editor (e.g., Notepad++) or command line (for large file) to look at the content of the dbNSFP output. The output file contains all nsSNVs and ssSNVs that have a match in the database, each of which occupies one line (*see* Fig. 2). The number of columns

```
22   15528159 A    G
22   15528159 A    C    M    L
22   15528161
Ensembl:ENST00000252835:M1K
Ensembl:ENSP00000252835:M1T
MT-ND1
Ensembl:ENSG00000198763
Ensembl:ENST00000361624
Ensembl:ENSP00000354876
Uniprot:NU5M_HUMAN
Uniprot:P00846
Entrez:4541
```

Fig. 1 Various input formats supported by dbNSFP

Table 2
Search program options. Examples can be found in the instruction file for the database

Option	Value	Description
-c	List of chromosome numbers to search	By default the program searches all chromosomes. Use this option to restrict the search to the chromosomes of interest. Multiple chromosome numbers are separately by commas
-i	Input file name	If the input file is not in the database directory, path to the file should be specified
-o	Output file name	If users want to write the output to other directory, path should be specified
-p	Not applicable	This option is for VCF input only. By default the output file does not contain any additional columns in the VCF file. Use this option to write all VCF columns
-s	scSNV output file name	By default the program does not search dbscSNV. Use this option to output entries with dbscSNV annotation to a separate file
-v	hg19 or hg18	By default human genome reference sequence version hg38 is used to interpret the genomic position. Use this option to specify other build versions matching the users' own data
-w	List of column numbers to output	By default all columns in the database are outputted. Use this option to specify the columns of interest. Column numbers can be found in the database readme file. Multiple column numbers are separately by commas. Hyphen can be used to specify a range of columns

#chr	pos(1-based)	ref	alt	aaref	aaalt	rs_dbSNP142	hg19_chr	hg19_pos(1 based)	hg18_chr	hg18_pos(1 based)	genename	cds_strand	refcodon	codonpos	codon_degeneracy	Ancestral_allele
22	15528159	A	C	M	L	.	14	19377594	14	18447594	OR11H1	+	ATG	1	0	a
22	15528159	A	G	M	V	.	14	19377594	14	18447594	OR11H1	+	ATG	1	0	a
22	15528160	T	A	M	K	.	14	19377595	14	18447595	OR11H1	+	ATG	2	0	t
22	15528160	T	C	M	T	.	14	19377595	14	18447595	OR11H1	+	ATG	2	0	t
22	15528161	G	A	M	I	.	14	19377596	14	18447596	OR11H1	+	ATG	3	0	g
22	15528161	G	C	M	I	.	14	19377596	14	18447596	OR11H1	+	ATG	3	0	g
22	15528161	G	T	M	I	.	14	19377596	14	18447596	OR11H1	+	ATG	3	0	g

Fig. 2 First several columns of an example dbNSFP output

in the output file is controlled by the option -**w** (*see* Table 2). If all columns are outputted, there will be 183 columns in the dbNSFP v3.0 output with all annotations and predictions in the database. All non-matches will be written to an **.err** file. If -**s** is used to search dbscSNV, a separate output file with 18 columns will be written containing all matched scSNVs. Description of each column can be found in the database readme file. For suggested usage of the output (*see* **Note 6**).

4   Notes

1. Download the zipped dbscSNV file from various sources provided by the developer and unzip the file to the same directory where the dbNSFP database files are saved. The size of unzipped dbscSNV file is ~2.5 GB.

2. The header line is optional in the input file. The search program queries the database to find records that match the input entries line by line. Therefore no records will match the header line and it will be written to the error file. Columns in the input file should be separated by whitespace (tab- or space-delimited). The search program is case-insensitive, so either uppercase or lowercase can be used to represent alleles or amino acids. For the four-column format (chr pos ref alt), each line should represent only one variant. Multiple alternative alleles for the same reference allele at a specific position (multi-allelic) shall be separated into multiple rows.

3. Besides the already mentioned custom format, users may specify an nsSNV based on amino acid change referring to an Ensembl transcript ID or Ensembl protein ID. Users may even query all potential nsSNVs for certain genes by specifying gene name (HGNC symbol), Entrez ID, Uniprot ID or accession number, or Ensembl gene ID or transcript ID. The database name is needed for Entrez, Uniprot, and Ensembl. Examples can be found in the sample input files.

4. A typical VCF (variant call format) file has the **.vcf** extension in its file name and is usually not generated by end users of the database. It contains several meta-information lines (starting with a ##), followed by one header line (starting with a #), and then many data lines. For more information about VCF please refer to http://www.1000genomes.org/wiki/Analysis/variant-call-format. The search program automatically recognizes the extension of the input file name and queries the database by the first, second, fourth, and fifth column of each data line (chr pos ref alt), respectively. In a VCF file, each data line contains information about a position instead of a variant. For this reason, one line may have more than one variant (multi-allelic), which requires preprocessing before annotation. Please refer to http://annovar.openbioinformatics.org/en/latest/articles/VCF/ for preprocessing methods.

5. The search program must be run in the directory (working directory) to which the database files were unzipped. If the input/output file is not in the working directory the path to the file should be specified in the command. When the input file is large (i.e., a VCF file), Java may report memory insufficiency and a larger memory is then needed. To achieve this, just add a Java option **–Xmx[memory value]** following the **java** command to specify the maximum memory allocation for Java.

6. Strategies for filtering and prioritizing SNVs using the output depend on different research needs. Basically one is able to filter SNVs according to: (a) user-defined minor allele frequency threshold in certain populations; (b) user-defined number of "deleterious" predictions among all methods outputted; and (c) gene expression, gene-by-gene interaction, disease association, or disease pathway information outputted.

References

1. Liu X, Jian X, Boerwinkle E (2011) dbNSFP: a lightweight database of human nonsynonymous SNPs and their functional predictions. Hum Mutat 32(8):894–899

2. Liu X, Jian X, Boerwinkle E (2013) dbNSFP v2.0: a database of human non-synonymous SNVs and their functional predictions and annotations. Hum Mutat 34(9):E2393–E2402

3. Liu X, Wu C, Li C, Boerwinkle E (2015) dbNSFP v3.0: a one-stop database of functional predictions and annotations for human nonsynonymous and splice site SNVs. Hum Mutat 37(3):235–241. doi:10.1002/humu.22932

4. Jian X, Boerwinkle E, Liu X (2014) In silico prediction of splice-altering single nucleotide variants in the human genome. Nucleic Acids Res 42:13534–13544

5. Ng PC, Henikoff S (2001) Predicting deleterious amino acid substitutions. Genome Res 11:863–874

6. Adzhubei IA, Schmidt S, Peshkin L, Ramensky VE, Gerasimova A, Bork P, Kondrashov AS, Sunyaev SR (2010) A method and server for predicting damaging missense mutations. Nat Methods 7:248–249

7. Chun S, Fay JC (2009) Identification of deleterious mutations within three human genomes. Genome Res 19:1553–1561

8. Schwarz JM, Cooper DN, Schuelke M, Seelow D (2014) MutationTaster2: mutation prediction for the deep-sequencing age. Nat Methods 11:361–362

9. Reva B, Antipin Y, Sander C (2011) Predicting the functional impact of protein mutations: application to cancer genomics. Nucleic Acids Res 39(17):e118

10. Shihab HA, Gough J, Cooper DN, Stenson PD, Barker GL, Edwards KJ, Day IN, Gaunt TR (2013) Predicting the functional, molecular, and phenotypic consequences of amino acid substitutions using hidden Markov models. Hum Mutat 34:57–65

11. Choi Y, Sims GE, Murphy S, Miller JR, Chan AP (2012) Predicting the functional effect of amino acid substitutions and indels. PLoS One 7(10):e46688

12. Carter H, Douville C, Stenson PD, Cooper DN, Karchin R (2013) Identifying Mendelian disease genes with the variant effect scoring tool. BMC Genomics 14(Suppl 3):3

13. Kircher M, Witten DM, Jain P, O'Roak BJ, Cooper GM, Shendure J (2014) A general framework for estimating the relative pathogenicity of human genetic variants. Nat Genet 46:310–315

14. Quang D, Chen Y, Xie X (2015) DANN: a deep learning approach for annotating the pathogenicity of genetic variants. Bioinformatics 31:761–763

15. Shihab HA, Rogers MF, Gough J, Mort M, Cooper DN, Day IN, Gaunt TR, Campbell C (2015) An integrative approach to predicting the functional effects of non-coding and coding sequence variation. Bioinformatics 31:1536–1543

16. Dong C, Wei P, Jian X, Gibbs R, Boerwinkle E, Wang K, Liu X (2015) Comparison and integration of deleteriousness prediction methods for nonsynonymous SNVs in whole exome sequencing studies. Hum Mol Gene 24:2125–2137

17. Gulko B, Hubisz MJ, Gronau I, Siepel A (2015) A method for calculating probabilities of fitness consequences for point mutations across the human genome. Nat Genet 47:276–283

18. Davydov EV, Goode DL, Sirota M, Cooper GM, Sidow A, Batzoglou S (2010) Identifying a high fraction of the human genome to be under selective constraint using GERP++. PLoS Comput Biol 6(12):e1001025

19. Pollard KS, Hubisz MJ, Rosenbloom KR, Siepel A (2010) Detection of non-neutral substitution rates on mammalian phylogenies. Genome Res 20:110–121

20. Siepel A, Bejerano G, Pedersen JS, Hinrichs AS, Hou M, Rosenbloom K, Clawson H, Spieth J, Hillier LW, Richards S, Weinstock GM, Wilson RK, Gibbs RA, Kent WJ, Miller W, Haussler D (2005) Evolutionarily conserved elements in vertebrate, insect, worm, and yeast genomes. Genome Res 15:1034–1050

21. Garber M, Guttman M, Clamp M, Zody MC, Friedman N, Xie X (2009) Identifying novel constrained elements by exploiting biased substitution patterns. Bioinformatics 25(12):i54–i62

Chapter 14

In Silico Methods for Analyzing Mutagenesis Targets

Troy C. Messina

Abstract

Molecular dynamics of complex biological and chemical systems is possible using personal computers due to increased computer performance and improved software design. Here we describe molecular dynamics methods using Not Another Molecular Dynamics (NAMD) and Visual Molecular Dynamics (VMD) programs that aid in understanding the structural effects a mutation has on a protein. We describe in silico methods for site-specific mutation to standard and phosphorylated amino acids. Molecular dynamics equilibrations are used to provide a means for measuring structural fluctuations. These fluctuations assist in defining a distance coordinate, or reaction coordinate, that is relevant to the function of the protein. Adaptive biasing force molecular dynamics are then demonstrated to evaluate the energy landscape, or potential of mean force, along the chosen reaction coordinate. The potential of mean force identifies variations of the predominant structures among mutants that may affect function.

Key words Molecular dynamics simulation, Adaptive biasing force, Mutagenesis

1 Introduction

Improvements in desktop computer and molecular dynamics software performance have made it feasible to run simulations of complex biological systems without the need for high-performance computing clusters (HPCC). While there are methods that necessitate an HPCC, in silico studies designed to support and/or validate wet laboratory experiments are possible within reasonable timeframes using desktop and laptop computers. In fact, some software developers have focused on making their software work well as a desktop application, e.g., NAMD and gromacs [1–3]. Many other software packages are available for molecular dynamics simulations. See, for example, https://en.wikipedia.org/wiki/List_of_software_for_molecular_mechanics_modeling. The methods that follow are specific to NAMD/VMD. NAMD is an acronym for Not (just) Another Molecular Dynamics program, and VMD stands for Visual Molecular Dynamics. This software was chosen for several reasons: (1) It is freely available for a variety of operating systems and simple to install; (2) It has very developed

Andrew Reeves (ed.), *In Vitro Mutagenesis: Methods and Protocols*, Methods in Molecular Biology, vol. 1498, DOI 10.1007/978-1-4939-6472-7_14, © Springer Science+Business Media New York 2017

tutorials, mailing list, and other user resources; and (3) It can be used at different levels of sophistication, e.g., from a graphical user interface, from the command line, and through scripting. VMD allows users to go from start to finish without extensive programming or command line interfacing. However, VMD and NAMD also offer experienced users a great amount of flexibility and opportunity for customization. All of the activities we describe here could be performed with other modeling packages and viewers if there are reasons to use them.

For protein systems of approximately 15–200 amino acids one can obtain details about structural and functional changes related to mutations in a timeframe of 1 week to 1 year. The actual timescale depends on computer, size of the molecular system, and the number of molecular system variations being studied. In most molecular dynamics software, standard and phosphorylated amino acids can be substituted site-specifically. Multiple mutations can be made by sequential mutagenesis within the software. Mutated molecules can be subsequently simulated using a few different strategies. Here we detail the methods for running molecular dynamics to obtain an initial guess for a reaction coordinate, i.e., a structural change relevant to the mutation. Then, steered dynamics are used to drive the protein along the reaction coordinate to determine whether a mutation results in free energy changes that would affect the function of the protein. The methodology presented here presents previously published methods in more detail and extends beyond those methods to obtain the free energy profile and subsequent structural analysis [4].

2   Materials

1. A computer running Windows®, Mac OS®, or Linux®. Minimal computer requirements change with time, and the reader should consult the software distributor's most recent requirements (http://www.ks.uiuc.edu/Development/Download/download.cgi?PackageName=NAMD). The NAMD/VMD software repository dates backward a few years such that older computers running, for example, Windows XP® can still be used. Computational speed will be significantly reduced for older computers. The methodology presented here assumes the reader is using Windows 7® or a more recent version of Windows®.

2. NAMD and VMD software available from the link above.

3. Protein Databank molecular structure (pdb) file for the molecular system of interest.

3   Methods

All methods presented here use a tryptophan zipper, beta hairpin peptide with Protein Databank (pdb) identifier 1LE0. This peptide consists of 12 amino acids. It simulates quickly and has serine and lysine terminal residues that make it amenable to mutagenesis work. It works well for becoming familiar with the software and mutagenesis methods.

The software should be installed as described on each machine to be used. VMD is installed through a typical Windows installer. By default, it will be located in the directory C:\Program Files\ University of Illinois\VMD. It will also appear in the Start menu list of programs. NAMD extracts as a folder containing the primary executable, namd2.exe. There are other executables, dynamic-link libraries (.dll), and a folder of libraries for performing various types of molecular dynamics in the NAMD folder. The folder will be titled something similar to namd_2.10_Win64-multicore. The actual title depends on the NAMD version (2.10 2.11, etc.) and computer being used (64-bit vs. 32-bit). The entire NAMD folder can be placed in the University of Illinois subdirectory.

NAMD runs via a command prompt or via VMD. If the Windows® PATH variable is not manually updated it is necessary to provide the full path to namd2.exe upon every run. The system PATH variable can be edited to make the namd2.exe executable from a command prompt or from within VMD without typing the full path to namd2.exe. To do this in Windows® 7, open Windows Explorer, right-click on "Computer" and select "Properties"; click "Advanced system settings"; click "Environment variables…"; Select "PATH" and click "Edit". At the end of the variable value, add the path to the namd2.exe, which should be similar to ";C:\Program Files\University of Illinois\NAMD_2.10_Win64-multicore\". The preceding semicolon is necessary for separating a previous software path in the variable. Earlier versions of Windows® have a slightly different method, and Linux and Mac OS should update the PATH variable during the installation process. Many of the introductory topics discussed here are also covered in Chiang [4] and the accompanying tutorial videos (https://goo.gl/R9YlFe).

A "working directory" should be created for storing all of the files associated with a particular protein. In Windows Explorer navigate to "Documents". Windows identifies this directory with the path C:\Users\username\Documents, where username is the login name of the current computer user. In this directory create a folder titled "NAMD-simulations". Next create a subfolder in NAMD-simulations titled with the pdb identifier. For example, in this methods chapter we use 1LE0, a tryptophan zipper peptide, as a model system. We thus have a working directory with the path C:\ Users\messinat\Documents\NAMD-simulations\1LE0. An image of the 1LE0 structure is shown in Fig. 1.

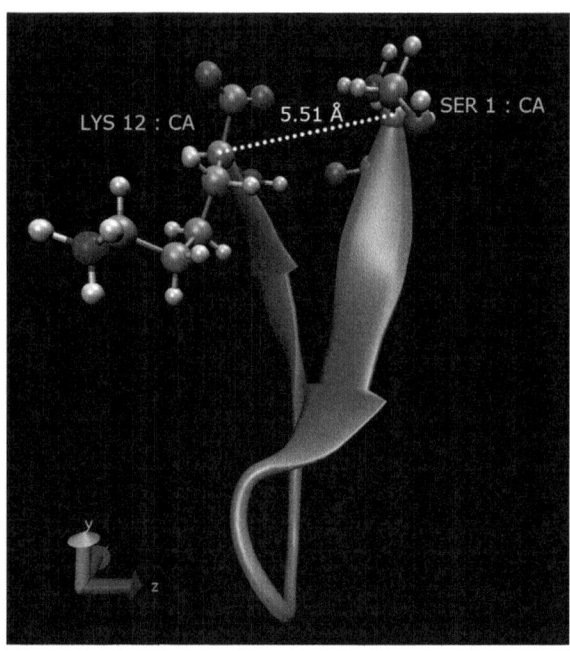

Fig. 1 The structure of the tryptophan zipper with PDB ID 1LE0. The structure is a beta-hairpin with a serine at the N-terminus and a lysine at the C-terminus. The peptide is 12 amino acids. The end-to-end distance between the terminal alpha carbons is shown to be 5.51 Å

NAMD does not include many of the force fields that are necessary for the molecular dynamics simulations. Force fields are defined in Charmm topology parameter files and define properties such as equilibrium bond lengths and angles, bond and angle force constants, and dihedrals. These files can be created by advanced users; however, a fairly extensive set of files can be obtained from the author's public repository at https://goo.gl/bpL9yz. Alternatively, these files can be obtained from the source, Alex MacKerrell, a the University of Maryland (http://mackerell. umaryland.edu/ [5]). Place the topology files in a subfolder titled "topology" (C:\Users\username\Documents\NAMD-simulations\ topology). It is advised for those new to molecular dynamics simulations to use a protein structure with no ligands, ATP, metal ions, stabilizing salts, etc. This is because the force fields for such molecules are not always readily available. A small peptide is advisable for the learning phase because test simulations will complete quickly. Working through this methodology with the 1LE0 structure to learn the process, then, switching to the protein system of interest is advised.

3.1 Save a Clean pdb (No Water, Salts, Ligands, etc.)

Open the VMD software, which should have a shortcut in the Windows® Start menu. Three windows will open (VMD Main, VMD OpenGL display, and vmd.exe terminal). In VMD Main, click the Extension menu and select TkConsole. This opens a terminal window. Many future actions will require opening files in the working directory. To simplify these actions we want to place VMD's reference to this folder. Do this by typing

`cd C:/Users/username/Documents/NAMD-simulations/pdbfoldername` where "`username`" should be the specific computer user login name and "`pdbfoldername`" should be the specific name of the pdb folder, e.g., 1LE0. *Note*: All console commands and scripting in TkConsole will be written in the `courier font`. Notice in TkConsole one uses forward slashes (`/`). In the command prompt or other Windows® environments, one uses backward slashes (\). This is a difference between programs that evolved from Unix (Tk, NAMD, VMD), and programs that evolved from DOS (Windows®). The use of these different slashes can be a cause of errors. Another issue that may arise is files being save to the incorrect folder. If files are not saved to the working directory as expected (*see* **Note 1**).

Return to the VMD Main window, and load your pdb file from the New Molecule menu. It should appear as lines color coded by atom type, e.g., carbon is cyan, oxygen is red, and nitrogen is blue. It is important to remove any solvent molecules that are part of the pdb structure because we will fully solvate the system later. To do this, return to TkConsole and enter

<code>set prot [atomselect top protein]</code>

`Set` is a command for creating a new variable. `Prot` is the name of the variable. `Atomselect` is a command for selecting a portion of a protein structure. `Top` makes the selection from the topmost structure, indicated by the "T" on the VMD Main window. Assuming only one structure is open in VMD, top will use that structure. If more than one structure is open, care must be exercised to select the structure of interest. Any open molecular structure can be set to top by double-clicking the T column of that structure in the VMD Main window. Alternatively, the structures in VMD are given ID numbers that can be seen in the VMD Main window, and these can be used in place of `top`. The command above will create a selection with the variable name `prot` that stores all of the pdb information for the protein only. If the protein has non-protein molecules that are important to the simulation, the above command will need to be modified to include these molecules by using Boolean operators. For example, an enzyme with ATP bound to it would be selected using the command

`set protATP [atomselect top "protein or resname ATP"].`

In this example, "or" adds either protein-related molecules or ATP molecules to the selection. A list of residue names in the loaded protein structure can be viewed. Click "Graphics → Representations" in the VMD Main window. In the "Graphical Representations" window click the "Selections" tab.

Then, under "Keywords", choose "resname". A list of residues, both standard amino acids and other molecules, will be listed in the bottom, right box (*see* **Note 2**).

The selection stored in the variable prot (or protATP) can be written to a new pdb file using the command

```
$prot writepdb 1LE0_prot.pdb,
```

where the $ indicates that the information stored in the variable prot is to be written to a pdb file titled 1LE0_prot.pdb. The file will be saved to the working directory. The original pdb molecule can be closed by clicking to select it in the VMD Main window and then select "Delete Molecule" from the "Molecule" menu.

3.2 Generating a Protein Structure File

Load the pdb file that was just saved containing only protein, 1LE0_prot.pdb. The protein structure data must be stored in a protein structure file (.psf) to inform NAMD about charges, bonds, angles, etc. The psf creation process also will add hydrogen atoms to the structure if not already present. The autopsf generator can be used to generate these files. In VMD Main, click "Extensions→ Modeling→Automatic PSF Builder". Manual generation of pdb and psf files is possible if necessary or desired (*see* the NAMD tutorial [6]). In the Automatic PSF Builder, the input molecule should default to the top open structure, 1LE0_prot. The output can be left as the default 1LE0_prot_autopsf. Click "Load input files". Next, click "Guess and split chains". VMD attempts to determine the number and identifiers for each segment, monomer, etc. It will list segments identified. For most pdb files, VMD is successful (*see* **Note 3**). Click "Create chains". A message box will appear prompting for patches. Click OK. Another message box will appear indicating pdb and psf files have been created. Click "OK". The new structures will automatically load into VMD. Hydrogen atoms will be visible on the structure as white lines. This is another indicator that the psf generation was successful.

3.3 Mutating the Protein

Open the Mutator from the VMD Main window by selecting "Extensions→Modeling→Mutate Residue." The PSF and PDB fields will automatically fill with the structure that was automatically loaded into VMD (1LE0_prot_autopsf). If this structure is not loaded, close all open molecules using the "Delete Molecule" menu item, and open the psf then pdb file. In VMD, one should open the psf file first. Then, open the pdb file. In VMD Main, only the psf file shows; however, the pdb file is loaded on top of the psf. The psf acts as supporting structural information. Return to the Mutator window and see that the structures are properly shown. The third box filled with "MUTATED" may be changed to the desired file output. As an example, we will mutate residue 12 from lysine to alanine in order to eliminate the positive charge on the lysine. Replace "MUTATED" with 1LE0_K12A to indicate the point mutation. The optional segment name may be left blank.

"ID target of residue" should be filled with the residue number 12. "Mutation (three letter residue name)" should be filled with ALA. Click "Run Mutator". The Mutator will generate the mutation in a pdb file and create a corresponding psf file. The mutated structure will be automatically loaded into VMD. If not, open it now, psf first. Check the mutation by opening the "Graphical Representations" window and creating a representation. In "Selected Atoms" enter "resid 12 and resname ALA". This selects residue 12 only if it has been mutated to alanine due to the logical and. The single amino acid showing in the OpenGL window indicates the mutation was successful. Any amino acid can be mutated in this way by entering the corresponding residue number and any of the 18 standard amino acid three letter identifiers available in VMD (*see* Table 1). Mutations from or to glycine are not recommended for novice users of VMD due to some issues with bond angles. Proline is not supported at all.

Table 1
Eighteen standard amino acids, three-letter identifiers, and single-letter identifiers available in the VMD Mutator without the need for advanced experience using VMD

Amino acid	Three-letter identifier	One-letter identifier
Alanine	ALA	A
Arginine	ARG	R
Asparagine	ASN	N
Aspartic acid	ASP	D
Cysteine	CYS	C
Glutamine	GLN	Q
Glutamic acid	GLU	D
Histidine	HIS	H
Isoleucine	ILE	I
Leucine	LEU	L
Lysine	LYS	K
Methionine	MET	M
Phenylalanine	PHE	F
Serine	SER	S
Threonine	THR	T
Tryptophan	TRP	W
Tyrosine	TYR	Y
Valine	VAL	V

The Mutator in VMD accepts only the three-letter identifiers

3.4 Mutating to Phosphorylated Amino Acids

It is possible to manually use patches available in topology files to make changes to nonstandard amino acids such as a phosphorylated amino acid. Patches SP2, THP2, and TP2 mutate serine, threonine, and tyrosine to phosphoserine, phosphothreonine, and phosphotyrosine, respectively. The 2 in the patch name indicates the patch results in a dianionic form for a charge of negative two on the phosphate. This charge corresponds to physiological pH. There are also monoanionic patches corresponding to low pH. These are invoked by replacing the 2 with 1. Below are the file contents for a script to mutate a serine at residue 1 to phosphoserine. The first two lines source the force field information from the topology directory. It is important these two files are in the appropriate directories. The first file "top_all27_prot_na.rtf" is in the topology folder. The second file "toppar_prot_na_all.str" is in a subfolder of topology titled "stream". The "../" indicates that the topology folder is a directory one level higher than the working directory, i.e., in the NAMD-simulations folder. The third line aliases atom types that may have a different atom identifier for isoleucine and glycine. The fifth line indicates the segment name "A" is created from the structural information of the pdb file indicated. The segment to be mutated is from a pdb file with the filename 1LE0_prot_autopsf.pdb. This filename should be edited to reflect the pdb file for the structure being mutated when working on a protein other than 1LE0. The seventh line indicates the patch SP2 to be applied to the new segment A on residue 1. The patch and residue number should be edited to reflect the appropriate amino acid being phosphorylated when working on a protein other than 1LE0. The eighth line recalculates bond and dihedral angles because the insertion of the phosphate likely created some structural changes that need reorienting. The coordinates are then read from the pdb file into segment A. Hydrogens are added if they are not present, and coordinates will be guessed for any unknown or added atoms, such as hydrogen. Output is written to pdb and psf files on the last two lines. These filenames should be edited to reflect the structure being studied when working on a protein other than 1LE0. *See* **Note 4** for troubleshooting. Copy the script below into a file in Wordpad. Save the file as phosphoserine.pgn in the working directory.

```
topology  ../topology/top_all27_prot_na.rtf
topology  ../topology/stream/toppar_prot_na_all.
str
alias atom ILE CD1 CD
alias atom GLY OXT OT1
segment A {pdb 1LE0_prot_autopsf.pdb
}
patch SP2 A:1
regenerate angles dihedrals
coordpdb 1LE0_prot_autopsf.pdb A
```

```
guesscoord
writepdb LE0_S1pS.pdb
writepsf LE0_S1pS.psf
```

In the TkConsole, enter the command

```
source phosphoserine.pgn
```

`Source` is a command to open, evaluate, and/or incorporate the contents of a script. In this case, the source command will run the script and save the mutated protein files to the working directory. Open the mutated files in VMD (psf first), and check that the phosphoserine mutation occurred. To check the mutation use the Graphical Representation window found in the Graphics menu to enter Selected Atoms "resid 1 and resname SER". Click "Apply". The phosphate should be apparent as a tan phosphorous atom and four, red oxygen atoms. The script above can be modified for the other phosphorylation patches by simply changing the patch from SP2 to TP2 or THP2 and selecting an appropriate residue that has tyrosine or threonine, respectively. The input and output filenames should be edited as indicated above.

3.5 Solvating the Protein

To most accurately simulate the protein behavior, one should solvate the protein structure in water. Solvation in explicit solvent significantly increases the computation time required to obtain results. Simulating proteins in vacuum will provide results that can be used to better understand the dynamic behavior; however, these simulations will result in protein structures that favor a more tightly folded or more closed configuration than a simulation in water. Proteins appear to be more rigid in vacuum, too. This is observed in the histogram of distance measurements that are described later in this chapter. The trade-off between time and accuracy is one that the practitioner must make. Implicit solvent modeling is also possible. These models use an average solvent dielectric. In practice, this implementation is as complex as explicit solvent; simulations take less time than explicit solvent; however, the results are less accurate. For the sake of accuracy, we detail methods for simulating in explicit water as a solvent.

Download the script file "wat_sphere.tcl" for solvating a structure in a sphere of water from https://goo.gl/4PTLUc. Save this file to the NAMD-simulations folder. In TkConsole enter `source ../wat_sphere.tcl`. This call to the `source` command causes the console to incorporate the procedures contained in the script, which is located one directory higher than the working directory. The contents of the wat_sphere.tcl script are not described in this methods chapter; however the original script is described at http://www.ks.uiuc.edu/Training/Tutorials/namd/namd-tutorial-unix-html/node7.html. Once the script is sourced, the procedure "addsphere" can be called to execute. The addsphere procedure requires one input parameter, which is the name of the psf and pdb files for the protein structure. For this script, the filenames of both the psf and pdb files must be the same except for the extensions .psf and .pdb. In TkConsole, enter the command addsphere 1LE0_prot_autopsf. Running the script will generate quite a bit of output to the console. When complete, the center of mass and radius of the sphere will appear as well as two lines reading "Autopsf: Updating structures". It is very important to store the center of mass and radius for use in upcoming simulation steps. Write the values in a laboratory notebook, copy them into a text file in the working directory, or do both. Keeping track of 4 or 5 significant digits is sufficient. Two new files are written to the working directory. They will have the original filename with "_ws" added to the end, which is short for water sphere. It is recommended that the experimenter open the new psf and pdb files to check that they are compatible with one another, i.e., no errors arise loading the pdb file onto the psf file in VMD. It is also a good practice to check all generated files to ensure each step goes as planned, especially due to the nature of this work using user-generated files to generate files in a subsequent step. In checking the files, make sure mutated residues maintain their three-dimensional structure and water solvation looks spherical. Another good practice is to create folders for each structure to be simulated. For example, if wild-type and mutant S1pS are to be simulated, create folders in NAMD-simulations titled "1LE0_ws" and "1LE0_S1pS_ws". Repeat this process for each mutant (wild-type, phosphorylated mutant, standard amino acid mutant, etc.). Put the solvated psf and pdb files corresponding to each mutant structure in their own folders. In the next step, we will be placing simulation configuration files in these folders. To see examples of folder contents and hierarchy, go to https://goo.gl/4TCxLU.

3.6 Molecular Dynamics

The experimenter must determine how many simulations to run. As with experiments, it is necessary to perform control simulations for comparison to mutated protein simulation. For the tryptophan zipper model used in this chapter, one may be interested in the rigidity of the hairpin. One measurement that might capture this rigidity is to measure the end-to-end distance between the N- and C- terminal amino acids, residues 1 (serine) and 12 (lysine). A set of simulations to determine the effects of mutagenesis on this distance might consist of the wild-type, an S1pS mutant, and a double mutant S1pS and K12A. Here, we only show results of simulating wild-type and S1pS. Other simulation experiments that novices could easily adapt from this tutorial include substrate binding site mutagenesis to measure structural changes such as collapse, expansion, or change in distance between docking residues, activation loop mutagenesis to determine loop flexibility or locking that may hinder or enhance activation.

With the solvation in a water sphere, molecular dynamics simulations must be run using configuration files and the command line (either TkConsole or the Windows® Command Prompt. An example configuration script follows. This script can be copied and pasted into a file saved as LE0_ws_eq.conf, or it can be downloaded from https://goo.gl/BnUs2s. The script should be saved to the directory 1LE0_ws. The pound (#) signs indicate comments in the script. The script is commented for readers to understand the components.

```
###################################################
#############
## JOB DESCRIPTION ##
###################################################
#############
# Minimization and Equilibration of
# 1LE0 wild-type in a Water Sphere at 310 K
###################################################
#############
## ADJUSTABLE PARAMETERS ##
###################################################
#############
# "./" indicate files in the same folder as this
script.
# The following 2 lines are the starting files (psf
and pdb).
# The filenames should be changed appropriately.
structure ./1LE0_prot_autopsf_ws.psf
coordinates ./1LE0_prot_autopsf_ws.pdb
# The following two lines set the temperature in
Kelvin and
# create a name for output files.
set temperature 310
set outputname LE0_ws_eq
# Start the simulation at zero. There is no need
to modify.
firsttimestep 0
###################################################
#############
## SIMULATION PARAMETERS ##
## Imports topologies. Sets force field cut-offs.
##
## For new users, this section should not be modi-
fied ##
## other than the parameters when simulating non-
standard ##
## amino acids. ##
###################################################
#############
# Input
```

```
# The "../" indicates files in a folder one
directory
#higher than this script's directory
paraTypeCharmm on
parameters ../topology/par_all27_prot_lipid.inp
temperature $temperature
# Force-Field Parameters
exclude scaled1-4
1-4scaling 1.0
cutoff 12.0
switching on
switchdist 10.0
pairlistdist 14.0
# Integrator Parameters
timestep 2.0;# 2 fs/step
rigidBonds all;# needed for 2 fs steps
nonbondedFreq 1
fullElectFrequency 2
stepspercycle 10
# Constant Temperature Control
langevin on;# do langevin dynamics
langevinDamping 1;# damping coefficient (gamma) of
1/ps
langevinTemp $temperature
langevinHydrogen off;# don't couple langevin bath
to hydrogens
# Output
outputName $outputname
restartfreq 500;# 500steps = every 1 ps if
timestep=2
dcdfreq 250
outputEnergies 100
outputPressure 100
#################################################
#############
## EXTRA PARAMETERS ##
#################################################
#############
# Spherical boundary conditions (sphericalBC)
# The center and r1 need to be changed to the val-
ues obtained during
# the water sphere solvation step.
# The last two lines of this section are force
parameters
# to keep the sphere intact. New users should
leave these two
# lines as the defaults, 10 & 2.
sphericalBC on
```

```
sphericalBCcenter -1.71360 0.93285 -0.37772
sphericalBCr1 12.24
sphericalBCk1 10
sphericalBCexp1 2
#####################################################
#############
## EXECUTION SCRIPT ##
#####################################################
#############
# Minimization to alleviate any stresses in the
system from
# mutation or solvation. Increase if analysis
using
# namdstats.tcl shows it is not minimized.
minimize 100
reinitvels $temperature
# 5,000,000 = 10 ns, each timestep is 2 fs.
run 5000000;# number of steps to run.
#####################################################
#############
## END CONFIGURATION SCRIPT ##
#####################################################
################
```

Each mutant will require a configuration file that indicates the appropriate input files. The lines to modify for input files are

```
structure ./1LE0_prot_autopsf_ws.psf
coordinates ./1LE0_prot_autopsf_ws.pdb
```

These should be changed to the appropriate psf and pdb filenames. As shown here, it is a good idea to have the same filename to keep track of which files correspond to one another. The temperature can be changed to any desired temperature in Kelvin. The output filenames can be modified to reflect the mutant being simulated. Because new psf and pdb files will not be generated during the simulation, the `set outputname` can be the same as the input `structure` and `coordinates` filenames.

```
set outputname LE0_ws_eq
```

The parameters line may need to be modified depending on the protein system. The line in the script above is for most protein systems with common amino acids. If a phosphorylated amino acid is in the structure the parameter files should be

```
parameters ../topology/par_all27_prot_na.inp
parameters ../topology/par_all27_prot_lipid.inp
parameters    ../topology/stream/toppar_prot_na_
all.str
```

If ATP is included, the configuration file will need the following parameters.

```
parameters     ../topology/stream/toppar_all27_na_
nad_ppi.str
```

It is important to make sure all of these parameter files are in the topology folder. Notice the files with .str file extension are in a subdirectory "stream" that is found within the topology directory. The extra parameters section contains information about the spherical water solvation. The `sphericalBCcenter` should contain the *x*, *y*, and *z* values saved from the TkConsole during the solvation step. The `sphericalBCr1` should contain the radius value saved from the TkConsole during the solvation step. In the last section labeled EXECUTION SCRIPT, the number of minimization and simulation steps can be modified. For small peptide systems, 1000 minimization steps is adequate. This may need to be increased to as high as 10,000 for larger proteins and enzymes. The minimization part of the procedure removes any unwanted energetic strains that may have been introduced during mutation or solvation. It will also help eliminate artifacts due to low temperature crystallization if that was the method used to obtain the original protein structure.

The configuration script can be run from TkConsole or the Windows Command Prompt. Because the TkConsole is already pointed to the working directory, it may be simpler to use it for running the molecular dynamics. However, the NAMD/VMD tutorials recommend the Command Prompt To run the configuration script, go to TkConsole or Command Prompt (pointed to the working directory via "cd" commands) and enter

```
namd2 LE0_ws_eq.conf > LE0_ws_eq.log
```

This command executes the NAMD program (namd2.exe) to run the configuration script titled "LE0_ws_eq.conf" and output all related execution information to a file in the working directory titled "LE0_ws_eq.log". When complete the cursor will return in TkConsole (*see* **Note 5** if issues arise). The log file will contain various energy contributions at each saved iteration step. This file can be used to determine whether more minimization steps should be used. Figure 2 is a plot that shows a minimized structure is obtained within 1000 minimization steps as indicated by the asymptotic behavior of the total energy over time.

The data for this graph may be obtained as follows: download the namdstats.tcl script from http://www.ks.uiuc.edu/Research/vmd/mailing_list/vmd-l/att-11735/namdstats.tcl. Copy and save the script file to the working directory as "namdstats.tcl". In TkConsole, the following commands will incorporate the script procedures and run the procedure `data_time` to extract the total energy of the system for time steps 0–999 (1000 minimization steps).

```
source namdstats.tcl
data_time TOTAL LE0_ws_eq.log 0 999
```

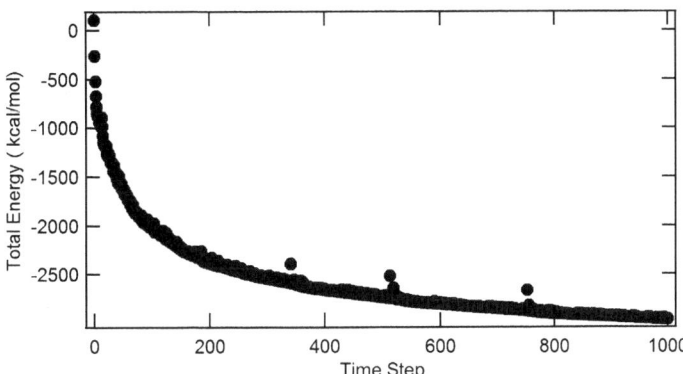

Fig. 2 A plot of the total energy of a protein during an energy minimization. The graph shows that the energy is minimized within 1000 time steps by the asymptotic behavior, reaching an energy minimum

A file will be created called "total.dat". This file has two columns of data corresponding to time step and total energy in kcal/mol. The data is tab-separated and may be opened and plotted in Excel or other analysis software of choice. If the minimization did not reach a minimum asymptote, restart the simulation using more minimization steps. The chosen increase in number of steps will depend on how close to minimization the result appears. For example, if the trend is still a steep negative slope, try 10,000 minimization steps. If the trend has begun to curve towards a somewhat horizontal slope, increase the number of steps to 2500 or 5000. To analyze the results of these restarts, the reader will need to change the second command above so that 999 is the number of minimization steps minus 1. The 5,000,000 step simulation on 1LE0 solvated in water will take a few hours depending on the computer being used. Mutants will take a similar amount of time. When simulating larger proteins of 100–300 amino acids, the experimenter can expect the simulation to take approximately 24 h. Once minimization and dynamics simulation are complete for each mutant, proceed to the next section.

3.7 Distance Measurement

Analysis of the molecular dynamics simulation of 1LE0 will be performed by measurement of the end-to-end distance, that is, the distance between the N- and C-terminal amino acids. For other protein systems, the reader will need to identify a coordinate that has meaning for that protein's activity, e.g., the distance between two purportedly interacting amino acids, one of which is the mutated residue. In VMD, load the psf file for the protein. Then, load the simulation file onto the psf file. The simulation file will have the output name indicated in the configuration script with a file extension dcd. In the 1LE0 wild-type example that is solvated in a water sphere this file is "LE0_ws_eq.dcd".

A script for analyzing the distance between any two selections of a protein can be downloaded from https://goo.gl/LlntcV. This script should to be saved to the NAMD-simulations directory. The script filename is "distance.tcl". It should be sourced in the TkConsole

```
source ../distance.tcl
```

The procedure "distance" is then available for execution. The distance procedure will calculate the distance between the centers of mass of two selections. It will save the distance calculated at every time step in the simulation file. The configuration was set to save to the dcd file every 250 steps. For a simulation of 5,000,000 steps, this will be 20,000 frames. The actual number may be higher depending on the number of minimization steps, e.g., 250 minimization steps will add one frame to the dcd simulation file. The distance procedure requires five input parameters. The general form for execution is

```
distance select1 select2 num_bins distance_file
histogram_file
```

where `select1` is a protein selection (e.g., N-terminus); `select2` is a protein selection (e.g., C-terminus); `num_bins` is the number of distance bins to create for a histogram of the data; `distance_file` is the name of a file to save the distance measured at each saved time step; and `histogram_file` is the name of a file to save a histogram of the distance data. Both files will be comma–separated variables and should have .csv file extensions. For the 1LE0 example, the command to execute the distance measurements should be

```
distance "resid 1 and name CA" "resid 12 and name
CA" 50 dist_1LE0_ws.csv hist_1LE0_ws.csv
```

This will analyze the topmost structure that is open in VMD. Therefore, make sure the simulation file is the only open file in VMD or that it has the "T" indicator for top in the VMD Main window. The selections made above are the alpha carbon of the N- and C-terminal amino acids. The reason for this selection is that mutations will change the location of the center of mass of the amino acids based upon the size of the sidechain. Measuring the distance between alpha carbons gives more robust measure of end-to-end distance. Note that the experimenter is able to make comparisons of different selections to see what effect the selections have on the results.

The histogram file can be opened in Excel or another analysis software package to plot the results. Figure 3 shows the results of molecular dynamics simulations of wild-type and S1pS mutant of 1LE0. The graph shows that the phosphorylation appears to cause the end-to-end distance to decrease from 6.0 to 5.4 Å, which is an expected outcome given the positively charged lysine at the C-terminus. *See* Fig. 1 for a depiction of the measured C-alpha to C-alpha distance. Further analysis could be pursued to determine

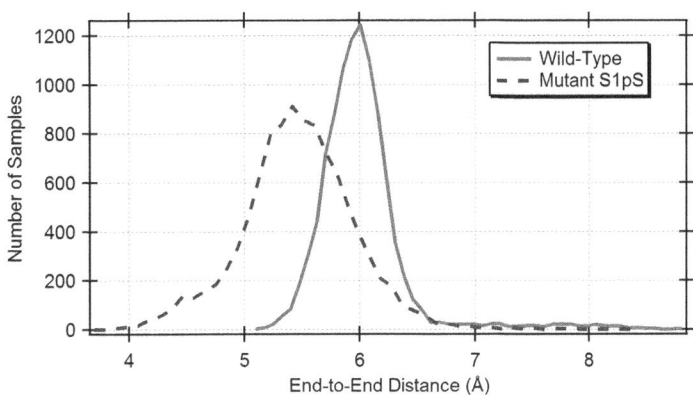

Fig. 3 Histograms of the wild-type and S1pS mutant end-to-end distances from molecular dynamics simulations of the tryptophan zipper with pdb identifier 1LE0. The distance is calculated between the alpha-carbons on the N- and C-terminal amino acids

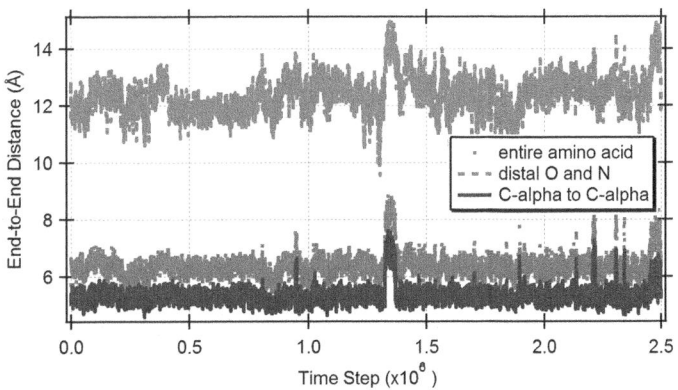

Fig. 4 The wild-type N- to C-terminus end-to-end distance calculated using three different references, alpha carbons, distal sidechain atoms, and entire amino acid. The data is for simulations with 2,500,000 time steps. Each time step corresponds to two femtoseconds of simulation. The total simulation is thus 5 ns. Structural information was saved every 250 time steps giving 10,000 data points for structural analysis of the end-to-end distance

the relative side chain positions by using the distance procedure to measure distances between the phosphorous on residue 1 (phosphoserine) and the side chain nitrogen on residue 12. Figures 4 and 5 show the comparison of calculating center of mass distances between the alpha carbons, the distal atoms (oxygen and nitrogen in the wild-type and phosphorous and nitrogen in the mutant), and the entire amino acids. One should note that using the entire amino acid or the distal atoms on the S1pS mutant would have resulted in a very different histogram due to the large end-to-end distances at the beginning of the simulation. The conclusions drawn from this analysis would be quite different from the alpha carbon analysis. When comparing the three different calculations,

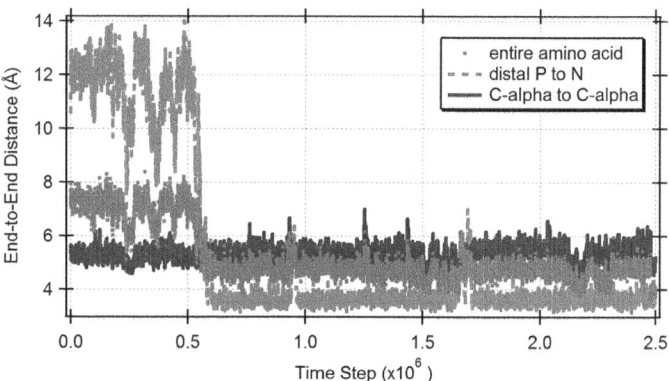

Fig. 5 The S1pS mutant N- to C-terminus end-to-end distance calculated using three different references, alpha carbons, distal atoms, and entire amino acid. The data is for simulations with 2,500,000 time steps. Each time step corresponds to 2 fs of simulation. The total simulation is thus 5 ns. Structural information was saved every 250 time steps giving 10,000 data points for structural analysis of the end-to-end distance. The change in end-to-end distance that occurs after 0.5 ns for the distal and entire amino acid calculations may be due to a sidechain reorientation that does not appear in the alpha carbon calculation

a possible conclusion is that the mutation caused an unstable side chain orientation that reoriented during the simulation. The different structures can be viewed in VMD by going through the frames of the simulation (dcd file). The VMD Main window has tools for scrolling or choosing particular frames at the bottom of the window. The distance.csv file can be used to know the distance at a particular frame number. The commands below can be used to save pdb and psf files of representative distance measurements once the frame is known. The example below shows the selection and saving of frame zero once a dcd file has been loaded. To reiterate, it is critical that the psf file be loaded first and the dcd file second for this to work properly.

```
set LE0 [atomselect top protein frame 0]
$LE0 writepdb LE0_frame0.pdb
$LE0 writepsf LE0_frame0.psf
```

From this short simulation it is not possible to draw accurate conclusions about the energetic differences, if they exist, between these orientations. The randomness of the molecular dynamics simulations means that many replicas should be run using the methods described above. Then, the replicas could be averaged to obtain a reliable result. Even doing this is not as reliable as using a more robust method such as adaptive biasing force simulations.

3.8 Adaptive Biasing Force (abf) Simulations

In many protein systems, there will be conformational energy barriers, and often they will be large enough that thermal energy at physiological temperatures is not enough to drive frequent transitions to these conformations. This will prohibit the protein from exploring a large enough range of motion to be meaningful. There is no single way to know when this is the case. If the experimenter believes a low energy conformation exists that is not observed by molecular dynamics it is possible to steer the protein across large energy barriers that may hinder the access to these states. One technique for doing this is adaptive biasing force (abf) simulations [7, 8]. An abf simulation will steer the protein along a given reaction coordinate, while calculating the applied force needed to eliminate conformational energy barriers. When a minimal number of samples have been accrued for each point along the defined coordinate, the algorithm applies the spatially dependent, calculated force. This allows the protein to sample evenly along the reaction coordinate. The result is an accurate calculation of the protein conformational free energy (or potential of mean force) along the coordinate. More information about abf can be found in the NAMD tutorials (http://www.ks.uiuc.edu/Training/Tutorials/namd/ABF/tutorial-abf.pdf).

The distance histogram in Fig. 3 that was produced from the molecular dynamics simulation provides the information needed for creating a reaction coordinate. For example, during the molecular dynamics simulation the wild-type 1LE0 fluctuates over an end-to-end distance range of 5–9 Å. Thus, one could perform abf simulations over a range of 4–10 Å. This would extend to small enough distances to include the S1pS mutant end-to-end distance at 5.4 Å. It is important to choose a coordinate range that is broad enough to cover all simulated mutants' distance coordinates. Background literature (NMR [9], crystallography [10], fluorescence resonant energy transfer—FRET [11], etc.) on the protein may reveal additional information about conformational flexibility that could be used in defining a reaction coordinate. The coordinate range can always be expanded by repeating the adaptive biasing force simulations over a coordinate range that overlaps a previously simulated range. To do this, one must find a structure from a previous simulation that has a distance within the range being simulated. This is done using the frame-saving method discussed above. These psf and pdb files are then used as the input structure and coordinates of an abf simulation. An alternative to creating overlapping regions is to restart the adaptive biasing force simulations using a broader coordinate range. Unfortunately, one cannot easily expand the coordinate range in the adaptive biasing force technique without restarting the abf simulation from the beginning.

To perform adaptive biasing force simulations two files are used. The first file, "abf1.conf" is a configuration file similar to that of the molecular dynamics simulations. The second file, "Distance. in" defines the reaction coordinates and adaptive biasing force parameters. Both files can be copied and saved from this chapter or downloaded from https://goo.gl/bpL9yz. The following is the configuration file contents.

```
################################################
##
################################################
##
#
# ABF calculation
#
# Wild-type 1LE0 4-10 A
#
################################################
##
################################################
##
################################################
##
# MD SECTION
################################################
##
# NUMBER OF MD-STEPS
numsteps 5000000
# TOPOLOGY
# The following line should correspond to the cor-
rect psf
structure 1LE0_ws_frame1000.psf
# FORCE FIELD
parameters ../../topology/par_all27_prot_na.inp
parameters   ../../topology/par_all27_prot_lipid.
inp
parameters   ../../topology/stream/toppar_all27_
na_nad_ppi.str
parameters  ../../topology/stream/toppar_prot_na_
all.str
paraTypeCharmm on
# 1-4 TERMs
exclude scaled1-4
1-4scaling 1.0
# INPUT FILES
# The following line should correspond to the cor-
rect pdb
coordinates 1LE0_ws_frame1000.pdb
#bincoordinates abf_1.coor
#binvelocities abf_1.vel
temperature 310.0
# OUTPUT FILES
binaryoutput no
binaryrestart yes
outputname abf_1_0
```

```
restartname abf_1
# DCD FILE
dcdFile abf_1.dcd
# FREQUENCY FOR DUMPING OUTPUT DATA
outputenergies 1000
outputtiming 1000
outputpressure 1000
restartfreq 1000
XSTFreq 1000
dcdFreq 1000
# CUT-OFFs
hgroupcutoff 2.8
switching on
switchdist 10.0
cutoff 12.0
pairlistdist 14.0
# CONSTANT-T
langevin on
langevintemp 310.0
langevindamping 10.0
# MULTIPLE TIME-STEP PROPAGATOR
timestep 1.0
# SHAKE/RATTLE
rigidbonds all
# PARALLELISM
stepspercycle 16
splitpatch hydrogen
margin 2.0
#################################################
#############
## EXTRA PARAMETERS ##
#################################################
#############
# Spherical boundary conditions
sphericalBC on
sphericalBCcenter -1.71360 0.93285 -0.37772
sphericalBCr1 12.24
sphericalBCk1 10
sphericalBCexp1 2
# ABF SECTION
colvars on
colvarsConfig Distance.in
#colvarsInput abf_1.colvars.state
#######################
## End File ##
#######################
```

The lines beginning with structure, parameters, and coordinates should have the appropriate filenames. In the script above, files saved from frame 1000 are used for the protein structure and coordinates. This is an energy minimized molecular structure from which the simulation starts. All parameters files discussed in this chapter are listed. Not all are required for every simulation and should be commented out if unnecessary (*see* details in Subheading 3.5). The spherical boundary conditions section should be edited to be correct for the protein structure being simulated. All other lines may be left as they are unless a temperature other than 310 K is desired. The temperature line should be modified if this is the case. The second file is "Distance.in". The contents follow.

```
colvarsTrajFrequency 500
colvarsRestartFrequency 10000
colvar {
name COMDistance
width 0.1
lowerboundary 4.0
upperboundary 10.0
lowerwallconstant 10.0
upperwallconstant 10.0
distance {
group1 {
atomnumbers {5 6}
}
group2 {
atomnumbers {198 199}
}
}
}
abf {
colvars COMDistance
fullSamples 500
hideJacobian
}
```

Lines that may be modified by novices are `width`, which indicates the separation of each point along the reaction coordinate. The `width` value can be as large as twice the standard deviation of the list of distances obtained in the Distance Measurement section. Increasing the `width` value lowers the resolution of the results and speeds the computation. A trade-off the user must manage. The `lowerboundary` value is the smallest distance to evaluate, and the `upperboundary` value is the largest distance to evaluate. In the context of 1LE0, the values of these three variables are 0.1, 4.0, and 10.0, respectively. This will create a reaction coordinate with values 4.0, 4.1, 4.2, ..., 9.8, 9.9, 10.0 Å. The `atomnumbers` lines are space-separated values for the atom numbers in the pdb file that will be used to calculate a center of mass. `Group1` and `group2` are the two sets of atoms separated by some distance to be calculated.

The numbers 5 6 and 198 199 correspond to the alpha carbons and alpha hydrogens of the N- and C-terminal amino acids of the wild-type 1LE0. The `atomnumbers` can be obtained by opening the pdb file in Microsoft® Word or Wordpad. The line beginning `fullsamples` may be modified. This number corresponds to the number of samples required at each distance before the biasing force is applied to the protein simulation. A larger number will result in more accurate force calculations. However, a larger number can increase simulation time dramatically. This is another instance where time vs. accuracy is a decision to be made. Reasonable values for this variable range from 500 to 10,000. Examples for the low and high ends are simulations of peptides in vacuum on the low end to solvated, large proteins of several hundred amino acids on the high end. It is possible that every distance bin will be sampled in excess of the `fullsamples` before the last bin is sampled even once. The diffusive aspect of molecular dynamics simulations make it impossible to know the total computational time required to sample the reaction coordinate fully.

To run the adaptive biasing force simulation, return to TkConsole and enter the command

```
namd2 abf1.conf > abf1.log
```

The simulation of 1LE0 will take several hours. For proteins with hundreds of amino acids, the simulation may take several days. After the simulation completes, it is necessary to determine whether the reaction coordinate was fully sampled. Open the file with extension ".count". Word or Excel can be used to read these files. This file contains the number of simulation counts (second column) in each bin of the reaction coordinate (first column). When all of the values in the count column are above the `fullsamples` value, the adaptive biasing force has been applied and the simulation can be considered finished. The file with extension ".pmf" has the potential of mean force. The potential of mean force will only be valid after all sample counts are above `fullsamples`. The pmf file may also be opened in Excel to create a graph. Examples of the count and pmf data are shown in Figs. 6 and 7, respectively. These data are for two 5,000,000 step runs that were performed as a continuation described below. Ideally, the sample counts will be uniform across the reaction coordinate. A factor of 10 variation in the number of samples across the reaction coordinate seems to be as uniform as can be expected. Running the abf simulation for long enough simulation time will eventually result in uniform sampling. This is an infinite-simulation-time result. Increasing the simulation time will improve the sample uniformity. Below we discuss a secondary method for determining when a simulation is complete in a statistical sense.

The potential of mean force can be seen in Fig. 7 to have a common low energy conformation for both the wild-type and the S1pS mutant at an end-to-end distance of 5.4 Å. The wild-type has

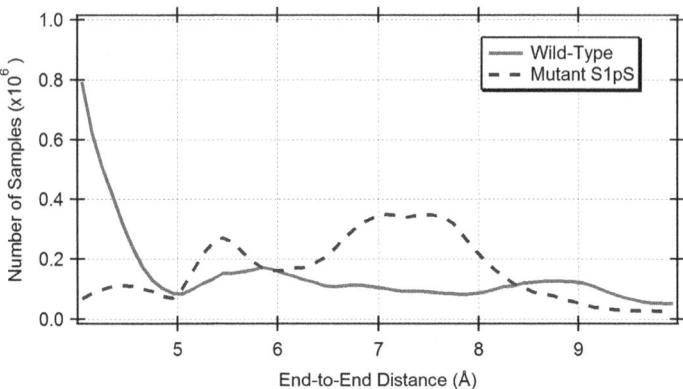

Fig. 6 The number of samples is plotted versus the reaction coordinate. The reaction coordinate corresponds to the N-terminal to C-terminal distance calculated using the alpha carbon and alpha hydrogen center of mass. Ideally, an adaptive biasing force simulation will result in uniform sampling of the reaction coordinate. In this example the wild-type is somewhat more heavily sampled at short distances. The mutant S1pS is reasonably uniform. Both have uniform sampling within an order of magnitude across the coordinate of interest

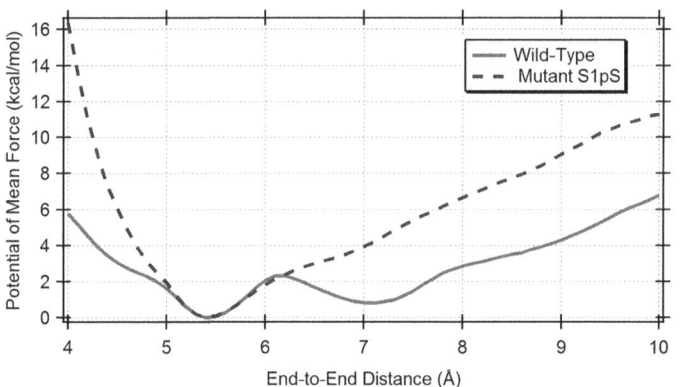

Fig. 7 The potential of mean force vs. reaction coordinate for wild-type and S1pS mutant of 1LEO. The energy landscape shows a common low energy conformation with an end-to-end distance of 5.4 Å. The wild-type has an additional low energy end-to-end distance of 7.1 Å. This conformation should be thermally accessible due to a barrier less than RT separating it from the lower energy conformation

a second low energy conformer at 7.1 Å. This second conformer is separated from the closer, 5.4 Å end-to-end distance by a barrier that is 2.3 kcal/mol. The thermal energy, $RT = 0.008315$ kJ/ mol K $\times$ 310 K $= 2.58$ kJ/mol, gives a statistical measure for evaluating conformational accessibility. Energies within RT of the global energy minimum are accessible with a probability greater than 30 %, that is, these conformations should be accessed frequently. Thus, the barrier should be accessed approximately 30 % of the timesteps before the adaptive force is applied. The energy minima will be accessed with higher frequencies that are related to their

Bolztmann factors, $e^{-E/RT}$, where E is the energy of a position on the reaction coordinate given by the potential of mean force.

In the event that an abf simulation does not sample the entire reaction coordinate adequately, the simulation can be continued from the endpoint of the last simulation. To continue running an abf simulation the configuration file should be saved as "abf2.conf". In general, increment the number as more continuations are made. In the file abf2.conf, the lines beginning `#bincoordinates` and `#binvelocities` should be uncommented by removing the # symbol. The line beginning `temperature` should be commented by adding a # symbol at the beginning. The temperature will be calculated from the coordinate and velocity information of the preceding simulation, which is read from the lines mentioned above. The lines beginning `outputname`, `restartname`, and `dcdFile` need to have the number 1 incremented to 2 because these are now output files for abf2.conf. Finally, the last line of the file beginning `#colvarsInput` needs to be uncommented. Future continuations of an abf simulation can be saved from abf2.conf. These files will only need to have the numbers incremented on the input and output files, i.e, $1 \rightarrow 2$, $2 \rightarrow 3$, $3 \rightarrow 4$, etc. To run the continuation, go to TkConsole and enter `namd2 abf2.conf > abf2.log`. Possible issues continuing an abf simulation are detailed in **Note 6**.

Each abf simulation continuation will create a new set of output files with file names given by `outputname`, `restartname`, and `dcdFile`. The count file will be updated to contain the total sample counts from the entire simulation, including all continuations. The pmf file will be updated to represent the best estimate given all of the simulation continuation data. The abf simulations should be continued until `fullsamples` is achieved for all points along the reaction coordinate. The simulation can be continued beyond this point to refine the potential of mean force for more accuracy. Best practice would be to continue the simulation and compare the potential of mean force until the desired resolution is met, i.e., continuation does not change the potential of mean force significantly. The dcd files that contain structural snapshots of the simulation can be analyzed further. For example, the distance.tcl script can be used to calculate different relevant distances, which might be put into a histogram for side by side comparison to the reaction coordinate. An example might be side chain distances. There are other measurements that might be made using the `measure` command that is the basis of the distance script. Some examples are energy, bond and dihedral angles, solvent accessible surface area (SASA), and hydrogen bonds. All of these parameters might be pertinent to the effects of a mutation. More information on the `measure` command can be found at http://www.ks.uiuc.edu/Research/vmd/vmd-1.8.6/ug/node124.html. The developers of NAMD and VMD at the University of Illinois have a repository of VMD scripts at http://www.ks.uiuc.edu/Research/vmd/script_library/.

4   Notes

1. When actions are performed in VMD, the output files are not going to the working directory. This is likely a result of VMD not pointing to the proper working directory. Make sure TkConsole has been redirected through "cd" commands to the working directory. Entering the command "ls" will list files in the current directory. This can be used as an additional check of referencing the proper directory. When sourcing script files or accessing files through the TkConsole, using correct directory hierarchy is critical. Any error that indicates a file cannot be located is usually a directory hierarchy issue.

2. Sometimes errors occur when making molecular selections in VMD using the atomselect command. This is often due to syntax, improper use of Booleans, or an invalid selection such as indicating a residue number and a residue name that are not self-consistent. Check that these aspects are correct. For full details of the atomselect command refer to http://www.ks.uiuc.edu/Research/vmd/vmd-1.8.6/ug/node109.html.

3. Guess and split chains can result in errors in the AutoPSF builder. This is not a very common issue, but it is typically due to a chain named something uncommon. Check the pdb file in a text editor (Word or Wordpad) to ensure that the fifth and/or the last column have a chain identifier, typically A, B, C, D. Protein Databank files with multiple chains (quaternary structure or multiple conformers) can cause issues. It may also be due to a missing termination line that begins TER at the end of the protein chain in the pdb file. See http://deposit.rcsb.org/adit/docs/pdb_atom_format.html#TER for more information if this is suspected.

4. When mutating to a phosphorylated amino acid, the segment name sometimes causes issues. Most commonly A, B, C, D… are used to identify chains. This is more commonly a problem when multiple structures exists. For example NMR pdb structures with multiple conformers or structures where quarternary structure exists. In these files there are multiple chains. The chain identifier is typically the fifth column of the pdb file. More details about pgn files for generating psf and pdb files can be found at http://www.ks.uiuc.edu/Training/Tutorials/namd/namd-tutorial-win-html/node7.html.

5. The necessary files for handling ATP, phosphoserine, phosphothreonine, and phosphotyrosine were mentioned above. There are many nonstandard atoms and molecules in protein structure pdb files. To list the force fields for every possibility would be impossible. In general, one can search the Internet for "namd moleculename topology" or "charmm moleculename

topology", where moleculename should be the name of the particular molecule for which a force field is required. Many of the molecules that appear in pdb structures have force fields that can be found. It is possible to construct a topology file from scratch. This is an advanced method, and readers should consult with Charmm or NAMD developers on this topic.

Errors are sometimes encountered when running the namd2 executable. Often it is a directory hierarchy issue discussed in **Note 1**. To find what errors occurred when running namd2, open the .log file and scroll to the end. There will be a statement about the first error encountered that caused the program to halt. Often copying this error into a Web search engine will find a solution. An example error is "FATAL ERROR: UNABLE TO OPEN CHARMM PARAMETER FILE". This error is due to a directory hierarchy issue in the configuration parameter file list. Sometimes there are multiple errors. If there are multiple errors, fixing the first error causes the program to halt on the next error. It may be an iterative process to resolve all errors.

6. Continuing simulations requires attention to detail. The input file names must all be correct, or the files will not be found. The output file names must be changed, or files will be overwritten. This will cause loss of data. There are reasons for running simulations in increments. Power outages will stop simulations and may cause a loss of data. Incremental simulations also allow the experimenter to check that the simulations are running as expected. It is frustrating to wait a week for a simulation to complete only to find that something in the setup was incorrect, e.g., the atom selection for the reaction coordinate was incorrect, or the span of the reaction coordinate was too small/big.

References

1. Phillips JC, Braun R, Wang W, Gumbart J, Tajkhorshid E, Villa E, Chipot C, Skeel RD, Kalé L, Schulten K (2005) Scalable molecular dynamics with NAMD. J Comput Chem 26:1781–1802

2. Pronk S, Páll S, Schulz R, Larsson P, Bjelkmar P, Apostolov R, Shirts MR, Smith JC, Kasson PM, van der Spoel D, Hess B, Lindahl E (2013) GROMACS 4.5: a high-throughput and highly parallel open source molecular simulation toolkit. Bioinforma (Oxford) 29:845–854

3. Adcock SA, McCammon JA (2006) Molecular dynamics: survey of methods for simulating the activity of proteins. Chem Rev 106: 1589–1615

4. Chiang H, Robinson LC, Brame CJ, Messina TC (2013) Molecular mechanics and dynamics characterization of an in silico mutated protein: a stand-alone lab module or support activity for in vivo and in vitro analyses of targeted proteins. Biochem Mol Biol Educ 41:402–408

5. Best RB, Zhu X, Shim J, Lopes PE, Mittal J, Feig M, Mackerell AD Jr (2012) Optimization of the additive CHARMM all-atom protein force field targeting improved sampling of the backbone ϕ, ψ and Side-Chain $\chi 1$ and $\chi 2$ dihedral angles. J Chem Theory Comput 8:3257–3273

6. Schulten K. NAMD tutorials. http://www.ks.uiuc.edu/Training/Tutorials/

7. Darve E, Rodríguez-Gómez D, Pohorille A (2008) Adaptive biasing force method for scalar and vector free energy calculations. J Chem Phys 128:144120

8. Hénin J, Chipot C (2004) Overcoming free energy barriers using unconstrained molecular dynamics simulations. J Chem Phys 121: 2904–2914

9. Mayor U, Johnson CM, Daggett V, Fersht AR (2000) Protein folding and unfolding in microseconds to nanoseconds by experiment and simulation. Proc Natl Acad Sci U S A 97: 13518–13522

10. Noonan RC, Carter CW, Bagdassarian CK (2002) Enzymatic conformational fluctuations along the reaction coordinate of cytidine deaminase. Protein Sci Publ Protein Soc 11:1424–1434

11. Schuler B, Lipman EA, Eaton WA (2002) Probing the free-energy surface for protein folding with single-molecule fluorescence spectroscopy. Nature 419:743

Chapter 15

Methods for Detecting Critical Residues in Proteins

Nurit Haspel and Filip Jagodzinski

Abstract

In proteins, certain amino acids may play a critical role in determining their structure and function. Examples include flexible regions, which allow domain motions, and highly conserved residues on functional interfaces, which play a role in binding and interaction with other proteins. Detecting these regions facilitates the analysis and simulation of protein rigidity and conformational changes, and aids in characterizing protein–protein binding. We present a protocol that combines graph-theory rigidity analysis and machine-learning-based methods for predicting critical residues in proteins. Our approach combines amino-acid specific information and data obtained by two complementary methods. One method, KINARI, performs graph-based analysis to find rigid clusters of amino acids in a protein, while the other method relies on evolutionary conservation scores to find functional interfaces in proteins. Our machine learning model combines both methods, in addition to amino acid type and solvent-accessible surface area.

Key words Docking, Evolutionary conservation, Machine learning, Protein binding interfaces, Protein–protein interaction

Abbreviations

PDB   Protein Data Bank
VdW   van der Waals
lRMSD   Least root mean square deviation
SVM   Support vector machine
AI   Artificial intelligence

---

1 Introduction

Proteins and protein complexes play a central role in cellular organization and function, ion transport and regulation, signal transduction, protein degradation, and transcriptional regulation [1]. Since the three dimensional structure of proteins is closely related to their function, analyzing the structural and dynamical properties of proteins is crucial for understanding their role in the cell.

Andrew Reeves (ed.), *In Vitro Mutagenesis: Methods and Protocols*, Methods in Molecular Biology, vol. 1498,
DOI 10.1007/978-1-4939-6472-7_15, © Springer Science+Business Media New York 2017

Some specific regions or amino acids in a protein may play a critical role in its structural, dynamical, and functional properties. For example, proteins usually bind to one another through a specific site on their surfaces, which tends to be highly conserved [2]. Proteins also have flexible regions or residues that act as hinges, which allow the protein to undergo small- or large-scale domain motions. Finding these critical regions can facilitate the analysis of protein flexibility and improve the performance of docking algorithms.

In this chapter we describe our work in analyzing the relative importance of amino acids in proteins using the combination of graph-based rigidity analysis and evolutionary conservation. These two methods use different input data, and thus combining them might allow us to infer locations of critical residues, which either of the methods alone could not accomplish.

1.1 Related Work

One way in which a residue can be identified as critical is by performing a mutation in a physical protein, and then measuring the effect of the substitution. Matthews et al. [3, 4] have designed and analyzed many mutants of lysozyme from bacteriophage T4, and from their work they were able to infer many interesting effects of the mutations. For example, they concluded that the unoccupied volume that is caused by some mutations induces a collapse of that region, while in other cases the cavity remains empty. Also, some residues of T4 lysozyme with high mobility or high solvent accessibility were shown to be much less susceptible to destabilizing substitutions. Although such mutagenesis studies provide precise, experimentally verified insights into the role of a residue based on its mutation, they are time consuming and often cost prohibitive.

Gilis et al. [5] estimated the folding free energy changes upon mutations using database-derived potentials. They concluded that hydrophobic interactions contribute most to stabilizing the protein core, and thus residues that do not engage readily in hydrophobic interactions are not as critical as those that do. In other work, Guerois et al. have developed force fields to help predict protein stability, and to provide a fast and quantitative estimation of the importance of the interactions contributing to the stability of molecules and protein complexes [6]. They concluded that packing density around each atom is a suitable parameter that can be used to predict the flexibility of proteins, and that ranking of residues by their involvement in hydrophobic interactions may provide information about the importance of each residue in maintaining the protein's stability.

Thus, progress has been made in predicting whether a residue is critical. However, many such methods rely on performing and measuring the effect of mutations in the physical protein, or rely on techniques that are computationally intensive, which makes their use on large datasets of proteins prohibitive.

To complement these existing methods, we apply rigidity concepts and conservation scoring techniques to help efficiently identify critical residues. In the following two subsections, we describe these two methods.

1.2 Rigidity Analysis and KINARI

Rigidity analysis [7, 8] is a graph-based method that gives information about the flexibility of biomolecules. Atoms and their chemical interactions are used to construct a mechanical model of a molecule. A graph is constructed from the mechanical model, and efficient algorithms based on the pebble game paradigm [9, 10] are used to analyze the rigidity of the graph. More details are given in the Subheading 2. Figure 1 shows a schematic summarizing how a biomolecule is represented as a graph on which the pebble game algorithm is run.

The rigidity results are used to infer the rigid and flexible regions of the mechanical model, and hence the protein. KINARI-Web [11] is a Web server for rigidity analysis of molecular structures. In Fig. 2a, we show the cartoon rendering of staphylococcal nuclease (PDB ID 1stn). The visualization of its rigidity properties calculated using KINARI-Web is shown in Fig. 2b, where colored bodies indicate clusters of atoms that are rigid.

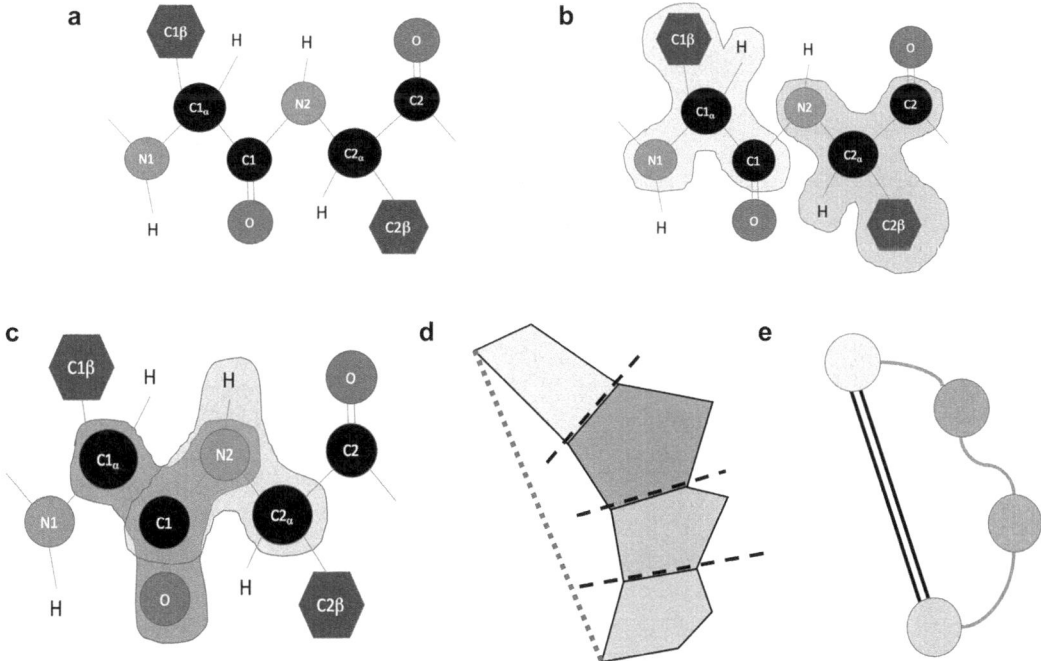

Fig. 1 Rigidity analysis using KINARI software. In a hypothetical two-amino acid polypeptide (**a**) *solid lines*, *spheres*, and *octagons* designate covalent bonds, atoms, and the beta carbons of two residues. Rigidity analysis using the KINARI software identifies four rigid units (**b**, **c**) from the biochemical properties of the polypeptide chain. The mechanical model (**d**) of the polypeptide is constructed by joining rigid units by hinge joints where two rigid units overlap. A sample hydrophobic interaction is also shown. The pebble game algorithm is run on the associated graph (**e**) of the mechanical model, where nodes represent rigid bodies and edges represent degrees of freedom between rigid units

a

b

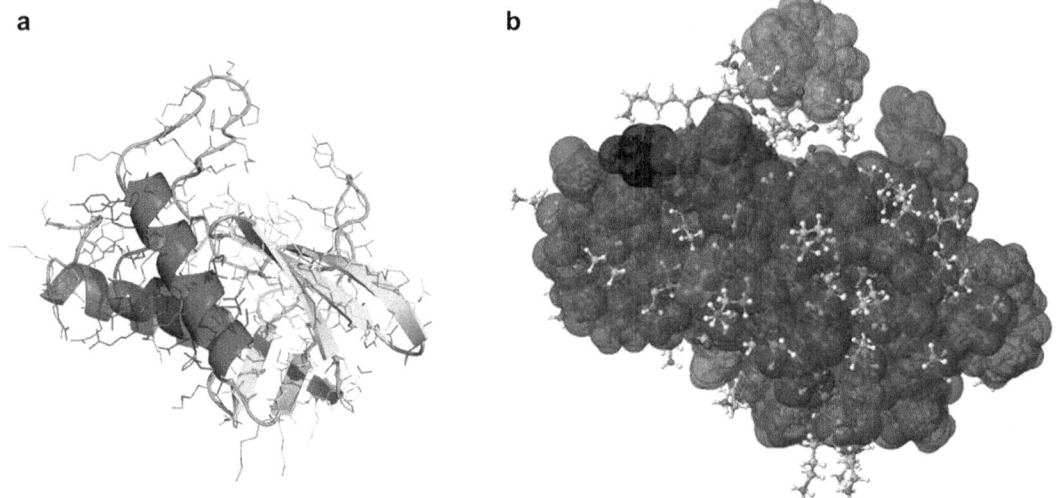

Fig. 2 A schematic rendering of the crystal structure of staphylococcal nuclease. The crystal structure of the nuclease (PDB ID 1stn), refined at 1.7Å resolution, is shown in (**a**). KINARI-Web was used to calculate the protein's rigidity properties, visualized in (**b**). *Color bodies* represent clusters of atoms that are rigid (only the largest clusters are shown)

1.3 KINARI-Mutagen

KINARI-Mutagen [12] relies on a rigidity-theoretical approach for fast evaluation of the effects of mutations that may not be easy to perform in vitro, because it is not always possible to express a protein with a specific amino acid substitution. It tests the effect(s) of the loss of hydrogen bonds and hydrophobic interactions upon a residue's change to glycine, to predict the effects of a mutation. KINARI-Mutagen simulates a mutation by removing the side-chain hydrogen bonds and hydrophobic interactions from the molecular model, and identifies critical residues based on the degree to which the mutation affects the protein's rigidity. Jagodzinski et al. [12] showed that predictions made by KINARI-Mutagen successfully correlate with experimental stability measurements such as $\Delta\Delta G$.

1.4 Identifying Critical Residues via Evolutionary Trace-Based Conservation Scores

Proteins usually bind to one another at a specific site on their interfaces, through a combination of geometric complementarity and specific chemical interactions. In many cases the binding site is not known experimentally and therefore docking algorithms, which try to predict the way proteins bind, have to scan the entire protein surface for possible binding sites on the protein interface, or use methods that try to detect the binding location. Identifying functional interfaces in interacting proteins can greatly reduce the search time for correct rigid-body transformations, as the only transformations that need to be considered are those that match features on predicted interfaces, while the rest of the monomeric interface is not considered. A powerful way to estimate the relative importance of amino acids in a protein is to inspect the degree to

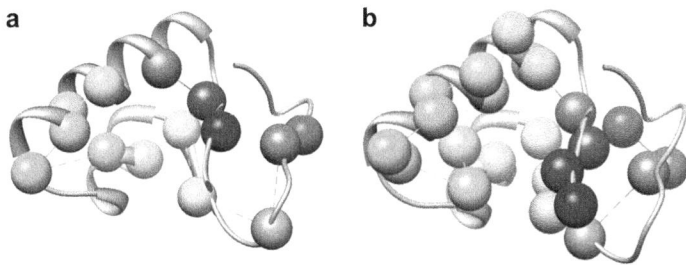

Fig. 3 Barnase structure (PDB:1bni) with critical residues highlighted in *spheres*. (**a**) Critical residues detected by evolutionary conservation analysis. (**b**) Critical residues detected experimentally by the Protherm database

which they are conserved among similar protein structures. Some residues in a protein, such as those on binding interfaces, tend to be highly conserved. The Evolutionary Trace (ET) method [13, 14] ranks residues in proteins based on their functional importance by a sequence conservation analysis in homologues. Proteins belonging to the same family perform similar functions and show lower mutation rates in the residues that contribute the most to the functionality.

Using this conservation value, Akbal-Delibas et al. [15] defined a scoring scheme to identify structures that have clusters of functionally or structurally important residues around binding interfaces. Figure 3 shows crambin (PDB:1crn) with its critical residues as calculated based on known conservation data (right) and our ET score [16] (left). The evolutionary conservation score was used to refine coarsely docked protein structures and demonstrated to significantly improve the input structures both in terms of geometry (lower RMSD with respect to the native structure compared to the input) and lower energy, which may indicate improved interface packing. In a more recent work, we combined evolutionary conservation with other features such as potential energy (electrostatic interactions and VdW), amino acid types and interface size into an artificial neural network scheme to predict the RMSD of a docking candidate with respect to the native structure, as well as refining docked complexes [17, 18].

2   Methods

2.1   Detecting Critical Regions in Proteins

Important amino acids do not include only the binding interface. Other regions in a protein, such as flexible hinges, also play an important role in protein structure and function. Detecting various critical regions in proteins facilitates the analysis and simulation of protein rigidity and conformational changes, and aids in characterizing protein–protein binding. We developed a machine-learning based method to analyze and predict critical residues in proteins

[16, 19]. We combined residue specific information and data obtained by two complementary methods: KINARI-Mutagen [12], which performs graph-based analysis to find rigid clusters of amino acids in a protein, and evolutionary conservation scores to find functional interfaces in proteins, similar to the docking refinement work discussed above. We devised a machine learning model that combined both methods and other features: amino acid type and solvent-accessible surface area. We applied the method to a dataset of proteins with experimentally known critical residues, and were able to achieve a 77% prediction rate, more than either of the methods separately. The ET Server [20] provided the residue rank files for a large number of proteins. The range of rank values in ET files varies from protein to protein, which makes evaluating the relative conservation of a residue in one protein chain with respect to another a difficult task. We used the normalized score devised in [15] and described as:

$$c_i = (\mu - \text{residueRank}) / \sigma$$

where residueRank is the ET rank value of the residue, μ is the mean of ET rank values of residues in the chain, and σ is the standard deviation of ET rank values of residues in the chain. Lower ET rank values represent lower mutation rates and higher conservation rates. Similarly, larger ET rank values will have negative conservation values. Atoms with positive conservation values are considered critical.

2.2 Rigidity Analysis and KINARI

Rigidity analysis [7, 8] is a graph-based method that detects rigid and flexible regions in proteins. A mechanical model of the molecule is built in which covalent bonds are modeled as hinges, and other interactions, such as hydrogen bonds and hydrophobic interactions, are represented as hinges or bars. A graph is constructed from the mechanical model such that each body is associated to a node, a hinge between two bodies is associated to five edges between two nodes, and a bar is associated to an edge. Efficient algorithms based on the pebble game paradigm [9, 10] are used to analyze the rigidity of the graph and infer the rigid and flexible regions of the mechanical model and hence the protein. KINARI-Web [11] is a Web server for rigidity analysis of molecular structures. KINARI-Mutagen [12] relies on a rigidity-theoretical approach that evaluates the effects of mutations that may not be easy to perform in vitro, because it is often impossible to experimentally express a protein with a specific amino acid substitution. KINARI-Mutagen simulates the loss of hydrogen bonds and hydrophobic interactions. Because glycine does not have a beta carbon, removing any hydrogen bonds or hydrophobic interactions that a residue engages in is equivalent to mutating that amino acid to a glycine.

The tool identifies critical residues based on the degree to which an in silico mutation to glycine affects the protein's rigidity. For the sake of the work described here, KINARI-Mutagen was expanded to allow in silico mutating residues to alanine. This new features was developed to permit using datasets of the effects of mutations to alanine.

2.3 Towards a Combined Approach

While evolutionary conservation and rigidity analysis use different approaches that measure different properties, they have one important thing in common—both aim to discover highly important residues in a biomolecule. Therefore, combining them can give us richer, more accurate information about the relative importance of residues in a protein. Therefore, in [16] we applied the evolutionary conservation-based score and used KINARI-Mutagen on a large dataset of proteins to test whether combining these two methods could provide more information about the importance of residues than either of the methods separately. Later, in [19], our aim was to use machine learning to smoothly integrate the two approaches into a combined method that can provide accurate and robust prediction of the importance of residues in proteins. We tested our data against the Protherm dataset [21] that contains information about single point mutations, and a dataset of interaction partners, PiSite [22]. PiSite searches the PDB for different protein complexes that include the same protein, and returns information about that protein's interaction sites and partners, at the residue level.

Classification Using Machine Learning: Machine learning is a branch of artificial intelligence, which aims to classify, group, and learn from data. The classification generally contains the following three stages:

1. Representing a set of known data points (training data) as a set of feature vectors labeled by classes. Often there are two classes—positive and negative, but there can be more than two.

2. Training the set to construct a model that best explains the data, and

3. Using the model to classify a set of unknown data points (test data).

Support Vector Machines (SVM) [23] are a type of machine learning model which constructs a high-dimensional hyperplane that best separates the two classes of data and defines a kernel function to map the data onto the plane. There are many different types of kernels, and the most popular ones are linear, polynomial, radial basis function (RBF) or sigmoid. Many machine learning and statistical methods have been developed to help predict the effects of mutations and to infer which residues are critical [24–27].

Extraction of Experimental Data: We searched the Protherm Database [21] for single point mutations to Glycine or Alanine with known $\Delta\Delta G$ values. $\Delta\Delta G$ is the change to the protein's free energy value (ΔG) following the mutation. A negative $\Delta\Delta G$ value means that the mutation has a destabilizing effect on the protein.

Feature Selection: We used an SVM library, libsvm [23], to train and test our data. The features we selected were (single letter abbreviations refer to amino acids):

1. Amino acid type: Charged (D,E,K,R), Polar (N,Q,S,T), Aromatic (F,H,W,Y) or Hydrophobic (A,C,G,I,L,M,P,V).

   Evolutionary Trace score, normalized according to the equation $ci = (\mu - \text{residueRank})/\sigma$.

2. Rigidity score, expressed as the size of the largest rigid cluster obtained by KINARI-Mutagen.

3. Solvent-Accessible Surface Area (SASA) of the residue.

The feature vectors were labeled as +1 (destabilizing mutation according to Protherm) or −1 (not destabilizing). Our threshold for destabilization was a $\Delta\Delta G$ value of −0.5 or less. We conducted a threefold cross-validation by conducting three tests where the roles of the test and training set rotated between three equal-sized, randomly selected samples. The data was scaled to the [−1,+1] range in order to have all the features in the same order of magnitude. Grid-based cross-validation was used to select the optimal penalty C the RBF kernel parameter γ for the training set. The training set was trained to build the model using the RBF kernel and the test set was classified using the obtained model.

2.4 Case Study: Barnase

Figure 4a shows the native barnase protein structure and Fig. 5 shows the prediction of critical residues using the SVM classifier (Fig. 5a), evolutionary traces (Fig. 5b) and rigidity only (Fig. 5c), respectively, with respect to the experimental data from the Protherm database for barnase (PDB:1bni). The Protherm database contains experimental data for 47 residues; 38 of them (80.1%) are critical and 9 are not critical according to our criteria outlined

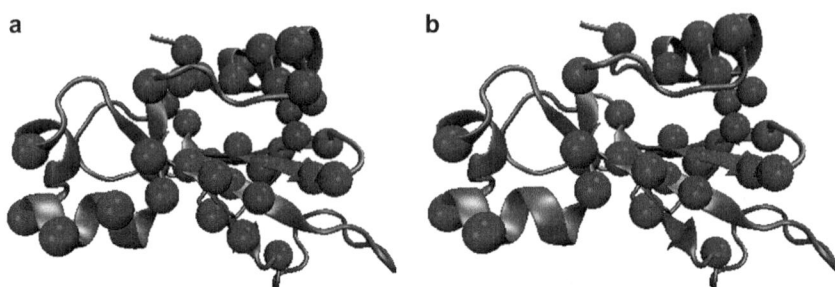

Fig. 4 Barnase structure (PDB:1bni) with critical residues highlighted in *spheres*. (**a**) Critical residues detected experimentally by the Protherm database. (**b**) Critical residues detected by the SVM classifier

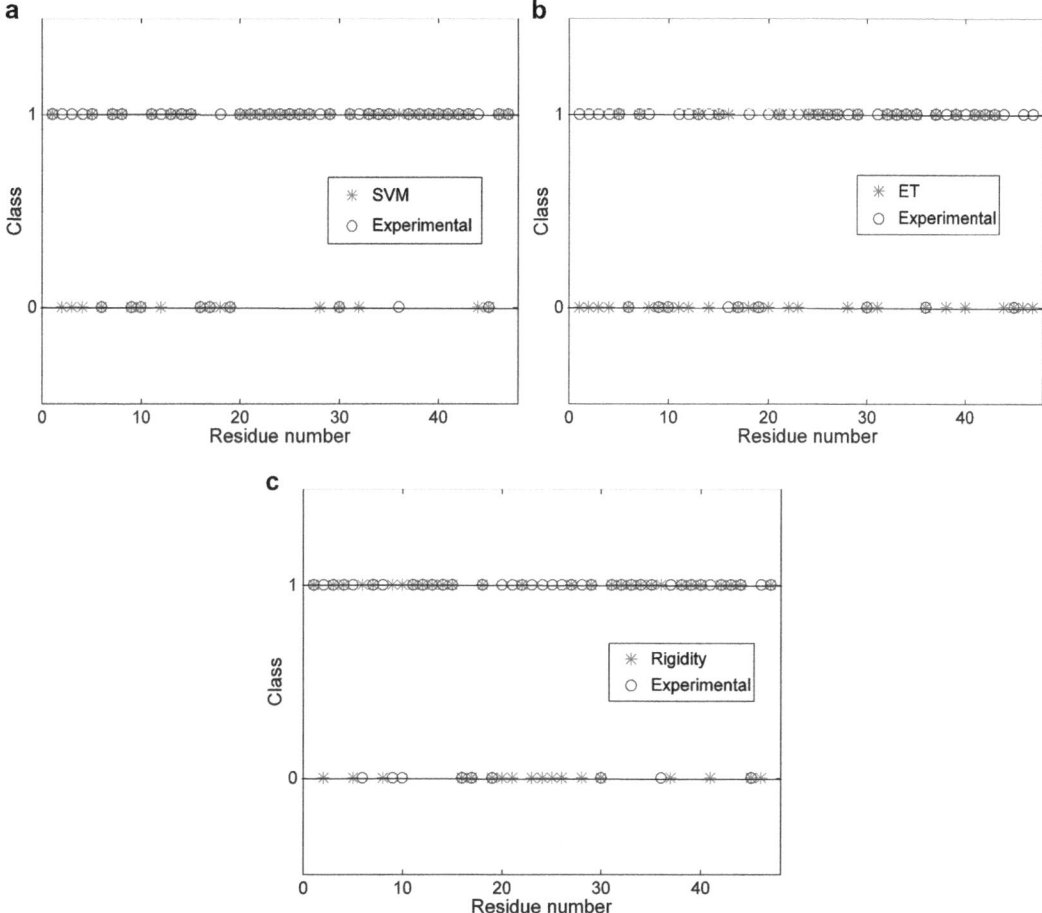

Fig. 5 Comparison of experimentally available criticality data with our SVM based approach (**a**) conservation (**b**) and rigidity analysis (**c**) for barnase (PDB: 1bni). The *bottom line* (*0*) indicates noncritical residues and the *top line* (*1*) indicates critical residues. An *x* and a *circle* at the same position show an agreement between computational and experimental data. The *x* axis is a serial residue number and is not necessarily the residue number in the protein

above (*see* Table 1). The SVM approach correctly predicted 38 out of 47 residues (80.1%) to be critical or noncritical. Thirty out of the 38 were true positives and 8 true negatives. Out of the rest, eight were false positives and one false negative (*see* Table 1).

The ET-based score correctly predicted 27 residues (57.5%) and the rigidity analysis correctly predicted 30 residues (63.8%). There is only partial overlap between the residues identified by the three methods, and the SVM classifier had a much better prediction ability than any of the methods separately. Both the conservation and the rigidity-based approach showed weaker positive correlation with the experimental data (correlation coefficients of 0.31 and 0.18, respectively), but the SVM classifier showed a much

Table 1
Comparison of experimental and computational data for PDB:1bni (barnase)

Res.	$\Delta\Delta G^a$	SVM label[b]	ET score[c]	Δ(LRB)[d]
I4	−1.4	1 (TP)	−0.08	29
N5	−1.9	0 (FN)	−0.40	0
T6	−2.5	0 (FN)	−0.82	4
D8	−0.7	0 (FN)	−1.27	5
V10	−3.4	1 (TP)	0.18	0
D12	1.1	0 (TN)	−1.18	5
Y13	−3.7	1 (TP)	0.28	32
L14	−4.3	1 (TP)	−0.14	0
Q15	0.0	0 (TN)	−1.08	10
T16	−0.3	0 (TN)	−1.66	4
Y17	−2.2	1 (TP)	−0.69	13
H18	−1.9	0 (FN)	−1.24	9
N23	−2.2	1 (TP)	0.05	7
I25	−3.5	1 (TP)	−0.11	6
T26	−1.9	1 (TP)	0.63	4
K27	−0.4	0 (TN)	1.28	0
S28	0.4	0 (TN)	−1.53	0
E29	−1.2	0 (FN)	−0.79	8
Q31	0.1	0 (TN)	−1.34	0
A32	−0.9	1 (TP)	−1.60	0
G34	−3.1	1 (TP)	1.18	0
V36	−1.3	1 (TP)	−1.37	15
V45	−1.8	1 (TP)	−0.21	0
I51	−4.7	1 (TP)	0.89	0
G52	−5.3	1 (TP)	1.35	0
G53	−3.4	1 (TP)	1.35	0
D54	−3.0	1 (TP)	0.93	5
I55	−1.1	0 (FN)	−0.56	0
N58	−2.7	1 (TP)	1.61	51
K66	0.2	0 (TN)	−1.21	0

(continued)

Table 1
(continued)

Res.	$\Delta\Delta G^a$	SVM label[b]	ET score[c]	Δ(LRB)[d]
R72	−2.5	1 (TP)	−0.72	19
E73	−2.5	1 (TP)	1.54	8
I76	−1.9	1 (TP)	0.47	40
N77	−1.6	0 (FN)	0.02	15
N84	−2.0	1 (TP)	−0.01	10
S85	−0.1	1 (FP)	−0.37	3
I88	−4.5	1 (TP)	0.12	0
L89	−7.0	1 (TP)	−0.30	12
S91	−1.9	1 (TP)	1.15	4
S92	−2.8	1 (TP)	−0.34	4
L95	−4.7	1 (TP)	0.25	0
I96	−4.0	1 (TP)	0.21	6
Y97	−6.6	1 (TP)	0.51	13
T100	−2.8	0 (FN)	−0.89	6
Q104	−0.2	0 (TN)	−1.11	0
I109	−0.5	1 (TP)	0.35	0
R110	−0.6	1 (TP)	−0.55	27

TP true positive, *TN* true negative, *FP* false positive, *FN* false negative
[a]We considered values below −0.5 as critical, due to possible measurement errors
[b]0—non critical, 1—critical
[c] The data is represented as a normalized score shown in the Subheading 2 above. A positive score indicates a critical residue
[d]The change in size of the LRB (largest rigid body) following the mutation. A number greater than 0 indicates a critical residue

higher correlation with the experimental (correlation coefficient of 0.57). It should be noted that the majority of residues for which experimental information exists are critical, so the test set as well as the SVM-based model is biased towards critical, rather than non-critical residues. *See* [19] for more details.

Figure 4 shows the protein structure with the critical residues highlighted, both for the experimentally detected residues (Fig. 4a) and the true positive critical residues detected by the SVM classifier (Fig. 4b). It can be seen that the classifier missed a few residues but was able to detect most of them, especially on the surface.

2.5 Case Study: Critical Residues on Binding Sites

Analysis of experimental data shows that known critical residues may have different percentages of solvent accessibility. This is not surprising since buried critical residues play an important role in maintaining the protein's structure, while critical residues on the surface are related to binding sites. To further validate this assumption, we searched the PiSite Database [22] for binding sites and interaction partners. Using the PiSite database, we found that Bovine Pancreatic Trypsin Inhibitor (PDB:1bpi) has six different binding partners and ten binding states. The number of binding partners for each known critical residue is shown in the last column of Table 2. Out of 13 solvent-accessible critical residues that have $\Delta\Delta G$ less than −1.0, 11 residues had at least one binding partner, which means they are on the binding interface. The SVM classifier correctly predicted most of these residues as critical or not, and as seen in the table, most of the incorrect predictions are associated with borderline $\Delta\Delta G$ values (as mentioned above, we defined a residue as critical if its $\Delta\Delta G$ was −0.5 or lower).

These results are very promising since detecting critical residues on the interface would be very helpful for scientists working on the docking problem. Halperin et al. [28] mention that binding sites are typically part rigid and part flexible, with far greater extent of movements in the interface than in any other exposed parts of the structure. Hence, information about critical residues on the surface would not just help in reducing the search space but also in detecting residues that are critical for flexibility on the surface. Protein binding can then be modeled more realistically with the flexible residues on the binding site for a more compact docking.

3 Towards Multiple Mutations

The function of many proteins is altered only if multiple—sometimes more than ten—mutations are present. For example it has been shown for HIV-1 protease that 28 is a median number of mutations in the protease gene, which confers drug-associated resistance to protease inhibitors duranavir and tipranavir [29]. Thus, there is a need for new methods to infer the effect of multiple mutations because that is beyond the scope of existing tools. In our ongoing work we are generating in silico mutant structures with two or more mutations. The rigidity and conservation score values associated with these mutant structures with multiple mutations are being used as features of our SVM method. It is being trained against Protherm $\Delta\Delta G$ values of mutants with multiple amino acid substitutions. Our combined machine learning approach in combination with in silico mutants with multiple mutations will permit reasoning about which multiple residues are critical in affecting the structure or function of a protein.

Table 2
Rigidity analysis and conservation score analysis for protein bovine pancreatic trypsin inhibitor (PDB:1bpi) with residue mutations to alanine

Mutation	WT Residue SASA (Å^2)	$\Delta\Delta G$	No. binding partners	SVM label
K46A	177.11	0.1	2	–1 (TN)
R53A	174.71	–0.1	2	–1 (TN)
T54A	68.66	–0.1	2	1 (FP)
T32A	114.38	–0.1	2	–1 (TN)
E49A	116.65	–0.2	1	–1 (TN)
G56A	20.42	–0.2	2	–1 (TN)
G57A	39.32	–0.2	0	–1 (TN)
R17A	211.65	–0.3	5	–1 (TN)
K15A	196.87	–0.4	5	–1 (TN)
K41A	105.59	–0.4	2	–1 (TN)
D50A	51.92	–0.4	1	1 (FP)
R42A	167.75	–0.5	2	1 (TP)
Q31A	79.04	–1.0	1	–1 (FN)
G28A	41.29	–1.0	1	1 (TP)
Y35A	14.74	–1.1	2	1 (TP)
P13A	70.66	–1.2	4	1 (TP)
Y10A	73.8	–1.2	1	–1 (FN)
V34A	117.65	–1.2	3	1 (TP)
I18A	98.24	–1.5	4	1 (TP)
S47A	35.24	–1.6	1	1 (TP)
M52A	122.96	–1.7	2	1 (TP)
G12A	16.54	–1.8	4	1 (TP)
R20A	36.99	–1.8	2	1 (TP)
F22A	21.02	–2.0	0	1 (TP)
G36A	0.25	–2.1	4	1 (TP)
I19A	158	–2.1	3	1 (TP)
N24A	35.71	–2.2	0	1 (TP)
G37A	36.14	–2.3	4	1 (TP)
N44A	19.98	–3.3	2	1 (TP)

WT wild type, *TP* true positive, *TN* true negative, *FP* false positive, *FN* false negative
The table rows are ordered by $\Delta\Delta G$; the mutations that are least destabilizing are at the top of the table, while the mutations that are most destabilizing are towards the bottom of the table

4   Notes and Conclusions

Specific amino acids play a critical role in a protein's structural stability and dynamics. Being able to detect these amino acids is very useful, as it can help in structural analysis, the simulation of protein motions, and the discovery of protein–protein and protein–drug interactions and binding modes. There is increasing evidence that binding interfaces in proteins are highly conserved and there are many experimental and computational methods that detect clusters of conserved residues, or "hotspots" on protein surfaces. In this chapter we introduce our work in discovering amino acids that may be critical for protein structure and binding. First, we show a method for protein–protein docking and refinement using a combination of geometric complementarity, physicochemical interactions, and evolutionary conservation. Our goal was to bias the docking search and ranking stages towards clusters of conserved residues on the protein surface. We show that this approach indeed helps reduce the computational cost and improve the prediction of binding interfaces. In a subsequent work we devised a machine learning classifier to predict the importance of amino acids in proteins. The features we used were based on a graph-based method to detect rigid and flexible regions in proteins, evolutionary conservation, amino acid type, and solvent-accessible surface area. We were able to achieve high levels of prediction, higher than each one of the features separately. More recently, we devised an artificial intelligence (AI) based method to predict and refine docked complexes [17, 30]. The AI-based method uses more features and seems to give very good results in predicting protein–protein interactions. While the work described here focuses primarily on protein–protein interactions, predicting binding interface and incorporating binding site knowledge into docking methods has many useful applications in drug design and in analyzing protein dynamics.

Acknowledgements

The work described here was partially funded by NSF grant CCF-1116060. The authors thank Dr. Bahar Akbal-Delibas for her collaboration.

References

1. Goodsell DS, Olson AJ (2000) Structural symmetry and protein function. Annu Rev Biophys Biomol Struct 29(1):105–153
2. Chambers KA, Pavletich NP, Pabo CO (1993) The DNA-binding domain of p53 contains the four conserved regions and the major mutation hot spots. Genes Dev 7:2556–2564
3. Xu J, Baase WA, Baldwin E, Matthews BW (1998) The response of T4 lysozyme to large-to-small substitutions within the core and its

relation to the hydrophobic effect. Prot Sci 7(1):158–177

4. Bell JA, Becktel WJ, Sauer U, Baase WA, Matthews BW (1992) Dissection of helix capping in T4 lysozyme by structural and thermodynamic analysis of six amino acid substitutions at Thr 59. Biochemistry 31:3590–3596

5. Gilis D, Rooman M (1997) Predicting protein stability changes upon mutation using database-derived potentials: solvent accessibility determines the importance of local versus non-local interactions along the sequence. J Mol Biol 272(2):276–290

6. Guerois R, Nielsen RE, Serrano L (2002) Predicting changes in the stability of proteins and protein complexes: a study of more than 1000 mutations. J Mol Biol 320:369–387

7. Jacobs DJ, Rader AJ, Thorpe MF, Kuhn LA (2001) Protein flexibility predictions using graph theory. Proteins 44:150–165

8. Jacobs DJ, Thorpe MF (1995) Generic rigidity percolation: the pebble game. Phys Rev Lett 75:4051–4054

9. Lee A, Streinu I (2008) Pebble game algorithms and sparse graphs. Discrete Mathematics 308(8):1425–1437

10. Jacobs DJ, Hendrickson B (1997) An algorithms for two-dimensional rigidity percolation: the pebble game. J Comp Phys 137:346–365

11. Fox N, Jagodzinski F, Li Y, Streinu I (2011) KINARI-Web: a server for protein rigidity analysis. Nucleic Acids Res 39(Web Server Issue):W177–W183

12. Jagodzinski F, Hardy J, Streinu I (2012) Using rigidity analysis to probe mutation-induced structural changes in proteins. J Bioinform Comput Biol 10(3):1242010

13. Lichtarge O, Bourne HR, Cohen F (1996) An evolutionary trace method defines binding surfaces common to protein families. J Mol Biol 257(2):342–358

14. Mihalek I, Res I, Lichtarge O (2004) A family of evolution-entropy hybrid methods for ranking of protein residues by importance. J Mol Biol 336(5):1265–1282

15. Akbal-Delibas B, Hashmi I, Shehu A, Haspel N (2012) An evolutionary conservation based method for refining and re-ranking protein complex structures. J Bioinform Comput Biol 10(3):1242002

16. Akbal-Delibas B, Jagodzinski F, Haspel N (2013) A conservation and rigidity based method for detecting critical protein residues. BMC Struct Biol 13(Suppl 1):S6

17. Akbal-Delibas B, Pomplun M, Haspel N (2014) AccuRMSD: a machine learning approach to predicting structure similarity of docked protein complexes. In: Proc. of ACM-BCB (5th ACM International conference on Bioinformatics and Computational Biology). pp 289–296

18. Akbal-Delibas B, Pomplun M, Haspel N (2015) Accurate prediction of docked protein structure similarity. J Comp Biol 22(9):892–904

19. Jagodzinski F, Akbal-Delibas B, Haspel N (2013) An evolutionary conservation & rigidity analysis machine learning approach for detecting critical protein residues. In: CSBW (Computational Structural Bioinformatics Workshop), in proc. of ACM-BCB (ACM International conference on Bioinformatics and Computational Biology), pp 780–786

20. Lichtarge O-Evolutionary trace server. http://mammoth.bcm.tmc.edu/ETserver.html

21. Kumar MD, Bava KA, Gromiha MM, Prabakaran P, Kitajima K, Uedaira H, Sarai A (2005) Protherm and pronit: thermodynamic databases for proteins and protein–nucleic acid interactions. Nucleic Acids Res 34(suppl 1):D204–D206

22. Higurashi M, Ishida T, Kinoshita K (2009) Pisite: a database of protein interaction sites using multiple binding states in the PDB. Nucleic Acids Res 37(suppl 1):D360–D364

23. Chang CC, Lin CJ (2011) LIBSVM: a library for support vector machines. ACM Trans Intel Syst Technol 2(3)

24. Cheng J, Randall A, Baldi P (2006) Prediction of protein stability changes for single-site mutations using support vector machines. Proteins 62:1125–1132

25. Lise S, Buchan D, Pontil M, Jones DT (2011) Predictions of hot spot residues at protein-protein interfaces using support vector machines. PLoS One 6(2):e16774

26. Worth CL, Preissner R, Blundell L (2011) SDM-a server for predicting effects of mutations on protein stability and malfunction. Nucleic Acids Res 39(Web Server Issue):W215–W222

27. Suresh MX, Gromiha MM, Suwa M (2015) Development of a machine learning method to predict membrane protein-ligand binding residues using basic sequence information. Adv Bioinform 2015:7

28. Halperin I, Ma B, Wolfson H, Nussinov R (2002) Principles of docking: an overview of search algorithms and a guide to scoring functions. Proteins 47(4):409–443

29. Rhee S-Y, Taylor J, Fessel WJ, Kaufman D, Towner W, Troia P, Ruane P, Hellinger J,

Shirvani V, Zolopa A, Shafer RW (2010) Hiv-1 protease mutations and protease inhibitor cross-resistance. Antimicrob Agents Chemother 59(8):4253–4261

30. Akbal-Delibas B, Pomplun M, Haspel N (2015) AccuRefiner: a machine learning guided refinement method for protein-protein docking. In: proceedings of BICoB (7th international conference on Bioinformatics and Computational Biology)

Chapter 16

A Method for Bioinformatic Analysis of Transposon Insertion Sequencing (INSeq) Results for Identification of Microbial Fitness Determinants

Nengding Wang and Egon A. Ozer

Abstract

Transposon insertion sequencing is a process whereby microbial fitness determinants can be identified on a genome-wide scale. This process uses high-throughput next generation sequencing to screen for changes in the composition of a pool of transposon mutants after exposure to selective conditions. One commonly used process for generating transposon insertion sequencing libraries is called INSeq that works with mutant pools produced using a modified *Mariner* transposon. Libraries produced using the INSeq process are sequenced on the Illumina platform. In this chapter, we describe our method for processing the raw Illumina sequencing reads, aligning the reads to a reference sequence to determine read counts, and using the online transposon insertion sequencing data analysis server, ESSENTIALS, to interpret the results.

Key words Transposon insertion sequencing, INSeq, Bioinformatics, Gene function, Next-generation sequencing, Microbial fitness

1 Introduction

Transposon insertion sequencing offers a high-throughput screening tool to identify genes and pathways important for microbial fitness in a particular set of conditions being examined [1]. A population of mutants is generated using a tagged, randomly inserting transposon. This population is then exposed to selective conditions in vivo or in vitro and changes in the representation of mutants are identified by next generation sequencing (NGS) techniques. Mutants that decrease in relative abundance after selection are likely to have disruptions in genes that contribute to microbial fitness in the experimental condition. Conversely, mutants that increase in relative abundance are likely to have disruptions in genes that are deleterious in the condition. This technique is widely applicable to rapidly and accurately identify genes important under a wide variety of selective conditions and in a number of different microbes provided they are amenable to genetic manipulation.

Andrew Reeves (ed.), *In Vitro Mutagenesis: Methods and Protocols*, Methods in Molecular Biology, vol. 1498, DOI 10.1007/978-1-4939-6472-7_16, © Springer Science+Business Media New York 2017

Several methods for generating transposon insertion mutant pools, library preparation, and sequencing have been described. These include transposon sequencing (Tn-seq) [2], high-throughput insertion tracking by deep sequencing (HITS) [3], transposon-directed insertion site sequencing (TraDIS) [4], and insertion sequencing (INSeq) [5]. These techniques differ in the choice of transposons used and in the lengths of sequencing libraries produced. The INSeq protocol uses the Himar I *Mariner* transposon to generate mutant libraries. The transposon has been modified to include *Mme*I restriction sites in the inverted repeats. Digestion of DNA from a library of mutants results in 2-bp staggered cuts 20 or 21 bp downstream of the restriction sites, capturing 16 or 17 bp of genomic sequence flanking the transposon insertion site which is subsequently sequenced. A published protocol for INSeq library preparation is available [6].

Analysis of sequence data from transposon insertion sequencing methods, including INSeq, consists of identifying the portion of the reads that represents the transposon insertion site flanking sequences, aligning them to a reference sequence to determine the gene that has been disrupted by the transposon and counting the number of aligned reads, then comparing relative read abundances between the mutant pool prior to exposure to the selective condition and pools collected after selection. In practice, the analysis can be more complicated with the potential for errors to be introduced during library generation and alignment to reference sequences potentially resulting in inaccurate read counts and incorrect conclusions being drawn from relative read abundance comparisons. There are, as of yet, no standard methods for analysis of transposon insertion sequencing data. Goodman et al. provided a suite of software with their INSeq protocol for read alignment and pool comparison [6], but this lacks more advanced statistical evaluations of the results. The online analysis server ESSENTIALS [7] offers more robust statistical analysis tools, but its requirement for raw read or alignment data to be provided as a direct link can be difficult for bioinformatics novices. Given the importance of accurate data processing for interpreting transposon insertion sequencing data, we have prepared this simplified data analysis protocol.

2 Materials

1. File(s) of raw sequencing reads generated from libraries produced using the INSeq protocol [6] (*see* **Note 1**). The reads should be in fastq format and can either be compressed with gzip or uncompressed (*see* **Note 2**).

2. Desktop or laptop computer running Mac OSX or Linux. Perl and `gzip` must be installed, but these come preinstalled on most OSX and Linux systems. To check, type 'perl -v' in the

Terminal application. If there is a response showing Perl version, then the program is installed. To check if gzip is installed, type 'gzip -V' in the Terminal application. If there is a response showing the gzip version, then the program is installed.

3. Application *bowtie* (http://bowtie-bio.sourceforge.net/index.shtml) downloaded and installed on your computer system. For best results, the directory with the application should be available in your PATH. To check if it is installed correctly in your PATH, type 'bowtie --version' in your Terminal application. If you get a response showing the bowtie version, then the program is installed and available. Similarly, check that *bowtie-build* is in your PATH by typing 'bowtie-build --version' in your Terminal application.

4. Scripts INSeq_read_preprocess.pl and INSeq_ESSENTIALS_filter.pl, available for download from https://github.com/egonozer/TIS_tools. Scripts can be directly downloaded via the website or, if *git* is installed, by entering in Terminal:

```
git clone https://github.com/egonozer/TIS_tools.git
```

5. File containing genomic sequence of bacterial strain used to produce the transposon mutant library. Sequence should be in "fasta" format and all known sequences (chromosomal and plasmid) should be combined in the same file. Even if you are only interested in genes in, for example, the chromosome and not the plasmids, all known sequences in the strain should still be included in this file to minimize misalignments during processing. Sequence files can be downloaded from NCBI (http://www.ncbi.nlm.nih.gov/ or ftp://ftp.ncbi.nlm.nih.gov/ (*see* **Note 3**).

6. Account on dropbox.com (*see* **Note 1**).

7. Microsoft Excel (or similar spreadsheet program).

3 Methods

3.1 Sequence Read Processing

1. If you have not already done so, download and install *bowtie*, INSeq_read_preprocess.pl, and INSeq_ESSENTIALS_filter.pl (*see* **Materials**)

2. Create a directory that will contain all of the analysis output. Example:

```
mkdir /Users/ResearcherX/INSeq_experiment_1.
```

3. Move, copy, or make an alias to the Illumina sequencing reads file downloaded from your sequencing provider (*see* **Note 2**).

4. Create a text file containing the library names followed by their corresponding adapter barcodes separated by a tab. Save this

file in the directory created in **step 2** of Subheading 3.1. For example:

```
Input_1 ATCG
Input_2 CATG
Output_1 GGAA
Output_2 GGGG
```

5. Move, copy, or make an alias to the transposon mutant reference strain sequence (*see* **step 5** in Subheading 2) in the directory created in **step 2** in Subheading 3.1. If you also sequenced your transposon mutant reference strain (*see* **Note 3**), then move, copy, or make an alias to this sequence in the directory as well.

6. In your Terminal application, navigate to the analysis directory you created in **step 2** in Subheading 3.1. Example:

```
cd /Users/ResearcherX/INSeq_experiment_1
```

7. Run the read processing script INSeq_read_preprocess.pl. For example (with minimum *required arguments*):

```
perl        /path/to/script/INSeq_read_preprocess.
pl  -r<compressed  or  uncompressed  reads  file
from  Subheading  3.1,  step 3>-c<barcode  file  from
Subheading  3.1,  step 4>-g<reference  sequence  file
from  Subheading  3.1,  step 5>
```

To see all program options, run the script without arguments or see the README file included with the script.

8. Read processing usually takes 1–3 h, depending on the number of reads being analyzed.

9. When processing is completed, there will be several files output in the directory. Each of these files will be named in following format:

<library name>_<barcode>.<reference sequence>.wiggle

Each line of each file will contain the position of a transposon insertion site in the sequence and the number of reads aligning at that position after filtering. Files will range in size depending on the length of the reference sequence, but for bacterial chromosomes the files are generally 2–4 Mb each.

3.2 ESSENTIALS Analysis

1. Open Microsoft Excel and create a new spreadsheet.

2. Type the following in the indicated columns in the first row:

   (a) <Leave blank for now>.

   (b) N.

   (c) N.

   (d) <Leave blank for now>.

   (e) lib1.

(f) wiggle.

(g) none.

(h) <Leave blank for now>.

3. Copy the first row as many times as you have input pool library replicates in your read sets plus the number of times as you have output pool library replicates in your read sets. For example, if you prepared and sequenced three libraries from your input transposon mutant library pool and prepared and sequenced one library each from three separate infected animals under the same conditions, then you should copy the cell entries in **step 3** in Subheading 3.2 into the next five rows.

4. Log into your Dropbox account via a web browser (*see* **Note 5**).

5. Create a new folder in Dropbox and click the newly created folder.

6. Drag and drop all of the ".wiggle" files created in **step 9** in Subheading 3.1 onto the web browser window to upload the files. You do not need to upload any files that do not end with ".wiggle".

7. For each uploaded file, perform the following steps:

   (a) Click the "Share" button in the row with file name.

   (b) Copy the link URL given in pop-up window under "Link to file".

   (c) Paste this link URL into one of the cells left empty in column A.

   (d) Change "https://www.dropbox.com/" at the start of the link URL to "https://dl.dropboxusercontent.com/" (*see* **Note 6**).

   (e) In the empty cell in column D, if this is an input pool library, type "control." If this is an output pool library, type "target".

   (f) In the empty cell in column H, type the name of the library. For example "Input_1" or "Mouse4", etc.

8. Save the file as a "Windows Formatted Text (.txt)" tab-delimited text file named "configuration_target.txt" in the directory you created in **step 2** in Subheading 3.1 (*see* **Note 7**).

9. Copy all the rows corresponding to the input pool(s), i.e., those with "control" in column D, to a new spreadsheet. Save this new spreadsheet as a "Windows Formatted Text (.txt)" file named "configuration_input.txt" in the directory that was created in **step 2** in Subheading 3.1.

10. Navigate to the ESSENTIALS web server (http://bamics2. cmbi.ru.nl/websoftware/essentials/essentials_start.php)

11. Under the first heading "Template genome," select "Finished genome" and select the bacterial genus, species, and strain

name corresponding to the transposon mutant library parent strain from the drop-down menu.

12. Under the second heading "Run mode," under "Upload configuration file," click the "Choose file" button and select the "configuration_input.txt" file from **step 9** in Subheading 3.2.

13. Click the "Next" button.

14. On the next page, change the following options from their default settings:

    (a) Library size: Change this number to the expected number of transposon mutants in your library.

    (b) Create ZIP archive: Select "Yes" from the drop-down menu.

15. Click the "Proceed" button to start the run and load the run progress window. This window will automatically update every 10 s until the run is completed. Leave the window open until complete or bookmark the page to return to it later. The analysis usually completes within about 10 min (*see* **Note 8**).

16. On the results page, under the "Essential genes" section, click on the "Plot of signal versus $\log_2$ fold change on essential genes (*see* **Note 9**).

17. The figure should have two major peaks representing essential (left-most peak) and nonessential genes (right-most peak). A local minimum between the two peaks will be plotted with a red dot and number. See example in Fig. 1. Make note of this number as the optimal cutoff between essential and nonessential genes.

18. Also under the "Essential genes" section, download the table under the link "Combined table for essential genes (.tsv)". Save this file in the directory that was created in **step 2** in Subheading 3.1.

19. Download the archive file under the link "Zipped file of output files excluding sequences and raw alignment output" for future reference. Save this file in the directory that was created in **step 2** in Subheading 3.1 and rename it as "input.zip".

20. Repeat **steps 10–15** in Subheading 3.2, but using the "configuration_target.txt" produced in **step 9** in Subheading 3.2 as the uploaded file in **step 12** in Subheading 3.2.

21. On the results page, under the "Conditionally essential genes" section, click the link "PCA plot of normalized read counts on genes". This figure will give an indication of how similar read count distributions within condition replicates are, if provided (*see* Fig. 2).

22. On the results page, under the "Conditionally essential genes" section, download the table under the link "Combined table of conditionally essential genes (.tsv)". Save this file in the directory that was created in **step 2** in Subheading 3.1.

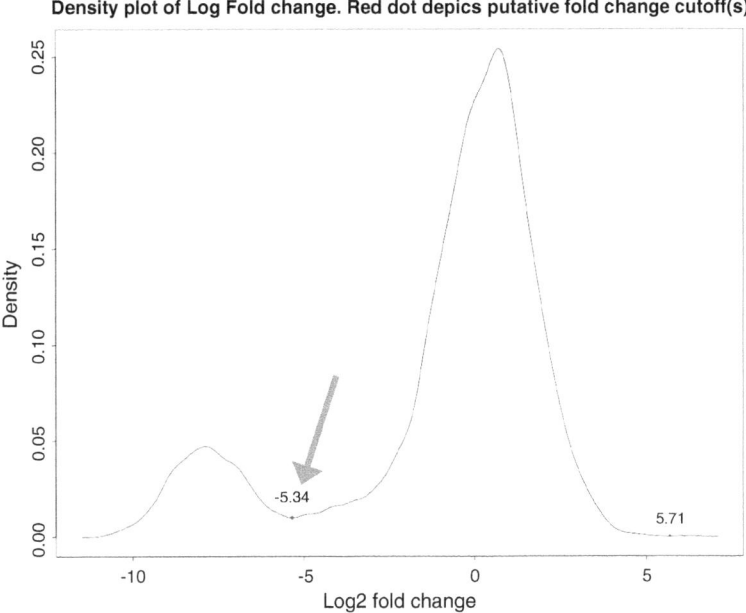

Fig. 1 Example of density plot of log fold change produced by ESSENTIALS. Local minimum between two major peaks is marked by ESSENTIALS with a red dot and log$_2$ fold change value, highlighted here with *orange arrow*

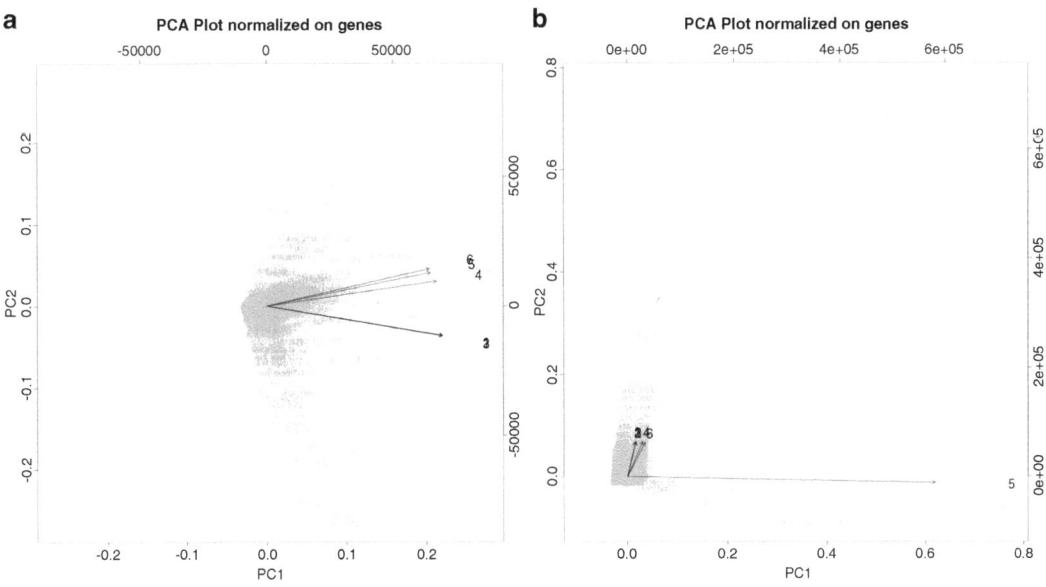

Fig. 2 Examples of PCA plots of genes in analysis replicates produced by ESSENTIALS. *Arrows* represent individual libraries. In these figures, libraries *1*, *2*, and *3* are replicates of the input pool and libraries *4*, *5*, and *6* are biological replicates of the treatment condition. (**a**) Example of closely related samples with libraries of the same conditions clustering closely together. (**b**) Example of control samples clustering closely together, but one of the treatment samples (#5) deviates considerably from the other two (#4 and #6) suggesting that a search for potential problems in the treatment, preparation, sequencing, or analysis of these samples is warranted

23. Download the archive file under the link "Zipped file of output files excluding sequences and raw alignment output" for future reference. Save this file in the directory that was created in **step 2** in Subheading 3.1 and rename it as "target.zip".

24. In the Terminal application, run the results processing script INSeq_ESSENTIALS_filter.pl. For example, with minimum *required arguments*:

```
perl ../INSeq_ESSENTIALS_filter.pl -m<local minimum
cutoff from Subheading 3.2, step 17>-c essenti-
algenes_alloutputmerged.tsv<from Subheading 3.2,
step      18>-t      gene_alloutputmerged.tsv<from
Subheading 3.2, step 22>
```

To see all program options, run the script without arguments or see the README file included with the script.

25. There will be three output files in the directory that the script was run in. They can be opened for viewing and sorting in Excel or a similar spreadsheet program:

(a) <prefix>_essentialgenes_filtered.tsv: A list of all of the genes that had $\log_2$ fold changes less than local minimum cutoff, i.e., the genes most likely to be essential for growth in culture medium. Column headings and descriptions, in order:

  • [untitled]: Unique ID number.

  • "Row.names": Locus ID of gene.

  • "<library name>_<sample number>_control.x" (one or more columns): Read counts before between-sample normalization.

  • "unique_flanking_sequences": Number of unique insertion site flanking sequences in the gene.

  • "<library name>_<sample number>_control.y" (one or more columns): Read counts after normalization (pseudocounts).

  • "expected_reads": Expected number of reads in the gene.

  • "logConc": The log fraction of the number of reads of a gene divided by the total number of reads.

  • "logFC": The $\log_2$ of the sample reads divided by the expected number of reads.

  • "P.Value": The *p*-value of the comparison between the observed number of reads and the expected number of reads.

  • "adj.P.Val": *p*-Value adjusted for multiple testing.

  • "Start", "Stop", "Strand", "Length", "PID", "Gene", "Product": Gene information.

(b) <prefix>_genes_filtered.tsv: A list of all genes like the file downloaded in Subheading 3.2, **step 22**, but now with the genes identified above as essential for growth in media, as well as genes with less than ten reads per sample or no insertion sites, marked with "NA" in all measurements. Column headings and descriptions are the same as above with the following exceptions:

- No "unique_flanking_sequences" or "expected_reads" columns.

- "logFC": The log2 of the reads of the target samples divided by the reads of the control samples.

- "P.Value": The p-value of the comparison between the number of reads in the target samples to the number of reads in the control samples.

(c) <prefix>_genes_treatment_related.tsv: A list of genes that are not essential for growth in media and have $\log_2$ fold changes above or below the given thresholds (defaults are 1 and −1) and adjusted p-values less than the given threshold (default is 0.05). Genes in the file are sorted from lowest negative $\log_2$ fold change to highest positive $\log_2$ fold change. Genes with negative $\log_2$ fold changes represent those genes with the greatest decrease in read counts at transposon insertion sites in the treatment pool(s) compared read counts in the control pool(s) and hence are genes most likely to be important for persistence in the treatment condition. Column headings and descriptions are the same as in the "<prefix>_genes_filtered.tsv" file.

4   Notes

1. We perform INSeq library sequencing as described in the Goodman, Wu, and Gordon's protocol [6] with the following modifications/notes:

   (a) Isolation of crude DNA (**steps 1–16** in [6]) can be performed using an automated DNA purification platform such as the Promega Maxwell 16 or similar instrument.

   (b) We find that gel electrophoresis of library fragments at 200 V for 30 min (step 72 in [6]) results in overheating of the gel. We instead run the gel at 100 V for 1 h.

   (c) Library mixture should be sequenced on Illumina GA-IIx HiSeq using the SR cluster kit. We have attempted to sequence libraries on the HiSeq PE cluster kit as well as the MiSeq platform using a V3 600 cycle kit without success. Only the HiSeq SR kit has successfully sequenced the libraries in our experience.

(d) This protocol replaces **steps 79** through **88** of Goodman et al.'s protocol.

2. Sequencing providers often provide reads compressed in the gzip format. If so, the sequencing read file name will end with a ".gz" suffix. The decompressed file is often very large. This file does NOT need to be decompressed for processing.

3. Optional: Consider separately performing whole-genome sequencing of the parent strain used to produce the transposon mutant library and either de novo assemble the sequencing reads (i.e., using *velvet* or another assembler) or align reads to the reference sequence and produce a corrected consensus (i.e., using *bwa* with *samtools mpileup*). INSeq_read_pre-process.pl can be use this sequence to compare with the available reference strain sequence and account for possible genetic drift in the strain used to produce the mutant library.

4. Using Dropbox is one of the simplest methods of making the alignment files available for remote analysis by ESSENTIALS, but any similar cloud sharing service will also be sufficient as long as the service allows direct linking of files, i.e., links provided by the service go directly to the file and not to an intermediate web page from which the file can then be downloaded. Setting up web sharing on a desktop or laptop computer is also an option, but may not work if the computer is connected to the Internet behind a firewall.

5. If using a method other than Dropbox to make files available for direct download, replace **steps 4, 5, 6**, and 7a, b, d in Subheading 3.2 with the method being used.

6. Modifying the link URL will allow the file to be directly downloaded by ESSENTIALS rather than linking to an intermediary download web page.

7. Create a separate configuration file for each unique selection condition examined and include all replicates performed under this condition. For example, if three libraries are prepared from the input mutant pool, three libraries from the pool after infection in wild-type mice and three libraries from the pool after infection in immunocompromised mice, make one configuration file that includes the reads from the three input pool libraries and the three wild-type mouse libraries and make another configuration file that includes the reads from the three input pool libraries and the three immunocompromised mouse libraries. Perform separate ESSENTIALS runs with these two files. Separate configuration files should be created and separate ESSENTIALS runs performed for each reference sequence of interest. For example, if in **step 5** in Subheading 3.1 reads were aligned to a reference sequence file with one chromosomal sequence and one plasmid sequence, ESSENTIALS should be

performed first on the alignments against the chromosomal sequence, then on the alignments against the plasmid sequence.

8. The ESSENTIALS website estimates that runs will take about 90 min to complete, but the run time is usually far shorter when providing prealigned read counts as is being done here.

9. For more detail about the formatting and data in the ESSENTIALS output files, see the ESSENTIALS manual: http://bamics2. cmbi.ru.nl/websoftware/essentials/manual.pdf.

Acknowledgments

This work was supported by a Mentored Research Scholar Grant in Applied and Clinical Rsearch, MRSG-13-220-01—MPC from the American Cancer Society (EAO).

References

1. van Opijnen T, Camilli A (2013) Transposon insertion sequencing: a new tool for systems-level analysis of microorganisms. Nat Rev Microbiol 11:435–442

2. van Opijnen T, Bodi KL, Camilli A (2009) Tn-seq: high-throughput parallel sequencing for fitness and genetic interaction studies in microorganisms. Nat Methods 6:767–772

3. Gawronski JD, Wong SM, Giannoukos G, Ward DV, Akerley BJ (2009) Tracking insertion mutants within libraries by deep sequencing and a genome-wide screen for *Haemophilus* genes required in the lung. Proc Natl Acad Sci U S A 106:16422–16427

4. Langridge GC, Phan MD, Turner DJ, Perkins TT, Parts L, Haase J, Charles I, Maskell DJ, Peters SE, Dougan G, Wain J, Parkhill J, Turner AK (2009) Simultaneous assay of every *Salmonella Typhi* gene using one million transposon mutants. Genome Res 19:2308–2316

5. Goodman AL, McNulty NP, Zhao Y, Leip D, Mitra RD, Lozupone CA, Knight R, Gordon JI (2009) Identifying genetic determinants needed to establish a human gut symbiont in its habitat. Cell Host Microb 6:279–289

6. Goodman AL, Wu M, Gordon JI (2011) Identifying microbial fitness determinants by insertion sequencing using genome-wide transposon mutant libraries. Nat Protoc 6:1969–1980

7. Zomer A, Burghout P, Bootsma HJ, Hermans PW, van Hijum SA (2012) ESSENTIALS: software for rapid analysis of high throughput transposon insertion sequencing data. PLoS One 7:e43012

Section IV

In Vitro Transposon Mutagenesis Methods in Diverse Prokaryotic Systems

Chapter 17

Application of In Vitro Transposon Mutagenesis to Erythromycin Strain Improvement in *Saccharopolyspora erythraea*

J. Mark Weber, Andrew Reeves, William H. Cernota, and Roy K. Wesley

Abstract

Transposon mutagenesis is an invaluable technique in molecular biology for the creation of random mutations that can be easily identified and mapped. However, in the field of microbial strain improvement, transposon mutagenesis has scarcely been used; instead, chemical and physical mutagenic methods have been traditionally favored. Transposons have the advantage of creating single mutations in the genome, making phenotype to genotype assignments less challenging than with traditional mutagens which commonly create multiple mutations in the genome. The site of a transposon mutation can also be readily mapped using DNA sequencing primer sites engineered into the transposon termini. In this chapter an in vitro method for transposon mutagenesis of *Saccharopolyspora erythraea* is presented. Since in vivo transposon tools are not available for most actinomycetes including *S. erythraea*, an in vitro method was developed. The in vitro method involves a significant investment in time and effort to create the mutants, but once the mutants are made and screened, a large number of highly relevant mutations of direct interest to erythromycin production can be found.

Key words *S. erythraea*, In vitro transposon mutagenesis, Erythromycin, Strain improvement, Gene replacement, Actinomycetes

---

1 Introduction

The commercial production of antibiotics and other natural products from microbial sources using large-scale submerged fermentations is often economically challenging due to low production efficiencies of the compound of interest by the wild-type organism. Since the 1940s, strain improvement has been performed by random mutation and screening which often led to much better strains but the mutational site(s) were not identified. With today's technology and with access to the improved strains of the past it is now possible, by DNA sequence analysis, to catalog the mutations that make up improved commercial strains [1, 2]. However, taking this catalog and applying it to rational strain engineering has not been

Andrew Reeves (ed.), *In Vitro Mutagenesis: Methods and Protocols*, Methods in Molecular Biology, vol. 1498,
DOI 10.1007/978-1-4939-6472-7_17, © Springer Science+Business Media New York 2017

straightforward. This is because many mutations in commercial strains may be silent or deleterious. Therefore, identifying the beneficial mutations and determining their order of introduction using this "top-down" approach is still a laborious task.

One way of addressing the limitations of the "top-down" approach is to start with the wild type strain and work from the "bottom up," that is, to identify the mutations after each step of the strain improvement process, thus not only uncovering the identity of the most important mutations but also learning the order that they must be engineered into the strain to exert their effect. This could be performed using transposon mutagenesis combined with a small amount of DNA sequencing as described in this chapter, or it can be done using traditional mutagenesis and performing whole genome sequencing on each improved mutant. Since the transposon approach will not work in all strains of interest (for technical reasons), the traditional mutagenesis approach combined with whole genome sequencing is a plausible alternative. As information from "bottom-up" approaches is accumulated, assembled, and analyzed, the need for further bottom-up analyses such as this will diminish and general strain improvement principles will emerge that can be directly applied to new strain improvement programs.

This chapter describes a method for performing transposon-based strain improvement in *Saccharopolyspora erythraea* that leads to the identification of the genes influencing the production of erythromycin. A similar set of materials and methods have been previously published in Reeves and Weber [3], however the earlier chapter was directed towards *Aeromicrobium erythreum*, whereas this chapter is directed towards *Saccharopolyspora erythraea*. Some of the materials and methods overlap between these two procedures and many of the methods used in this procedure are modifications from methods originally described in Hopwood et al. [4] or Keiser et al. [5] for *Streptomyces coelicolor* and *Streptomyces lividans*.

Random mutagenesis and screening has a long history of success in the strain improvement of actinomycetes [6]; however, very little has been done when it comes to using transposons as the mutagenic agent despite the many advantages they can bring to the process such as creating single mutations that can be easily identified and mapped in the genome [7]. The development of this technology has been hindered by the fact that no generally applicable in vivo transposition tools have been found or developed for *Actinomyces*. The development of in vitro transposition by Goryshin and Resnikoff [8] however, opened up a new route for using transposons in actinomycete research. The general approach is to isolate genomic DNA from the actinomycete of interest, and then mutate that DNA in vitro with a transposon. The transposon-mutated DNA is then put back into the host in its natural location in the genome. A strain mutated this way is indistinguishable from a mutant made by in vivo transposon mutagenesis and the information gained is equally as powerful.

A technical requirement, that may be a limitation for many actinomycetes, is that to perform this procedure the strain must have an efficient method for acquiring exogenous DNA. For the wild type strain of *Saccharopolyspora erythraea* (ATCC 11635, NRRL 2338) a high transformation efficiency can be difficult to obtain, but we were able to mitigate this limitation through adjustments to the time of harvest of the mycelium. Another route around the transformation problem is to use strain variants or mutants that are more highly transformable. For *S. erythraea* the red-variant strain can be transformed with plasmid DNA at a much higher frequency than the wild type. The "red variant" is a deeply red pigmented strain on certain media and is a spontaneous variant (mutant) of the wild-type strain. Not all red variants carry the same genomic rearrangement, but most are significantly more transformable than the wild type strain. One issue with the red variant strains is that (in our hands) they produce less (~5-fold less) erythromycin; however, this limitation may be addressed through manipulation of the growth medium and conditions.

Another technical requirement involves a strain's ability to perform efficient gene replacements with incoming DNA. Gene replacement occurs in *S. erythraea* naturally once homologous DNA enters the cell cytoplasm but only about 10% of transformants actually carry out a double crossover event (depending on the size of the homologous fragment), as opposed to about 50% of the transformants that undergo a single crossover event. In the remainder of the transformants the plasmid DNA does not integrate at all and is segregated away during sporulation of the strain.

Despite the effort required for setting up an in vitro transposon mutagenesis experiment, the work performed is rewarded with the unambiguous identification of mutations that influence erythromycin production. Also, when the wild type strain is used (as opposed to a highly mutated production strain), these mutations appear at a high frequency (~3%) [8] and therefore many mutations can be found relatively easily. Despite the fact that intense interest is currently being focused on the many "top-down" -omics technologies available to study strain improvement, the contributions that transposons can offer has never been fully exploited and this "bottom-up" approach offers a unique and beneficial way to study this complex process.

2 Materials

2.1 Chemicals and Reagents

1. Lysozyme solution: 4 mg/ml lysozyme in a solution of 25 mM Tris buffer containing 25 mM EDTA, pH 8.0 (*see* **Note 1**).

2. Pronase E (*see* **Note 2**).

3. 5×-Tris–borate–EDTA buffer (TBE): (*see* **Note 3**) for noncommercial production.

4. 250 mM EDTA (ethylenediaminetetraacetic acid).

5. 10.3% sucrose solution. Prepared in distilled water and sterilized by autoclaving or 0.45 mm filter.

6. Phenol–chloroform.

7. Rnase A.

8. 3 M sodium acetate, pH 5.2.

9. Corex® glass centrifuge tubes.

10. TE buffer, pH 8.0.

11. 70% ethanol.

12. 100% ethanol.

13. Ethidium bromide, 1 mg/ml solution.

2.2 Media

1. TSB (Tryptic soy broth): Add 30 g of commerically bought TSB powder per liter of water. TSB is a general growth medium for actinomycetes.

2. Lysogeny broth (LB): tryptone 10 g/l, yeast extract 5 g/l, NaCl 10 g/l. pH to 7.0 with 5 N NaOH. For agar plates, add agar to 1.5%. Sterilize by autoclaving. For antibiotic supplementation ampicillin was added to liquid medium or plates at a final concentration of 100 μg/ml. X-gal and IPTG (isopropyl β-D-1-thiogalactopyranoside) was added as inducer (*see* **Note 4**).

3. E20A agar: Bacto Soytone, 5.0 g/l, soluble starch, 5.0 g/l, calcium carbonate, 3.0 g/l, MOPS buffer, 2.1 g/l, Bacto agar, 20.0 g/l. Add components to 900 ml of distilled water. After complete mixing raise final volume to 1 l. Sterilize by autoclaving.

4. SGGP medium: Bacto peptone, 4.0 g, Bacto yeast extract, 4.0 g, magnesium sulfate: heptahydrate ($MgSO_4 \cdot 7H_2O$), 0.5 g, distilled water to 1000 ml. pH adjust to 7, sterilize by autoclaving, then add sterile solutions of 50% D-(+)-glucose in distilled water, 20.0 ml and Potassium phosphate monobasic (0.5 M), 20.0 ml.

5. R2T2 plate medium: Before autoclaving, the following are added to a total of 1 l: sucrose, 103.0 g, potassium sulfate, 0.25 g, Bacto yeast extract 6.5 g, Bacto tryptone, 5.0 g, Bacto agar, 22.0 g. After autoclaving the following sterile solutions are added: D-(+)-glucose (50% w/v in distilled water) 20.0 ml, Trizma base (pH 7.0, 1 M), 25.0 ml, potassium phosphate monobasic (0.5% in distilled water) 5.0 ml, sodium hydroxide (NaOH, 1 N in distilled water) 2.5 ml, calcium chloride dihydrate ($CaCl_2 \cdot 2H_2O$, 1 M in distilled water) 50.0 ml, magnesium chloride hexahydrate ($MgCl_2 \cdot 6H_2O$, 1 M in distilled water) 50 ml, trace elements, 2.0 ml.

6. Trace element solution: zinc chloride (ZnCl$_2$) 0.040 g, ferric chloride hexahydrate (FeCl$_2$·6H$_2$O) 0.200 g, Cupric chloride dihydrate (CuCl$_2$·2H$_2$O) 0.010 g, manganese chloride tetrahydrate (MnCl$_2$·4H$_2$O) 0.010 g, sodium borate decahydrate (Na$_2$B$_4$O$_7$·10H$_2$O) 0.010 g, ammonium molybdate tetrahydrate (NH$_4$)$_6$Mo$_7$O$_{24}$·4H$_2$O) 0.010 g, distilled water to 1000 ml. Plates work in transformations best when dried to ~85 % of their original weight. This can be performed in a sterile laminar flow hood.

7. SOC medium: For SOB medium add tryptone 20 g, yeast extract 5 g, NaCl 0.5 g, KCl, 0.186 g. Add ddH$_2$O to 1000 ml. After all the components are dissolved add MgCl$_2$, anhydrous 0.952 g, and MgSO$_4$ (heptahydrate) 2.5 g. pH to 7.0 with NaOH. Add glucose 3.6 g (20 mM final concentration) to the SOB medium to make SOC. Sterilize by autoclaving.

2.3 Molecular Biology Reagents

1. EZ-Tn5™ < R6Kγori/KAN-2 > Insertion Kit (Epicentre Biotechnologies). Epicentre's EZ-Tn5 Transposon Tools (kits and reagents) are based on the hyperactive Tn5 transposition system developed by Goryshin and Reznikoff [9]. Reagents included in the transposition kit include: transposase, transposon, 10× reaction buffer, stop solution, forward and reverse primers.

2. Primers KanFP-1 and KanRP-1 to sequence from the ends of the transposon into the adjacent chromosomal DNA (Epicentre Technologies).

3. 2×PT buffer: sucrose, 200.0 g, potassium sulfate, 0.50 g, magnesium chloride hexahydrate, 10.17 g, trace elements solution [3], 2.0 ml, distilled water to 750 ml.

   At time of use complete by adding to 150 ml of above 2× PT solution: calcium chloride dihydrate (CaCl$_2$·2H$_2$O, 1 M in distilled water), 10.0 ml, TES buffer, pH 7.2, 0.25 M, 40.0 ml.

4. 1× PT Buffer: 2× PT Buffer (Protoplast Transformation Buffer), 50.0 ml, distilled water, 50.0 ml.

5. Thiostrepton overlay solution: mix 6.25 µl of a 25 mg/ml thiostrepton stock solution (in DMSO: ethanol, 50:50) into 1 ml distilled water.

6. Kanamycin overlay solution: Mix 7.5 µl of a 100 mg/ml stock solution of kanamycin in 1 ml distilled water.

7. 50 % PEG 10,000: polyethylene glycol (average molecular weight 10,000), 50 g, distilled water to 100 ml.

8. QIAquick gel purification kit.

9. Plasmid purification kit.

10. Restriction enzymes.

11. T4 DNA ligase and buffer.

3   Methods

Carry out all procedures at room temperature unless otherwise specified.

3.1 S. erythraea Genomic DNA

1. Use spores of *S. erythraea* (we used *S. erythraea mutB* FL2302) to inoculate 250 ml shake flasks containing 30 ml of TSB medium. Incubate while shaking at 375 rpm for 48–72 h at 32 °C (*see* **Note 5**).

2. Harvest the cells from the culture by centrifugation (10 min at 13,700×g).

3. Resuspend cells in 10 ml of lysozyme solution (4 mg/ml). Incubate at 37 °C for 15 min or until cells become translucent.

4. Add 5 ml of 250 mM EDTA, 250 μl of pronase E (10 mg/ml), 1.4 ml of 10% SDS and incubate at 37 °C for 30 min.

5. Add 5 ml of phenol–chloroform and mix thoroughly. Separate the liquid phases by centrifugation at 13,700×g for 10 min. Remove viscous upper aqueous phase (10–12 ml) using a wide-bore pipet tip.

6. Add 40 μl of 10 mg/ml RNase and incubate at 37 °C for 1 h.

7. Extract once more with 2.5 ml of phenol–chloroform, separate the liquid phases by centrifugation at 13,700×g for 10 min.

8. Transfer supernatant (12 ml) to a 30 ml Corex® glass centrifuge tube, add 1.5 ml of 3 M sodium acetate, pH 5.2 and 11 ml of isopropanol. Mix gently but thoroughly to extract DNA.

9. A compact white DNA globule will appear that can be removed from solution with a hooked pasteur pipet. Dip the DNA globule into 10 ml of 70% ethanol to wash off the salts. Repeat this step, then transfer the globule to 1 ml of TE pH 8.0.

10. Put the DNA solution on a gentle rocker to mix. The DNA globule will be completely dissolved in the buffer after overnight rocking incubation at 37 °C.

3.2 Fragmentation of S. erythraea DNA

1. Fragmentation is performed through partial enzymatic digestion of the *S. erythraea* chromosomal DNA using DNA restriction enzyme pairs: (1) *Bam*HI, *Eco*RI; (2) *Eco*RI, *Hind*III; and (3) *Hind*III, *Sac*I. These pairs of enzymes were chosen based on the restriction sites available for forced cloning into pFL2073 [9] (*see* **Note 6**) (Fig. 1).

2. Although each reaction for the partial digestion of the *S. erythraea* chromosomal DNA has to be optimized by trial and error, we found that excellent results were obtained by mixing 140 μl of water, 25 μl of chromosomal DNA (from **step 5** in Subheading 3.1), 20 μl of 10× restriction enzyme buffer, 5 μl of the first restriction enzyme (e.g., *Bam*HI), and 10 μl of a

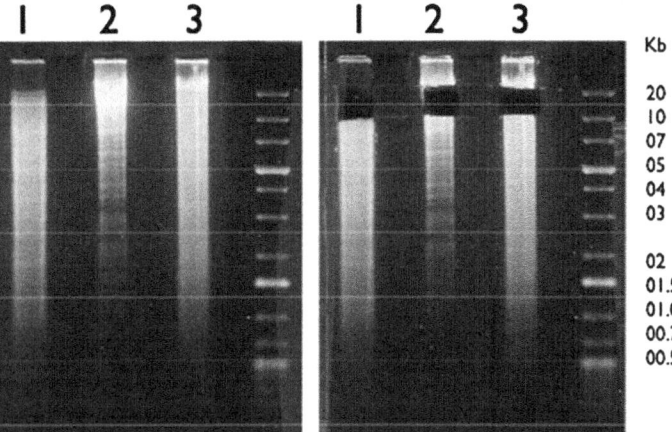

Fig. 1 Digestion of *S. erythraea* DNA by double restriction enzyme method and physical removal of DNA in the desired size range by scalpel. *Lane 1, BamHI + EcoRI; Lane 2, EcoRI + HindIII; Lane 3, HindIII + SacI*

second enzyme (e.g., *Eco*RI). The reactions were incubated at 37 °C for 60 min at which point 0.5 M EDTA and 20 µl of 6× blue DNA loading dye were added.

3. The partially digested DNA was immediately electrophoresed through a 100 ml 0.8% agarose gel in 1× TBE buffer containing ethidium bromide (2 µl of a 1 mg/ml solution). Gel is run at 15 V overnight. Visualization of the DNA was performed under UV light illumination and a DNA size ladder was run on the gel to estimate the size of DNA fragments.

4. DNA in the desired size range was sliced from the agarose gel with a scalpel and extracted using the QIAquick Gel Extraction Kit Protocol using a microcentrifuge.

3.3   Cloning of S. erythraea DNA

1. Plasmid pFL2073 [9] was digested with pairs of restriction endonucleases matching the enzymes used to create the chromosomal DNA fragments as in **step 1** in Subheading 3.2. The digested vector DNA was purified using a QIAquick gel purification procedure (*see* **Note 7**).

2. An example ligation was performed by combining pFL2073 digested with *Bam*HI and *Eco*RI with an excess of 10 kb fragments of *S. erythraea* chromosomal DNA partially digested with *Bam*HI and *Eco*RI. This mixture of the two DNA preparations was heated at 65 °C for 10 min then put on ice for 5 min. 1 µl of 10× ligation buffer was added along with 1 µl of ligase and the mixture was held at room temperature for 18 h at which point it was diluted 1:5 with water.

3.4   E. coli DNA Transformation

1. 1 µl of the reaction mixture from **step 2** in Subheading 3.3 was used for the transformation of *E. coli*. Transformants were selected on LB amp-100 plates with X-gal and IPTG inducer added.

2. Approximately 85 % of the transformants from the ligation in **step 1** displayed the "white" phenotype on X-gal plates indicating successful cloning. The white transformants were analyzed for plasmid content and 70 % were found to have inserts with an average size of over 10 kb. Insert size was determined by digesting the plasmid with *Pst*I and adding up the size of the fragments that were not part of the vector DNA. Plasmids carrying fragments larger than 9 kb and with similar copy numbers were combined in pools of approximately 20 plasmids each. Numerous different plasmid pools were prepared in this manner from the different ligation mixtures and were used in separate in vitro transposition reactions.

3.5 In Vitro Mutagenesis

1. The mutagenesis reaction mixture was performed using the EZ-Tn5 < R6Kγ ori/Kan-2 > kit (Epicentre Biotechnologies) [8] that was comprised of 10× reaction buffer (2 μl), the target plasmid pool DNA (16 μl, see above), transposon oriV-kan2 (1 μl), and transposase (0.5 μl). The reaction was held at 37 °C for 2 h, then terminated by the addition of 10× stop buffer (2 μl) and incubation of the mixture at 70 °C for 10 min.

2. A 1 μl portion of the in vitro mutagenesis reaction mixture containing the mutagenized plasmids was used to transform *E. coli* DH5α cells (50 μl). Following transformation, the cells were grown in SOC medium at 37 °C for 1 h then plated on LB supplemented with kanamycin at 40 μg/ml, ampicillin sulfate at 50 μg/ml and inducer. The plates were then transferred to room temperature and grown out another 48 h at which point all the cells were harvested and the DNA isolated using a DNA plasmid purification kit such as Genejet or Qiagen DNA extraction and purification methods.

3.6 S. erythraea Protoplasts

1. Inoculate 25 ml of SGGP broth in a 250 ml shake flask with spores from a 10–14 day old E20A agar plate of *S. erythraea* FL2302 (*mutB* knockout).

2. Inoculate shake flasks at five different time points so that at harvest time the cultures fall within a range of 40–96 h old (e.g., 45, 53, 64, 72, 88 h) (*see* **Note 8**).

3. Incubate the shake flasks at 375 rpm (1-in. orbit) at 32 °C.

4. At time of harvest, pellet cells by centrifugation at $3220 \times g$ for 10 min (*see* **Note 9**). Use disposable tubes or tubes not washed with detergent throughout. Discard supernatant.

5. Wash pellet with 20 ml of distilled water. Pellet the cells using the same procedure as above and discard supernatant.

6. Wash cells a second time with 10 ml of 10.3 % sucrose. Pellet the cells as before and discard supernatant.

7. Resuspend cells in 10 ml of 1× PT buffer containing 4 mg/ml of lysozyme.

8. Transfer cells to a 15 ml conical tube, and incubate at 32 °C for at least 1 h.

9. Promote the release of protoplasts from the mycelium by gently inverting the sealed tube at ~10 min intervals during the 1-h incubation. At the end of the incubation period perform a final mixing by drawing the mycelium/protoplast mixture into and out of a 10 ml pipet one time. Allow the undigested mycelium to settle to the bottom of the tube which takes at least 15 min.

10. Transfer supernatant containing protoplasts into a fresh 15 ml conical tube using a pipette. Be careful not to draw up any undigested cells at the bottom of the tube.

11. Gently pellet protoplasts (suggested: $1500 \times g$ for 30 min on a benchtop centrifuge). Draw off the supernatant with a disposable pipet and discard.

12. Add an equal volume of 1× PT to the volume of the protoplasts to resuspend them as a dense milky suspension. Perform transformation procedure immediately (see below), or place protoplasts on ice and freeze them at –80 °C for long-term storage (*see* **Note 10**).

3.7 S. erythraea DNA Transformation

1. Transfer 50 μl of protoplasts suspended in 1× PT into 1.5 ml Eppendorf tubes (*see* **Note 11**).

2. Add 10 μl of highly purified and concentrated plasmid DNA (~10 μg) (*see* **Note 12**).

3. Add 200 μl of 25 %-polyethylene glycol (PEG, 10,000 mw) in 1× PT and mix gently.

4. Spread the mixture evenly over a R2T2 plate that has been dried with an even surface to 85 % of its original weight.

5. Include a control where no DNA is added but drug selection is applied to the protoplasts to check for leak-through of spontaneous drug-resistant mutants.

6. Place the plates in the sterile laminar flow hood, if necessary, to speed the drying of the protoplast/PEG puddles on the plates.

7. Incubate plates overnight at 32 °C and low (20 %) humidity, if possible.

8. After protoplasts have regenerated (24–28 h), overlay with drug by covering each protoplast sector individually with the drug solution (~250 μl). This will be either a thiostrepton or a kanamycin selection (*see* Subheading 2.3).

9. On each sector (except the no DNA control sector), gently spread drug solution over the entire sector of regenerated cells (125 μl for a 1/8 sector).

10. For the no-DNA control, place a small pool of 100 μl of the drug solution onto the center of the sector, leaving a perimeter of the sector uncovered to serve as a regrowth control.

11. Twenty-four hours after the overlay, perform a sterile transfer of the agar pad with the regenerated cells on top onto a fresh uninoculated E20A agar plate to promote growth and sporulation of the primary transformants. This also keeps the plate from drying out too quickly in the desiccated incubator.

12. Spores from transformants can be harvested 7–10 days later. Plate diluted (10^4) spores on E20A-Kan25 agar plates (kanamycin at 25 μg/ml in E20A agar) and screen the resulting single colonies by replica plating to find those that have undergone gene replacement and are therefore kanamycin-resistant and thiostrepton-sensitive. Ten percent of the colonies screened this way should have undergone a gene replacement event and these are the colonies that will be screened in the next step.

3.8 Erythromycin Production Screen

1. Pour E20A agar (about 250 μl/well) into 96-well flat-bottom plates using a multichannel pipettor. Allow plates to cool and solidify.

2. Using a sterile sharp-end toothpick, take a small representative quantity of spores from the thiostrepton-15 plate kanamycin-resistant/thiostrepton-resistant patch or replica plate and inoculate one well into the 96-well plate. Using a fresh sterile toothpick, repeat for a duplicate sample.

3. Inoculate two wells with the parent *S. erythraea* FL2302 and two wells with wild-type strain FL2267.

4. Incubate plate upside down (to avoid condensation), slightly askew from cover to allow circulation at 30 °C. Incubate for 5 days (*see* **Note 13**).

5. Prepare an erythromycin standards 96-well plate one day ahead of the bioassay plate inoculation.

6. Transfer plugs using a sterile flat toothpick to slide the plug out onto the surface of a *Bacillus subtilis* prepared bioassay plate (*see* **Note 14**). Alternatively, a forced air procedure can be used to lift the agar plug from the microtiter well (*see* **Note 15**).

7. After incubation of the bioassay plate, measure the clear zones and calculate the differences.

3.9 Plasmid Rescue

1. Prepare chromosomal DNA from strains containing an inserted transposon mutation.

2. Digest 17 μl DNA with *Nar*I or *Sal*I in a 20 μl reaction. Add 70 μl water, heat at 70 °C, 10 min. Cool on ice.

3. Add 10 μl of 10× ligase buffer and 1 μl ligase, incubate at room temperature for 1 h.

4. Add 10 μl tRNA (10 mg/ml), 10 μl of 3 M sodium acetate (pH 5.2), 250 μl of 100 % ethanol. Cool on ice for 20 min.

5. Centrifuge (15 min), pour off ethanol, revealing small white pellet of DNA at bottom of tube.

6. Wash with 500 μl of cold 70% ethanol, repeat wash once. Remove all traces of liquid.

7. Resuspend the DNA pellets in 25 μl water, add 5 μl of DNA to 20–35 μl of cold cells in a cold microcentrifuge tube.

8. Transfer the cold cells and DNA to a cold electro-cuvette. Electroporate cells at 2.5 kV.

9. Add 1 ml of SOC broth and incubate the broth containing cells at 37 °C for 1 h.

10. Plate 20 μl of cells directly on LB plates with kanamycin (40 μg/ml).

11. Centrifuge the broth and cells at 3220×g for 5 min to gently pellet cells. Decant supernatant and resuspend cells in 50 μl of 20% glycerol. Plate 20 μl of cells in a sector of a LB plate supplemented with kanamycin (40 μg/ml).

12. Incubate at 37 °C overnight, colonies appear by morning (*see* **Note 16**).

13. Prepare high-purity plasmid DNA from the colonies that appear and submit for DNA sequence analysis.

14. Use primers KanFP-1 and KanRP-1 to sequence from the ends of the transposon and the DNA sequence information will provide the transposon insertion site in *S. erythraea* chromosomal DNA (*see* **Note 17**) (*see* Fig. 2).

4 Notes

1. Lysozyme digests the peptidoglycan cell wall of the actinomycete mycelium.

2. Pronase E is used to degrade proteins during nucleic acid purification.

3. TBE Recipe (1 l of 5× stock solution): 54 g of Tris base, 27.5 g of boric acid, 20 ml of 0.5 M EDTA (pH 8.0). Adjust pH to 8.3 with HCl. TBE can be diluted to 1× or 0.5× prior to use in electrophoresis.

4. In gene cloning, X-gal (bromo-chloro-indolylgalactopyranoside) is a visual indicator of expression of β-galactosidase in blue/white screening, a method for distinguishing a successful cloning product. X-gal was added to a final concentration of 32 μM. IPTG (isopropyl-β-D-thiogalactopyranoside) was used at a final concentration of 160 μM and is an analog of a lactose metabolite that triggers transcription of the *lac* operon, including the β-galactosidase gene.

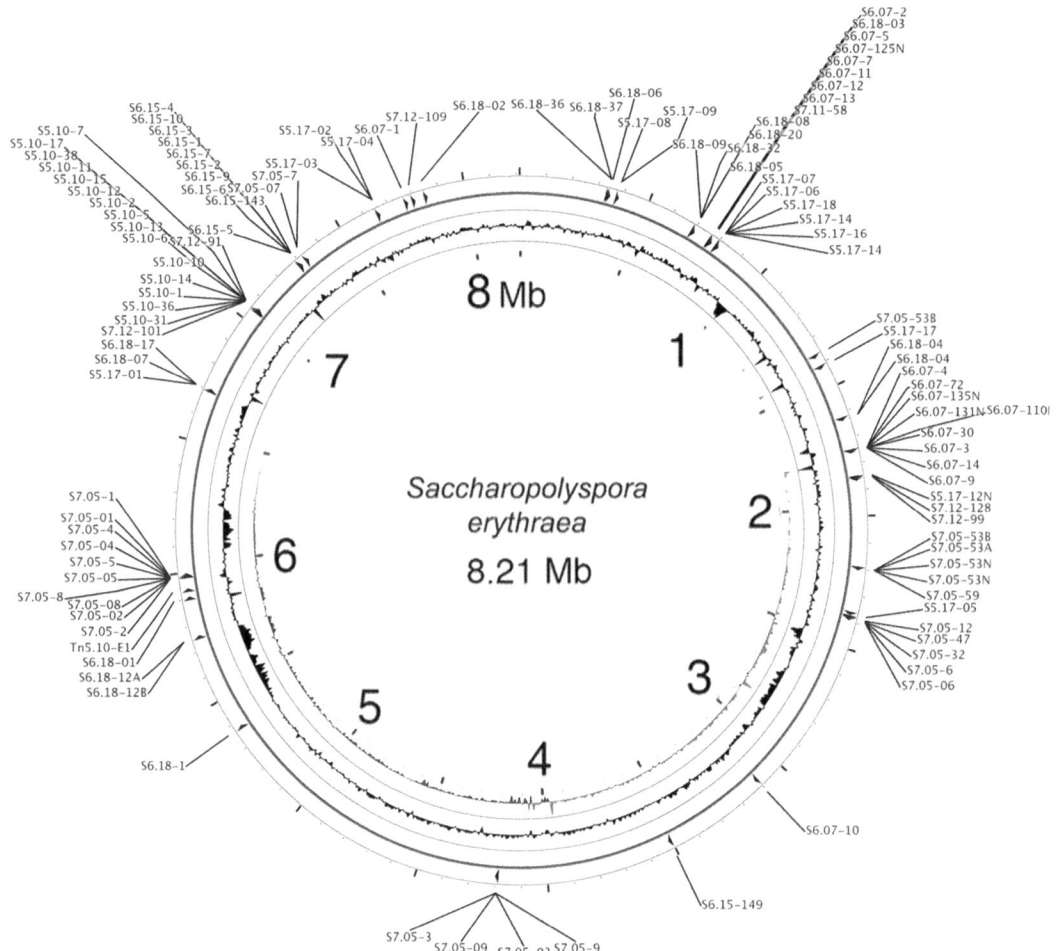

Fig. 2 Genome map of *S. erythraea* showing transposon insertion sites of gene-replacement mutants generated in this study. Map is based on data generated by Oliynyk et al. [10]

5. The choice of strain is based on the goals of the experiment. In the ideal situation a wild type strain is used to find the greatest number of genetic targets for erythromycin strain improvement.

6. Multiple combinations of two restriction enzymes were used (for forced cloning) rather than using a single frequent cutting enzyme (such as *Mbo*I or *Sau*3A) to improve the cloning efficiency for large fragments of DNA into pFL2073 (*see* Fig. 1).

7. An important consideration for this procedure is the choice of cloning vector. pFL2073 was chosen because it can stably and autonomously replicate in *E. coli* even with 10–15 kb inserts of *S. erythraea* chromosomal DNA. *E. coli*-derived pFL2073, containing 10–15 kb inserts of *S. erythraea* chromosomal DNA, can then be transformed into *S. erythraea* where it will

integrate into the chromosome through homologous recombination. The *tsr* (thiostrepton-resistance) gene allows for selection of transformants in *S. erythraea*.

8. Transformation of mutagenized plasmid DNA into *S. erythraea* requires careful timing of the harvesting of mycelium in order to obtain high efficiency transformations, but the timing varies greatly from one experiment to the next.

9. Avoid packing protoplasts into tight pellets. A slower, gentler centrifugation will help preserve the viability of the protoplasts.

10. Protoplasts of *S. erythraea* can withstand multiple freeze–thaw cycles and still maintain their high transformation efficiency.

11. If thawed protoplasts are being used for the transformation, gently spin the tube containing the protoplasts (e.g., $845 \times g$ for 7 min in a microcentrifuge) to pellet the protoplasts. Discard the supernatant which may contain nucleases, add PEG and DNA as before. For the transformation use highly purified and highly concentrated DNA (Qiagen or similar plasmid purification kits are recommended). Gentle handling of protoplasts may be an important factor in achieving high transformation efficiencies. At no point should protoplasts be subjected to high-speed centrifugation or vortex mixing.

12. Highly purified gDNA preparations must be used at high concentration (20 μg of DNA per reaction). Protoplasts should ideally be produced from cultures that were inoculated from freshly harvested spores.

13. The most practical method for high throughput screening of *S. erythraea* mutants was in solid-phase microtiter fermentations. Shake flask fermentations are still the gold standard for strain improvement determinations and can also be used in the primary testing as well as repeat testing phases. Different mutations will be found depending upon the screening method.

14. Preparation of Bioassay plates. TSB agar is prepared by adding 30 g/l tryptic soy broth (Difco) to distilled water, and adding 15 g/l Fisher agar with a stir bar included in the container (2 l flask). After 50 min of sterilization at 121 °C, 18 psi, agar is placed in a preheated water bath to prevent solidification. Sterile bioassay plates are placed on a level surface, and a 100-ml bottom layer of TSB agar is evenly poured. Plates are gently tilted to ensure even coverage. The bottom layer is allowed to solidify completely, until the agar is at room temperature (10–15 min). The remaining agar for the second layer is kept in a water bath on a stirring heat block at approximately 90–100 °C to prevent solidification while the first layer cools. Just prior to pouring the second layer, add 2.5 ml of a 1% tetrazolium red solution/l and 100–200 μl/l *B. subtilis* spores

to the remaining agar. TSB is mixed thoroughly by swirling and stir bar. The second layer of 100 ml TSB agar containing dye and spores is then poured evenly over the first layer, then allowed to cool completely (30–60 min). Cooled plates were dried in a circulating biohood for 10 min to prevent condensation due to excess moisture in the agar.

15. Compressed air removal of agar plugs from mictrotiter wells. Materials needed are a microtiter plate with agar plugs to be removed, a short Pasteur Pipette attached to rubber tubing, a source of compressed air, forceps, and bioassay plates (see above). The method is: (1) turn on compressed air to lowest pressure with tubing and Pasteur pipette attached; (2) carefully aim pipette down central hole of agar plug all the way until it is flush with bottom of well; (3) carefully move tip to loosen plug and then slightly raise pipette tip; (4) air pressure will force plug up the outside of the shaft on the pipette while making sure no agar fragments adhere to the bottom of the plug preventing it from being flush with the surface of the bioassay plate; and (5) place pipette over assay site on bioassay plate and slide plug down to plate with forceps making sure the bottom of the plug is flush with the surface of the bioassay plate.

16. When the plasmid is transformed into a *pir+ E. coli* strain, it propagates using the R6Kγ origin of replication contained within the transposon. The sequence of chromosomal DNA flanking the transposon is determined using the primers that read off both ends of the transposon. Since the procedure is simple and rapid, many mutant hits can be analyzed this way, including mutants that have a decrease in erythromycin production. Yield-lowering mutants will also provide insights into secondary metabolic processes.

17. Primers can be purchased directly from Epicentre Technologies or as part of the EZ:Tn in vitro transposon mutagenesis kit. The primers are designed to read off the ends of the Tn5-based transposon to identify the transposon insertion site. They can be used to sequence directly off the transposons from the chromosomal templates, or more preferably with rescued plasmid DNA (*see* Fig. 2).

References

1. Peano C, Talà A, Corti G, Pasanisi D, Durante M, Mita G, Bicciato S, De Bellis G, Alifano P (2012) Comparative genomics and transcriptional profiles of *Saccharopolyspora erythraea* NRRL2338 and a classically improved erythromycin over-producing strain. Microb Cell Fact 11:32

2. Li YY, Chang X, Yu WB, Li H, Ye ZQ, Yu H, Liu BH, Zhang Y, Zhang SL, Ye BC, Li YX (2013) Systems perspectives on erythromycin biosynthesis by comparative genomic and transcriptomic analyses of *S. erythraea* E3 and NRRL23338 strains. BMC Genomics 14:523

3. Reeves AR, Weber JM (2012) Metabolic engineering of antibiotic-producing actinomycetes using *in vitro* transposon mutagenesis. Methods Mol Biol (Clifton, NJ) 834:153–175

4. Hopwood DA, Bibb MJ, Chater KF, Kieser T, Bruton CJ, Kieser HM, Lydiate DJ, Smith CP, Ward JM, Schrempf H (1985) Genetic manipulation of *Streptomyces*. A laboratory manual. The John Innes Foundation, Norwich, England

5. Kieser T, Bibb MJ, Buttner MJ, Chater KF, Hopwood DA (2000) Practical *Streptomyces* genetics. The John Innes Foundation, Norwich, England

6. Vinci VA, Byng G (1999) Strain improvement by nonrecombinant methods. In: Demain AL, Davies JE (eds) Manual of industrial microbiology and biotechnology, 2nd edn. ASM Press, Washington, DC, pp 103–113

7. Griffiths AJF, Miller JH, Suzuki DT, Lewontin RC, Gelbart WM (2000) An introduction to genetic analysis. Review of transposable elements in prokaryotes, 7th edn. W. H. Freeman, New York, http://www.ncbi.nlm.nih.gov/books/NBK21805/

8. Goryshin IY, Reznikoff WS (1998) Tn5 *in vitro* transposon mutagenesis. J Biol Chem 273:7367–7374

9. Fedashchin A, Cernota WH, Gonzalez MC, Leach BI, Kwan N, Wesley RK, Weber JM (2015) Random transposon mutagenesis of the *Saccharopolyspora erythraea* genome reveals additional genes influencing erythromycin biosynthesis. FEMS Microbiol Lett 362:22

10. Oliynyk M, Samborskyy M, Lester JB, Mironenko T, Scott N, Dickens S, Haydock SF, Leadlay PF (2007) Complete genome sequence of the erythromycin-producing bacterium *Saccharopolyspora erythraea* NRRL23338. Nat Biotechnol 25:447–453

Chapter 18

Engineering Gram-Negative Microbial Cell Factories Using Transposon Vectors

Esteban Martínez-García, Tomás Aparicio, Víctor de Lorenzo, and Pablo I. Nikel

Abstract

The construction of microbial cell factories *à la carte* largely depends on specialized molecular biology and synthetic biology tools needed to reprogram bacteria for modifying their existing functions or for bestowing them with new-to-Nature tasks. In this chapter, we document the use of a series of broad-host-range mini-Tn5 vectors for the delivery of gene(s) into the chromosome of Gram-negative bacteria and for the generation of saturated, random mutagenesis libraries for studies of gene function. The application of these tailored mini-transposon vectors, which could also be used for chromosomal engineering of a wide variety of Gram-negative microorganisms, is demonstrated in the platform environmental bacterium *Pseudomonas putida* KT2440.

Key words Mini-transposon, Tn5 transposon, *Pseudomonas putida*, *Escherichia coli*, Synthetic biology, Metabolic engineering, Microbial cell factory, Genome editing

1 Introduction

Mini-transposon vectors allow for the stable insertion of foreign DNA into the chromosome of many types of Gram-negative bacterial targets [1, 2]. Tn5-derived elements [3] present clear advantages over the use of their plasmid-based counterparts for the random interruption of gene(s), or for the introduction and expression of heterologous genes into several bacterial species. These features include, but are not limited to: (1) maintenance of the corresponding transgenes without antibiotic selective pressure; (2) long-term stability of the constructs and re-usability of the functional DNA parts; and, furthermore, (3) mini-Tn5-based vectors enable the cloning and chromosomal delivery of considerably long DNA fragments (which would be cumbersome to manipulate in other DNA delivery tools). As the transposase gene *tnpA* is lost following each transposition event [4, 5], one added value of mini-Tn5 vectors is the possibility to use them recursively in the same

Andrew Reeves (ed.), *In Vitro Mutagenesis: Methods and Protocols*, Methods in Molecular Biology, vol. 1498, DOI 10.1007/978-1-4939-6472-7_18, © Springer Science+Business Media New York 2017

microbial host, provided that they bear different selection markers. Since the TnpA transposase tends to act in *cis* [6], it promotes the insertion of DNA sequences borne by the plasmid irrespective of any previous DNA insertions in a given chromosome. These features allow for the delivery and integration of various DNA cargoes into the same target genome. However, the original layout of such mini-transposon vectors was not exempt of downsides. One of them is the unavoidable inheritance of long, nonfunctional DNA fragments stemming from the intricate cutting-and-pasting DNA methods available at the time when the original vectors were constructed. These procedures were also afflicted by the presence of an excessive and inconvenient number of non-useful restriction sites scattered along the plasmids, and the suboptimal transposition machinery encoded therein.

Martínez-García et al. [7] thoroughly revisited the original mini-Tn5 transposon vector concept. The most attractive features of the mini-Tn5-aided mutagenesis procedure have been enhanced while each of its drawbacks (identified along >20 years of use in many independent laboratories worldwide) has been eliminated [8]. The functional modules that constitute the vector (including the mosaic elements, MEs) have been edited to minimize the length of the corresponding DNA fragments, improving their functionality and making them entirely modular and exchangeable. The final product was the entirely synthetic plasmid construct termed pBAM1 (*b*orn-*a*gain *m*ini-transposon). This design was soon followed by a series of synthetic, modular broad-host-range mini-Tn5 plasmids derived from pBAM1. These vectors, termed pBAMDs vectors [9], enable the possibility of easy cloning and subsequent chromosomal insertion of functional DNA cargoes with three different and interchangeable antibiotic resistance markers. Another set of pBAM1-derived plasmids, termed pBELs and pBEXs vectors [10], were designed to exploit the possibility of delivering DNA cargoes under the control of regulated gene expression modules (i.e., LacI^Q/*Ptrc* in pBELs vectors or XylS/*Pm* in pBEXs vectors). Furthermore, the antibiotic-resistance determinants in the mini-transposon modules of the pBELs and pBEXs vectors can be removed by means of the FLP recombinase from *Saccharomyces cerevisiae* [11]. In all the cases presented above, the functional parts of the mini-transposon vectors can be easily swapped by digestion with the appropriate restriction enzymes, allowing for the easy shuffling of each DNA element as needed. Finally, the multiple cloning site of all the mini-transposon vectors share the same set of restriction sites, which eases the subcloning of DNA cargoes by making them compatible with plasmids from the *S*tandard *E*uropean *V*ector *A*rchitecture (SEVA) initiative [12, 13].

The expansion of the available mini-transposon tools is a step forward in our efforts to purposely engineer microbial cell factories, based on environmental bacteria. *Pseudomonas putida* KT2440 is a robust host for strong oxidative bioreactions [14–16], it exhibits the GRAS (i.e., *g*enerally *r*ecognized *a*s *s*afe) status [17], and it has the inherent ability to grow on a wide range of (often, difficult-to-degrade) substrates [18–21]. Rewiring its extant genetic features to broaden its metabolic potential—or even introducing new-to-Nature functions—is a task continuously undergoing in our laboratory. In the present chapter, we detail all the experimental steps needed to either (1) construct random mutant libraries by mini-Tn5 insertions to explore gene-function relationships, or (2) deliver a DNA cargo into a target chromosome, with the option of FLP-catalyzed removal of the antibiotic resistance determinant.

2 Materials

2.1 Strains and Plasmids

1. The bacterial strains, plasmids and vectors used in this protocol are described in Tables 1 and 2, respectively. The sequences of the following vectors have been deposited in the GenBank

Table 1
Bacterial strains used in this protocol

Bacterial strain	Description and genotype	Relevant characteristics[a]	Reference
Escherichia coli			
CC118 λpir	Cloning host for plasmids containing an R6K origin of replication; Δ(*ara-leu*) *araD* Δ*lac*X174 *galE galK phoA20 thi-1 rpsE rpoB argE* (Am) *recA1*, λ*pir* lysogen	Sp^R, Rif^R, Thi⁻, Leu⁻	[42]
SM10 λpir	Cloning and mobilizing host for plasmids containing an R6K origin of replication; F⁻ *thi-1 thr leu tonA lacY glnV recA*::RP4-2-Tc::Mu, λ*pir* lysogen	Km^R, Thi⁻, Thr⁻, Leu⁻	[43]
S17-1 λpir	Cloning and mobilizing host for plasmids containing an R6K origin of replication; F⁻ *recA1 endA1 thiE1 pro-82 creC510 hsdR17* RP4-2-Tc::Mu-Km::Tn7, λ*pir* lysogen	Sm^R/Sp^R, Tp^R, Thi⁻, Pro⁻	[44]
MFD λpir	Cloning and mobilizing Mu-free host for plasmids containing an R6K origin of replication; F⁻ λ⁻ *ilvG rfb-50 rph-1* RP4-2-Tc::[ΔMu1::*aac* [3] IV Δ*aphA* Δ*nic35* ΔMu2::*zeo*] Δ*dapA*::(*erm-pir*) Δ*recA*	$Apra^R$,Zeo^R,Erm^R,DAP⁻	[45]
HB101	Mating helper strain; F⁻ λ⁻ *hsdS20*(rB⁻ mB⁻) *recA13 leuB6*(Am) *araC14* Δ(*gpt-proA*)*62 lacY1 galK2*(Oc) *xyl-5 mtl-1 thiE1 rpsL20 glnX44^B*(AS)^B	Sm^R, Thi⁻, Leu⁻, Pro⁻	[46]
Pseudomonas putida			
KT2440	Wild-type strain; derivative of strain mt-2 [47] cured of the TOL plasmid pWW0	Prototroph	[48]

[a]Antibiotic and auxotrophy markers: Apra, apramycin; Erm, erythromycin; Km, kanamycin; Rif, rifampicin; Sp, spectinomycin; Sm, streptomycin; Tp, trimethoprim; Zeo, zeocin; DAP, diaminopimelic acid; Leu, leucine; Thi, thiamine (vitamin B1); Thr, threonine; and Pro, proline. Please note that not all these features are used in the experiments described in the present protocol

Table 2
Plasmids used in this protocol

Plasmid	Description and relevant characteristics[a]	Reference
pRK600	Helper plasmid used for conjugation; $oriV$(ColE1), RK2(mob^+ tra^+); derivative of plasmid pRK2013 [49]; Cm^R	[50]
pBAM1[b]	Mini-Tn5 delivery plasmid; $oriV$(R6K), $oriT$; Ap^R, Km^R	[7]
pBAM1-GFP[b]	Mini-Tn5 delivery plasmid to create random $gene$::gfp fusions by insertion; $oriV$(R6K), $oriT$; Ap^R, Km^R	[7]
pBAMD1-2[b]	Mini-Tn5 delivery plasmid; $oriV$(R6K), $oriT$; Ap^R, Km^R	[9]
pBAMD1-4[b]	Mini-Tn5 delivery plasmid; $oriV$(R6K), $oriT$; Ap^R, Sm^R/Sp^R	[9]
pBAMD1-6[b]	Mini-Tn5 delivery plasmid; $oriV$(R6K), $oriT$; Ap^R, Gm^R	[9]
pBELK	Mini-Tn5 delivery plasmid for inserting a DNA cargo under the control of the LacI^Q/$Ptrc$ expression system; oriV(R6K), oriT; Ap^R, Km^R	[10]
pBELG	Mini-Tn5 delivery plasmid for inserting a DNA cargo under the control of the LacI^Q/$Ptrc$ expression system; $oriV$(R6K), $oriT$; Ap^R, Gm^R	[10]
pBEXK	Mini-Tn5 delivery plasmid for inserting a DNA cargo under the control of the XylS/Pm expression system; $oriV$(R6K), $oriT$; Ap^R, Km^R	[10]
pBEXG	Mini-Tn5 delivery plasmid for inserting a DNA cargo under the control of the XylS/Pm expression system; $oriV$(R6K), $oriT$; Ap^R, Gm^R	[10]
pFLP2	Helper plasmid used to eliminate antibiotic markers flanked by FRT sequences; $oriV$(pRO1600), RK2(mob^+tra^+), $oriT$, λP_R::FLP, $\lambda cI857$, $sacB$; Cb^R	[39]

[a]Antibiotic markers: Ap ampicillin, Km kanamycin, Cm chloramphenicol, Sm streptomycin, Sp spectinomycin, Gm gentamicin, Cb carbenicillin
[b]Plasmids pBAM1 (ID 60487), pBAMD1-2 (ID 61564), pBAMD1-4 (ID 61565), and pBAMD1-6 (ID 61566) have been deposited in AddGene (Cambridge, MA, USA; http://www.addgene.org). Plasmid pBAM1-GFP (ID BAA-2426) is also available at the American Type Culture Collection (ATCC™; Manassas, VA, USA; http://www. lgcstandards-atcc.org)

database (http://www.ncbi.nlm.nih.gov/genbank) with the indicated GenBank accession numbers: pBAM1 (HQ908071), pBAM1-GFP (HQ908072), pBADM1-2 (KM403113), pBAMD1-4 (KM403114), pBAMD1-6 (KM403115), pBELG (JQ663863), pBELK (JQ663864), pBEXG (JQ663865), and pBEXK (JQ663866).

2.2 Media Preparation

Unless otherwise stated, all the culture medium components and chemicals described below were purchased from Sigma-Aldrich Co. (St. Louis, MO, USA) or other major suppliers of molecular biology reagents and kits. Whenever appropriate, please follow all waste disposal regulations when discarding waste materials.

1. Lysogeny broth (LB) medium: tryptone 10 g, yeast extract 5 g, and NaCl 10 g. These are dissolved and brought up to 1 l with deionized H_2O, the pH is adjusted to 7.0 with 5 N NaOH

and the broth is sterilized by autoclaving (20 min at 121 °C and 1.05 kg/cm^2). This culture medium can be indefinitely stored at room temperature when protected from light.

2. M9 minimal medium: A 10× stock of M9 salts is prepared by dissolving 42.5 g of Na$_2$HPO$_4$·2H$_2$O, 15 g of KH$_2$PO$_4$, 2.5 g of NaCl, and 5 g of NH$_4$Cl in deionized H$_2$O up to a final volume of 500 ml. Sterilize the 10× salts medium by autoclaving as described above. To make M9 medium, mix the following sterile components: ~800 ml of distilled H$_2$O, 100 ml of 10× M9 salts, 2 ml of 1 M MgSO$_4$, 20 ml of 20 % (w/v) sodium citrate, and 100 μl of 1 M CaCl$_2$. Adjust to 1000 ml with sterile distilled H$_2$O. Nutritional selection is employed as a general strategy to counter-select for *P. putida*. The M9 minimal medium is supplemented in this study with sodium citrate at 0.2 % (w/v) as the sole carbon source (*see* **Note 1**) and is used for this purpose since *E. coli* cannot grow on citrate. A 20 % (w/v) sodium citrate solution is prepared by dissolving 20 g of anhydrous sodium citrate in deionized H$_2$O up to 100 ml. All these solutions are separately sterilized by autoclaving as described above, and can be stored indefinitely at room temperature. Components are mixed and diluted as appropriate with deionized H$_2$O to prepare M9 minimal medium immediately prior to use.

3. To prepare solid LB medium containing 1.5 % (w/v) agar, add 15 g of bacteriological agar to 1 l of LB medium and autoclave it as indicated above. In the case of M9 minimal medium plates, prepare a 1.6 % (w/v) agar suspension in deionized H$_2$O, autoclave it separately from the other medium stock solutions, and then mix an adequate amount of this suspension with the rest of the M9 minimal medium components to reach a final agar concentration of 1.4 % (w/v). Antibiotics and other supplements are added when the molten agarized medium reaches about 50 °C. Distribute the agarized culture media in plastic petri dishes (25 ml of molten culture medium per 90-mm petri dish), and let the medium solidify at room temperature. Culture medium plates are freshly prepared immediately prior to use, but they can be stored at 4 °C (ideally for no longer than 1 week, especially if antibiotics were added to the agarized culture medium).

4. All the antibiotics needed for bacterial selection in this protocol are prepared as concentrated stock solutions in deionized H$_2$O at the concentrations indicated below, sterilized by filtration (0.45 μm), and stored at –20 °C for several months (*see* **Note 2**). The concentration of the stock solutions is as follows: ampicillin (Ap), 150 mg/ml; kanamycin (Km), 50 mg/ml; streptomycin (Sm), 50 mg/ml; gentamicin (Gm), 10 mg/ml; and carbenicillin (Cb), 500 mg/ml. The chloramphenicol (Cm) stock solution is prepared at 30 mg/ml in 100 % (v/v) ethanol.

Unless otherwise indicated, all the antibiotic stock solutions are considered to be 1000× concentrated.

5. 1× phosphate-buffered saline (PBS): 8 mM Na_2HPO_4, 1.5 mM KH_2PO_4, 3 mM KCl, and 137 mM NaCl, pH 7.0. Dissolve 1.44 g of Na_2HPO_4, 0.24 g of KH_2PO_4, 0.2 g of KCl, and 8 g of NaCl in 800 ml of deionized H_2O, adjust the pH to 7.0 by dropwise addition of 6 N HCl, and bring the volume to 1 l with deionized H_2O. Sterilize this buffer by autoclaving as indicated above, and store it at room temperature.

6. 300 mM sucrose solution: Dissolve 10.27 g of sucrose in deionized H_2O up to a final volume of 100 ml, sterilize by filtration (0.45 μm), and keep at room temperature.

7. 20% (v/v) glycerol in LB medium: add 118 ml of 85% (v/v) glycerol to 382 ml of LB medium, sterilize by filtration (0.45 μm), and keep this solution at room temperature protected from light.

2.3 Molecular Biology Reagents

1. Plasmid purification kit such as the QIAprep Spin Miniprep™ kit (Qiagen, Valencia, CA).

2. Commercial genomic DNA isolation kit (*see* **Note 3**).

3. DNA polymerase for PCR.

4. 10 mM stock solution of deoxynucleotide triphosphates (dNTPs) containing equimolar amounts of dATP, dCTP, dGTP, and dTTP in MilliQ water (resistivity ≥18 MΩ cm at 25 °C). Store the solution at –20 °C for up to 8–12 months.

5. Agarose gel and PCR product cleanup kit (e.g., NucleoSpin™ Gel and PCR cleanup kit or the ExoSAP-IT™ PCR kit).

6. Agarose gel electrophoresis system using TBE or TAE buffer is routinely used to identify, quantify, and purify DNA fragments. Basic protocols are provided in Makovets [22], and running parameters must be adjusted according to the specific application.

2.4 Primers

All the oligonucleotides needed for arbitrary PCR amplifications are indicated below, along with the mini-transposon vectors used for the chromosomal insertions. Primers are purchased from Sigma-Aldrich Co., as desalted, lyophilized DNA, and resuspended in the appropriate volume of MilliQ (≥18 MΩ cm) or equivalent H_2O to obtain 5 μM primer solutions. These solutions are aliquoted and stored at –20 °C for several months.

1. Primers for arbitrary PCR amplifications. Common to all vectors (a) ARB6: 5′-GGC ACG CGT CGA CTA GTA C*NN NNN NNN NN*A CGC C-3′. This primer is used in arbitrary PCR, round 1 [23]. In this sequence, *N* represents any nucleotide. (b) ARB2: 5′-GGC ACG CGT CGA CTA GTA C-3′. This primer is used in arbitrary PCR, round 2 [23].

2. Primers specific to the ME-O end of each mini-transposon. pBAMD1-2: (a) ME-O-Km-Ext-F: 5′-CGT CTG TTT CAG AAA TAT GGC AT-3′. This primer is used in arbitrary PCR, round 1 [9]. (b) ME-O-Km-Int-F: 5′-ATC TGA TGC TGG ATG AAT TTT TC-3′. This primer is used in arbitrary PCR, round 2, and for sequencing to map the integration point [9].

   pBAMD1-4: (a) ME-O-Sm-Ext-F: 5′-CTT GGC CTC GCG CGC AGA TCA G-3′. This primer is used in arbitrary PCR, round 1 [9]. (b) ME-O-Sm-Int-F: 5′-CAC CAA GGT AGT CGG CAA AT-3′. This primer is used in arbitrary PCR, round 2, and for sequencing to map the integration point [9].

   pBAMD1-6: (a) ME-O-Gm-Ext-F: 5′-GCA CTT TGA TAT CGA CCC AAG T-3′. This primer is used in arbitrary PCR, round 1 [9]. (b) ME-O-Gm-Int-F: 5′-TCC CGG CCG CGG AGT TGT TCG G-3′. This primer is used in arbitrary PCR, round 2, and for sequencing to map the integration point [9].

   pBELG and pBELK: (a) pBEL-ME-O-Ext-F: 5′-CTG CGA CAT CGT ATA ACG TTA CTG GTT TC-3′. This primer is used in arbitrary PCR, round 1 [10]. (b) pBEL-ME-O-Int-F: 5′-GGG CGC TAT CAT GCC ATA CCG-3′. This primer is used in arbitrary PCR, round 2 [10].

   pBEXG and pBEXK: (a) pBEX-ME-O-Ext-F: 5′-CTT CTT ACA TTT GGG ACG CTT CGC TG-3′. This primer is used in arbitrary PCR, round 1 [10]. (b) pBEX-ME-O-Int-F: 5′-CCT TCC GAC ACC CTG CGT CAA TG-3′. This primer is used in arbitrary PCR, round 2 [10].

3. Primers specific to the ME-I end of each mini-transposon. pBAMD1-2: (a) pBAM-ME-I-Ext-R: 5′-CTC GTT TCA CGC TGA ATA TGG CTC-3′. This primer is used in arbitrary PCR, round 1 [7]. (b) pBAM-ME-I-Int-R: 5′-CAG TTT TAT TGT TCA TGA TGA TAT A-3′. This primer is used in arbitrary PCR, round 2, and for sequencing to map the integration point [7].

   pBAMD1-4: (a) ME-I-Sm-Ext-R: 5′-ATG ACG CCA ACT ACC TCT GAT A-3′. This primer is used in arbitrary PCR, round 1 [9]. (b) ME-I-Sm-Int-R: 5′-TCA CCG CTT CCC TCA TGA TGT T-3′. This primer is used in arbitrary PCR, round 2, and for sequencing to map the integration point [9].

   pBAMD1-6: (a) ME-I-Gm-Ext-R: 5′-GTT CTG GAC CAG TTG CGT GAG-3′. This primer is used in arbitrary PCR, round 1 [9]. (b) ME-I-Gm-Int-R: 5′-GAA CCG AAC AGG CTT ATG TCA-3′. This primer is used in arbitrary PCR, round 2, and for sequencing to map the integration point [9].

4. Primers specific to the mini-transposon vector backbone. Oligonucleotides annealing to specific sequences in SEVA vectors

[12, 13] are routinely used to detect the presence of the Tn5-bearing mini-transposon vector backbone in transconjugant cells. If the plasmid is present in such cells, a PCR amplification using primers PS4 (5'-CCA GCC TCG CAG AGC AGG-3') and PS5 (5'-CCC TGC TTC GGG GTC ATT-3') generates a DNA amplicon of 225 bp.

5. Primers to confirm elimination of antibiotic resistances by ectopic expression of the FLP recombinase. When employing the specialized pBELs or pBEXs mini-Tn5 vectors [10], the presence or the absence of antibiotic resistance genes can be easily assessed by colony PCR with primers c*FRT*-Ab-R (5'-GAG AAT AGG AAC TTC GGA ATA GG-3') in combination with either cKm-F (5'-CGG AAT GCT ATG CAG ACG-3', when using pBELK or pBEXK) or cGm-F (5'-CCC GTA TGC CCA ACT TTG-3', when using pBELG or pBEXG). The corresponding expected amplicon lengths are 736 bp (for pBELG and pBEXG) and 535 bp (for pBELK and pBEXK).

2.5 Additional Materials

Other laboratory materials, equipment and standard procedures used in this protocol are as indicated below.

1. 10-ml plastic test tubes containing 3 ml of the corresponding liquid culture medium for cell growth. *P. putida* KT2440 is incubated at 30 °C and *E. coli* strains are incubated at 37 °C with rotary agitation at 170 rpm.

2. Electroporation system.

3. 2-mm gap width cuvettes for electroporation.

4. 0.45-µm filter disks, mixed cellulose esters.

5. 25-mm diameter and blunt-end filter forceps.

6. Thermocycler.

7. Sterile, round plastic petri dishes either of regular size (i.e., diameter 90 mm) or larger plates (i.e., diameter 140 mm).

8. Glass beads (5–10, sterilized), 3-mm diameter.

9. 50 ml Falcon tubes.

10. Spatula.

11. Cryogenic tubes and vials.

3 Methods

The mini-transposon vectors described in the present protocol have two principal uses: (a) the generation of libraries of random mutants to correlate a particular and observable phenotype to a specific gene [1, 7, 8, 24]; or (b) the introduction of heterologous gene(s) randomly inserted into the chromosome of a target

Gram-negative bacterium [2, 9, 10]. In the first application, a typical random mutagenesis protocol involves three steps: (1) *delivery* of the non-replicative plasmid bearing the mini-transposon into a recipient strain, (2) *selection* of transconjugants carrying the transposon, and (3) *storing* the library for future uses or directly *identifying* the insertion point of the mini-transposon in the genome of selected recipient cells. In the second case, when the objective is to introduce heterologous DNA into a bacterial genome, a previous step is included in which the gene or genes of interest have to be cloned into the multiple cloning site of the mini-Tn5 delivery plasmid. Steps (1), (2), and (3) are then followed as described in the specific procedure below. Finally, when using any of the pBELs or pBEXs mini-Tn5 plasmids [10] to deliver cargo DNA under a controllable expression system, a final, extra step could be added *ad libitum* to remove the antibiotic marker of the mini-Tn5 cassette.

The first step of the protocol (i.e., introduction of the mini-transposon plasmid into the recipient strain), could be done either by mating or electroporation. Specific protocols to perform either delivery technique are described below, and a specific application example, in which plasmid pBAMD1-2 [9] was employed to generate a library of random insertion mutants of *P. putida* KT2440, is discussed. The main steps of the protocol are summarized in Fig. 1.

3.1 Mini-Tn5 Delivery: Conjugation   Whenever possible, we recommend using the conjugation delivery method since it is more efficient than the electroporation method. Conjugation requires cell contact to transfer DNA from a donor cell to a recipient strain. To establish such intimate contact, donor bacteria produce the conjugative pilus (i.e., a type IV secretion system) that ultimately retracts, bringing both cells together. A number of proteins of the donor bacterial cell form a bridge between both

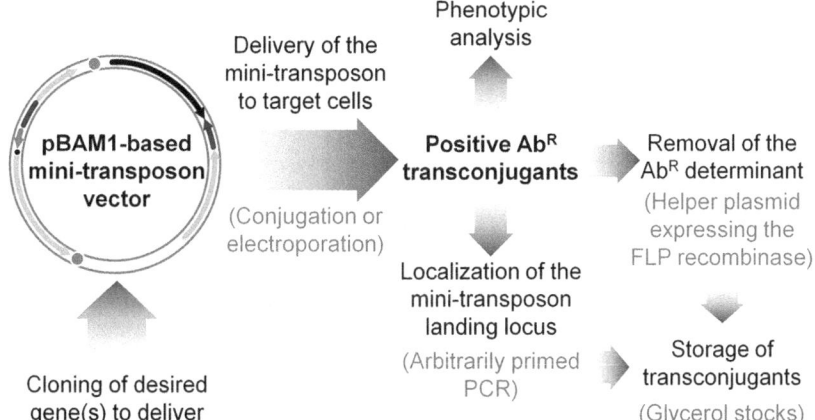

Fig. 1 Outline of the procedure described in this protocol. Mini-transposon vectors can be used for delivering gene(s) into a target chromosome in virtually any Gram-negative bacterium, as well as to obtain random mutant libraries. *Ab^R* antibiotic resistance

donor and recipient cells forming a mating pair (i.e., Mpf proteins, for *mating pair formation*). Then, the *relaxosome* (i.e., a complex formed by a relaxase and auxiliary proteins) recognizes the origin of transfer (*oriT*) sequence and moves one strand of the target DNA to the recipient cell (for a review on the biology behind this process, please *see* Zechner et al. [25] and Ilangovan et al. [26] and references therein).

To perform a conjugation experiment, one just needs to bring together the donor cell (i.e., the one bearing the mini-transposon plasmid), the recipient cell (i.e., the target bacterium), and a helper bacterial strain to assist and catalyze the mating process. The mating helper is an *E. coli* strain that provides the conjugation machinery. Typically, this molecular machinery is derived from the IncPα plasmid RP4 (also known as RK2 or RP1), and involves the mobilization (*mob*) and transfer (*tra*) functions [27, 28], supplied *in trans*. There are two basic types of helper strains, which express the *mob/tra* functions either (1) in a plasmid (e.g., *E. coli* HB101 carrying plasmid pRK600) (*see* Tables 1 and 2), or (2) integrated into the genome (e.g., *E. coli* S17-1 λ*pir*, *E. coli* SM10 λ*pir*, or *E. coli* MFD λ*pir*) (*see* Table 1). When using the former type of helper *E. coli* strain, the user needs to include three bacterial strains in the mating process (i.e., a triparental mating). A triparental mating offers the possibility of changing the donor *E. coli* strain (which should contain the *pir* gene as a λ*pir* lysogen, e.g., *E. coli* CC118 λ*pir* or DH5α λ*pir*) in order to favor counter selection of transconjugants as needed. In the case of performing a biparental mating, the protocol is exactly the same as in the triparental mating procedure, but uses only two bacterial strains in the mixture (i.e., the recipient strain and the mobilizing donor cell, that in addition to the mini-Tn5 plasmid also has the *mob/tra* functions integrated into the genome).

In the present protocol, we describe a triparental mating using *P. putida* KT2440 as the recipient strain and pBAMD1-2 as the mini-Tn5 delivery plasmid [9]. In order to properly generate a random mutagenesis library when working with other recipient bacterial species, it is important to perform several previous tests to determine the optimal experimental conditions for successful DNA transfer, since the expected number of transconjugant colonies depends on several factors such as the nature of the recipient species, the initial amount of recipient cells, the mixing ratio of recipient to donor cells, and the mating incubation time. With the help of these prior experiments, the user should be able to set the appropriate experimental conditions and to estimate the number of plates needed to obtain a saturated library.

1. To prepare the mating mixture, grow the following strains overnight as indicated:

   (a) Donor: *E. coli* CC118 λ*pir* (carrying plasmid pBAMD1-2) grown in LB medium supplemented with Ap at 150 μg/

ml. Incubate the cells for 18 h at 37 °C with rotary agitation (~250 rpm). These cells bear the mobilizable and non-replicative plasmid with the Tn5 mini-transposon (*see* **Note 4**).

(b) Mating helper: *E. coli* HB101 (carrying plasmid pRK600) grown in LB medium supplemented with Cm at 30 μg/ml. Incubate for 18 h at 37 °C with rotary agitation. This bacterium provides the plasmid with the mobilization (*mob*) and transfer (*tra*) functions, encoded on plasmid pRK600.

(c) Recipient: *P. putida* KT2440 grown in LB medium at 30 °C with rotary agitation (*see* **Note 5**).

2. Measure the optical density at 600 nm (OD_{600}) of the bacterial cultures and adjust the bacterial suspensions to an OD_{600} of 1 with PBS in a final volume of 1 ml in a 1.5-ml Eppendorf tube.

3. Centrifuge the cultures at $7200 \times g$ for 2 min at room temperature, discard the supernatant, and resuspend the pellet in 1 ml of 10 mM $MgSO_4$ to wash the cells.

4. Mix the three bacterial suspensions in a 1:1:1 ratio (i.e., 150 μl of each suspension) in a test tube containing 4.55 ml of 10 mM $MgSO_4$. The final OD_{600} should be ~0.03 (*see* **Note 6**).

5. Pass the 5-ml cell suspension through a filter disk (0.45-μm pore-size, 25-mm diameter) using a 20-ml sterile syringe (*see* **Note 7**). Discard the flow-through and place the filter, aseptically, onto an LB medium agar plate (cells facing up). Incubate the plate containing the filter (lid facing up) at 30 °C during the desired mating time (1 h, 6 h, or up to 24 h) (*see* **Note 8**).

6. Gently lift the filter from the agar plate with sterile tweezers (blunt-end filter forceps, previously sterilized by quickly dipping them in 70% (v/v) ethanol and flaming) and place it in a 10-ml test tube containing 5 ml of 10 mM $MgSO_4$.

7. Resuspend the cells in the mating mixture from the filter by vigorous vortexing (at least 1 min) and plate appropriate dilutions (*see* **Note 9**) onto M9 minimal medium plus sodium citrate at 0.2% (w/v) and Km at 50 μg/ml (i.e., selective culture medium for transconjugant *P. putida* cells harboring the mini-transposon) (*see* **Notes 10** and **11**).

3.2 Mini-Tn5 Delivery: Electroporation

If no other choice is available (or in cases when the cargo DNA is to be integrated into a target genome, where higher frequencies are not that important as they are for the construction of mutant libraries) then electrotransformation is the preferred DNA delivery method, mainly due to its ease of use. This technique is based on the transient permeabilization of the cell membrane, which allows for the entry of DNA after applying a high electric field [24, 29, 30].

1. Inoculate a 100-ml Erlenmeyer flask containing 20 ml of LB medium with *P. putida* KT2440 from a fresh LB medium agar plate (or directly from a frozen stock, by scraping the surface of the stock with a sterile toothpick). Let the cells grow overnight (e.g., 18–24 h) aerobically (170 rpm) at 30 °C.

2. Transfer the saturated culture to a 50-ml Falcon tube and centrifuge it at $3220 \times g$ for 10 min at room temperature.

3. Discard the supernatant, add 10 ml of 300 mM sucrose and softly resuspend the cell pellet. Then, centrifuge the suspension at $3220 \times g$ for 10 min at room temperature.

4. Remove the supernatant and add 1 ml of 300 mM sucrose, resuspend the cells, and transfer the suspension to a 2-ml sterile Eppendorf tube. Centrifuge at $7200 \times g$ for 3 min at room temperature.

5. Remove the supernatant, add 800 µl of 300 mM sucrose, resuspend the cells, and centrifuge the suspension at $7200 \times g$ for 3 min at room temperature. Repeat this washing step once more.

6. Remove the supernatant and add 500 µl of 300 mM sucrose to resuspend the pellet and to obtain a concentrated cell suspension (after the final resuspension step, the density of electrocompetent bacteria should be $\sim 5 \times 10^{10}$ cells/ml.)

7. Transfer 100 µl of the electrocompetent cell suspension to a sterile 1.5-ml Eppendorf tube and add ~500 ng of plasmid pBAMD1-2 (in <10 µl total volume). Pipet the plasmid DNA–cell suspension mix to a 2-mm gap width electroporation cuvette (care has to be taken to avoid the formation of bubbles at this step, which would reduce the overall efficiency of the electroporation process.).

8. Place the cuvette in the electroporation (e.g., MicroPulser™) apparatus, set the electroporation program to EC2, and proceed to electroporate the cells. With these working conditions and using an optimum electric pulse (a single pulse of 2.5 kV with a field strength of 12.5 kV/cm), a time constant (τ) between 4 and 5 ms should be obtained.

9. Immediately after the electric shock, add 900 µl of LB medium to the cuvette and transfer the cells to a sterile 1.5-ml Eppendorf tube. Incubate the cells aerobically for 3 h at 30 °C with gentle shaking.

10. Spread dilutions of the cell suspension obtained in the step above onto LB medium agar plates containing Km at 50 µg/ml. Since no *E. coli* cells are used in this procedure, there is no need for nutritional selection as performed in the mating experiments.

3.3 Mini-Transposon Insertion Sites

1. If specifically looking for particular phenotypes, select interesting colonies based on a trait (e.g., morphology or color) different from that observed in the wild-type cells, and streak them with a sterile toothpick onto both (a) M9 minimal medium plates containing 0.2% (w/v) citrate and 50 μg/ml Km, and (b) M9 minimal medium plates containing 0.2% (w/v) citrate and 500 μg/ml Ap. This process is aimed at differentiating between genuine transposition events (i.e., Km^R colonies) from spurious mini-Tn5 plasmid co-integration incidents (i.e., Km^R and Ap^R colonies). Incubate the plates overnight at 30 °C.

2. Select Km^R and Ap^S clones. Also, use colony PCR amplifications with oligonucleotides PS4 and PS5 to confirm the absence of the delivery plasmid backbone.

3. As a strain purity check, restreak selected colonies several times onto M9 minimal medium plates containing 0.2% (w/v) citrate and 50 μg/ml Km.

4. Make a frozen stock in 20% (v/v) glycerol in LB medium of the selected mutants and store the resulting stocks at –80 °C. Bacterial frozen stocks can be prepared by growing the cells of interest on LB medium plates (with appropriate antibiotics, if necessary) overnight, and adding 2 ml of 20% (v/v) glycerol in LB medium. Cells are gently scraped from the surface by using a sterile glass rod (i.e., a Drigalski spatula). One milliliter of the resulting suspension is then transferred into a cryotube (e.g., a 1.8-ml Nunc™ CryoTubes™ cryogenic vial, round bottom). Cells can be stored at –80 °C under these conditions for several years without significant loss of viability, provided that the bacterial stock is not repeatedly freeze-thawed.

5. Take mutant clones from the frozen stock and streak the cells onto LB medium agar plates containing 50 μg/ml Km. Grow the cells overnight at 30 °C.

6. In order to genetically analyze the transconjugants, first choose one of the mini-transposon ends (i.e., ME-I or ME-O) to determine its insertion site in the genome and then perform arbitrarily primed colony PCR [31]. The DNA sequence of the primers needed to perform arbitrarily primed PCR when using different mini-Tn5 plasmids is described in **item 1** in Subheading 2.4 (*see* **Note 12**).

7. Prepare a PCR reaction mix on ice with the following recipe. Note that most of the components indicated in the recipe are provided along with the commercial *Taq* DNA polymerase kit. Thoroughly vortex each concentrated solution before pipetting into the PCR reaction mix.

   – 5 μl of 5× Green or Colorless Go*Taq*™ reaction buffer.

   – 1.5 μl of 25 mM MgCl$_2$.

- 0.5 µl dNTPs (10 mM).
- 0.5 µl of dimethyl sulfoxide (when performing amplifications from high G + C DNA templates).
- 1 µl of 5 µM arbitrary primer.
- 1 µl of 5 µM mini-transposon primer (i.e., ME-I or ME-O).
- 0.2 µl of 5 U/µl Go*Taq*™ Flexi DNA polymerase.

8. Aliquot 15.3 µl of sterile deionized H_2O into each PCR tube.

9. Transfer fresh colonies from agar plates directly into the PCR reaction tube with a sterile toothpick.

10. Distribute 9.7 µl of the PCR reaction mix into each PCR tube.

11. The primers needed for round 1 of the arbitrary primed PCR amplification are ARB6 together with the external ME-I or ME-O primers (i.e., either ME-I-Ext or ME-O-Ext).

12. The settings for round 1 of the arbitrarily primed PCR amplification are as follows:
    - 5 min at 95 °C, 1 cycle.
    - 30 s at 95 °C, 30 s at 30 °C, and 1.5 min at 72 °C for 6 cycles.
    - 30 s at 95 °C, 30 s at 45 °C, and 1.5 min at 72 °C for 30 cycles.

13. Directly take 1 µl of the PCR after running round 1 (i.e., no need to check for positive amplifications in an agarose gel) and use it as the template for round 2 of arbitrarily primed PCR. In this round, use primer ARB2 together with the internal ME-I or ME-O primers (i.e., either ME-I-Int or ME-O-Int) (*see* **Note 13**). Prepare the PCR reaction mix for round 2 as indicated in **step 8** above, this section.

14. The settings for round 2 of the arbitrarily primed PCR amplification are as follows:
    (a) 1 min at 95 °C, 1 cycle.
    (b) 30 s at 95 °C, 30 s at 52 °C, and 1.5 min at 72 °C for 30 cycles.
    (c) 4 min at 72 °C.

15. Clean up the PCR product from the second round of the arbitrary PCR amplification using a PCR product cleanup kit (such as either the NucleoSpin™ Gel and PCR cleanup kit or the ExoSAP-IT™ PCR product cleanup kit).

16. Sequence the PCR product [32, 33] using the ME internal primer used in round 2 of the arbitrary PCR.

17. Analyze the sequencing results. Start by identifying the DNA sequence of the mini-transposon end (i.e., either ME-I or ME-O) (*see* **Note 14**), and then trim that part and select the

rest of the DNA sequence. Use the BlastN program [34], available online at: www.pseudomonas.com/blast/set [35], to map the precise genomic coordinates of the mini-transposon insertion (*see* **Note 15**).

18. Once an interesting mutant is spotted, in which the phenotype-gene relationship has been identified, it is always recommended to complement that mutant back with the identified gene(s) to rule out the occurrence of polar effects, since mini-Tn5 insertions are known to alter the expression of neighboring genes [36, 37].

3.4 Removing Antibiotic Resistance

When using any of the pBELs or pBEXs mini-Tn5 vectors (*see* Table 2) to introduce heterologous DNA under the control of an expression system (i.e., LacI^Q/P*trc* or XylS/P*m*) [10], the genes conferring resistance to Km (*aphA*) or Gm (*aacC1*) in these transposons can be removed as they are flanked by FLP recombinase target (*FRT*) sequences [38]. This layout offers the possibility to the user of eliminating that marker by means of ectopic expression of the FLP recombinase from *Saccharomyces cerevisiae* using plasmid pFLP2 (*see* Table 2). The expression of the FLP recombinase in plasmid pFLP2 is driven by the strong, rightward λ promoter (located within the *FLP-cI857* intergenic region) and is regulated by the temperature-sensitive, *cI857*-encoded λ repressor [39].

1. Select a transconjugant *P. putida* clone in which the insertion site of the mini-transposon has been successfully localized.

2. Introduce plasmid pFLP2 (*see* **Note 16**) into this selected clone by either mating or electroporation as described above.

3. Plate the cells on M9 minimal medium plates added with sodium citrate at 0.2% (w/v) and Cb at 500 μg/ml. Incubate the plates overnight at 30 °C. If no discernible colonies are observed after this incubation period, try lowering the Cb concentration to 350 μg/ml.

4. Select two or three independent colonies and re-streak them on M9 minimal medium plates supplemented with sodium citrate at 0.2% (w/v) and Cb at 500 μg/ml. Incubate the plates overnight at 30 °C.

5. Pick single colonies and check for Km or Gm sensitivity and Cb resistance on LB medium plates containing these antibiotics. Double check for the removal of the antibiotic gene by colony PCR using the primers described in **item 5** in Subheading 2.4 (i.e., cFRT-Ab-R and either cKm-F or cGm-F). Such PCR should give no amplification. If possible, use primers annealing within the gene(s) delivered in the mini-transposon cassette to conduct a colony PCR of the antibiotic-sensitive clone to make sure that the gene(s) of interest have been stably inserted into the target chromosome.

6. Cure plasmid pFLP2 from the selected clone by performing several (at least three) cycles of growth in LB medium without any antibiotic.

7. Plate cells onto M9 minimal medium plates supplemented with 0.2% (w/v) sodium citrate.

8. Pick single colonies and restreak two times onto M9 minimal medium plates supplemented with 0.2% (w/v) sodium citrate containing Cb at 500 μg/ml.

9. Select Cb^S clones and store them as frozen stocks at –80 °C.

3.5 Mutant Library Storage

After obtaining a random mutagenesis library, it is always useful to save it for downstream analyses. The steps below describe the procedure to store the library after introduction of plasmid pBAMD1-2 in strain KT2440 as explained in the preceding sections.

1. Spread dilutions of the triparental mating mixture onto selective agar plates in order to obtain approximately 3000 transconjugant colonies per plate in a standard-sized (i.e., 90 mm) petri dish. Estimate the number of plates needed to obtain a non-saturated mutant library as indicated by Liberati et al. [37].

2. Add 2.5 ml of LB medium containing 20% (v/v) glycerol to each overnight-incubated plate and with the aid of an inoculation loop or a Drigalski spatula, gently scrap the cells from the agar surface. Tilt the plate and collect 1-ml of the bacterial suspension with a micropipet (*see* **Note 17**).

3. Mix the liquid content collected from all the plates, aliquot the resulting suspension into several cryotube vials, and store the library as a series of frozen stocks at –80 °C.

4 Notes

1. The appropriate culture medium composition has to be defined to select against *E. coli* donor/mating helper cells when using other bacterial species as the target strain. As a general rule, try to make use of specific carbon sources in which only the recipient strain grows by taking advantage of the auxotrophies of the *E. coli* donor/mating helper cells (e.g., most laboratory *E. coli* strains need thiamine·HCl to grow [40]).

2. Avoid repeated freezing and thawing of antibiotic solutions as that may cause a loss of potency. We routinely distribute antibiotic stock solutions in 0.5-ml working aliquots that are used only a couple of times before discarding them.

3. If no amplification is obtained through colony PCR, genomic DNA can be isolated with a commercial kit (e.g., the UltraClean™ Microbial DNA isolation kit, MoBio Laboratories Inc., Carlsbad, CA) and used as the template for amplifications.

4. It is very important to grow the donor bacterial strain in the presence of the antibiotic for which the plasmid backbone carries a resistance gene (e.g., Ap) to avoid inadvertent selection of transposed donor cells. Note that there is a vector derived from pBAM1 which carries a promoterless and leaderless *gfp* gene (plasmid pBAM1-GFP), which allows for the visual inspection of successful *gene::gfp* fusions after the transposition event.

5. In some cases, incubating the recipient strain at high temperatures (40–42 °C) for a few hours before mating is known to increase the efficiency of the process by inactivating its endogenous DNA restriction machinery.

6. Different ratios of the bacterial strains to be included in the triparental mating could also be tested if needed (e.g., by increasing the amount of donor cells). To do this, simply adjust the volume of each bacterial suspension appropriately with 10 mM MgSO$_4$ to bring the final volume to 5 ml, and proceed as indicated. In the case of integrating a cargo DNA into the bacterial genome, where there is no need for the high numbers of transconjugant colonies usually required in random mutant libraries, one can use only 100 μl of each overnight culture (i.e., without adjusting the OD$_{600}$ of the individual cultures).

7. If a filter system for bacterial matings is not available, one can simply mix the three bacterial strains in a 1.5-ml Eppendorf tube (e.g., 150 μl of each bacterial suspension adjusted at OD$_{600}$ of 1), centrifuge the cells at 7200×g, discard the supernatant, and resuspend the pellet in 25 μl of 10 mM MgSO$_4$ (i.e., a small buffer volume to maximize cell contact). The 25-μl mating mix can be placed onto a 0.45-μm filter disk previously laid onto a LB medium plate, or be directly spotted onto the surface of an LB medium plate. In the latter case, cells can easily be recovered using an inoculation loop and resuspended in 10 mM MgSO$_4$ before plating on a selective culture medium.

8. When creating non-saturated random mutant libraries it is better to use shorter incubation times to minimize the number of cell divisions of the transconjugants.

9. Depending on the purpose of the experiment, different size petri dishes could be used to recover more transconjugants per plate. Adjust the plating volume accordingly.

10. It is also important to plate (a) the donor strain, (b) the mating helper strain, and (c) the recipient strain onto the selective

culture medium used to recover transconjugants. These three bacterial strains should not grow in the selective culture medium (i.e., they are used as negative controls).

11. Take into account that different mini-Tn5-bearing plasmids need other antibiotics (e.g., Sm at 80 μg/ml or Gm at 10 μg/ml in the case of plasmids pBAMD1-4 and pBAMD1-6, respectively) to select for positive transconjugants.

12. Note that the specific ME primers have to be chosen depending on the mini-Tn5 plasmid used for insertions and on the selected ME-end.

13. Clean up the PCR products obtained after the first round of arbitrary-primed PCR with a commercial kit to eliminate unbound primers to avoid potential problems in the second round of arbitrary-primed PCR (e.g., no amplification observed in the second round).

14. If the insertion of the transposon cannot be unequivocally mapped using the protocol and primers suggested here, select another set of arbitrary primers, such as primer ARB1 [23] or even others as described by Das et al. [31]. Alternatively, a new custom arbitrary primer could be designed by changing the five nucleotides at the 3'-end of the oligonucleotide sequence to match the G + C content of the recipient bacterial strain, thereby increasing the frequency of appearance of that motif in the target genome.

15. For other target bacterial species, use the BlastN tool against the genome of the desired recipient strain. If the complete genome sequence of your favorite microorganism is not available, perform a more general BlastN search in order to identify homologous genes or sequences in related species. Specific primers could then be designed on the basis of these results to sequence the exact locus in which the mini-Tn5 insertion has occurred.

16. Other plasmids can be used for the ectopic expression of the FLP recombinase, such as plasmid pBBFLP [41]. The procedure to be followed in this case is essentially the same as the one described in the main protocol, but using tetracycline (at 15 μg/ml) instead of Cb to select for the presence of the pBBFLP helper plasmid. If the insertion of pBELs or pBEXs vectors is carried out in *E. coli*, plasmid pCP20 [38] is recommended for the FLP-dependent removal of antibiotic-resistance determinants after transposition.

17. It is a good procedure to perform several independent matings in order to yield a representative random mutant library.

Acknowledgments

The work described in this protocol was supported by the CAMBIOS Project of the Spanish Ministry of Economy and Competitiveness (RTC-2014-1777-3), the ST-FLOW (FP7-KBBE-2011-5-289326), EVOPROG (FP7-ICT-610730), ARISYS (ERC-2012-ADG-322797), and EmPowerPutida (EU-H2020-BIOTEC-2014-2015-6335536) Contracts of the European Union, and the PROMPT Project of the Autonomous Community of Madrid (CAM-S2010/BMD-2414). The authors declare that there is no conflict of interest. All the bacterial strains and plasmids described in the text are available upon request.

References

1. de Lorenzo V, Herrero M, Jakubzik U, Timmis KN (1990) Mini-Tn5 transposon derivatives for insertion mutagenesis, promoter probing, and chromosomal insertion of cloned DNA in Gram-negative eubacteria. J Bacteriol 172:6568–6572

2. de Lorenzo V, Timmis KN (1994) Analysis and construction of stable phenotypes in Gram-negative bacteria with Tn5- and Tn10-derived minitransposons. Methods Enzymol 235:386–405

3. Reznikoff WS (2008) Transposon Tn5. Annu Rev Genet 42:269–286

4. Berg DE (1989) Transposon Tn5. In: Berg DE, Howe MM (eds) Mobile DNA. American Society for Microbiology Press, Washington, D.C., pp 185–210

5. Reznikoff WS (2006) Tn5 transposition: a molecular tool for studying protein structure-function. Biochem Soc Trans 34:320–323

6. Phadnis SH, Sasakawa C, Berg DE (1986) Localization of action of the IS50-encoded transposase protein. Genetics 112:421–427

7. Martínez-García E, Calles B, Arévalo-Rodríguez M, de Lorenzo V (2011) pBAM1: an all-synthetic genetic tool for analysis and construction of complex bacterial phenotypes. BMC Microbiol 11:38

8. de Lorenzo V, Herrero M, Sánchez JM, Timmis KN (1998) Mini-transposons in microbial ecology and environmental biotechnology. FEMS Microbiol Ecol 27:211–224

9. Martínez-García E, Aparicio T, de Lorenzo V, Nikel PI (2014) New transposon tools tailored for metabolic engineering of Gram-negative microbial cell factories. Front Bioeng Biotechnol 2:46

10. Nikel PI, de Lorenzo V (2013) Implantation of unmarked regulatory and metabolic modules in Gram-negative bacteria with specialised mini-transposon delivery vectors. J Biotechnol 163:143–154

11. Schweizer HP (2003) Applications of the Saccharomyces cerevisiae Flp-FRT system in bacterial genetics. J Mol Microbiol Biotechnol 5:67–77

12. Martínez-García E, Aparicio T, Goñi-Moreno A, Fraile S, de Lorenzo V (2014) SEVA 2.0: an update of the Standard European Vector Architecture for de-/re-construction of bacterial functionalities. Nucleic Acids Res 43:D1183–D1189

13. Silva-Rocha R, Martínez-García E, Calles B, Chavarría M, Arce-Rodríguez A et al (2012) The Standard European Vector Architecture (SEVA): a coherent platform for the analysis and deployment of complex prokaryotic phenotypes. Nucleic Acids Res 41:D666–D675

14. Nikel PI, Martínez-García E, de Lorenzo V (2014) Biotechnological domestication of pseudomonads using synthetic biology. Nat Rev Microbiol 12:368–379

15. Nikel PI, de Lorenzo V (2014) Robustness of Pseudomonas putida KT2440 as a host for ethanol biosynthesis. New Biotechnol 31:562–571

16. Benedetti I, de Lorenzo V, Nikel PI (2016) Genetic programming of catalytic Pseudomonas putida biofilms for boosting biodegradation of haloalkanes. Metab Eng 33:109–118

17. Timmis KN (2002) Pseudomonas putida: a cosmopolitan opportunist par excellence. Environ Microbiol 4:779–781

18. Nikel PI, Chavarría M, Fuhrer T, Sauer U, de Lorenzo V (2015) Pseudomonas putida

KT2440 strain metabolizes glucose through a cycle formed by enzymes of the Entner-Doudoroff, Embden-Meyerhof-Parnas, and pentose phosphate pathways. J Biol Chem 290:25920–25932

19. Nikel PI, Kim J, de Lorenzo V (2014) Metabolic and regulatory rearrangements underlying glycerol metabolism in *Pseudomonas putida* KT2440. Environ Microbiol 16:239–254

20. Nikel PI, Romero-Campero FJ, Zeidman JA, Goñi-Moreno A, de Lorenzo V (2015) The glycerol-dependent metabolic persistence of *Pseudomonas putida* KT2440 reflects the regulatory logic of the GlpR repressor. mBio 6:e00340-00315

21. Nikel PI, Silva-Rocha R, Benedetti I, de Lorenzo V (2014) The private life of environmental bacteria: pollutant biodegradation at the single cell level. Environ Microbiol 16:628–642

22. Makovets S (2013) Basic DNA electrophoresis in molecular cloning: a comprehensive guide for beginners. Methods Mol Biol 1054:11–43

23. Pratt LA, Kolter R (1998) Genetic analysis of *Escherichia coli* biofilm formation: roles of flagella, motility, chemotaxis and type I pili. Mol Microbiol 30:285–293

24. Martínez-García E, de Lorenzo V (2012) Transposon-based and plasmid-based genetic tools for editing genomes of Gram-negative bacteria. Methods Mol Biol 813:267–283

25. Zechner EL, Lang S, Schildbach JF (2012) Assembly and mechanisms of bacterial type IV secretion machines. Philos Trans R Soc Lond B Biol Sci 367:1073–1087

26. Ilangovan A, Connery S, Waksman G (2015) Structural biology of the Gram-negative bacterial conjugation systems. Trends Microbiol 23:301–310

27. Álvarez-Martínez CE, Christie PJ (2009) Biological diversity of prokaryotic type IV secretion systems. Microbiol Mol Biol Rev 73:775–808

28. Babic A, Guérout AM, Mazel D (2008) Construction of an improved RP4 (*RK2*)-based conjugative system. Res Microbiol 159:545–549

29. Iwasaki K, Uchiyama H, Yagi O, Kurabayashi T, Ishizuka K et al (1994) Transformation of *Pseudomonas putida* by electroporation. Biosci Biotechnol Biochem 58:851–854

30. Choi KH, Kumar A, Schweizer HP (2006) A 10-min method for preparation of highly electrocompetent *Pseudomonas aeruginosa* cells: application for DNA fragment transfer between chromosomes and plasmid transformation. J Microbiol Methods 64:391–397

31. Das S, Noe JC, Paik S, Kitten T (2005) An improved arbitrary primed PCR method for rapid characterization of transposon insertion sites. J Microbiol Methods 63:89–94

32. Zimmermann J, Voss H, Schwager C, Stegemann J, Ansorge W (1988) Automated Sanger dideoxy sequencing reaction protocol. FEBS Lett 233:432–436

33. Shendure JA, Porreca GJ, Church GM, Gardner AF, Hendrickson CL et al (2011) Overview of DNA sequencing strategies. Curr Prot Mol Biol 96:7.1.1–7.1.23

34. Altschul SF, Gish W, Miller W, Myers EW, Lipman DJ (1990) Basic local alignment search tool. J Mol Biol 215:403–410

35. Winsor GL, Lam DK, Fleming L, Lo R, Whiteside MD et al (2011) *Pseudomonas* Genome Database: improved comparative analysis and population genomics capability for *Pseudomonas* genomes. Nucleic Acids Res 39:D596–D600

36. Berg DE, Weiss A, Crossland L (1980) Polarity of Tn5 insertion mutations in *Escherichia coli*. J Bacteriol 142:439–446

37. Liberati NT, Urbach JM, Miyata S, Lee DG, Drenkard E et al (2006) An ordered, nonredundant library of *Pseudomonas aeruginosa* strain PA14 transposon insertion mutants. Proc Natl Acad Sci U S A 103:2833–2838

38. Cherepanov PP, Wackernagel W (1995) Gene disruption in *Escherichia coli*: Tc^R and Km^R cassettes with the option of Flp-catalyzed excision of the antibiotic-resistance determinant. Gene 158:9–14

39. Hoang TT, Karkhoff-Schweizer RR, Kutchma AJ, Schweizer HP (1998) A broad-host-range Flp-FRT recombination system for site-specific excision of chromosomally-located DNA sequences: application for isolation of unmarked *Pseudomonas aeruginosa* mutants. Gene 212:77–86

40. Bachmann BJ (1996) Derivations and genotypes of some mutant derivatives of *Escherichia coli* K-12. In: Neidhardt FC, Curtiss R III, Ingraham JL, Lin ECC, Low KB Jr et al (eds) *EcoSal–Escherichia coli* and *Salmonella*: cellular and molecular biology. American Society for Microbiology Press, Washington, DC, pp 2460–2488

41. de las Heras A, Carreño CA, de Lorenzo V (2008) Stable implantation of orthogonal sensor circuits in Gram-negative bacteria for environmental release. Environ Microbiol 10:3305–3316

42. Herrero M, de Lorenzo V, Timmis KN (1990) Transposon vectors containing non-antibiotic resistance selection markers for cloning and

stable chromosomal insertion of foreign genes in Gram-negative bacteria. J Bacteriol 172: 6557–6567

43. Miller VL, Mekalanos JJ (1988) A novel suicide vector and its use in construction of insertion mutations: osmoregulation of outer membrane proteins and virulence determinants in *Vibrio cholerae* requires *toxR*. J Bacteriol 170:2575–2583

44. de Lorenzo V, Cases I, Herrero M, Timmis KN (1993) Early and late responses of TOL promoters to pathway inducers: identification of postexponential promoters in *Pseudomonas putida* with *lacZ-tet* bicistronic reporters. J Bacteriol 175:6902–6907

45. Ferrières L, Hémery G, Nham T, Guérout AM, Mazel D et al (2010) Silent mischief: bacteriophage Mu insertions contaminate products of *Escherichia coli* random mutagenesis performed using suicidal transposon delivery plasmids mobilized by broad-host-range RP4 conjugative machinery. J Bacteriol 192:6418–6427

46. Boyer HW, Roulland-Dussoix D (1969) A complementation analysis of the restriction and modification of DNA in *Escherichia coli*. J Mol Biol 41:459–472

47. Worsey MJ, Williams PA (1975) Metabolism of toluene and xylenes by *Pseudomonas putida* (*arvilla*) mt-2: evidence for a new function of the TOL plasmid. J Bacteriol 124:7–13

48. Bagdasarian M, Lurz R, Rückert B, Franklin FC, Bagdasarian MM et al (1981) Specific-purpose plasmid cloning vectors. II. Broad host range, high copy number, RSF1010-derived vectors, and a host-vector system for gene cloning in *Pseudomonas*. Gene 16: 237–247

49. Ditta G, Stanfield S, Corbin D, Helinski DR (1980) Broad host range DNA cloning system for Gram-negative bacteria: construction of a gene bank of *Rhizobium meliloti*. Proc Natl Acad Sci U S A 77:7347–7351

50. Kessler B, de Lorenzo V, Timmis KN (1992) A general system to integrate *lacZ* fusions into the chromosomes of Gram-negative eubacteria: regulation of the *Pm* promoter of the TOL plasmid studied with all controlling elements in monocopy. Mol Gen Genet 233:293–301

Chapter 19

PERMutation Using Transposase Engineering (PERMUTE): A Simple Approach for Constructing Circularly Permuted Protein Libraries

Alicia M. Jones, Joshua T. Atkinson, and Jonathan J. Silberg

Abstract

Rearrangements that alter the order of a protein's sequence are used in the lab to study protein folding, improve activity, and build molecular switches. One of the simplest ways to rearrange a protein sequence is through random circular permutation, where native protein termini are linked together and new termini are created elsewhere through random backbone fission. Transposase mutagenesis has emerged as a simple way to generate libraries encoding different circularly permuted variants of proteins. With this approach, a synthetic transposon (called a permuteposon) is randomly inserted throughout a circularized gene to generate vectors that express different permuted variants of a protein. In this chapter, we outline the protocol for constructing combinatorial libraries of circularly permuted proteins using transposase mutagenesis, and we describe the different permuteposons that have been developed to facilitate library construction.

Key words Circular permutation, Library, Protein engineering, Transposase, Transposon

1 Introduction

Mutational processes that alter protein length (fission and domain insertion) and contact order (circular permutation) are frequently used to develop biotechnologies for systems and synthetic biology [1, 2]. However, our understanding of sequence–structure–function relationships is not yet sufficient to predict the effects of these mutational lesions on protein structure and function [3]. Because it is hard to anticipate how the structure of marginally stable proteins responds to mutational lesions, combinatorial libraries of vectors encoding mutants are typically generated and screened (or selected) to discover proteins with the desired functional properties [4]. Libraries of vectors encoding circularly permuted proteins have traditionally been generated by manipulating protein-coding sequences using nonspecific nucleases, such as DNAse I [5, 6]. While these efforts have led to the discovery of functional proteins, nuclease-based protocols can be arduous to learn and implement

Andrew Reeves (ed.), *In Vitro Mutagenesis: Methods and Protocols*, Methods in Molecular Biology, vol. 1498,
DOI 10.1007/978-1-4939-6472-7_19, © Springer Science+Business Media New York 2017

because nuclease incubation times must be optimized, and low efficiency blunt cloning is used within the library construction workflow. In addition, vectors created in these libraries express protein variants that often have deletions and duplications of DNA sequence [7–9], which may not be desired. One way to avoid deletions and duplications is to use methods that leverage transposases [10]. Unlike nucleases, transposases generate well-defined DNA modifications that do not result in sequence deletions.

Transposase methods have emerged as a simple strategy to create libraries encoding proteins with random amino acid deletions [11], backbone cleavages [12–14], peptide and domain insertions [15, 16], affinity tag insertions [17], and truncations [18]. With these library construction approaches, the MuA transposase is used to insert synthetic transposons at different locations within a gene encoding the protein of interest. Recently, a transposase approach was described for creating circularly permuted protein libraries that is called PERMutation Using Transposase Engineering, PERMUTE [19]. In PERMUTE, a new class of transposons (called *permuteposons*) are used with MuA transposase to generate an expression vector library. A permuteposon is linear DNA that contains all of the attributes of an expression vector (*see* Fig. 1a), including an origin of replication (ori), a selectable marker (Ab^R-1), and regulatory sequences required to initiate the transcription and translation of permuted proteins. Permuteposons additionally contain MuA transposase binding sites located at the ends of the transposon (R1R2 and R2R1), which allow for MuA recognition and insertion into the protein coding sequence being permuted [10]. In the first step of constructing a circularly permuted protein library, MuA is used to insert a permuteposon into a target vector (*see* Fig. 1b) that contains the gene of interest (GOI), i.e., the gene being permuted. To isolate gene–permuteposon hybrids from the resulting MuA library, the product of the MuA reaction is incubated with a restriction enzyme (RE1) that cuts at sites flanking the gene of interest in the target vector. The gene–permuteposon hybrids are then separated from genes lacking a permuteposon using gel electrophoresis, and those genes containing a single permuteposon are purified and circularized by ligation to yield the final library of vectors that express different permuted variants of a protein [19]. PERMUTE creates well-defined sequence diversity equal to the gene length times two, since permuteposons can be inserted at any location and in two orientations.

2 Materials

All reactions should use molecular biology grade water that is DNAse- and protease-free and has been filtered through a 0.2 μm filter and autoclaved (*see* **Note 1**).

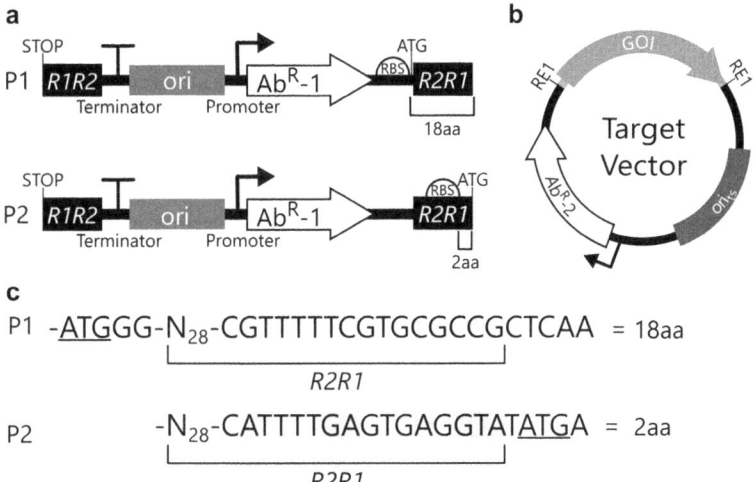

Fig. 1 DNA constructs required to perform PERMUTE. (**a**) The permuteposon serves as an expression vector for permuted proteins in the final PERMUTE libraries. This linear DNA must contain an origin of replication (ori) that functions at any *E. coli* growth temperature and a selectable marker (Ab^R-1) that is distinct from the target vector. Additionally, the permuteposon must contain MuA transposase recognition sequences (R1R2 and R2R1) at each end and regulatory sequences that allow for transcription and translation of the permuted ORFs created by PERMUTE, including a promoter, RBS, and terminator. The RBS can precede R2R1 as shown for permuteposon 1 (P1) or be embedded within R2R1 as illustrated for permuteposon 2 (P2). (**b**) The gene of interest (GOI) encoding the protein being permuted is cloned into a target vector, which must contain a temperature-sensitive origin of replication (ori$_{ts}$). In this vector, the GOI should be flanked on both sides by the same restriction site (RE1) and should lack a stop codon. Extra base pairs should be included between RE1 and the GOI, which encode the linker used for permutation. The number of base pairs in the linker can vary, but the number of base pairs in the linker and restriction site must add up to a number that is divisible by 3. The target vector must also contain a selectable marker (Ab^R-2), which differs from that in the permuteposon. (**c**) R2R1 sequences can be used that lack or contain an RBS, which differ in the location of the first codon used for translation initiation (ATG). When the RBS sequence precedes R2R1 (P1), the number of residues added to the termini of permuted proteins is large. However, this design allows flexibility in RBS sequence. In contrast, R2R1 containing mutations (*highlighted*) that incorporate an RBS into the transposase binding site minimize peptide additions (P2)

2.1 Nucleic Acids

1. Temperature-sensitive target vector containing the gene being permuted (*see* **Note 2**): 600 ng (*see* **Note 3**). As illustrated in Fig. 1b, the gene being permuted must be flanked by the sequence that encodes a peptide linker and restriction sites (RE1) that allow for gene fragment ligation in the last step of PERMUTE (*see* **Note 4**). In the final library, the linker encodes the region that covalently connects the original N- and C-termini of the protein.

2. Permuteposon: 200 ng (*see* **Note 5**). As shown in Fig. 1a, per-muteposons require an origin of replication (*see* **Note 6**), a selectable marker (*see* **Note 7**), two transposase binding sites (*see* **Note 8**), a promoter for transcribing permuted genes (*see* **Note 9**), a stop codon followed by a transcriptional termina-tor for permuted genes (*see* **Note 10**), and a ribosome bind-ing site (RBS) for initiating translation of permuted proteins (*see* **Note 11**).

2.2 Transposase Reaction

1. PCR tube: one sterile tube.

2. Target vector: 600 ng per 40 μL reaction (*see* **Note 12**).

3. Linearized permuteposon: 200 ng per 40 μL reaction (*see* **Note 13**).

4. MuA transposase: 0.44 μg per 40 μL reaction (*see* **Note 14**).

5. MuA reaction buffer: 8 μL of a 5× stock containing 125 mM Tris–HCl, pH 8.0, 50 mM $MgCl_2$, 550 mM NaCl, 0.25% Triton X-100, and 50% glycerol.

6. Thermal cycler (*see* **Note 15**).

7. DNA purification kit: Zymo DNA Clean and Concentrator kit or equivalent kit from another manufacturer that yields high-quality, purified DNA (*see* **Note 16**).

2.3 Cellular Materials

1. Electrocompetent *Escherichia coli*: 50 μL of cells (*see* **Note 17**) and 1 mL of recovery medium (*see* **Note 18**).

2. Electroporation cuvette and electroporator.

3. Culture tube: sterile 14 mL tube.

4. Luria broth (LB) agar plates: 10 g/L tryptone, 5 g/L yeast extract, and 10 g/L NaCl containing 1.5% agar. Adjust to pH 7.0 using 5 N NaOH. Five petri plates (70 mm diameter) containing 25 mL LB-agar and two antibiotics, one that selects for the permuteposon and one that selects for the tar-get vector (*see* **Note 19**).

5. Shaking incubator: set to 37 °C and 250 rpm (*see* **Note 20**).

6. Gravity incubator: set to 37 °C or higher (*see* **Note 21**).

2.4 Vector Harvesting

1. LB medium: 10 mL for harvesting cells from five LB-agar plates (*see* **Note 22**).

2. Sterile spreader.

3. Tubes: sterile 14 mL tube (*see* **Note 23**) and a 0.5 mL PCR tube.

4. DNA purification kit such as the Qiagen Miniprep Kit (*see* **Note 24**).

5. Restriction enzyme (RE1) and appropriate buffer (*see* **Note 25**).

6. Agarose gel and casting unit. Typically run agarose gels in Tris–acetate–EDTA (TAE) buffer (*see* **Note 26**) and 1 μg/mL ethidium bromide (*see* **Note 27**).

7. Gel loading dye: 4 μL of 6× stock containing 15 % Ficoll®-400, 66 mM EDTA, 20 mM Tris–HCl, pH 8.0, 0.1 % SDS, and 0.09 % bromophenol blue (*see* **Note 28**).

8. Gel box and transilluminator (*see* **Note 29**).

9. DNA recovery kit such as Zymo Gel DNA Recovery kit (*see* **Note 30**).

10. Incubator or heat-block: set to 37 °C.

2.5 DNA Ligation

1. PCR tube: one sterile tube.

2. Gene–permuteposon hybrids: 100 ng for 20 μL ligation reaction (*see* **Note 31**).

3. T4 DNA ligase: 400 U per 20 μL reaction (*see* **Note 32**).

4. T4 DNA ligase buffer: 2 μL of 10× stock containing 500 mM Tris–HCl, pH 7.5, 100 mM $MgCl_2$, 100 mM DTT, 10 mM ATP (*see* **Note 33**).

5. Thermal cycler: set to 16 °C.

6. DNA purification kit such as Zymo DNA Clean and Concentrator kit (*see* **Note 16**).

2.6 Quality Control

1. Electrocompetent *Escherichia coli*: 50 μL of cells (*see* **Note 17**) and 1 mL of recovery medium (*see* **Note 18**).

2. Electroporation cuvette and electroporator.

3. Culture tube: sterile 14 mL tube.

4. LB agar plates: five petri plates (70 mm diameter) containing 25 mL LB-agar and antibiotic (Ab^R-1) that selects for the final library.

5. Shaking incubator: set to 37 °C and 250 rpm.

6. Gravity incubator: set to 37 °C.

7. LB medium: 10 mL for harvesting cells from five LB-agar plates (*see* **Note 22**).

8. Sterile spreader.

9. Culture tube: sterile 14 mL tube (*see* **Note 23**).

10. DNA purification kit such as the Qiagen Miniprep kit (*see* **Note 24**).

11. Two restriction enzymes, which cut at sites RE1 and RE2 and buffers (*see* **Note 34**).

12. Agarose gel: cast using TAE buffer and 1 μg/mL ethidium bromide.

13. Gel-loading dye: 12 μL of 6× stock.

14. Gel box and transilluminator.

3 Methods

The protocol described below requires 2 days to produce a library of vectors that expresses the different possible circularly permuted variants of a protein (*see* Fig. 2).

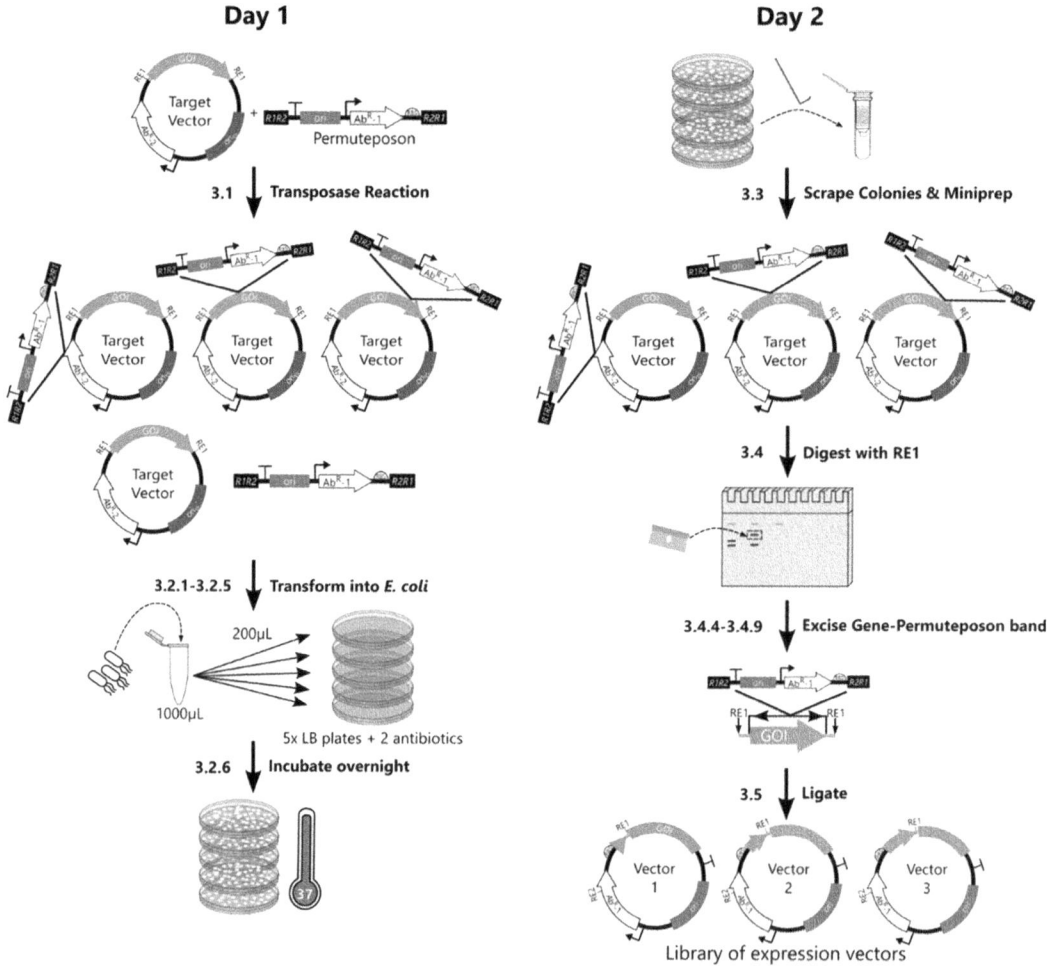

Fig. 2 Scheme for PERMUTE. First, a permuteposon that will ultimately serve as an expression vector for each permuted protein is inserted into a temperature-sensitive target vector containing a gene of interest (GOI) using MuA transposase. The vectors generated by this reaction have the permuteposon inserted at different locations. Second, the DNA product of the MuA reaction (which includes permuteposon, target vector and vector–permuteposon hybrids) is transformed into bacteria and plated on LB-agar containing a pair of antibiotics (Ab-1 and Ab-2) that select for cells containing a permuteposon and target vector. Third, to amplify cells with the vector–permuteposon hybrids, these plates are incubated at a high temperature where the target vector origin of replication (ori$_{ts}$) cannot function. Fourth, colonies are harvested from plates after a day of growth, and plasmid DNA is purified from this cell mixture. Fifth, the purified DNA is digested using an enzyme (RE1) that cuts at sites adjacent to the gene of interest. Gene–permuteposon hybrids are separated by agarose gel electrophoresis and the gene–permuteposon hybrids are purified using a DNA recovery kit. Finally, the gene–permuteposon hybrids are circularized through ligation to generate the final library of vectors that express different circularly permuted variants

**3.1 MuA
Transposase Reaction**

1. In the PCR tube, make up a 38 μL reaction using 8 μL 5×
   MuA Reaction Buffer, 600 ng target vector, 200 ng linear
   permuteposon, and water.

2. Add 0.44 μg MuA Transposase (2 μL) and mix gently by pipetting (*see* **Note 35**).

3. Incubate the reaction in a thermal cycler for 16 h at 37 °C
   followed by 10 min at 75 °C to inactivate the enzyme.

4. Purify the DNA from the reaction using a DNA cleanup kit
   (e.g., Zymo DNA Clean and Concentrator kit) and elute with
   6 μL of water (*see* **Note 16**).

**3.2 Vector–
Permuteposon Hybrids**

1. Add all of the DNA purified from the MuA reaction from **step 4**
   in Subheading 3.1 to 50 μL *E. coli* cells on ice and mix gently
   by pipetting.

2. Transfer the cell-DNA mixture to a chilled electroporation
   cuvette and electroporate using 2 kV (*see* **Note 36**).

3. Add 1 mL of recovery medium (e.g., SOC medium) to the
   cuvette, mix by pipetting, and transfer the slurry to a sterile
   14 mL culture tube.

4. Incubate culture for 1 h at 37 °C while shaking at 250 rpm
   (*see* **Note 20**).

5. Spread 200 μL aliquots of the cells on each LB-agar plate
   (*see* **Note 37**).

6. Incubate plates overnight at 37 °C to obtain colonies (*see*
   Note 21).

**3.3 Vector–
Permuteposon Library**

1. Add 1 mL of LB medium to each agar plate (*see* **Note 22**).

2. Generate a bacterial slurry by gently scraping plates with a sterile spreader.

3. Angle plates so that the cell slurry pools on one side (*see* **Note 38**).

4. Pool the cell slurries from each plate into a 14 mL culture tube
   (*see* **Note 39**).

5. Cap the 14 mL tube containing the cell slurry and invert to
   mix.

6. Purify vectors from the pooled cells using a Qiagen Miniprep
   kit. Only 20% of the recommended culture volume (400 μL
   slurry) should be used in each miniprep (*see* **Note 40**).

7. Determine the concentration of purified DNA (*see* **Note 41**).

**3.4 Gene–
Permuteposon Hybrids**

1. Digest 500 ng of the purified DNA from **step 6** in
   Subheading 3.3 in a PCR tube using 1 unit of a restriction
   enzyme (RE1) that cuts at sites flanking both sides of the GOI
   (*see* **Note 42**). This reaction should be performed in a 20 μL
   volume using water and the buffer provided with the restriction
   enzyme.

2. Incubate the restriction digest reaction at 37 °C for 1 h.

3. Heat-inactivate the restriction enzyme as recommended by the manufacturer.

4. Add 4 μL of 6× gel loading dye to the digested product.

5. Separate the DNA products of the restriction digest using agarose gel electrophoresis (*see* **Note 43**).

6. Visualize the gel using a transilluminator.

7. Use a clean razor blade to excise the band corresponding to the size of the gene–permuteposon hybrid.

8. Recover the gene–permuteposon DNA using a DNA clean up kit such as the Zymo Gel DNA Recovery kit and elute the DNA using 6 μL sterile water two times.

9. Quantify the gene–permuteposon hybrid concentration (*see* **Note 44**).

3.5 Circularize Gene–Permuteposon Hybrids

1. In a sterile PCR tube, make a 19 μL reaction containing 2 μL 10× T4 DNA ligase buffer, all of the DNA recovered from the gel in **step 8** in Subheading 3.4, and water to 19 μL.

2. Mix the reaction thoroughly and then add 1 μL of T4 DNA ligase (400 U/μL).

3. Incubate the ligation reaction at 16 °C for 16 h.

4. Purify the DNA with a DNA purification kit such as Zymo DNA Clean and Concentrator kit and elute with 6 μL of water (*see* **Note 45**).

3.6 Quality Control

1. Add 6 μL of the ligated DNA from **step 4** in Subheading 3.5 into an 1.5 mL Eppendorf tube containing 50 μL electrocompetent *E. coli* cells on ice and mix by pipetting.

2. Transfer the cell-DNA mixture to a chilled electroporation cuvette and electroporate at 2 kV (*see* **Note 36**).

3. Add 1 mL of recovery medium (e.g., SOC medium) to the cuvette, mix by gently pipetting, and transfer the slurry to a sterile 14 mL culture tube.

4. Incubate the culture for 1 h at 37 °C while shaking at 250 rpm.

5. Spread 200 μL aliquots of the transformation mix onto each LB-agar plate.

6. Incubate plates overnight at 37 °C to obtain colonies.

7. Use the protocol outlined in Subheading 3.3 above to harvest the final plasmid library, which encodes expression vectors for different permuted variants of the gene of interest.

8. Digest 200 ng of the final library with a restriction enzyme that cuts the restriction site (RE1) that was ligated to create the final library using a 20 μL reaction (*see* **Note 46**).

9. Digest 200 ng of the final library with a restriction enzyme (RE2) that cuts at a single site within the permuteposon using a 20 μL reaction (*see* **Note 47**).

10. Digest 200 ng of the final library with the two restriction enzymes (RE1 and RE2) in parallel using a 20 μL reaction (*see* **Note 48**).

11. Add 4 μL of 6× gel-loading dye to each reaction.

12. Separate the DNA products of the restriction digests using agarose gel electrophoresis (*see* **Note 43**).

13. Visualize the gel using a transilluminator.

4   Notes

1. Commercial molecular biology grade water is recommended.

2. The gene of interest must lack a stop codon so that permuted open reading frames can be fully translated.

3. A vector with a temperature-sensitive origin of replication (ori$_{ts}$) allows for selection of vector–permuteposon hybrids at high temperatures where the ori$_{ts}$ cannot function. A method for using a circularized gene as an alternative was recently described [20].

4. Structural information should be used to estimate a linker length that is sufficiently long enough to enable proper folding of permuted proteins. In addition, glycine/serine-rich sequences encoding flexible linkers are recommended [2].

5. Permuteposon sequence information was previously described [19]. All aspects of a permuteposon can be customized, provided that functional transposase binding sites are included in the design (*see* Fig. 1a).

6. Avoid using the same origin of replication in the permuteposon and target vector to minimize undesired recombination events.

7. Use distinct selectable markers in the permuteposon and target vector to allow for selection of vector–permuteposon hybrids using two antibiotics (Ab^R-1 and Ab^R-2).

8. Two sequences of transposase binding sites have been reported (*see* Fig. 1c). In the first transposon described [19], the RBS used to initiate translation is separated from the open reading frame of permuted proteins by the transposase binding site. This separation results in the addition of a large 18 amino acid peptide (MGFRIYRETLSRFSCAAQ) at the N-terminus of each permuted variant, which is encoded by the transposase binding site. More recently, a transposase binding site was described that allows for translation initiation closer to the gene sequence encoding the protein variant [13]. With this

transposon, only two extra residues are amended to the N-terminus of proteins expressed from the transposon after insertion into the gene of interest, a methionine followed by a residue whose identity varies depending on the sequence at the site of permuteposon insertion.

9. Regulated or constitutive promoters can be used to control transcription of the permuted genes.

10. A terminator should be incorporated into the region of the permuteposon that follows the stop codon. This terminator reduces the plasmid burden on cell growth caused by unnecessary transcription.

11. When using transposase binding sites (*see* Fig. 1c) that contain mutations that introduce an RBS [13], it is important to use a strong promoter because translation initiation will vary with each permuted protein due to changes in the genetic context of the RBS.

12. Temperature-sensitive vectors frequently acquire mutations, so it is important to perform all cloning at temperatures that avoid selective pressure on this phenotype. Additionally, it is important to run controls to verify that the vector preparation used for cloning retains the temperature sensitive phenotype.

13. Permuteposons can be amplified like vectors. This can be achieved by flanking the terminal transposase binding sites with a unique restriction enzyme [19] and circularizing the permuteposons through ligation. However, circular permuteposons must be linearized by restriction digestion to serve as a substrate for MuA.

14. MuA transposase can be purchased with the 5× reaction buffer from Thermo Scientific (Catalog No. F-750).

15. A thermal cycler is recommended to provide fine control over the reaction temperature and the heat inactivation of the enzyme.

16. To elute DNA from the spin columns, molecular biology grade water is recommended rather than the DNA elution buffer provided with the kit. This allows for electroporation of the eluted DNA without further manipulation.

17. Library-quality competent cells (Invitrogen MegaX DH10B; Catalog No. C6400-03) are recommended to ensure that sequence diversity is not limited by the transformation.

18. Recovery medium provided with commercial cells is recommended to maximize transformation efficiency.

19. Multiple plates are recommended to allow for sufficient separation of the colonies and counting. The DNA sequence diversity depends on the number of transformants obtained at each step and can be limited by poor transformations.

20. The shaking incubator should be set to a temperature that selects against the temperature-sensitive vector.

21. The gravity incubator temperature must be sufficient to select against a temperature-sensitive vector lacking an inserted permuteposon. Control experiments that examine the colony counts obtained from transforming a target vector alone are recommended to verify the temperature is sufficiently high to select against the temperature-sensitive vector.

22. Prior to scraping colonies from plates, sterile liquid should be added to each plate. The addition of liquid makes it easier to remove the scraped cells by pipetting. If cells are not harvested immediately after the incubation, colonies can harden and become challenging to scrape off plates.

23. Cells harvested from each plate should be pooled into one tube.

24. Water should be used rather than the buffer provided with the kit to minimize the salt concentration in the eluted DNA, since high salt levels can inhibit the activity of some restriction enzymes.

25. A restriction enzyme (RE1) is needed that cuts at sites adjacent to the 5′ and 3′ ends of the gene being permuted. An enzyme should be used that does not cut anywhere else in the target vector, gene, or permuteposon.

26. An agarose percentage is recommended that enables separation of the different DNA products generated by a digest with the restriction enzyme (RE1).

27. Alternative dyes can be used, such as SYBR® Stains.

28. Commercial gel loading dye is recommended (New England Biolabs; Catalog No. 7021S).

29. Gel visualization tools should be used that are compatible with the DNA visualization dye.

30. Water is used rather than the buffer provided by the manufacturer to minimize the salt concentration in the eluted DNA, which can affect the ligation reaction.

31. Lower concentrations of gene–permuteposon hybrids can be used, but this could limit the sequence diversity sampled.

32. T4 DNA ligase from New England Biolabs (Catalog No. M0202S) is recommended, which is provided at a concentration of 400 U/μL.

33. To maximize ligation efficiency, a ligase buffer should be used that has not been subjected to multiple freeze-thaw cycles.

34. Two enzymes are required for quality control (*see* Fig. 3): (1) RE1, which cuts at a single unique site in the gene being

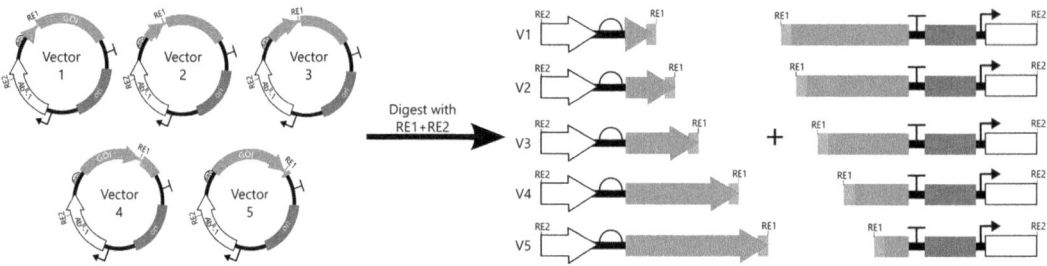

Fig. 3 A scheme illustrating how to assess PERMUTE library quality. All of the variants contain two restriction sites: (1) RE1, which is found at a unique location within each of the permuted genes, and (2) RE2, which is found at one location within the permuteposon backbone. RE2 should be identified before beginning the library construction. A double digest of the final PERMUTE library with RE1 and RE2 will yield DNA smears that vary by a length that corresponds to the number of base pairs in the GOI. In contrast, digestion of the final PERMUTE library with each restriction site individually will yield linear DNA, which has a size corresponding to the length of the GOI plus the length of the permuteposon

permuted, and (2) RE2, which cuts at a single unique site in the permuteposon.

35. The activity of MuA can be calibrated by analyzing the activity using a commercially available transposon (Entranceposons from Thermo Scientific) with a selectable marker that is not present in the target vector.

36. The kV setting on the electroporator should match the value recommended by the electrocompetent cell manufacturer.

37. LB agar plates should be dried sufficiently to remove any excess water on the surface, which can cause colony smearing upon incubation.

38. Elevating one end of the plate 1 cm above the other end should be sufficient.

39. If cells are not fully removed by the protocol, then add 1 mL of LB medium to each plate, scrape, and remove additional liquid.

40. Scraped cells are typically more concentrated than cells from liquid cultures, so cell volumes should be reduced to those as recommended by the manufacturer.

41. DNA concentration can be quantified using a spectrophotometer (e.g., a NanoDrop) or by gel electrophoresis.

42. Permuteposons are randomly inserted into the target vector. In some vectors, they will be found within the GOI (*see* Fig. 2), while other vectors will contain a permuteposon within the vector backbone. Because transposons are randomly inserted, restriction enzyme digestion using RE1 will yield four DNA products: (1) a gene lacking a permuteposon, (2) a gene containing a permuteposon, (3) a target vector backbone lacking a

permuteposon, and (4) a target vector backbone containing a permuteposon.

43. A DNA ladder should be used to provide a frame of reference when performing the size-selection step. The goal of this step is to purify the band corresponding to the gene–permuteposon hybrid.

44. The desired yield is 100 ng of gene–permuteposon.

45. The DNA obtained from this purification can be screened or selected for variants with desired functional properties, although quality control is recommended to ensure the desired sequence diversity was generated.

46. Any restriction enzyme can be used for RE1, provided that it only cuts at one unique site within the gene that was permuted. This digest should yield a single product.

47. Any restriction enzyme can be used for RE2, provided that it cuts at one unique site within the permuteposon and does not cut within the gene that was permuted. This digest should yield a single product.

48. The double digest at RE1 and RE2 should yield products of varying size, resulting in a smear on the gel, because the distance between the two restriction sites will vary with each of the permuted genes (*see* Fig. 3).

Acknowledgments

This work was supported by the National Science Foundation (1150138). AMJ and JTA were supported by the National Science Foundation Graduate Research Fellowship Program (NSF GRFP) under grant number (R3E821).

References

1. Ostermeier M (2009) Designing switchable enzymes. Curr Opin Struct Biol 19:442–448

2. Yu Y, Lutz S (2011) Circular permutation: a different way to engineer enzyme structure and function. Trends Biotechnol 29:18–25

3. Liberles DA, Teichmann SA, Bahar I, Bastolla U, Bloom J, Bornberg-Bauer E, Colwell LJ, de Koning APJ, Dokholyan NV, Echave J et al (2012) The interface of protein structure, protein biophysics, and molecular evolution. Protein Sci 21:769–785

4. Zhao H, Arnold FH (1997) Combinatorial protein design: strategies for screening protein libraries. Curr Opin Struct Biol 7:480–485

5. Graf R, Schachman HK (1996) Random circular permutation of genes and expressed polypeptide chains: application of the method to the catalytic chains of aspartate transcarbamoylase. Proc Natl Acad Sci U S A 93:11591–11596

6. Hennecke J, Sebbel P, Glockshuber R (1999) Random circular permutation of DsbA reveals segments that are essential for protein folding and stability. J Mol Biol 286:1197–1215

7. Guntas G, Ostermeier M (2004) Creation of an allosteric enzyme by domain insertion. J Mol Biol 336:263–273

8. Hida K, Won SY, Di Pasquale G, Hanes J, Chiorini JA, Ostermeier M (2010) Sites in the

AAV5 capsid tolerant to deletions and tandem duplications. Arch Biochem Biophys 496:1–8

9. Judd J, Wei F, Nguyen PQ, Tartaglia LJ, Agbandje-McKenna M, Silberg JJ, Suh J (2012) Random insertion of mCherry into VP3 domain of adeno-associated virus yields fluorescent capsids with no loss of infectivity. Mol Ther Nucleic Acids 1:e54

10. Haapa S, Taira S, Heikkinen E, Savilahti H (1999) An efficient and accurate integration of mini-Mu transposons in vitro: a general methodology for functional genetic analysis and molecular biology applications. Nucleic Acids Res 27:2777–2784

11. Jones DD (2005) Triplet nucleotide removal at random positions in a target gene: the tolerance of TEM-1 beta-lactamase to an amino acid deletion. Nucleic Acids Res 33:e80

12. Segall-Shapiro TH, Nguyen PQ, Dos Santos ED, Subedi S, Judd J, Suh J, Silberg JJ (2011) Mesophilic and hyperthermophilic adenylate kinases differ in their tolerance to random fragmentation. J Mol Biol 406:135–148

13. Segall-Shapiro TH, Meyer AJ, Ellington AD, Sontag ED, Voigt CA (2014) A 'resource allocator' for transcription based on a highly fragmented T7 RNA polymerase. Mol Syst Biol 10:742

14. Pandey N, Nobles CL, Zechiedrich L, Maresso AW, Silberg JJ (2014) Combining random gene fission and rational gene fusion to discover near-infrared fluorescent protein fragments that report on protein-protein interactions. ACS Synth Biol. 4:615–624. doi:10.1021/sb5002938

15. Poussu E, Vihinen M, Paulin L, Savilahti H (2004) Probing the alpha-complementing domain of E. coli beta-galactosidase with use of an insertional pentapeptide mutagenesis strategy based on Mu in vitro DNA transposition. Proteins 54:681–692

16. Edwards WR, Busse K, Allemann RK, Jone DD (2008) Linking the functions of unrelated proteins using a novel directed evolution domain insertion method. Nucleic Acids Res 36:e78

17. Hoeller BM, Reiter B, Abad S, Graze I, Glieder A (2008) Random tag insertions by Transposon Integration mediated Mutagenesis (TIM). J Microbiol Methods 75:251–257

18. Poussu E, Jäntti J, Savilahti H (2005) A gene truncation strategy generating N- and C-terminal deletion variants of proteins for functional studies: mapping of the Sec1p binding domain in yeast Mso1p by a Mu in vitro transposition-based approach. Nucleic Acids Res 33:e104

19. Mehta MM, Liu S, Silberg JJ (2012) A transposase strategy for creating libraries of circularly permuted proteins. Nucleic Acids Res 40:e71

20. Pandey N, Kuypers BE, Nassif B, Thomas EE, Alnahhas RN, Segatori L, Silberg JJ (2016) Tolerance of a Knotted Near-Infrared Fluorescent Protein to Random Circular Permutation. Biochem 55:3763–3773

Chapter 20

Transposon Insertion Mutagenesis for Archaeal Gene Discovery

Saija Kiljunen, Maria I. Pajunen, and Harri Savilahti

Abstract

Archaea constitute the third domain of life, but studies on their physiology and other features have lagged behind bacteria and eukarya, largely due to the challenging biology of archaea and concomitant difficulties in methods development. The use of genome-wide en masse insertion mutagenesis is one of the most efficient means to discover the genes behind various biological functions, and such a methodology is described in this chapter for a model archaeon *Haloferax volcanii*. The strategy successfully employs efficient in vitro transposition in combination with gene targeting in vivo via homologous recombination. The methodology is general and should be transferable to other archaeal species.

Key words Insertion mutant library, Gene discovery, *Haloferax volcanii*, Halophilic archaea, MuA transposase

1 Introduction

Archaea constitute the third domain of life, are ubiquitous in different types of environments, and often live in habitats with extremely harsh conditions. Thus, these organisms are biologically intriguing and potentially constitute an enormous resource for biotechnology applications. Yet, largely due to their challenging biology and concomitant difficulties in methods development, the biochemical pathways and genetic basis behind many unique archaeal features remain poorly characterized or entirely uncharacterized. Advanced methodologies are thus warranted for these organisms.

Insertion mutant libraries, which contain randomly distributed genomic alterations in each gene, provide a valuable resource for studies aimed at delineating molecular mechanisms behind biological functions. Such libraries have proven their immense usefulness for gene discovery, particularly in microbiological research. Until recently, archaea were underrepresented in such studies, as adequate insertion libraries had only been made for two

Andrew Reeves (ed.), *In Vitro Mutagenesis: Methods and Protocols*, Methods in Molecular Biology, vol. 1498,
DOI 10.1007/978-1-4939-6472-7_20, © Springer Science+Business Media New York 2017

archaeal species, the methanogens *Methanosarcina acetivorans* and *Methanococcus maripaludis* [1, 2].

Now, a recent study has widened the scope and applicability of insertion mutant libraries to halophilic archaea. The paper by Kiljunen et al. [3] describes a transposition-based method to generate a comprehensive insertion mutant library for the easily cultivable model archaeon, *Haloferax volcanii*, and moreover, the use of the library for gene discovery. The methodology used exploited a MuA-transposase-catalyzed in vitro transposition reaction and combined it with in vivo gene targeting by homologous recombination. As a result, a robust and widely applicable strategy was devised. The strategy entails the following steps: (1) Isolation of genomic DNA, its fragmentation, and tagging with a transposon; (2) Cloning of the tagged genomic fragments in *E. coli* to generate a plasmid library that covers the entire chromosome with overlapping fragments; (3) Amplification of the library, isolation of tagged chromosomal fragments, and their size selection; and (4) Transformation of the fragments into *H. volcanii* and en masse gene targeting via homologous recombination. The end product is a *H. volcanii* insertion-mutant library, in which each clone harbors a single transposon insertion in its genome. The library is an ideal resource for efficient gene discovery, and it facilitates the identification of nonessential genes behind any specific biochemical pathway. The strategy used for the construction of the library should readily be transferable to other archaeal species.

2 Materials

H. volcanii is sensitive to trace amounts of contaminants such as detergents as well as to impurities found in culture media. Therefore, it is advisable to rinse all glassware thoroughly with water and use for medium preparation *only* those commercial products specified in this protocol (*see* **Note 1**).

2.1 Equipment

1. Tabletop centrifuge that accommodates 50 ml centrifuge tubes.
2. 37 °C incubator and shaker.
3. Tabletop centrifuge (for 15 ml tubes).
4. Microcentrifuge.
5. Ultracentrifuge.
6. Heating block.
7. Water bath.
8. Equipment for ion-exchange chromatography.
9. Equipment for preparative agarose gel electrophoresis.
10. Electroporation apparatus and cuvettes.

11. 4.6×100 mm steel anion-exchange column, Gen-Pak FAX (Waters).

2.2 Media

1. 18% Modified Growth Medium (MGM): First, make a 30% salt water (SW) solution by dissolving 240 g NaCl, 30 g $MgCl_2 \cdot 6H_2O$, 35 g $MgSO_4 \cdot 7H_2O$, and 7 g KCl into ~800 ml water. Add slowly 5 ml 1 M $CaCl_2$. Adjust the pH to 7.5 with 1 M Tris–HCl, pH 7.5 and add water to a final volume of 1000 ml. For liquid 18% MGM, dissolve 5 g peptone (Oxoid) and 1 g yeast extract (Difco) into 600 ml SW and 367 ml water. Adjust the pH to 7.5 with 1 M Tris base and fill the volume to 1000 ml with water. For solid medium, add Bacto agar (Difco) 15 g/l [3] (*see* **Note 1**).

2. SOB medium: 2% Bacto tryptone, 0.5% Bacto yeast extract, 10 mM NaCl, 2.5 mM KCl. Autoclave.

3. SOC medium: Add to 100 ml of SOB solution 1 ml of 2 M $MgCl_2$ and 2 M glucose from stock solutions sterilized by filtration through a 0.22 μm filter.

4. Luria–Bertani medium (LB): 1% Bacto tryptone, 0.5% Bacto yeast extract, 1% NaCl. For solid medium, add Bacto agar to 1.5%. Supplement LB with ampicillin (100 μg/ml) and chloramphenicol (10 μg/ml) when needed.

2.3 Chemicals, Buffers

1. ST buffer: 1 M NaCl, 20 mM Tris–HCl, pH 7.5.
2. Saturated phenol, pH 7.9.
3. Chloroform.
4. Absolute ethanol.
5. 70% ethanol.
6. TE buffer: 10 mM Tris–HCl, pH 7.5, 1 mM EDTA.
7. 3 M sodium acetate, pH 7.0.
8. CsCl.

2.4 Molecular Biology Reagents

1. Lysis solution: 100 mM EDTA, pH 8.0, 0.2% SDS.
2. RNase A stock solution (10 mg/ml).
3. *Aci*I, *Hpa*I, *Taq*I, *Bgl*II, *Cla*I restriction enzymes.
4. Transposon carrier plasmid pMPH20 [4].
5. pBlueScriptSK+ plasmid.
6. Plasmid DNA isolation kit.
7. Alkaline phosphatase.
8. T4 DNA ligase.
9. *E. coli* electrocompetent DH10B cells. Stored at –70 °C.
10. *E. coli* DH5α electrocompetent cells. Stored at –70 °C.
11. Ethidium bromide.

2.5 DNA Transformation of H. volcanii

1. Buffered spheroplasting solution: 1 M NaCl, 27 mM KCl, 50 mM Tris–HCl, pH 8.5, 15 % (w/v) sucrose [3].

2. Buffered spheroplasting solution with 15 % glycerol: 1 M NaCl, 27 mM KCl, 50 mM Tris–HCl, pH 8.5, 15 % (w/v) sucrose, 15 % (v/v) glycerol [3].

3. 0.5 M EDTA, pH 8.0.

4. Unbuffered spheroplasting solution: 1 M NaCl, 27 mM KCl, 15 % (w/v) sucrose. Adjust pH to 7.5 with 1 M NaOH (~10 μl/100 ml) [3].

5. 60 % PEG 600 solution: 1500 μl PEG 600 (Merck) + 1000 μl unbuffered spheroplasting solution [3].

6. Spheroplast dilution solution: Dissolve 15 g sucrose in 76 ml 30 % SW, and add water up to 100 ml. After autoclaving, add 0.75 ml of 0.5 M CaCl$_2$ [3].

7. Regeneration solution: First, make a 10× YPC solution by dissolving 1.25 g Yeast Extract (Difco), 0.25 g Peptone (Oxoid), and 0.25 g casamino acids (Difco) in ~19 ml water. Adjust the pH to 7.5 with 1 M KOH. Adjust the volume to 25 ml with distilled water. For the regeneration solution, dissolve 37.5 g sucrose in 150 ml 30 % SW and 25 ml 10× YPC. Adjust the volume to 250 ml with distilled water. After autoclaving, add 1.5 ml of 0.5 M CaCl$_2$ [3].

8. Transformant dilution solution: Dissolve 37.5 g sucrose in 150 ml 30 % SW and adjust the volume to 250 ml with water. After autoclaving, add 1.5 ml of 0.5 M CaCl$_2$ [3].

9. 80 % glycerol-6 % SW: Mix 80 ml glycerol and 20 ml 30 % SW. After autoclaving, add 200 μl of 0.5 M CaCl$_2$ [3].

10. Hv-Ca: First, make a 10× Ca stock solution by dissolving 1.7 g casamino acids (Difco) in ~25 ml water. Add 800 μl of 1 M KOH and adjust the volume to 33 ml with water. To prepare Hv-Ca agar, measure into a 500 ml bottle 5 g agar (Difco), 100 ml of water and 200 ml of 30 % SW. Boil to dissolve and add 10× Ca (33 ml). Autoclave. Cool to ~60 °C and add slowly 2 ml of 0.5 M CaCl$_2$ and 300 μl of a mixture of thiamine (0.89 mg/ml) and biotin (0.11 mg/ml). For strain *H. volcanii* H295 add 340 μl of uracil (50 mg/ml in DMSO) [3].

2.6 Mu In Vitro Transposition

1. 2× Mix: 50 mM Tris–HCl, pH 8.0, 200 μg/ml BSA (bovine serum albumin), 30 % (w/v) glycerol. Use high-quality molecular biology grade BSA. Store at –70 °C.

2. Triton X-100: Make a 1.25 % solution from a 10 % stock solution by diluting with H$_2$O directly prior to use.

   • 2.5 M NaCl, 0.25 M MgCl$_2$

3. MuA transposase protein (Thermo Fisher Scientific) 220 ng/ μl in MuA dilution buffer: 0.3 M NaCl, 25 mM Hepes, pH 7.6, 0.1 mM EDTA, 1 mM DTT, 10% (w/v) glycerol.

3   Methods

3.1   Transposon-Tagged Plasmid Library

For *H. volcanii* chromosomal DNA isolation and fragmentation, all procedures must be carried out at room temperature (RT) unless otherwise indicated.

1. Culture *H. volcanii* cells on 18% MGM agar dishes at 45 °C for 3–5 days (*see* **Note 2**). Inoculate 1–4 colonies into 5 ml of 18% MGM liquid medium and culture with shaking at 45 °C for ~32 h, until the late exponential culture phase is reached.

2. Collect cells from 3 ml of the cell suspension by microcentrifuging at $3500 \times g$ for 8 min and resuspend the cells in 200 μl of ST buffer. Transfer the suspension into a 1.5 ml microtube. To lyse the cells, add 200 μl of lysis solution and mix carefully by gently inverting the tube (to avoid extra DNA shearing, do not vortex or pipette to mix).

3. Add 400 μl of phenol. Mix gently with a tabletop rotator for 30 min. Transfer the tube into a heat block (or water bath) and incubate at 60 °C for 1 h. To separate phases, use a microcentrifuge at maximum speed for 5 min, and transfer the viscous supernatant into another 1.5 ml microtube. Repeat the phenol extraction procedure but without the 60 °C incubation step. Extract the supernatant twice with 600 μl of chloroform.

4. Add 1 ml of ethanol and mix gently for a few minutes to precipitate DNA. Transfer the thread-like DNA precipitate (e.g., with a pipette tip) into a tube containing 1 ml of 70% ethanol. Microentrifuge at maximum speed for 10 min, remove the supernatant, and air-dry the DNA pellet. Dissolve DNA in 300–500 μl of TE buffer. Complete dissolution of the DNA may require up to 3 days (*see* **Note 3**).

5. Remove RNA by incubating with RNase A (at 0.2 mg/ml) for 30 min at 37 °C. Remove the enzyme by extracting with saturated phenol and twice with chloroform. Ethanol-precipitate the DNA with 1/10 volume of 3 M sodium acetate (pH 7.0). Dissolve the DNA in TE buffer as above.

6. Digest genomic DNA partially with three different enzymes (*Aci*I, *Hpa*II, and *Taq*I) in three separate reactions (*see* **Note 4**). Aim for a broad DNA fragment size distribution, in which a large proportion of the fragments falls within the size range of 2–4 kb. Remove the enzyme by phenol and chloroform extractions as in **step 5** above. Ethanol-precipitate and dissolve the DNA in water (*see* **Note 5**).

3.2 Transposon DNA (See Note 6)

1. TrpA-cat-Mu transposon DNA (*see* **Note 7**) is released from its carrier plasmid pMPH20 [4] by *Bgl*II digestion. The 2212 bp linear transposon can be purified in large quantities by the use of anion-exchange chromatography [6]. Preparative agarose gel electrophoresis is a convenient means to purify several micrograms of transposon DNA for a few experiments. Store the purified transposon DNA in TE buffer.

3.3 Mu In Vitro Transposition and Size-selection of the Reaction Products

1. Use DNA digested with each enzyme (*Aci*I, *Hpa*II, and *Taq*I) as a target in three separate transposition reactions. With each of them, assemble several standard in vitro DNA transposition reactions on ice as specified in Table 1 (*see* **Note 8**). Add MuA transposase as the last component just prior to incubation.

2. Incubate the reactions at 30 °C for 60 min and pool all the reaction products into one tube.

3. Extract once with saturated phenol and twice with chloroform. Ethanol-precipitate the DNA and dissolve in TE buffer.

4. Purify the 4–6 kb fragments using anion-exchange chromatography with a Gen-Pak FAX column (*see* **item 10** in Subheading 2.1 for the column). Use conditions described in the manufacturer's instructions.

5. As an alternative method, use preparative agarose gel electrophoresis for size selection and gel purification kits for DNA isolation.

3.4 Vector DNA

1. Isolate pBlueScript SK+ plasmid DNA (*see* **Note 9**) using any standard commercial plasmid isolation kit.

2. Digest 10 μg of pBlueScript SK+ plasmid DNA with *Cla*I restriction enzyme (4 U/μg) at 37 °C overnight.

Table 1
Mu in vitro transposition reaction mixture

Reagent	Standard reaction
2× mix	25 μl
Digested genomic DNA as target (800–1000 ng)	Typically 2–6 μl
TrpA-cat-Mu transposon as donor DNA (0.5 pmol/μl)	2 μl (1 pmol)
2.5 M NaCl	2 μl
1.25 % Triton X-100 (freshly diluted)	2 μl
0.25 M MgCl$_2$	2 μl
H$_2$O	Up to 48 μl
MuA (220 ng/μl)	2 μl (440 ng, 5.4 pmol)
	Total 50 μl

3. Dephosphorylate the digested vector DNA with calf intestinal alkaline phosphatase as recommended by the supplier.

4. Isolate the dephosphorylated linear vector DNA using preparative agarose gel electrophoresis.

3.5 Plasmid Library Generation

1. Ligate the *Cla*I-digested vector with size-selected in vitro transposition reaction products using T4 DNA ligase and reaction conditions recommended by the supplier. Typically 50 ng of digested plasmid DNA is ligated with a 3-fold molar excess of DNA inserts in a reaction volume of 15 μl.

2. Thaw competent *E. coli* DH10B cells on ice (*see* **Note 10**).

3. Add 1 μl of ligation mixture into 25 μl of electrocompetent cells in a cold microtube. Transfer the mixture into an ice-cold electroporation cuvette (0.1 cm electrode spacing) (*see* **Note 11**).

4. Electroporate using the following pulse settings: voltage 1.8 kV, resistance 200 Ω, and capacitance 25 μF (*see* **Note 12**).

5. Add 1 ml SOC (room temperature solution), transfer into a microcentrifuge tube and incubate at 37 °C by shaking (220 rpm) for 30 min.

6. Spread the cells on LB culture plates containing ampicillin (100 μg/ml) and chloramphenicol (10 μg/ml) and incubate at 37 °C overnight.

7. Collect a suitable number of colonies (*see* **Note 13**) by adding per plate 1 ml LB medium supplemented with ampicillin (100 μg/ml) and chloramphenicol (10 μg/ml) (LB-amp-cam), and scraping the colonies into a single pool (*see* **Note 14**). Add a suitable volume of fresh LB-amp-cam medium and grow at 37 °C by shaking (220 rpm) for 2.5 h (*see* **Note 15**).

8. Isolate plasmid DNA using a plasmid isolation kit of suitable capacity. This plasmid pool represents the primary plasmid library with the diversity defined by the number of the collected colonies used for its preparation (*see* **Note 16**).

3.6 Library Amplification and Purification (See Note 17)

1. Electrotransform aliquots from the primary plasmid library into *E. coli* DH5α, selecting for ampicillin and chloramphenicol resistance as above (*see* **Note 18**).

2. Collect a suitable number of colonies (*see* **Note 19**) and isolate plasmid DNA as above.

3. Purify supercoiled plasmid forms from the plasmid library by CsCl gradient ultracentrifugation (*see* **Note 20**). The published protocol is recommended [7].

3.7 H. volcanii Insertion Mutant Library

1. Use two different pairs of restriction endonucleases to release transposon-tagged *H. volcanii* DNA fragments from the plasmid library (*see* **Note 21**).

2. Isolate suitable-sized (4–6 kb) fragments using preparative agarose gel electrophoresis. Prepare enough purified DNA (several micrograms) to be used in the subsequent transformation step.

Perform the following steps at room temperature unless otherwise indicated.

3. To prepare *H. volcanii* competent cells (*see* **Notes 22** and **23**), inoculate 5 ml of 18% MGM medium with ~4 *H. volcanii* colonies and culture by shaking at 200 rpm, 45 °C for 24 h. Transfer 0.5 ml of the cell culture into 50 ml of fresh 18% MGM and culture as above for ~20 h until the absorbance (OD) at 600 nm is 0.8–1.0.

4. Divide the cell culture into two 25 ml aliquots in 50 ml centrifuge tubes and centrifuge in a table-top centrifuge ($4500 \times g$) for 10 min. Remove the supernatant and resuspend each pellet in 10 ml of buffered spheroplasting solution. Centrifuge as above and resuspend each pellet in 2.5 ml of buffered spheroplasting solution with 15% glycerol. Pool the suspensions and divide into suitable (e.g., 600 µl) aliquots. Flash-freeze the cells in liquid nitrogen and store at –80 °C. *H. volcanii* competent cells can be stored for several months. However, a decrease in the transformation efficiency is expected during prolonged storage (*see* **Note 24**).

5. Thaw 200 µl of competent *H. volcanii* cells for each transformation at room temperature and transfer them into a 2 ml centrifuge tube. Add 20 µl of 0.5 M EDTA, pH 8.0. Mix gently by inverting the tube and incubate for exactly 10 min to form spheroplasts.

6. Mix 10 µl (containing 1 µg) of DNA to be transformed, 15 µl unbuffered spheroplasting solution, and 5 µl 0.5 M EDTA, pH 8.0 (*see* **Note 25**).

7. Following the spheroplasting incubation (*see* **step 3** in Subheading 3.7), add the DNA mixture onto the tube wall and by gently rotating mix the solution with spheroplasts. Incubate for 5 min. Add 250 µl of freshly prepared 60% PEG 600 solution by slowly pipetting along the tube wall. Mix gently as above and incubate for 30 min.

8. Add 1.5 ml of spheroplast dilution solution and incubate for 2 min. Microcentrifuge at $3500 \times g$ for 10 min. Remove the supernatant. To regenerate, add 1 ml of regeneration solution gently onto the spheroplasts and incubate undisturbed at 45 °C for 1.5–2 h. Resuspend the cells and incubate with slow rotation (~100 rpm) at 45 °C for 4 h.

9. Microcentrifuge at $3500 \times g$ for 8 min. Remove the supernatant and resuspend into 1 ml of transformant dilution solution.

10. Plate the transformed cells directly onto appropriate selection plates or freeze the suspension for later use (*see* **Note 26**).

11. To store the cell suspension, add 335 μl of 80 % glycerol-6 % SW, divide into suitable (e.g., 200 μl) aliquots, flash-freeze in liquid nitrogen, and transfer for storage at –80 °C.

12. Spread the cells onto appropriate selective plates as follows (*see* **Note 27**). Thaw the frozen cells at room temperature and spread (with a suitable dilution) onto a selection plate. Use for example, 100 μl per standard petri dish and culture at 45 °C for 1 week. Note that some insertion mutants have a reduced growth rate, resulting in variable colony sizes among the member clones of the library.

3.8 Library Validation

1. Validate the library with regard to the insertion copy number (*see* **Note 28**).

2. Determine the insertion site of the transposon in the genome of a library clone by DNA sequencing (*see* **Note 29**).

4   Notes

1. It has been observed that the quality of media components and reagents may differ substantially among manufacturers, and certain impurities inhibit the growth of haloarchaeal cells. More details about the purity requirements can be found in the Halohandbook [3].

2. Other rich media can be used. More alternatives can be found in Halohandbook [3].

3. Alternatively, a genomic DNA isolation kit may be used. Most commercial kits should be suitable for *H. volcanii* DNA isolation.

4. We recommend partial digestions for DNA fragmentation and the usage of several enzymes. This guarantees an extensive coverage of the genome with overlapping fragments and in practice eliminates any potential bias caused by restriction site distribution. The enzymes recommended in this protocol (*Aci*I, *Hpa*II, *Taq*I) each recognize a 4-bp sequence and generate a protruding 5'-GC overhang that is compatible with a *Cla*I site in cloning. The conditions for partial digestions need to be adjusted for each restriction enzyme separately. It is advisable to use otherwise constant reaction conditions but different amounts of enzyme. The size distribution of DNA fragments can be analyzed by the use of standard agarose gel electrophoresis.

5. It is advisable to use 5–10 μg of genomic DNA for digestions. It is important to dissolve the digested DNA fragments in

water, as extra salt is inhibitory in the subsequent Mu in vitro DNA transposition reaction. The concentration of the fragmented DNA should preferably be ≥200 ng/μl.

6. Mini-Mu transposons utilized in in vitro reactions are linear DNA molecules that contain at each of their ends, in an inverted relative orientation, a 50-bp segment from the right end of the phage Mu genome. This so-called R-end segment contains a pair of MuA transposase-binding sites. The DNA between the R-ends can be of any origin and modified with regard to the needs of a particular application.

7. The TrpA-cat-Mu transposon [4] contains two selectable markers: *cat* for *E. coli* and *trpA* for *H. volcanii*. It is maintained within its carrier plasmid pMPH20 that can be obtained by request from *H. savilahti.*

8. Material from three to six reactions per enzyme digestion should yield enough DNA for the subsequent cloning step.

9. Other standard *E. coli* cloning vectors may be used. However, a suitable vector needs to harbor a unique *Cla*I site, which is flanked by at least two pairs of unique restriction sites not present in the TrpA-cat-Mu transposon (the transposon sequence is available upon request).

10. Although several *E. coli* strains may be used for cloning, DH10B is recommended as it yields good quality plasmid DNA and can be electroporated efficiently. High efficiency electrocompetent cells can be prepared using the protocol described [5], or they may be obtained from commercial vendors.

11. Perform several parallel electroporations to yield a desired number of transformants for the generation of a plasmid library. The number defines the maximum diversity of potential transposon insertion sites within this library.

12. The protocol has been developed for Gene Pulser II electroporation apparatus (Bio-Rad). If another brand is used the optimal pulse parameters may differ.

13. The desired colony number (*see* **Note 11**) varies according to the gene number of the target organism. A tenfold excess over the gene number is recommended, as it guarantees a comprehensive library, in which each gene will be tagged with a very high likelihood. For the calculation of the probability *see* Kiljunen et al. [4].

14. Standard (diameter 9 cm) plates can accommodate up to 1000 separate colonies, although this number may vary among different strains. If larger plates are used, the volume of the medium added should be increased proportionally. For example, 10 ml is suitable for large (25×25 cm) square plates.

15. The volume recommended depends on the number of collected colonies. For example, 200 ml of medium may be used for 40,000 colonies.

16. It is advisable to minimize the time used for liquid cultures, as the growth rate of the clones may differ. Accordingly, longer culture times may bias the original plasmid diversity.

17. Amplification of the plasmid library guarantees a sufficient amount of plasmid DNA for the purification by CsCl gradient ultracentrifugation.

18. Standard chemical transformation may also be used here given the efficiency is high enough to obtain the desired number of transformants. Although several *E. coli* strains may be used, DH5α is recommended. This strain yields very high quality plasmid DNA with the majority of the molecules being supercoiled.

19. To retain the original plasmid diversity with a very high probability, collect at least 20 times more colonies than what was collected for the primary plasmid library. Notice that you will need a substantial amount of DNA for the next step. We recommend isolating 1–2 mg plasmid DNA at this stage.

20. Alkaline plasmid preparation methods produce a fraction of collapsed supercoiled plasmid forms, which enter the cells efficiently and cannot be digested with restriction endonucleases. It is important to remove them, as they would generate false positive colonies upon transformation into *H. volcanii* (*see* ref. 4 for more details). CsCl gradient ultracentrifugation is a recommended means to remove collapsed supercoiled plasmid forms.

21. In the study of Kiljunen et al. [4], pBlueScript SKı plasmid was used as the cloning vector. The restriction enzyme pairs used for the successful fragment release in that study were *XhoI/HindIII* and *KpnI/EcoRV*. It is advisable to digest a substantial amount of DNA at this stage (e.g., 100–200 µg). Use conditions recommended by the enzyme supplier.

22. In principle, any *H. volcanii* strain could be used. We used the strain H295, which is devoid of Mre11 and Rad50. Due to these deficiencies, its homologous recombination activity is increased 100-fold [8].

23. More information about *Haloferax* transformation procedures can be found in Halohandbook [3].

24. Transformation efficiency of competent cells can be tested with any *H. volcanii* plasmid and appropriate selection plate.

25. DNA may be in any commonly used buffer (such as TE) or in water.

26. The expected transformation efficiency is 10^4–10^5 cfu per microgram of transformed DNA.

27. With the tryptophan auxotrophy marker gene (*TrpA*), the selection plate used is Hv-Ca. See Halohandbook for more information about selection plates and the required additives [3].

28. A genome-wide insertion mutant library used as a gene discovery tool should contain mutants with single-copy insertions. Southern hybridization or quantitative PCR can be used to evaluate this. In the protocol described, the stoichiometry in the transformation step favors single-copy genomic integrations, i.e., 1 out of 2000 target cells becomes transformed [4]. Thus, the vast majority of the clones in the library is expected to contain a single genomic insertion.

29. The transposon insertion site in the genome of a mutant can be determined by initially cloning the transposon with its genomic flanks from the chromosomal DNA using a restriction enzyme that does not cut within the transposon DNA. The sequences bordering the transposon DNA are then determined using standard Sanger sequencing [5]. However, with the fast development of next-generation sequencing (NGS) techniques, whole-genome sequencing by NGS may soon be the fastest and most cost-effective way to determine the insertion site in a microbial genome.

References

1. Zhang JK, Pritchett MA, Lampe DJ et al (2000) *In vivo* transposon mutagenesis of the methanogenic archaeon *Methanosarcina acetivorans* C2A using a modified version of the insect mariner-family transposable element Himar1. Proc Natl Acad Sci U S A 97:9665–9670

2. Sarmiento F, Mrazek J, Whitman WB (2013) Genome-scale analysis of gene function in the hydrogenotrophic methanogenic archaeon *Methanococcus maripaludis*. Proc Natl Acad Sci U S A 110(12):4726–4731. doi:10.1073/pnas.1220225110

3. Dyall-Smith M (2009) The Halohanbook. http://www.haloarchaea.com/resources/halohandbook/

4. Kiljunen S, Pajunen MI, Dilks K, Storf S, Pohlschroder M, Savilahti H (2014) Generation of comprehensive transposon insertion mutant library for the model archaeon, *Haloferax volcanii*, and its use for gene discovery. BMC Biol 12:103. doi:10.1186/s12915-014-0103-3

5. Lamberg A, Nieminen S, Qiao M, Savilahti H (2002) Efficient insertion mutagenesis strategy for bacterial genomes involving electroporation of *in vitro*-assembled DNA transposition complexes of bacteriophage Mu. Appl Environ Microbiol 68:705–712

6. Haapa S, Taira S, Heikkinen E, Savilahti H (1999) An efficient and accurate integration of mini-Mu transposons *in vitro*: a general methodology for functional genetic analysis and molecular biology applications. Nucleic Acids Res 27:2777–2784

7. Sambrook J, Russell DW (2001) Molecular cloning: a laboratory manual, 3rd edn. Cold Spring Harbor Laboratory Press, Cold Spring Harbor, NY

8. Delmas S, Shunburne L, Ngo HP, Allers T (2009) Mre11-Rad50 promotes rapid repair of DNA damage in the polyploid archaeon *Haloferax volcanii* by restraining homologous recombination. PLoS Genet 5(7):e1000552. doi:10.1371/journal.pgen.1000552

Chapter 21

Genome-Wide Transposon Mutagenesis in *Mycobacterium tuberculosis* and *Mycobacterium smegmatis*

Gaurav Majumdar, Rendani Mbau, Vinayak Singh, Digby F. Warner, Marte Singsås Dragset, and Raju Mukherjee

Abstract

TnSeq, or transposon (Tn) insertion sequencing, is a powerful method for identifying the essential—as well as *conditionally essential*—regions in a genome, both coding and noncoding. The advent of accessible massively parallel DNA sequencing technologies in particular has resulted in the increased use of TnSeq-based approaches to elucidate various aspects of bacterial physiology and metabolism. Moreover, the availability of detailed protocols has enabled even nonspecialist laboratories to adapt and develop TnSeq approaches to address specific research questions. In this chapter, we describe a recently modified experimental protocol used in our laboratory for TnSeq in the major human pathogen, *Mycobacterium tuberculosis*, as well as the related non-pathogenic mycobacterium, *M. smegmatis*. The method, which was developed in close consultation with pioneers in the field of mycobacterial genetics, includes the steps involved in preparing a phage stock, generating a mutant library, selection of the library under a specific experimental condition, isolation of genomic DNA from the pooled population of mutants, amplification of the sites of Tn insertion and, finally, determining the essential genomic regions by next-generation sequencing.

Key words Transposon mutagenesis, TnSeq analysis, MycoMarT7

1 Introduction

For almost 20 years since the genome of *Mycobacterium tuberculosis* was annotated [1], the biological function of nearly 30 % of the coding sequence remains unknown [2]. The terms transposon insertion sequencing (TnSeq), transposon-directed insertion site sequencing (TraDIS), high-throughput insertion tracking by deep sequencing (HITS) or, simply, insertion sequencing (INSeq), refer to a group of unbiased genetic methods that can provide genome-wide information about the regions—coding (genes) and noncoding (intergenic regions)—that are required for sustaining bacterial growth under a specific experimental condition, thereby enabling

---

Gaurav Majumdar and Rendani Mbau contributed equally with all other contributors.

Andrew Reeves (ed.), *In Vitro Mutagenesis: Methods and Protocols*, Methods in Molecular Biology, vol. 1498,
DOI 10.1007/978-1-4939-6472-7_21, © Springer Science+Business Media New York 2017

the functions of unknown gene products to be delineated at a much larger scale (and more rapidly) than can be achieved through individual, targeted gene deletions [3–5].

Most experimental protocols are based on the generation of a high-density transposon (Tn) insertion library through the use of the Himar1 *Mariner* transposon which inserts randomly at TA dinucleotides (a TA dinucleotide occurs approximately once every 50 nucleotides in a mycobacterial genome). In principle, all Tn mutants containing inactivating insertions in genes (or gene regions) which are essential for growth of the bacterium under the conditions used to generate the library will be eliminated ab initio. For this reason, selection of growth medium (including potential use of supplements) and strain background (wild-type versus another, defined genotype) are critical. The resulting Tn library, when grown under a specific experimental condition, is then able to identify additional genes that are essential for growth under that particular condition; these are the *conditional essentials*, whose differentiation from the remaining mutants in the library is enabled owing to the selective enrichment of those mutants with insertions in a non-essential genome region (coding and non-coding), versus the selective depletion of the essential genes or regions. Next-generation deep sequencing techniques are then used to map the Tn insertion sites. Since the number of reads counted at any individual insertion site is proportional to the frequency of that mutant in the library, this "read count" value provides a reliable (and statistically demonstrable) indication of the fitness of that particular mutant under the applied selective pressure [5, 6].

With many recent developments in statistical methods for data analysis, it has become possible to distinguish subtle differences in gene essentiality, so that gene insertions which result in growth defects can now be differentiated from "true" essentials [7–9]. This methodology has been applied extensively to *M. tuberculosis* and other organisms to identify functions of unknown genes, identify virulence-associated genes, and even to study genetic interactions using defined mutant strains [6, 10–13]. In this chapter, we detail the steps involved in preparing high-density Tn libraries in *M. tuberculosis*, as well as a recently optimized protocol to generate corresponding libraries in *M. smegmatis*. It is important to acknowledge that the methods described in this chapter are heavily reliant on an experimental protocol published in a previous edition in this series [14], but incorporate specific modifications which were adopted in response to recent experimental and analytical advances, or as a result of our inability to access specialist equipment. Where relevant, these modifications are highlighted in the text. It is hoped, therefore, that this protocol can be generally applied, including in other resource-limited settings.

2   Materials

For simplicity and completeness, all the requisite materials needed for successful implementation of a given procedure within the overall method are provided below. As such, some of the Materials may appear in more than one section.

2.1   High-Titer Phage Stock

1. *M. smegmatis* mc^{2}155 glycerol stock, MycomarT7 phage stock.

2. MP buffer: 50 mM Tris–HCl (pH 7.5), 150 mM NaCl, 10 mM MgSO$_4$, 2 mM CaCl$_2$. Filter-sterilize using a 0.22 μm filter.

3. Middlebrook 7H9 medium (Difco) supplemented with glycerol and ADC (no Tween 80 supplement): Add 4.7 g of Middlebrook 7H9 to 900 mL water, add 2 mL glycerol and 100 mL ADC supplement. Filter-sterilize using 0.22 μm filter (*see* **Note 1**).

4. Top agar: Add 0.6% agar (w/v) in distilled water and autoclave. After cooling, but before the medium solidifies, add CaCl$_2$ to a final concentration of 2 mM from a sterile 1 M stock solution.

5. Luria–Bertani (LB) agar plates: 10 g/L tryptone, 5 g/L yeast extract, and 10 g/L NaCl. pH to 7.0 with 5 N NaOH. Add agar to 1.5% (15 g/L).

6. Middlebrook 7H10 agar plates (prepared as per manufacturer's instructions).

2.2   Transduction

1. *M. tuberculosis* or *M. smegmatis* wild-type or an appropriate mutant strain (*see* **Note 2**).

2. Phage stock generated as described in Subheading 3.1.

3. Middlebrook 7H9 medium supplemented with 0.2% glycerol, 10% OADC, and 0.05% Tween 80 (*see* **Note 3**).

4. MP buffer as described above.

5. Phosphate-buffered saline (PBS) containing 0.05% Tween 80 (v/v): To 800 mL ddH$_2$O add NaCl, 8 g, KCl, 0.2 g, Na$_2$HPO$_4$, 1.44 g, KH$_2$PO$_4$, 0.24 g, Tween 80, 0.5 mL. Adjust the pH to 7.4 (or 7.2, if required) with HCl, and then add ddH$_2$O to 1 L. Dispense the solution into aliquots and sterilize them by autoclaving for 20 min at 15 psi (1.05 kg/cm^2) on liquid cycle. Store PBS at room temperature.

6. Middlebrook 7H10 plates supplemented with glycerol, OADC, and Tween 80 (0.05%) containing kanamycin (20 μg/mL) (*see* **Note 4**).

7. 15 cm (150×15) petri plates.

2.3   Genomic DNA Isolation

1. TE buffer: 10 mM Tris–HCl (pH 8.0), 1 mM EDTA. pH adjust with HCl. Autoclave to sterilize.

2. Lysozyme (10 mg/mL).

3. 10% SDS.

4. Proteinase K (10 mg/mL).

5. 5 M NaCl.

6. 10% CTAB (cetyltrimethyl ammonium bromide).

7. 24:1 isoamyl alcohol–chloroform.

8. Isopropanol.

9. 70% ethyl alcohol.

10. Agarose and submarine agarose gel electrophoresis apparatus.

2.4 Sequencing

1. Adapters and oligonucleotide primers for amplification of Tn insertion junctions (*see* Table 1). Note that one adapter has an amino modification at the 3′ end.

2. Primers for amplification of scaffold region of TM4 phage:

   TM4_ScaF ATGGCAGAACAAACTGAG and

   TM4_ScaR GAATTGGTGTTGCCGTTG.

3. Primers for amplification of capsid region of TM4 phage:

   TM4_CF CATCCAAGAGGCTTACTC and

   TM4_CR AGGTTGATCTGGTTCTCG.

4. PCR reagents: Taq DNA polymerase with compatible buffers, dNTPs, DMSO, $MgCl_2$.

5. NEBNext dsDNA Fragmentase (New England Biolabs).

6. End-It DNA repair kit (Epicentre Biotechnologies).

7. QIAquick PCR purification kit and gel extraction kit (Qiagen).

8. 10 mM dATP.

9. T4 DNA ligase and 10× ligase buffer.

3 Methods

3.1 High-Titer Phage Stock

The phage stock is prepared in wild-type *M. smegmatis* mc²155, for downstream use in any mycobacterial background.

1. Grow 100 mL culture of *M. smegmatis* to $OD_{600} \sim 3$–4. Wash cells twice with 7H9-ADC stock (no Tween) and resuspend in 100 μL of the same medium. Add the washed cells to 3.5 mL top agar (cooled to 42 °C) and pour the mixture into LB or 7H10 agar plates, wait until the plates are dry (*see* **Note 5**).

2. Acquire φMycoMarT7. Make tenfold serial dilutions of the phage in 50 μL MP buffer. Mix 100 μL of washed *M. smegmatis* cells with the respective dilutions of phage. Add this mixture containing cells and phage to top agar (as in **step 1** of Subheading 3.1)

Table 1
List of primers used for preparing "ready-for-sequencing" samples

Oligo No./Step in protocol	Primer name		Sequence	Function
1 and 2/step 4 in Subheading 3.6	Adapter 1		5′-TACCACGACCA-NH2	Barcoded adapter that ligates to genomic DNA after fragmentation
	Adapter 2		5′ATGATGGCCGGTGGATTTGTGNNANNANNNTGGTCGTGGTAT	
3 and 4/step 1 in Subheading 3.7	JEL_AP1		5′-ATGATGGCCGGTGGATTTGTG	Amplifies the sequence between barcoded genomic DNA fragment and inserted Tn
	T7		5′-TAATACGACTCACTATAGGGTCTAGAG	
5/step 3 in Subheading 3.7	Sol_AP1_tagged 930	A	CAAGCAGAAGACGGCATACGAGATTGTTCCGAGTGACTGGAGTTCAG ACGTGTGCTCTTCCGATCTGTCAATGATGGCCGGTGGATTTGTG	Multiplex sequencing adapter. Long primers which inserts specific sequence allowing identification of samples after multiplex sequencing
		B	CAAGCAGAAGACGGCATACGAGATTGTTCCGAGTGACTGGAGTTCAG ACGTGTGCTCTTCCGATCTCGTCCATGATGGCCGGTGGATTTGTG	
		C	CAAGCAGAAGACGGCATACGAGATTGTTCCGAGTGACTGGAGTTCAG ACGTGTGCTCTTCCGATCTACAGTCCCATGATGGCCGGTGGATTTGTG	
		D	CAAGCAGAAGACGGCATACGAGATTGTTCCGAGTGACTGGAGTTCAG ACGTGTGCTCTTCCGATCTTAGTGGATGATGGCCGGTGGATTTGTG	
6/step 3 in Subheading 3.7	Sol_Mar	A	AATGATACGGCGACCACCGAGATCTACACTCTTTCCCTACACGACGC \TCTTCCGATCTCGGGACTTATCAGCCAACC	Common sequencing primer which recognizes the Tn
		B	AATGATACGGCGACCACCGAGATCTACACTCTTTCCCTACACGACGCT CTTCCGATCTTCGGGACTTATCAGCCAACC	
		C	AATGATACGGCGACCACCGAGATCTACACTCTTTCCCTACACGACGCT CTTCCGATCTGATACGGGGACTTATCAGCCAACC	
		D	AATGATACGGCGACCACCGAGATCTACACTCTTTCCCTACACGACGCT CTTCCGATCTATCTACGGGACTTATCAGCCAACC	

Refer to Long et al. [14] for the list of other Sol_AP1_tagged primers that can be used for up to an 8-plex sequencing reaction. The *bold, italicized* part of the Sol_AP_1_tagged primers is the variable part of the sequence. Note the amide modification at the 3′ end of adapter 1

and mix well by vortexing before pouring onto an LB agar plate. Incubate the plates at 30 °C for about 48 h until small, distinct plaques appear (*see* **Note 6**).

3. Using a sterile micro-tip, pick phage from several individual plaques and resuspend in separate tubes containing 50 μL MP buffer. Spot 10 μL of this suspension onto two replicate LB agar plates with *M. smegmatis* in the top agar. Allow the spots to dry at room temperature. Incubate one plate at 30 °C and another at 37 °C for about 2 days (48 h). Most plaques should appear only at 30 °C. Select the clones which form plaques only at 30 °C and not at 37 °C.

4. Excise the top agar containing a plaque from a plate incubated at 30 °C and crush it in MP buffer. Centrifuge at $4000 \times g$ for 30 s to pellet the agar pieces and collect the supernatant.

5. Titer the supernatant (as in **step 2** in Subheading 3.1), and determine the dilution at which confluent plaques appear on 7H10 plates (*see* Fig. 1a).

6. Grow *M. smegmatis* to $OD_{600} \sim 3$–4. Wash 500 μL cells twice with 7H9-ADC and resuspend in 500 μL of the same medium. Apply sufficient phage stock (as determined in **step 5** in Subheading 3.1) to 500 μL of washed cells. Add 100 μL of this mixture to 3.5 mL top agar (cooled to 42 °C), mix well and pour the mixture into the 7H10 plates. Make about 3–5 plates in total. Incubate at 30 °C for about 2 days until confluent plaques appear.

7. Flood the plates with 2 mL MP buffer, and rock the plates gently at 20 rpm at room temperature for 5 h, or overnight at 4 °C. Pool the remaining MP buffer (approximately 1 mL) from all plates and pass it through a 0.22 μm syringe filter to prepare the final phage stock. Store at 4 °C.

8. Prepare a tenfold dilution of the final phage stock (as in **step 2** in Subheading 3.1) and spot 10 μL from the individual dilutions onto a lawn of *M. smegmatis* as prepared in **step 1** in Subheading 3.1. Allow the spots to dry at room temperature before incubating the plates at 30 °C for 48 h. Calculate the phage titer (*see* **Note 7**) (Fig. 1a).

3.2 Transductions

1. Grow 100 mL of the *M. tuberculosis* strain of interest to $OD_{600} \sim 0.8$ or *M. smegmatis* strain of interest to $OD_{600} \sim 2$.

2. Pre-warm MP buffer, phage stock, PBS–Tween 80 and the centrifuge chamber to 37 °C. This is important to prevent any cell lysis due to the initiation of the phage lytic cycle at temperatures nearing 30 °C.

3. Pellet 100 mL bacterial culture into two 50 mL conical tubes by spinning at $4000 \times g$ for 10 min at 37 °C. Remove superna-

tant and thoroughly resuspend the pellet in 10 mL MP buffer. Add 30 mL MP buffer, mix thoroughly and spin down. Repeat the steps twice to ensure that Tween 80 is completely removed from the culture medium (*see* **Note 8**).

4. Resuspend the culture in 9 mL MP buffer for *M. tuberculosis* and in 4 mL for *M. smegmatis*. Mix well and remove 100 μL of the culture for plating as a negative control in either case.

5. For transduction, add 1 and 2 mL of the phage stock (titer of 1×10^{11} pfu/mL), respectively, to the *M. tuberculosis* (9 mL) and *M. smegmatis* (4 mL) suspensions. Seal the centrifuge tubes and incubate at 37 °C with very gentle shaking (20 rpm) for 18–20 h in the case of *M. tuberculosis*, and 5 h in the case of *M. smegmatis*, to achieve efficient transduction (*see* **Note 9**).

6. Pellet the cells by centrifugation at, and wash once with PBS–Tween 80. Finally, resuspend the cells thoroughly in 10 mL PBS–Tween 80.

7. Remove 100 μL of the suspension and plate serial dilutions on 7H10 medium with kanamycin to determine the titer of the mutant library. Plate the remaining cells on pre-warmed (37 °C) 15 cm 7H10-OADC plates with Tween 80 and kanamycin. Use 500 μL of cells per 15 cm plate. Spreading with the help of glass beads (~20 beads per plate) leads to quick and uniform spreading. Alternatively, 200 μL of cells can be plated on each regular-sized (9 cm) petri dish. Double bag the plates and incubate at 37 °C until colonies appear (3–4 weeks for *M. tuberculosis* and 2–3 days for *M. smegmatis*). The final library should contain at least ~100,000 CFU to achieve a high coverage of Tn insertions across the genome (*see* **Notes 10** and **11**).

8. Harvest the mutant library by scraping the colonies from the plates and resuspending in 7H9-OADC broth supplemented with Tween 80 (0.05%) and glycerol up to a final concentration of 15% (v/v). If desired, resuspend the cells thoroughly by sonication using a bath sonicator for two cycles of 5 s each, and freeze the library in 1 mL aliquots at –80 °C (*see* **Note 12**).

3.3 Selection of Mutant Tn Library

To understand the function of gene(s) under a specific condition, the Tn library is grown under an appropriate selection pressure. Since the nature of the library, and the specific research question, will influence the experimental design—including the choice of control samples or conditions—it is possible to provide only some general guidelines here. To this end, a generic protocol might look something like that which is detailed below.

1. Thaw frozen vials of transduced cells and use approximately 10^6 cells per experimental condition as this ensures that an adequately complex library is used consisting of all possible (viable) mutants with unique insertions.

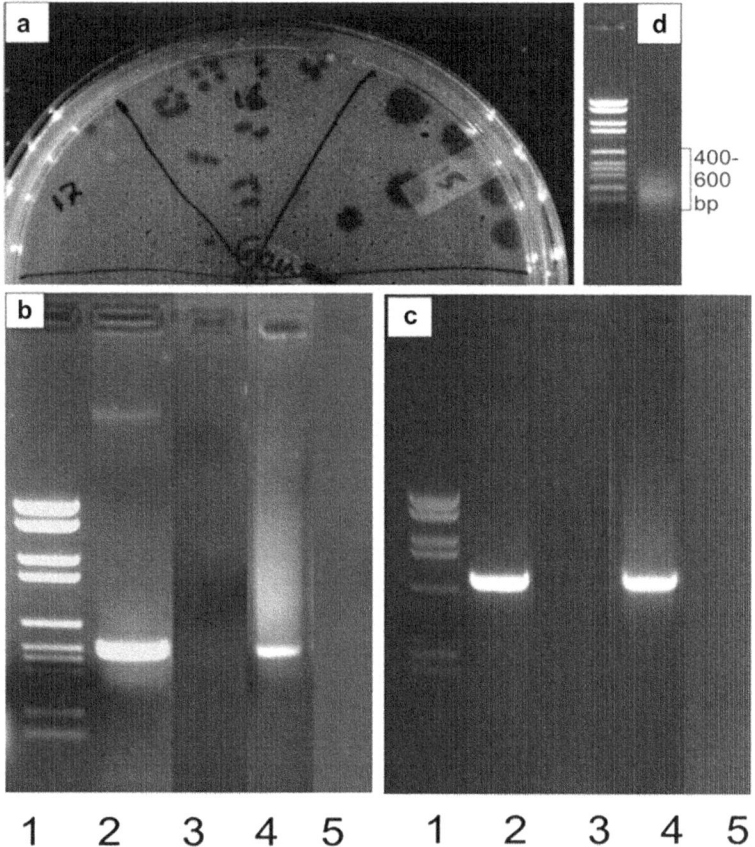

Fig. 1 (**a**) 7H10 plate showing plaques to calculate phage titer. (**b** and **c**) Gel images showing presence of phage DNA contamination in samples. Amplicons of size 517 bp (**b**) and 653 bp (**c**) for the scaffold and capsid regions in the phage genomes were amplified, respectively. *Lane 1*, DNA ladder. *Lane 2*, positive control of phage DNA (Phage stock). *Lane 3*, negative control of *M. tuberculosis* genomic DNA. *Lane 4*, a contaminated *M. tuberculosis* genomic DNA sample. *Lane 5*, a non-contaminated *M. tuberculosis* genomic DNA sample. (**d**) Image showing the amplification product after the first PCR of the Tn-insertion junctions

2. Grow the library under the presence of a selective pressure for an appropriate amount of time. It is necessary to include appropriate control conditions during the selection using the same ancestral library. For example, if performing a selection in a medium with an alternate carbon or nitrogen source it is prudent to include a control medium with zero carbon or nitrogen source in addition to a standard rich medium. After the selection, pellet the culture and isolate genomic DNA from the experimental and control samples.

   The magnitude and duration of the pressure depends upon the experimental conditions and the research question(s) asked. As noted above, part of the power of TnSeq vests in the capacity to perform both positive and negative modes of selection.

For example, a library grown in the presence of a drug will help to identify all the targets with which it will interact, as mutants with insertions in their respective genes will be positively selected and will have a growth advantage in the presence of the drug or in other words will be over-represented. Alternatively, during a negative selection under nutrient-limited or nutrient-starved conditions, mutants with inactivating insertions in genes that are now conditionally essential will fail to survive the selection pressure and will therefore be under-represented. With recent advances in data analysis methods it is now possible to differentiate conditionally essential genes from those that confer a growth defect, in turn reducing the extent of enrichment required below tenfold, as was required earlier [9]. Alternatively, if the goal is to only identify the regions of a minimal genome required for growth in vitro, a high-density library can be grown under a non-selective (optimal) condition to distinguish the essentials from the non-essentials.

3.4 Genomic DNA Isolation

CAUTION: For *M. tuberculosis*, **steps 1–8** must be performed inside a biosafety containment level 3 (BSL-3) laboratory.

1. For a 20 mL culture, resuspend the cell pellet in 500 μL TE buffer in a screw-capped tube.

2. Heat kill for 1 h at 80 °C using a water bath. Samples can be frozen at –80 °C at this point for extraction at a later stage.

3. Add lysozyme (50 μL of a 10 mg/mL stock) and incubate at 37 °C overnight.

4. Add 70 μL 10% SDS. Mix thoroughly and add 50 μL of proteinase K (10 mg/mL stock solution). Incubate for 1 h at 60 °C with shaking in a thermomixer.

5. Pre-heat 5 M NaCl and 10% CTAB to 60 °C.

6. While the samples are at 60 °C, stop the shaking, add 100 μL 5 M NaCl, and mix thoroughly by inverting.

7. Add 100 μL 10% CTAB and mix thoroughly by inverting.

8. Incubate at 60 °C for 15 min using a thermomixer with gentle shaking. Freeze samples for 15 min at –80 °C and re-incubate at 60 °C for 15 min with shaking after thawing.

   NOTE: Samples can be removed from the BSL-3 laboratory for the remaining steps below.

9. Allow the tubes to cool and add 700 μL chloroform–isoamyl alcohol (24:1). Mix by inverting the tubes several times. Make sure that the organic and aqueous components mix, forming a homogenous white solution.

10. Centrifuge for 10 min at $16,000 \times g$. Transfer upper (aqueous) phase to tube containing 700 μL cold isopropanol and mix by

inverting several times. It is sometimes possible to see a DNA precipitate forming at this point.

11. Incubate at –20 °C for 30 min to overnight.

12. Centrifuge at $16,000 \times g$ for 10 min at 4 °C. Drain the supernatant and wash pellet with 70% ethanol. Repeat centrifugation for 5–10 min.

13. Drain tube and dry in a speed vacuum concentrator for 10 min. Add 50 µL TE to resuspend the DNA before determining the DNA concentration using a spectrometer (e.g. Nanodrop) and running a small aliquot on a 0.8% agarose gel to check purity.

3.5 Detection of Phage Contamination by PCR

Detection of phage contamination by PCR is a required step before sequencing.

The extracted genomic DNA samples should be checked for phage DNA contamination at this stage as it is rare, but not unusual, for traces of bacteriophage DNA to be present in the genomic DNA preparations, and for this "contaminating" DNA to be inadvertently amplified during the subsequent stages. This additional step to check the quality of the genomic DNA represents a useful addition to the previously published protocol [14], and should ensure better results during sequencing.

Perform two reactions per sample, using primers: (1) TM4_ScaF and TM4_ScaR and (2) TM4_CF and TM4_CR.

1. Combine the following for each reaction (may be prepared as a master mix; 50 µL final volume per reaction): 5 µL 10× buffer, 2 µL of each primer (10 µM), 1 µL dNTP mix (10 mM dNTPs), 0.5 µL Taq DNA polymerase and 38.5 µL water.

2. Add 1 µL of extracted genomic DNA (10–40 ng), or water for the "no template" control, or phage DNA for the positive control, to the reaction mix and perform PCR as follows: One cycle of 95 °C for 3 min, then 40 cycles of 94 °C for 30 s, 55 °C for 30 s, 72 °C for 60 s, followed by a final extension (one cycle) of 72 °C for 10 min.

3. Run 10 µL of each PCR product on a 1.5% agarose gel to assess size and abundance of products (see Fig. 1b, c) (see **Note 13**).

3.6 Adapter Ligation

The preparation of DNA fragments usually relies on mechanical shearing utilizing a focused ultrasonicator [14]. However, here we describe an alternative, enzymatic method to generate short DNA fragments, for use in resource-limited settings where expensive technology is not available.

1. Mix 1–20 µL of DNA (3–5 µg of genomic DNA for each sample), 3 µL of 10× "NEBnext dsFragmentase" reaction buffer and water up to final volume of 28 µL. Add 2 µL "next dsFragmentase" enzyme and mix well. Incubate for 30 min at 37 °C

to generate DNA fragments in the range of 50–200 bp. Purify the fragmented DNA by ethanol precipitation or use directly for DNA clean up using column purification. Determine the DNA concentration and use 1.8 μg for end repair (see **Step 2**, Subheading 3.6).

2. Mix 1.8 μg DNA, 5 μL of 10× "End-it" DNA repair buffer, 5 μL dNTPs, 5 μL ATP and add sterile distilled water up to a final volume of 50 μL. Add 1 μL of enzyme, mix reaction mix with a micropipette and incubate at 37 °C for 45 min.

3. Purify the end-repaired DNA with QIAquick PCR purification column. Elute the DNA with 65 μL sterile distilled water to perform A-tailing of the end-repaired DNA. Add 10 μL 10× Taq DNA polymerase buffer, 20 μL 10 mM dATP, and 5 μL Taq DNA polymerase. Mix with a micropipette. Incubate at 72 °C in a thermocycler or water bath for 45 min. Purify the reaction products using QIAquick PCR purification column and elute with 50 μL of water.

4. Prepare a barcoded adapters mix before performing the next reaction. For this, mix equal volumes of the 100 μM adapter oligonucleotides 1 and 2 (*see* Table 1). To 50 μL of the adapter mix, add 2 μL of 50 mM MgCl$_2$ and heat the mix to 95 °C for 10 min in a thermocycler. Slowly reduce the temperature to 20 °C over a period of 2 h with ramping at 1% (*see* **Note 14**).

5. Add 4 μL of the above barcoded oligonucleotide mix to the 50 μL A-tailed DNA solution, 10 μL of the 10× ligase buffer, 8 μL of T4 DNA ligase, and sterile distilled water up to 100 μL final volume. Incubate at 16 °C overnight. Purify the ligation reaction using a QIAquick PCR purification kit. It is recommended to wash the columns four times before elution. For elution, add 100 μL sterile distilled water and incubate the column for 20 min at 37 °C before spinning (*see* **Note 15**).

3.7 PCR Amplification of Tn-DNA Junctions

This step involves two PCR amplifications: the first amplifies the Tn insertion junctions on the chromosome, and the second adds an adapter which will help in downstream identification after multiplex Illumina sequencing.

1. Set up eight PCR reactions for each sample obtained after purification of the ligation reaction in **step 5** in Subheading 3.6. For each PCR, use 100 ng of DNA and add 2.5 μL of DMSO, 5 μL of 10× PCR buffer, 5 μL of 2.5 mM dNTPs, 0.5 μL Taq DNA polymerase, 1.5 μL of 10 μM adapter primer (Oligo 3; *see* Table 1), and 1.5 μL of 10 μM Tn primer (Oligo 4; *see* Table 1). Amplify using the following PCR conditions: One cycle of 95 °C for 10 min, 20 cycles of 95 °C for 30 s, 58 °C for 30 s, 72 °C for 45 s and a final extension cycle of 72 °C for 5 min (*see* **Note 16**).

2. Pool all eight reactions and run the entire PCR reaction on a 2% agarose gel. A bright "smear" is expected between the region of 100 to 1000 bp (*see* Fig. 1d). Cut out the DNA band corresponding to fragment sizes of 400–600 bp, as estimated according to the DNA molecular weight marker ("DNA ladder") which should be run along with the samples. Elute the DNA using a Qiagen gel extraction kit. On the final elution step add 50 μL water and incubate at 37 °C for 20 min before eluting. Measure the DNA concentration using a (e.g. Nanodrop). Typically, the DNA concentration obtained is approximately 20–25 ng/μL (*see* **Note 17**).

3. Prepare oligonucletide mixes for a hemi-nested PCR. Prepare the four long oligonucleotide mix to prepare two sets of oligonucleotides. A Sol-ap-tag mix (Oligo mix 5; *see* Table 1) and the Sol-mar mix of four long primers (Oligo mix 6; *see* Table 1) for one sample. For sequence details of the oligonucleotides *see* Table 1.

4. Make final stocks of 1 μM of each primer mix (both Sol-mar mix and Sol-ap-tag mix). For multiplexing eight reactions in a single sequencing lane, make eight different Sol-ap-tag mixes of four primers as suggested in Long et al. [14].

5. Run a hemi-nested PCR using 4 μL of the gel extracted DNA from **step 2** in Subheading 3.7. Add 2 μL of 1 μM Sol-mar mix, 2 μL of 1 μM Sol-ap-tag mix, 2 μL of 10× PCR buffer, 2 μL of 2.5 mM dNTPs, 0.2 μL of Taq polymerase. Add water up to a final volume of 20 μL per PCR reaction. Perform PCR using these conditions: One cycle of 95 °C for 5 min, ten cycles of 95 °C for 30 s, 58 °C for 30 s, 72 °C for 45 s, and one final extension cycle of 72 °C for 5 min (*see* **Note 18**).

6. Ethanol precipitate the entire PCR product and resuspend it in 40 μL sterile distilled water. Run 2–5 μL on a gel to confirm appearance of a smear. Determine the concentration using a spectrophotometer. Subject 50 ng of DNA to next-generation (e.g. Illumina) sequencing.

3.8 Essentiality Analysis

1. Several TnSeq data analysis pipelines are now available which can be used to map Tn insertions and count reads at a particular insertion site. Among these, the methods which utilize a "Hidden-Markov Model" based analysis have been found to be the most robust. Importantly, these provide a higher resolution in distinguishing the genomic regions where an insertion leads to a growth defect or a growth advantage from those that are either essential or non-essential under the conditions tested. For a more detailed explanation of the theory and application of these statistical analysis methods, the interested reader is encouraged to read the following articles [7–9, 15].

4   Notes

1. It is important that this medium does not contain Tween 80.

2. A Tn library is prepared in a mutant background when relationships between interacting genes or pathways are under investigation. Since φMycoMarT7 inserts a kanamycin^R cassette, which is critical for selection of successful transductants, care should be taken to ensure that the mutant background is kanamycin sensitive.

3. This is a standard enriched medium for culturing all strains of mycobacterium. Tween and oleic acid components of the supplement were intentionally omitted from the washing steps (**step 1** in Subheading 3.1) as they affect viability of φMycoMarT7.

4. It is important to include Tween 80 in the solid medium as it helps to eliminate potential lysis owing to the presence of any residual phage.

5. Don't let the agar cool too much, otherwise it may not spread across the whole plate. Pouring the mixture in the middle of the plate, tilting and slowly rotating makes uniform plates.

6. Wait for the numerous small plaques to appear as the early-appearing larger plaques do not retain temperature sensitivity.

7. High-density transduction is achieved where phage stocks have a titer higher than 10^{11} plaque forming units (pfu)/mL as this increases the likelihood of insertion at most of the TA dinucleotides in the genome.

8. The first wash of MP buffer can be substituted with 7H9-ADC without Tween 80 medium. This minimizes the loss of cells since the pellet does not form well with MP buffer.

9. For *M. smegmatis*, a higher phage titer is added to adjust to the cell density of the initial culture (1 mL of 1×10^{11} pfu/mL for 1 OD_{600} unit).

10. Development of advanced statistical tools has made it possible to infer gene essentiality from a <50% coverage of TA dinucleotides; however, this is not ideal. This issue is more critical in the case of *M. smegmatis*, where transduction efficiency has been found to be lower than the slow-growing *M. tuberculosis*.

11. A higher concentration (0.1%) of Tween 80 can also be used to prepare plates to avoid contamination from residual activated phage.

12. In our experience, sterile disposable loops are preferable to cell scrapers for harvesting of colonies since scrapers tend to pick up some agar from plates.

13. Careful isolation of genomic DNA is important to avoid phage DNA contamination. If contamination is detected, all further processing is stopped.

14. Use of barcoded adaptors causes DNA fragments to be uniquely marked and thus can easily distinguish any erroneous overrepresentation of a fragment due to over-amplification of a Tn insertion junction.

15. Ethanol precipitation of the ligation reaction is an alternate option. Add 200 μL of alcohol in the reaction and 4 μL of 5 M sodium acetate (pH 5.2). Mix well and incubate at –20 °C for 2 h. Centrifuge at $16,000 \times g$ for 30 min and wash with 70% alcohol. Dry using a speed vacuum concentrator and finally resuspend in 100 μL water.

16. Alternatively, use 15 μL Thermo Scientific PCR master mix, 1.5 μL of the two primers, 100 ng of DNA and add water up to a final volume of 30 μL.

17. Make thick gels to avoid cross contamination between samples. Ensure that the whole gel melts when extracting DNA from the gel.

18. Alternatively, use 10 μL Thermo Scientific PCR master mix, 2 μL of the two primer mixes, 4 μL of DNA and add water to a final volume of 20 μL.

Acknowledgments

This work is funded by grants from the South African Medical Research Council (to DFW), the National Research Foundation (to DFW), and the Research Council of Norway through its Centres of Excellence funding scheme, project number 223255 (to MSD). RM is supported by the Howard Hughes Medical Institute through a Senior International Research Scholars grant to Valerie Mizrahi. GM is supported by a grant from the South African Medical Research Council (to DFW) with funds from National Treasury under the Economic Competitiveness and Support Package (MRC-RFA-UFSP-01-2013/CCAMP).

References

1. Cole ST, Brosch R, Parkhill J, Garnier T, Churcher C, Harris D, Gordon SV, Eiglmeier K, Gas S, Barry CE III, Tekaia F, Badcock K, Basham D, Brown D, Chillingworth T, Connor R, Davies R, Devlin K, Feltwell T, Gentles S, Hamlin N, Holroyd S, Hornsby T, Jagels K, Krogh A, McLean J, Moule S, Murphy L, Oliver K, Osborne J, Quail MA, Rajandream MA, Rogers J, Rutter S, Seeger K, Skelton J, Squares R, Squares S, Sulston JE, Taylor K, Whitehead S, Barrell BG (1998) Deciphering the biology of *Mycobacterium tuberculosis* from the complete genome sequence. Nature 393: 537–544

2. Lew JM, Kapopoulou A, Jones LM, Cole ST (2011) TubercuList-10 years after. Tuberculosis (Edinb) 91:1–7

3. Ates LS, Ummels R, Commandeur S, van de Weerd R, Sparrius M, Weerdenburg E, Alber M, Kalscheuer R, Piersma SR, Abdallah AM, Abd El Ghany M, Abdel-Haleem AM, Pain A, Jimenez CR, Bitter W, Houben EN (2015) Essential role of the ESX-5 secretion system in outer membrane permeability of pathogenic mycobacteria. PLoS Genet 11:e1005190

4. van Opijnen T, Camilli A (2013) Transposon insertion sequencing: a new tool for systems-level analysis of microorganisms. Nat Rev Microbiol 11:435–442

5. Chao MC, Abel S, Davis BM, Waldor MK (2016) The design and analysis of transposon insertion sequencing experiments. Nat Rev Microbiol 14:119–128

6. Zhang YJ, Ioerger TR, Huttenhower C, Long JE, Sassetti CM, Sacchettini JC, Rubin EJ (2012) Global assessment of genomic regions required for growth in *Mycobacterium tuberculosis*. PLoS Pathog 8:e1002946

7. DeJesus MA, Ambadipudi C, Baker R, Sassetti C, Ioerger TR (2015) TRANSIT–a software tool for Himar1 TnSeq analysis. PLoS Comput Biol 11:e1004401

8. Pritchard JR, Chao MC, Abel S, Davis BM, Baranowski C, Zhang YJ, Rubin EJ, Waldor MK (2014) ARTIST: high-resolution genome-wide assessment of fitness using transposon insertion sequencing. PLoS Genet 10:e1004782

9. DeJesus MA, Ioerger TR (2013) A Hidden Markov Model for identifying essential and growth-defect regions in bacterial genomes from transposon insertion sequencing data. BMC Bioinformatics 14:303

10. Nambi S, Long JE, Mishra BB, Baker R, Murphy KC, Olive AJ, Nguyen HP, Shaffer SA, Sassetti CM (2015) The oxidative stress network of *Mycobacterium tuberculosis* reveals coordination between radical detoxification systems. Cell Host Microbe 17:829–837

11. Kieser KJ, Baranowski C, Chao MC, Long JE, Sassetti CM, Waldor MK, Sacchettini JC, Ioerger TR, Rubin EJ (2015) Peptidoglycan synthesis in *Mycobacterium tuberculosis* is organized into networks with varying drug susceptibility. Proc Natl Acad Sci U S A 112:13087–13092

12. Griffin JE, Gawronski JD, Dejesus MA, Ioerger TR, Akerley BJ, Sassetti CM (2011) High-resolution phenotypic profiling defines genes essential for mycobacterial growth and cholesterol catabolism. PLoS Pathog 7:e1002251

13. Sassetti CM, Boyd DH, Rubin EJ (2003) Genes required for mycobacterial growth defined by high density mutagenesis. Mol Microbiol 48:77–84

14. Long JE, DeJesus M, Ward D, Baker RE, Ioerger T, Sassetti CM (2015) Identifying essential genes in *Mycobacterium tuberculosis* by global phenotypic profiling. Methods Mol Biol 1279:79–95

15. DeJesus MA, Zhang YJ, Sassetti CM, Rubin EJ, Sacchettini JC, Ioerger TR (2013) Bayesian analysis of gene essentiality based on sequencing of transposon insertion libraries. Bioinformatics 29:695–703

Section V

Site-Directed Mutagenesis: PCR and DNA Polymerase-Based Methods

Chapter 22

Multiple Site-Directed and Saturation Mutagenesis by the Patch Cloning Method

Naohiro Taniguchi and Hiroshi Murakami

Abstract

Constructing protein-coding genes with desired mutations is a basic step for protein engineering. Herein, we describe a multiple site-directed and saturation mutagenesis method, termed MUPAC. This method has been used to introduce multiple site-directed mutations in the green fluorescent protein gene and in the moloney murine leukemia virus reverse transcriptase gene. Moreover, this method was also successfully used to introduce randomized codons at five desired positions in the green fluorescent protein gene, and for simple DNA assembly for cloning.

Key words Site-directed mutagenesis, Saturation mutagenesis, Cloning, Protein engineering

1 Introduction

Site-directed mutagenesis, which introduces specific mutations at arbitrary positions of a gene, is a widely used technique in bioscience. Saturation mutagenesis is also frequently used in order to introduce random mutations into a gene, resulting in the generation of a set of all possible mutants, which are used in directed evolution to improve or modify protein properties [1–4]. Although various methods have been developed to introduce these mutations in a gene [5–12], development of more efficient and convenient methods is still demanded.

Recently, the ISO method (one-step isothermal in vitro recombination) [7] and its mutagenesis application, the MISO method (multi-change isothermal in vitro recombination) [13], have been reported. In the ISO method, DNA fragments, including 40-bp sequences homologous to the adjacent DNA fragments, are fused by the concerted action of three enzymes: (1) T5 exonuclease; (2) Phusion DNA polymerase; and (3) *Taq* DNA ligase. Mitchell et al. [13] demonstrated that the MISO method could assemble up to six DNA fragments and was used to introduce eight-point mutations in a gene, though the reaction conditions were not suitable

Andrew Reeves (ed.), *In Vitro Mutagenesis: Methods and Protocols*, Methods in Molecular Biology, vol. 1498,
DOI 10.1007/978-1-4939-6472-7_22, © Springer Science+Business Media New York 2017

for multiple site-directed mutagenesis (i.e., the ISO method was originally developed focusing on the assembly of long DNA fragments).

In this chapter, we describe an optimized DNA assembly and mutagenesis method, termed MUPAC (*Mu*ltiple *pa*tch *c*loning) [14] (*see* Fig. 1). MUPAC employs three common enzymes: (1) T5 exonuclease; (2) Klenow fragment (3′→5′ exo⁻): and (3) T4 DNA ligase. In the MUPAC reaction mixture, the DNA fragments are degraded from the 5′ to 3′ direction by T5 exonuclease. The exposed 3′-end overhanging DNA has a 16-mer complementary sequence to an adjacent DNA fragment, thus, these DNA fragments anneal to each other. The gaps, the result of the overdegradation by the T5 exonuclease, are filled by Klenow fragments (3′→5′ exo⁻), and the resulting nicked DNA strands are ligated by the T4 DNA ligase. This concerted reaction is achieved at 37 °C, which is lower than the temperature employed in the MISO method (50 °C). This feature makes the MUPAC method more suitable for shorter DNA assembly.

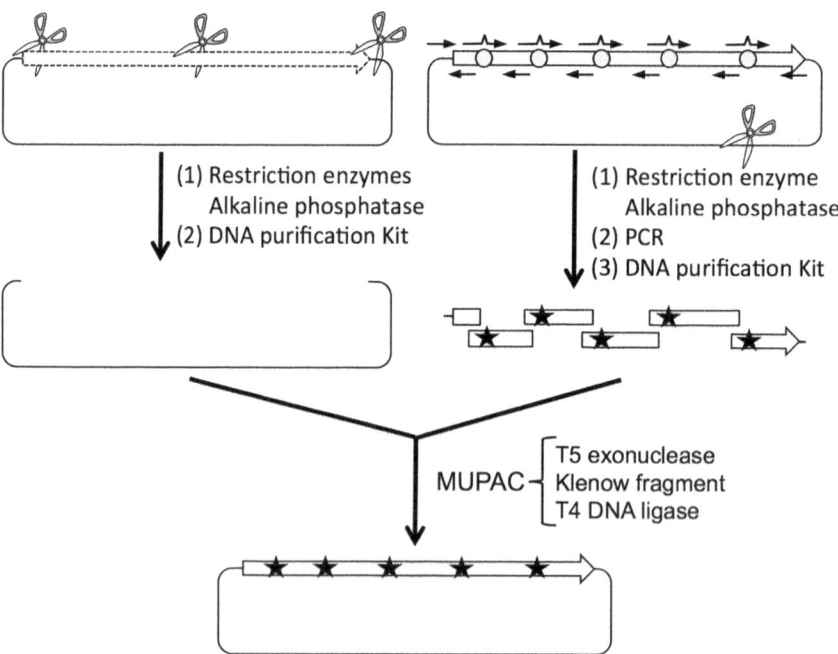

Fig. 1 Scheme of the multiple patch cloning (MUPAC) method. The plasmid vector is digested with restriction enzymes and treated with alkaline phosphatase. On the other hand, the template plasmid is also digested by a restriction enzyme and treated with alkaline phosphatase. The resulting template is used for amplification of DNA fragments by PCR. The resulting DNA fragments and digested vector containing homologous regions at each end were assembled at 37 °C by a concerted reaction with T5 exonuclease, Klenow fragment, and T4 DNA ligase. The *stars* indicate the mutation sites

2   Materials

2.1   Plasmids

1. pBAD-GFPuv [15] was purchased from Bio-Rad.

2. pBAD-GFPuv 5D mutant (V29D, Y66D, A110D, F165D, and L201D) was prepared by MUPAC using the procedure described herein.

3. pBAD-NXE was constructed by inserting the sequence 5′-<u>GCTAGC</u> ATTAA <u>CTCGA G</u>CTTAT <u>GAATTC</u>-3′ (*Nhe*I, *Xho*I, and *Eco*RI sites are underlined) into the *Nhe*I and *Eco*RI sites of pBAD-GFPuv.

4. pET16b+ was constructed by inserting the sequence 5′-<u>CATAT GTCCA ATGTC GACAT</u> TGCTG <u>GATCC</u>-3′ (*Nde*I, *Sal*I, and *Bam*HI sites are underlined) into *Nde*I and *Bam*HI sites of pET16b (Novagen).

2.2   Molecular Biology Reagents

1. KOD DNA polymerase (KOD-Plus-Neo).

2. Oligonucleotide primers.

3. DNA gel extraction kit.

4. Hi-fidelity restriction enzymes and buffers for: *Nde*I, *Sal*I, *Bam*HI, *Xho*I, *Eco*RI, *Eco*RV.

5. Calf intestinal alkaline phosphatase.

6. New England Biolabs buffer #4 with bovine serum albumin.

7. Electrocompetent *E. coli* cells (DH10B, JM109).

8. Electroporation apparatus.

9. Electroporation cuvettes (0.1 or 0.2 cm gap).

10. LB agar plates containing 0.2% arabinose and ampicillin at 50 μg/mL.

2.3   Reagents and Medium

All enzymes and buffers mentioned in this section were purchased from New England Biolabs. It is recommended to prepare all reagents on ice.

1. Luria–Bertani medium: 10 g/L tryptone, 5 g/L yeast extract, and 5 g/L NaCl. pH to 7.0 with 5 N NaOH. Add agar to 1.5% for agar plates.

2. Storage buffer: consists of 50 mM Tris–HCl (pH 7.5), 100 mM NaCl, 1 mM DTT, 0.1 mM EDTA, 50% Glycerol, 0.1% Triton X-100. Mix well in a 2 mL screw-cap cryotube, 414 μL of ultra-pure water, 500 μL of glycerol, 50 μL of 1 M Tris–HCl (pH 7.5), 33 μL of 3 M NaCl, 1 μL of 1 M DTT, 1 μL of 0.1 M EDTA and 1 μL of Triton X-100. Mix well and store at –20 °C.

3. Enzyme stock solution: consists of 4 U/μL T5 exonuclease, 0.2 U/μL Klenow fragment (3′ → 5′ exo⁻), and 40 U/μL T4 DNA ligase. Aliquot 10 μL of T5 exonuclease (10 U/μL),

1 μL of Klenow fragment (5 U/μL), 2.5 μL of T4 DNA ligase (400 U/μL), and 11.5 μL of the storage buffer (*see* previous step) into a sterile Eppendorf tube. Mix thoroughly but gently. Store at −20 °C (*see* **Note 1**).

4. Reaction buffer (2×): consists of 100 mM Tris–HCl (pH 7.5), 20 mM MgCl$_2$, 20 mM DTT, 2 mM ATP, 0.5 mM dNTPs, and 0.02 % BSA. Aliquot 40 μL of 10× T4 DNA ligase buffer (provided with the enzyme), 20 μL of 5 mM dNTPs, 4 μL of 100× BSA, and 136 μL of ultrapure water. Mix thoroughly but gently and store at −20 °C (*see* **Note 2**).

5. MUPAC reagent (2×): consists of 0.2 U/μL T5 exonuclease, 0.01 U/μL Klenow fragment (3′ → 5′ exo⁻), and 2 U/μL T4 DNA ligase in the reaction buffer made in the previous step. Take 1 μL of the enzyme stock solution and add 20 μL of the reaction buffer from **step 4** above. *Note*: the MUPAC reagent must be made fresh each time—it CANNOT be stored.

3   Methods

All procedures should be performed on ice unless otherwise noted. All restriction enzymes and buffers mentioned in this section were purchased from New England Biolabs but the same molecular biology reagents from other manufacturers can also be used.

3.1   Insert DNA fragments

1. Digest 20 ng of pBAD-GFPuv (use the 5D mutant for site-directed mutagenesis and the wild-type plasmid for saturation mutagenesis) with 2 U Hi-fidelity *Eco*RV followed by treatment with 1 U of calf alkaline intestinal phosphatase (CIP) in 4 μL of reaction mixture (NEBuffer 4 + BSA). Incubate for 1 h at 37 °C. This digestion step is performed in order to avoid contamination of the original plasmid in the PCR step (*see* **Note 3**).

2. Add 0.4 μL of the digested pBAD-GFPuv solution directly to 40 μL of each PCR mixture containing 0.8 U of KOD-Plus-Neo and each of the primer pairs listed in Table 1 (*see* **Notes 4** and **5**). Amplify the PCR fragment using the following cycling conditions: 94 °C for 3 min, followed by 25 cycles of 94 °C for 20 s, 50 °C for 30 s, and 68 °C for 40 s (*see* **Notes 6** and **7**).

3. Purify each PCR product using a DNA gel extraction kit. Elute the product with 20 μL of elution buffer provided with the kit (*see* **Note 8**).

3.2   Vector DNA Preparation

1. Digest 1 μg of pBAD-NXE with 10 U *Nhe*I-HF™ and 10 U *Eco*RI-HF™ and treat with 7.5 U CIP in a 40 μL reaction mixture containing NEBuffer 4 supplemented with BSA. Incubate at 37 °C for 2 h (*see* **Note 9**).

Table 1
List of primer pairs used for saturation mutagenesis

Primer pairs		Primer Names	Sequences (5′ → 3′)
1	Forward	pBADGFPmut2.F20	AGAAGGAGATATACATATGG
	Reverse	GFP29.R20	AGAAAATTTGTGCCCATTAA
2	Forward	GFP29X.F35	TGGGCACAAATTTTCT *NNK*-AGTGGAGAGGGTGAAG
	Reverse	GFP66.R20	AGAGAAAGTAGTGACAAGTG
3	Forward	GFP66X.F35	TGTCACTACT TTCTCT *NNK*-GGTGTTCAATG CTTTT
	Reverse	GFP110.R20	ACGCGTCTTGTAGTTCCCGT
4	Forward	GFP110X.F35	GAACTACAAGACGCGT *NNK*-GAAGTCAAGTTTGAAG
	Reverse	GFP165.R20	GTTAGCTTTGATTCCATTCT
5	Forward	GFP165X.F35	TGGAATCAAAGCTAAC *NNK*-AAAATTCGCCACAACA
	Reverse	GFP201.R20	GTAATGGTTGTCTGGTAAAA
6	Forward	GFP201X.F35	ACCAGACAACCATTAC *NNK*-TCGACACAATCTGCCC
	Reverse	pBADGFPmut4.R18	ATCCCCGGGTACCGAGCT

This table was reproduced from Taniguchi et al. [14]

2. Purify the digested product by using a DNA/gel extraction and purification kit. Elute the product with 20 μL of elution buffer provided with the kit.

3.3 MUPAC

1. Mix 0.2 μL of the digested vector and 0.8 μL of the insert DNA mixture (*see* **Note 10**).

2. Prepare the 2× MUPAC reagent as described in **step 5** in Subheading 2.3.

3. Mix 1 μL of the DNA mixture and 1 μL of the 2× MUPAC reagent.

4. Incubate the mixture at 37 °C for 30 min (*see* **Note 11**).

5. Purify the MUPAC reaction products by ethanol precipitation (*see* **Note 12**).

6. Use all the products to transform *E. coli* electrocompetent DH10B by electroporation at 18 kV/cm.

7. Add SOC medium, transfer cells to a 1.5 mL Eppendorf tube and incubate at 37 °C for 1 h with gentle shaking (~200 rpm).

8. Plate transformants onto LB agar plates containing 50 μg/mL ampicillin and 0.2 % arabinose.

9. Incubate the plate at 37 °C for 16 h (*see* **Note 13**).

---

4   Notes

1. The use of low-retention tubes is recommended. Add precise amounts of enzymes and mix them well. The concentration of each enzyme is very important for the concerted MUPAC reaction. The enzyme stock solution can be stored at –20 °C for 2 weeks.

2. Avoid repeated freeze–thaw cycles to prevent the oxidation of DTT.

3. The digestion step is critical to minimize false positive results as a consequence of intact template plasmids. The use of restriction enzymes that cut plasmid DNA at the region outside the gene and the use of multiple restriction enzymes to minimize the contamination of the template plasmid are recommended.

4. The lower limit of the insert DNA length seems to be between 50 and 70 bp. When the distance between two mutations is less than 50 bp, both mutations should be introduced by a pair of overlapping forward and reverse primers or by a single primer (*see* Fig. 2).

5. If the efficiency of the MUPAC reaction is low, extending the number of base pairs of the homologous DNA up to 40 bp is suggested. The efficiency of simple cloning of a gene into pET16b+ (digested by *Nde*I, *Sal*I, and *Bam*HI, and treated with CIP) could be increased by adding 40 bp of homologous DNA on both sides of the insert DNA.

6. The use of a DNA polymerase with $3' \rightarrow 5'$ exonuclease activity to generate DNA with blunt-ends is advised. We have also used Phusion DNA polymerase (New England Biolabs). DNA polymerase that does not possesses $3' \rightarrow 5'$ exonuclease activity, such as *Taq* DNA polymerase, adds extra nucleotides at the $3'$ end of the DNA, which prevents elongation of DNA by Klenow fragment ($3' \rightarrow 5$ exo⁻) during the MUPAC reaction.

7. The PCR conditions could vary according to the sequences and lengths of the DNA fragments and primers.

8. Alcohol contamination in the eluate will decrease the efficiency of the MUPAC reaction. Air-drying the column for at least 5 min before elution is recommended in order to evaporate any residual alcohol. Isopropanol precipitation can be used instead of the DNA extraction kit purification when the length

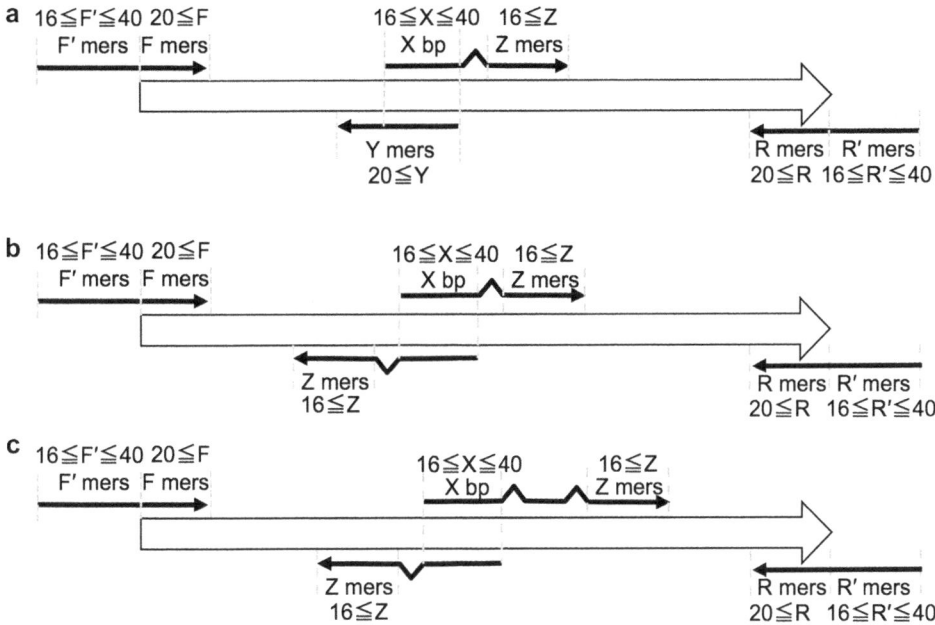

Fig. 2 Primer design for multiple patch cloning. (**a**) Introducing a single mutation with a single primer. (**b**) Introducing two mutations with overlapping forward and reverse primers. (**c**) Introducing two mutations with a single primer. F' and R' indicate the number of nucleotides of the homologous region toward the plasmid vector. F, R, Y, and Z indicate the annealing region of the primer, X indicates the region of the primer homologous to the adjacent DNA fragment

of the product DNA is more than 200 bp. This might improve the efficiency of the MUPAC reaction.

9. Additional restriction enzyme digestion of plasmids (e.g., *Xho*I for pBAD-NXE) between the cutting sites of the two restriction enzymes (e.g., *Nhe*I and *Eco*RI for pBAD-NXE) minimizes cross-contamination by the original plasmid. Transferring the reaction mixture to a new tube during the reaction also decreases carryover of the original plasmid.

10. Mixing equal amounts of DNA fragments to prepare the 0.8 μL of insert DNA mixture regardless of the number of insert fragments is suggested. Generally, from 2 to 10 ng of restriction enzyme-digested plasmid and 5–100 ng of total DNA fragments are sufficient for an efficient MUPAC reaction. It might be necessary to optimize the amounts of DNA in each case to increase the efficiency of the MUPAC reaction.

11. After mixing the reaction mixture, it is necessary to place the sample immediately into a water bath to avoid side reactions. It is suggested to keep the sample tubes under water in order to avoid evaporation.

12. Purification by ethanol precipitation is not necessary when chemical DNA transformation is performed. Chemical trans-

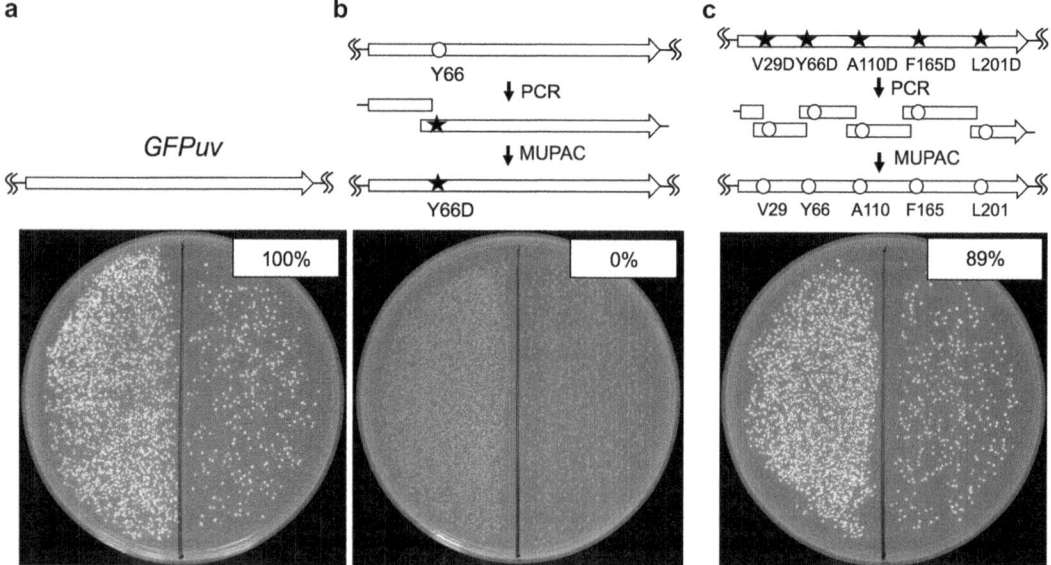

Fig. 3 Site-directed mutagenesis on the GFPuv gene. *E. coli* JM109 was transformed by a pBAD vector containing (**a**) wild-type GFPuv gene, (**b**) Y66D mutant gene generated by MUPAC, and (**c**) wild-type GFPuv gene generated by MUPAC from a mutant gene (V29D, Y66D, A110D, F165D, and L201D). The tenth aliquot of each transformant was spread on the right half of an LB agar plate, and the remainder on the *left half*. GFPuv fluorescence was observed through a Y48 filter (480 nm cutoff filter; HOYA, Tokyo, Japan) under 365 nm UV light. This figure was reproduced from Taniguchi et al. [14]

formation is effective enough for normal site-directed mutagenesis or cloning.

13. We have successfully introduced five mutations (D29V, D66Y, D110A, D165F, and D201L) in the GFPuv mutant gene with 89 % efficiency by assembling six DNA fragments (*see* Fig. 3). We have also introduced nine mutations (K69E, T147Q, D225P, F313W, N374I, G435L, K454N, N524D, and T605V) in the moloney murine leukemia virus reverse transcriptase gene with 30 % efficiency by assembling 11 DNA fragments. For saturation mutagenesis of the GFPuv gene, sequence analysis of the plasmid mixture extracted from all the resulting colonies showed DNA sequence chromatograms of apparent NNK mixture at the targeted sites (*see* Fig. 4).

Acknowledgments

This work was supported by the Funding Program for Next Generation World-Leading Researchers [LR011 to H.M.] and Grant-in-Aid for Scientific Research (A) [15H02006 to H.M.] from Japan Society for the Promotion of Science (JSPS).

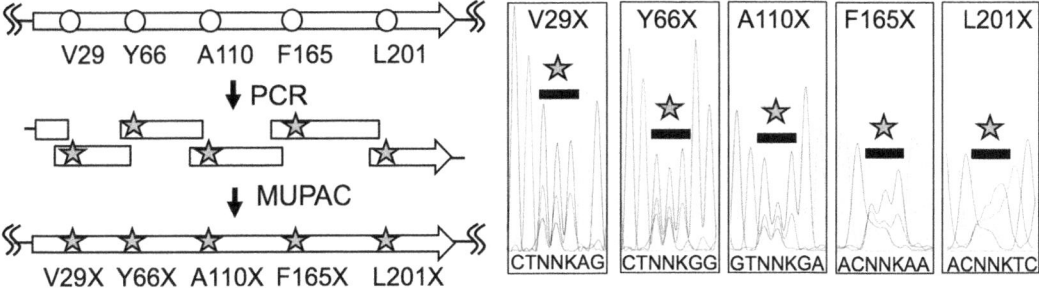

Fig. 4 Site-directed saturation mutagenesis on the GFPuv gene. DNA sequence chromatograms obtained by sequencing of a plasmid mixture from >10^4 clones. NNK codons, where *K* represents G or T, were simultaneously introduced into the GFPuv gene at five specific sites by using the MUPAC method. This figure was reproduced from Taniguchi et al. [14]

References

1. Sen S, Venkata-Dasu V, Mandal B (2007) Developments in directed evolution for improving enzyme functions. Appl Biochem Biotechnol 143:212–223

2. Brustad EM, Arnold FH (2011) Optimizing non-natural protein function with directed evolution. Curr Opin Chem Biol 15:201–210

3. Labrou NE (2010) Random mutagenesis methods for in vitro directed enzyme evolution. Curr Protein Pept Sci 11:91–100

4. Reetz MT (2011) Laboratory evolution of stereoselective enzymes: a prolific source of catalysts for asymmetric reactions. Angew Chem Int Ed Engl 50:138–174

5. Aslanidis C, de Jong PJ (1990) Ligation-independent cloning of PCR products (LIC-PCR). Nucleic Acids Res 18:6069–6074

6. Dennig A, Shivange AV, Marienhagen J, Schwaneberg U (2011) OmniChange: the sequence independent method for simultaneous site-saturation of five codons. PLoS One 6, e26222

7. Gibson DG, Young L, Chuang RY, Venter JC, Hutchison CA 3rd, Smith HO (2009) Enzymatic assembly of DNA molecules up to several hundred kilobases. Nat Methods 6:343–345

8. Hogrefe HH, Cline J, Youngblood GL, Allen RM (2002) Creating randomized amino acid libraries with the QuikChange multi site-directed mutagenesis Kit. Biotechniques 33:1158–1160, 1162, 1164–1165

9. Hsiao K (1993) Exonuclease III induced ligase-free directional subcloning of PCR products. Nucleic Acids Res 21:5528–5529

10. Li MZ, Elledge SJ (2007) Harnessing homologous recombination in vitro to generate recombinant DNA via SLIC. Nat Methods 4:251–256

11. Tillett D, Neilan BA (1999) Enzyme-free cloning: a rapid method to clone PCR products independent of vector restriction enzyme sites. Nucleic Acids Res 27, e26

12. Zhang Y, Werling U, Edelman W (2012) SLiCE: a novel bacterial cell extract-based DNA cloning method. Nucleic Acids Res 40, e55

13. Mitchell LA, Cai Y, Taylor M, Noronha AM, Chuang J, Dai L, Boeke JD (2013) Multichange isothermal mutagenesis: a new strategy for multiple site-directed mutations in plasmid DNA. ACS Synth Biol 2:473–477

14. Taniguchi N, Nakayama S, Kawakami T, Murakami H (2013) Patch cloning method for multiple site-directed and saturation mutagenesis. BMC Biotechnol 13:91

15. Crameri A, Whitehorn EA, Tate E, Stemmer WP (1996) Improved green fluorescent protein by molecular evolution using DNA shuffling. Nat Biotechnol 14:315–319

Chapter 23

Seamless Ligation Cloning Extract (SLiCE) Method Using Cell Lysates from Laboratory *Escherichia coli* Strains and its Application to SLiP Site-Directed Mutagenesis

Ken Motohashi

Abstract

Cell lysates from laboratory *Escherichia coli* strains endogenously exhibit homologous recombination activity, which can be utilized for seamless DNA cloning in vitro. This method, termed Seamless Ligation Cloning Extract (SLiCE) cloning, enables high cloning efficiency with simultaneous integration of two unpurified DNA fragments into a vector. In addition, the SLiCE method is highly cost-effective, as several laboratory *E. coli* strains may be utilized as sources of SLiCE. Previously, the SLiCE technique has been applied to site-directed mutagenesis to develop a novel technique termed SLiCE-mediated polymerase chain reaction (PCR)-based site-directed mutagenesis (SLiP site-directed mutagenesis). Two DNA fragments containing a mutation site can be simultaneously integrated into a vector while avoiding the introduction of undesirable mutations in the vector. Therefore, SLiP site-directed mutagenesis simplifies multiple procedures involved in PCR-based site-directed mutagenesis such as overlap extension method PCR or the Megaprimer method.

Key words Homologous recombination, Seamless DNA cloning, Site-directed mutagenesis, SLiCE, *Escherichia coli*

1 Introduction

DNA cloning is an indispensable technique in the field of molecular biology. Traditional cloning methods, which utilize restriction endonucleases and DNA ligases, have been widely applied for various protocols involving DNA manipulations such as vector construction and site-directed mutagenesis. However, gene manipulation using restriction endonucleases is restricted by the limited availability of unique recognition sites for restriction endonucleases. In order to overcome these limitations, various seamless DNA cloning methods, which do not require the use of restriction endonucleases, have been developed [1–8]. Recently, commercially kits for seamless DNA cloning have become available, e.g., the Clontech In-Fusion HD Kit [9, 10], GeneArt's Seamless Cloning and Assembly Kit (Thermo Fisher

Andrew Reeves (ed.), *In Vitro Mutagenesis: Methods and Protocols*, Methods in Molecular Biology, vol. 1498,
DOI 10.1007/978-1-4939-6472-7_23, © Springer Science+Business Media New York 2017

Scientific), and The Gibson Assembly Cloning Kit [1, 11] from New England Biolabs. While these commercially available kits have been widely used [12–16], they are, however, generally expensive.

Escherichia coli strains possess two endogenous RecA-dependent and RecA-independent homologous recombination pathways [17–21]. Laboratory *E. coli* RecA⁻ strains exhibit RecA-independent homologous recombination activity. Cell lysates from these RecA⁻ strains, which are termed Seamless Ligation Cloning Extract (SLiCE), possess efficient cloning activity in vitro [22]. SLiCE from laboratory *E. coli* strains have been demonstrated to efficiently integrate insert-DNA fragments into vectors with short homology regions at the 5′ and 3′ ends (15–19 bp) (*see* Fig. 1). This homologous recombination involves incubation for 10–60 min at 37 °C in a buffer containing ATP, MgCl₂, and dithiothreitol (DTT). As the SLiCE-based technique represents a highly efficient seamless cloning method, two insert DNA fragments can be simultaneously assembled into a vector. In addition, unpurified DNA fragments amplified by PCR can be also used as insert DNA [22]. SLiCE from laboratory *E. coli* strains is available for ultra-low-cost seamless cloning procedures at a cost of 0.4–0.5 Yen, or approximately US$0.003–0.004, per reaction [23], as these can be easily derived from cell lysates of laboratory *E. coli* strains via a simple protocol utilizing lysis buffer and centrifugation [22, 24]. Therefore, SLiCE from a laboratory *E. coli* strain can provide an effective alternative to commercially available seamless DNA cloning kits [23].

Recently, in vivo assembly cloning of DNA fragments in *E. coli* has been demonstrated [25–28]. This method is very simple as it does not require the incubation of DNA fragments with enzyme solutions; however, the use of 30–50 bp-overlap homology regions and high-efficiency competent cells (~10⁹ colony-forming units (CFU)/μg of pUC19 DNA) is recommended for efficient cloning [27]. In contrast, SLiCE from the laboratory *E. coli* JM109 strain

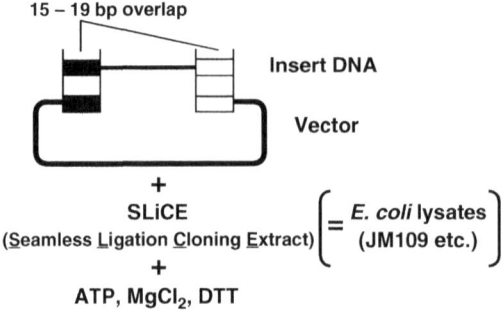

Fig. 1 Overview of SLiCE cloning. Primers were designed to contain 15–19 bp short-end homology regions. Either PCR-amplified vector or restriction enzyme-digested vector can be used. Several RecA⁻ strains are available for the preparation of SLiCE from laboratory *E. coli* strains [22]

exhibits 80- to 200-fold higher efficiency than the in vivo assembly cloning method, under the same conditions using 19-bp-overlap short-end homology regions [22]. Therefore, DNA cloning using SLiCE from this strain enables the use of "homemade" chemically competent cells (~10^7 CFU/μg of pUC19 DNA) for routine DNA cloning.

Numerous methods are available for site-directed mutagenesis such as QuickChange site-directed mutagenesis [29], the overlap extension method [30, 31], and the megaprimer method [32, 33]. The QuickChange method, which does not require a DNA assembly step, is commonly used for facile site-directed mutagenesis. However, long-PCR amplification of the entire plasmid may introduce undesirable mutations in the vector sequence. In order to avoid the introduction of such mutations in the vector, the overlap extension and megaprimer methods are used. These methods involve two sequential PCRs and purification steps of two insert DNA fragments; however, the possibility of generating undesirable mutations in the vector, which is associated with the use of restriction endonuclease-digested vectors, can be excluded (*see* Fig. 2a). The SLiCE method enables the simplification of complex procedures involved in the overlap extension method because it allows simultaneous cloning of multiple DNA fragments containing mutation sites (*see* Fig. 2b). *SLi*CE-mediated *P*CR-based site-directed mutagenesis (SLiP site-directed mutagenesis) can be widely applied to protocols requiring PCR-based site-directed mutagenesis methods.

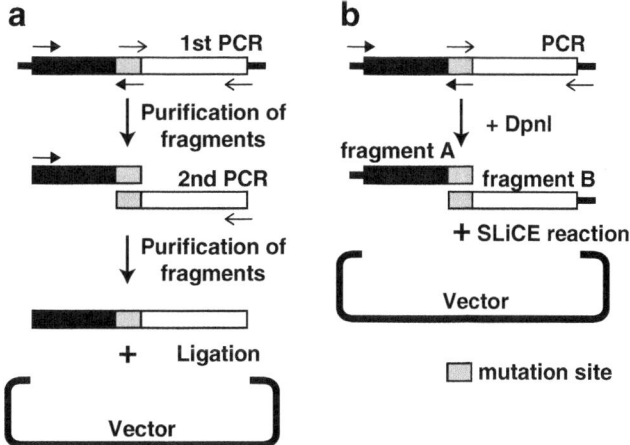

Fig. 2 PCR-based site-directed mutagenesis using SLiCE from laboratory *E. coli* strains. Arrows indicate PCR primers. (**a**) Standard overlap extension method for site-directed mutagenesis. (**b**) SLiCE-mediated PCR-based site-directed mutagenesis (SLiP site-directed mutagenesis. Reproduced from ref. [22]

2   Materials

2.1   Preparation of SLiCE from a Laboratory E. coli Strain

1. Luria broth (LB) medium: tryptone 10 g, yeast extract 5 g, NaCl 10 g. Dissolve components in 1 l of distilled H_2O. pH to 7.0 with 5 N NaOH. Sterilize by autoclaving at 121 °C, 15 psig, 20 min.

2. 2× YT medium: tryptone 16 g, yeast extract 10 g, NaCl 5 g. Dissolve components in 1 l of distilled H_2O. pH to 7.0 with 5 N NaOH. Sterilize by autoclaving at 121 °C, 15 psig, 20 min.

3. *E. coli* strain JM109 (*see* **Note 1**).

4. Round-bottom, long-neck Sakaguchi shaking flask (100 mL).

5. Lysis buffer: CelLytic B Cell Lysis Reagent (Sigma, B7435) (*see* **Note 2**).

6. Ice-cold sterile distilled water (50 mL).

7. Ice-cold 80 % (v/v) glycerol (1.2 mL, sterilized).

8. Liquid nitrogen.

2.2   SLiCE Reaction

1. SLiCE prepared from *E. coli* strain JM109.

2. High-fidelity PCR enzyme: PrimeSTAR max DNA polymerase (Takara) (*see* **Note 3**).

3. *Dpn* I (10 U/μL) restriction enzyme.

4. SLiCE buffer (10×): 500 mM Tris–HCl, pH 7.5, 100 mM $MgCl_2$, 10 mM ATP, 10 mM DTT) (*see* **Note 4**).

5. *E. coli* RecA⁻ chemically competent cells (*see* **Note 5**).

6. DNA purification kit (*see* **Note 6**).

3   Methods

3.1   Preparation of SLiCE from a Laboratory E. coli Strain

1. Inoculate *E. coli* JM109 into 1 mL of LB medium from a freezer stock kept at a –80 °C.

2. Incubate 1 mL of the culture with shaking at 37 °C until the OD_{600} reaches a value of ~1.0 (*see* **Note 7**).

3. Transfer *E. coli* JM109 precultured cells (1 mL) to 2× YT medium (50 mL) in a 100-mL round-bottom, long-neck Sakaguchi shaking flask.

4. Grow at 37 °C in a reciprocal shaker (160 rpm with 25 mm stroke) until the OD_{600} reaches a value of 2.0–3.0 (late log phase) (*see* **Note 8**).

5. Harvest by centrifugation at 5000×*g* for 10 min at 4 °C.

6. Wash with 50 mL of sterile distilled water (ice-cold), and centrifuge at 5000×*g* for 5 min at 4 °C (*see* **Note 9**).

7. Resuspend 0.3–0.4 g of cells (wet weight) gently in 1.2 mL of CelLytic B Cell Lysis Reagent (Sigma, B7435) (*see* **Note 2**), and leave to stand for 10 min at room temperature to proceed the lysis reaction.

8. Centrifuge at 20,000×*g* for 2 min at 4 °C. All subsequent procedures are to be performed on ice.

9. Transfer supernatants into 1.5-mL Eppendorf tubes to remove insoluble material.

10. Add an equal volume of ice-cold 80% (v/v) glycerol to the supernatant and mix gently.

11. Aliquot 40 μL of each SLiCE into 0.2-mL PCR tubes.

12. Snap freeze in a bath of liquid nitrogen, and transfer to –80 °C in 40% (v/v, final concentration) glycerol for long-term storage (*see* **Note 10**).

3.2 SLiCE-Mediated PCR-based Site-directed Mutagenesis (SLiP Site-directed Mutagenesis)

Design of primers for SLiP site-directed mutagenesis.

1. Design primers against the site-directed mutagenesis site(s) using PrimerX (*see* **Note 11**).

2. Set primer design parameters for "QuickChange site-directed mutagenesis kit by Stratagene" in the PrimerX program.

   Condition-1: melting temperature >78 °C.

   Condition-2: primer termination in G or C.

3. Design primers for insert DNA fragments containing 19-bp short sequences that overlap with the ends of the vector.

3.2.1 Standard SLiP-method

1. Amplify two DNA fragments containing a mutation site using a high-fidelity PCR enzyme such as PrimeSTAR max DNA polymerase (*see* **Note 12**).

2. Purify the PCR fragments using a DNA purification kit (via spin column). If multiple fragments are amplified by PCR, the fragments should be purified by agarose gel electrophoresis.

3. Prepare the linear vector by PCR amplification or digestion using the appropriate restriction enzymes.

4. Mix insert-DNA fragments and vector for SLiCE cloning under the following conditions (*see* Fig. 2b):

   (a) Insert DNA fragment A using purified insert–vector at a molar ratio of 1:1–3:1.

   (b) Insert DNA fragment B using purified insert–vector at a molar ratio of 1:1–3:1.

   (c) Vector (10–100 ng): *X*μL.

   (d) SLiCE buffer (10×): 1 μL.

(e) SLiCE: 1 μL.

(f) Sterile distilled water: up to 10 μL.

5. Incubate for 10–60 min at 37 °C (*see* **Note 13**).

6. Transform *E. coli* RecA⁻ chemically competent cells using the standard heat-shock procedure (*see* **Note 14**).

7. Plate the transformed *E.coli* cells on LB agar plates containing appropriate antibiotics for screening and incubate the plates at 37 °C for 12–16 h (overnight).

8. Screen for colonies harboring the correct insert by colony PCR (*see* **Note 15**) and confirm the insert sequence in the plasmid by DNA sequencing.

3.2.2 Rapid SLiP-method   This is the rapid protocol.

1. Amplify two DNA fragments containing a mutation site using a high-fidelity PCR enzyme (*see* **Note 12**).

2. Prepare linear vector by PCR-amplification or digestion using restriction enzymes.

3. Mix insert DNA fragments and the vector for SLiCE cloning under the following conditions (*see* **Note 16**):

(a) Unpurified insert DNA-A: 1 μL.

(b) Unpurified insert DNA-B: 1 μL.

(c) Vector (10–100 ng): XμL.

(d) SLiCE buffer (10×): 1 μL.

(e) SLiCE: 1 μL.

(f) Sterile distilled water: Up to 10 μL.

4. Incubate for 15 min at 37 °C.

5. Transform *E. coli* RecA⁻ chemically competent cells according to the standard heat-shock protocol (*see* **Note 14**).

6. Plate the transformed *E. coli* cells onto LB agar plates containing antibiotics for selection and incubate the plates at 37 °C for 12–16 h.

7. Screen for colonies harboring the correct insert by colony PCR (*see* **Note 15**), and confirm the insert sequence in the plasmid by DNA sequencing.

4   Notes

1. Other *E. coli* strains, such as DH5α, XL10-Gold, and Mach1-T1, can also be used [22].

2. Alternatively, buffers containing Triton X-100 can be used [24].

3. Other high-fidelity PCR enzymes, such as KOD-Plus-Ver.2 (Toyobo), can be used.

4. SLiCE buffer (10×) is filtered through a 0.2-μm filter, dispensed in 40-μL aliquots into 0.2-mL PCR tubes, and stored at –20 °C.

5. We generally use DH5α and XL10-Gold chemically competent cells.

6. For example, Gel extraction and PCR clean-up kit for DNA fragments (e.g., Nippon Genetics, Promega, Life Technologies).

7. This culture step usually takes about 3 h.

8. This culture step usually takes about 4–5 h. The OD_{600} value should be estimated using diluted samples as their absorbance readings should be in the linear range of the spectrophotometer.

9. At this stage, we are usually able to recover 0.3–0.4 g of wet cells.

10. The SLiCE can be stored at –20 °C short-term (about 2–3 months) without significant loss of activity. A SLiCE stock prepared from 50 mL of *E. coli* cells can be used in 2400 SLiCE cloning reactions.

11. The PrimerX program is available at: http://www.bioinformatics.org/primerx/.

12. The template DNA should be digested with *Dpn* I (5 U) for 30 min at 37 °C by adding the restriction enzyme into the PCR mixture following the amplification step. *Dpn* I treatment efficiently reduces background colony formation. When a restriction enzyme-digested vector is used, *Dpn*I should be inactivated for 15 min at 80 °C.

13. Prolonged incubation reduces the rate of colony formation [22].

14. Chemical transformation is performed as follows: (a) Gently thaw 50 μL of competent *E. coli* cells on ice. (b) Add 4 μL of the SLiCE reaction mixture into the chilled competent cells and gently flick the transformation mixture to disperse the DNA. (c) Incubate the transformation reaction mixture on ice for 30 min. (4) Heat the mixture at exactly 42 °C for 45 s and immediately cool it on ice for 2 min. (e) Add 800 μL of SOC medium into the transformation reaction mixture and incubate it at 37 °C for 30–60 min in a thermomixer or in a reciprocal shaker incubator at 200 rpm with 25 mm stroke.

   After the SLiCE reaction, the mixture should be immediately transformed into chemically competent cells in order to avoid DNA degradation due to nuclease contamination in the SLiCE mix. If the SLiCE mixture is not to be used immediately, it should be stored at –20 °C.

15. Each colony is picked using a sterile toothpick and placed into the bottom of a 0.2-mL, 8-strip PCR tube. After the toothpicks are removed from the PCR tube, 10 µL of PCR mix is added to each sample. Sample solutions are reacted according to standard protocols [34, 35].

16. Treatment of unpurified insert DNA fragments by ExoSAP-IT (Affymetrix) enhances cloning efficiency of the SLiCE by six- to sevenfold [22]. Ethanol precipitation of the unpurified insert DNA fragments also enhances cloning efficiency to the same extent.

Acknowledgments

I thank Yuki Okegawa for critically reading the manuscript. This work was supported by the MEXT-Supported Program for the Strategic Research Foundation at Private Universities (to K.M.).

References

1. Gibson DG, Young L, Chuang RY, Venter JC, Hutchison CA 3rd, Smith HO (2009) Enzymatic assembly of DNA molecules up to several hundred kilobases. Nat Methods 6:343–345

2. Klock HE, Koesema EJ, Knuth MW, Lesley SA (2008) Combining the polymerase incomplete primer extension method for cloning and mutagenesis with microscreening to accelerate structural genomics efforts. Proteins 71:982–994

3. Quan J, Tian J (2009) Circular polymerase extension cloning of complex gene libraries and pathways. PLoS One 4, e6441

4. Li MZ, Elledge SJ (2012) SLIC: a method for sequence- and ligation-independent cloning. Methods Mol Biol 852:51–59

5. Thieme F, Marillonnet S (2014) Quick and clean cloning. Methods Mol Biol 1116:37–48

6. Chino A, Watanabe K, Moriya H (2010) Plasmid construction using recombination activity in the fission yeast Schizosaccharomyces pombe. PLoS One 5, e9652

7. Matsuo Y, Kishimoto H, Horiuchi T, Tanae K, Kawamukai M (2010) Simple and effective gap-repair cloning using short tracts of flanking homology in fission yeast. Biosci Biotechnol Biochem 74:685–689

8. Goto K, Nagano Y (2013) Ultra-low background DNA cloning system. PLoS One 8, e56530

9. Hamilton MD, Nuara AA, Gammon DB, Buller RM, Evans DH (2007) Duplex strand joining reactions catalyzed by vaccinia virus DNA polymerase. Nucleic Acids Res 35:143–151

10. Zhu B, Cai G, Hall EO, Freeman GJ (2007) In-fusion assembly: seamless engineering of multidomain fusion proteins, modular vectors, and mutations. Biotechniques 43:354–359

11. Gibson DG, Smith HO, Hutchison CA 3rd, Venter JC, Merryman C (2010) Chemical synthesis of the mouse mitochondrial genome. Nat Methods 7:901–903

12. Baek CH, Chesnut J, Katzen F (2015) Positive selection improves the efficiency of DNA assembly. Anal Biochem 476:1–4

13. Lanyon-Hogg T, Masumoto N, Bodakh G, Konitsiotis AD, Thinon E, Rodgers UR, Owens RJ, Magee AI, Tate EW (2015) Click chemistry armed enzyme-linked immunosorbent assay to measure palmitoylation by hedgehog acyltransferase. Anal Biochem 490:66–72

14. Jacobs TB, LaFayette PR, Schmitz RJ, Parrott WA (2015) Targeted genome modifications in soybean with CRISPR/Cas9. BMC Biotechnol 15:16

15. Nakagawa Y, Sakuma T, Sakamoto T, Ohmuraya M, Nakagata N, Yamamoto T (2015) Production of knockout mice by DNA microinjection of various CRISPR/Cas9 vectors into freeze-thawed fertilized oocytes. BMC Biotechnol 15:33

16. Dovala D, Sawyer WS, Rath CM, Metzger LE (2016) Rapid analysis of protein expression

and solubility with the SpyTag-SpyCatcher system. Protein Expr Purif 117:44–51

17. Kuzminov A (1999) Recombinational repair of DNA damage in Escherichia coli and bacteriophage lambda. Microbiol Mol Biol Rev 63:751–813

18. Muyrers JP, Zhang Y, Buchholz F, Stewart AF (2000) RecE/RecT and Redalpha/Redbeta initiate double-stranded break repair by specifically interacting with their respective partners. Genes Dev 14:1971–1982

19. Lovett ST, Hurley RL, Sutera VA Jr, Aubuchon RH, Lebedeva MA (2002) Crossing over between regions of limited homology in Escherichia coli. RecA-dependent and RecA-independent pathways. Genetics 160: 851–859

20. Dutra BE, Sutera VA Jr, Lovett ST (2007) RecA-independent recombination is efficient but limited by exonucleases. Proc Natl Acad Sci U S A 104:216–221

21. Persky NS, Lovett ST (2008) Mechanisms of recombination: lessons from E. coli. Crit Rev Biochem Mol Biol 43:347–370

22. Motohashi K (2015) A simple and efficient seamless DNA cloning method using SLiCE from Escherichia coli laboratory strains and its application to SLiP site-directed mutagenesis. BMC Biotechnol 15:47

23. Okegawa Y, Motohashi K (2015) A simple and ultra-low cost homemade seamless ligation cloning extract (SLiCE) as an alternative to a commercially available seamless DNA cloning kit. Biochem Biophys Rep 4: 148–151

24. Okegawa Y, Motohashi K (2015) Evaluation of seamless ligation cloning extract preparation methods from an Escherichia coli laboratory strain. Anal Biochem 486:51–53

25. Li C, Wen A, Shen B, Lu J, Huang Y, Chang Y (2011) FastCloning: a highly simplified, purification-free, sequence- and ligation-independent PCR cloning method. BMC Biotechnol 11:92

26. Jacobus AP, Gross J (2015) Optimal cloning of PCR fragments by homologous recombination in *Escherichia coli*. PLoS One 10, e0119221

27. Kostylev M, Otwell AE, Richardson RE, Suzuki Y (2015) Cloning should be simple: *Escherichia coli* DH5alpha-mediated assembly of multiple DNA fragments with short end homologies. PLoS One 10, e0137466

28. Beyer HM, Gonschorek P, Samodelov SL, Meier M, Weber W, Zurbriggen MD (2015) AQUA cloning: a versatile and simple enzyme-free cloning approach. PLoS One 10, e0137652

29. Papworth C, Bauer JC, Braman J, Wright DA (1996) Site-directed mutagenesis in one day with >80% efficiency. Strategies 9:3–4

30. Higuchi R, Krummel B, Saiki RK (1988) A general method of *in vitro* preparation and specific mutagenesis of DNA fragments: study of protein and DNA interactions. Nucleic Acids Res 16:7351–7367

31. Ho SN, Hunt HD, Horton RM, Pullen JK, Pease LR (1989) Site-directed mutagenesis by overlap extension using the polymerase chain reaction. Gene 77:51–59

32. Sarkar G, Sommer SS (1990) The "megaprimer" method of site-directed mutagenesis. Biotechniques 8:404–407

33. Landt O, Grunert HP, Hahn U (1990) A general method for rapid site-directed mutagenesis using the polymerase chain reaction. Gene 96:125–128

34. Okegawa Y, Koshino M, Okushima T, Motohashi K (2016) Application of preparative disk gel electrophoresis for antigen purification from inclusion bodies. Protein Expr Purif 118:77–82

35. Okegawa Y, Motohashi K (2016) Expression of spinach ferredoxin-thioredoxin reductase using tandem T7 promoters and application of the purified protein for in vitro light-dependent thioredoxin-reduction system. Protein Expr Purif 121:46–51

Chapter 24

Facile Site-Directed Mutagenesis of Large Constructs Using Gibson Isothermal DNA Assembly

Isaac T. Yonemoto and Philip D. Weyman

Abstract

Site-directed mutagenesis is a commonly used molecular biology technique to manipulate biological sequences, and is especially useful for studying sequence determinants of enzyme function or designing proteins with improved activity. We describe a strategy using Gibson Isothermal DNA Assembly to perform site-directed mutagenesis on large (>~20 kbp) constructs that are outside the effective range of standard techniques such as QuikChange II (Agilent Technologies), but more reliable than traditional cloning using restriction enzymes and ligation.

Key words Site-directed mutagenesis, Gibson assembly, Polymerase chain reaction, *Escherichia coli*, DNA assembly

1 Introduction

Following the precedent set by Hutchison et al. [1], site-directed mutagenesis is traditionally performed using PCR with primers that install the desired mutation followed by traditional restriction-ligation cloning. This finicky process is dependent on operator technique, reagent quality, mutation location and identity, and sometimes luck. Other techniques, such as Kunkel mutagenesis, were developed, which depend on exotic cloning strains [2].

The widespread availability of cheap, long oligonucleotides and reliable polymerases made possible a second generation of site-directed mutagenesis methods, such as QuikChange and "Round-the-horn" site-directed mutagenesis [3, 4]. Although much easier, these techniques are limited by polymerase extension and, if applicable, minimization of unplanned mutations created by PCR; both of these parameters scale by plasmid size. Moreover, occasionally a difficult-to-diagnose problem (such as impaired *Dpn*I activity in the case of QuikChange) can ruin an attempt at mutagenesis and require starting from scratch. As enzymatic pathway- and genome-scale manipulations become more widespread, the inherent plasmid

Andrew Reeves (ed.), *In Vitro Mutagenesis: Methods and Protocols*, Methods in Molecular Biology, vol. 1498,
DOI 10.1007/978-1-4939-6472-7_24, © Springer Science+Business Media New York 2017

size limitations for mutagenesis are more relevant, especially the inability of polymerases to reliably amplify beyond 10–15 kbp.

We present an update of traditional site-directed mutagenesis with Gibson Isothermal DNA Assembly [5] with increased robustness and operator ease. The compartmentalized procedure also allows for easy diagnosis of problems, and for hierarchical and multistep manipulations, rollback to previous correct result, while still being reasonably rapid. With optimized ~1 h PCR thermocycling, the bulk of the process can be completed in <6 h and if a particularly aggressive timeline is needed, colonies can be recovered within 24 h of starting the procedure, resulting in a net 2-day turnaround for validated mutants.

In addition to standard site-directed mutagenesis, this strategy can be used to generate combinations of existing mutations. In our case, a panel of six single-point substitutions previously shown to be advantageous for enzyme activity were combined (2×4) to generate an 8-member defined library in a single step [6].

This is a facile technique for rapid generation of panels of mutants, especially advantageous in the case of large constructs and/or for use by operators with limited skill level. With only guidance for strategic pipelining of the process, double-checking design elements, and debugging failures, an undergraduate student with limited molecular biology experience was capable of producing 50 mutants of the hydrogenase enzyme over the course of 3 months, including training time [6]. The pipeline also afforded the undergraduate student time to complete biochemical experiments on half of the mutants produced.

1.1 Strategy

1. Choose the site(s) where sequence changes are desired.

2. Identify two restriction enzymes that each cut only once in the plasmid at sites flanking the site of desired change, or alternatively identify a single enzyme cutting more than once only near the change site (see **Note 1**) (see Fig. 1). If no such sites exist, pick a location to introduce a new restriction site (see **Note 2**). Use this procedure to introduce the new restriction

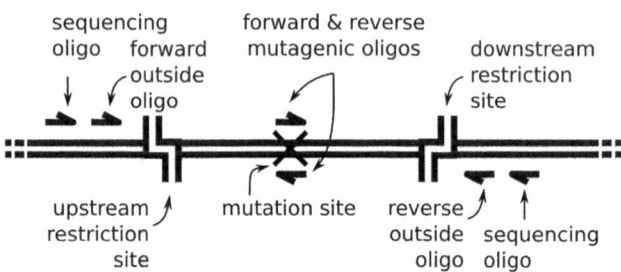

Fig. 1 A cartoon depicting the relative locations of key features of the Gibson isothermal DNA assembly procedure

site, before repeating the procedure for further mutagenesis. Alternatively, an enzyme cutting twice in the plasmid—once in a favorable location and once elsewhere—could be employed by eliminating the undesired site.

3. Select sequences for primers (i.e., "outside oligonucleotides") that match sequence 20–40 bp upstream of the first restriction site and downstream of the second restriction site. Match the melting temperatures of these primers to a desired PCR protocol.

4. Select sequences for primers (i.e., "sequencing oligonucleotides") that match sequence upstream of the forward outside oligonucleotide and downstream of the reverse outside oligonucleotides (*see* **Note 3**).

1.2 Site-Directed Mutagenesis

1. Design sequences for overlapping primers (i.e., "mutagenic oligonucleotides") that match ~20–40 bp of sequence surrounding the desired mutation, match the melting temperatures of these primers to the outside oligonucleotides, correcting for mismatch penalty.

1.3 Combinatorial Cloning

1. Choose the locations where genetic diversity has been introduced. Ideally these sites should be separated by at least 20 bp (*see* **Note 4**). Select an intervening primer between the sites of diversity, with a melting temperature matching the outside oligonucleotides.

2 Materials

2.1 Molecular Biology Reagents

1. Appropriate restriction endonucleases.

2. 2× Gibson isothermal assembly mix (*see* **Note 5**).

3. Synthetic oligonucleotide primers (e.g.,, Integrated DNA Technologies).

4. 2× High-fidelity PCR master mix such as Phusion (New England Biolabs) or PrimeSTAR Max (Clontech).

5. 2× Taq polymerase PCR master mix such as OneTaq (New England Biolabs).

6. Agarose.

7. 1× TAE buffer.

8. Agarose gel electrophoresis apparatus.

9. Ethidium bromide.

10. Gel imaging and documentation system.

11. Blue light gel illuminator.

12. Single-edge razor blade.

13. QIAquick Gel extraction kit (Qiagen).

14. QIAquick PCR cleanup kit (Qiagen).

15. Spectrometer (e.g., Nanodrop).

16. PCR thermocycler.

17. Electroporation apparatus with cuvettes.

18. Electrocompetent *E. coli* cells.

19. Luria–Bertani broth: tryptone 10 g, yeast extract 5 g, and NaCl 10 g to 1 l of distilled H_2O. pH to 7.2 with 5 N NaOH. Sterilize by autoclaving. For plating cells, add agar to a final concentration of 1.5 %. Add appropriate antibiotics for selection.

20. QIAprep miniprep kit (Qiagen).

3 Methods

3.1 Plasmid Backbone

1. Digest the template plasmid (~1–2 µg) to linearize with the identified restriction enzyme(s). Follow the recommended protocol from the restriction enzyme manufacturer. Ensure complete digestion by providing adequate enzyme, digestion time, or both.

2. Verify successful digestion using agarose gel electrophoresis. Usually a subset of the digest is separated by gel electrophoresis, stained with ethidium bromide, and imaged using UV illumination. The remainder is reserved for gel purification.

3. Separate the remaining digested plasmid by agarose gel electrophoresis. Stain with ethidium bromide but do not expose to UV light (*see* **Note 6**). Place the gel on a blue light illuminator and extract the plasmid band using a commercially available gel extraction kit (*see* **Notes 6** and **7**).

4. The digested, purified plasmid may be stored at –20 °C and reused for multiple rounds of mutagenesis.

3.2 Mutagenic Inserts

1. Perform two PCR reactions, each templated by the source plasmid using the following two primer pairs (*see* Fig. 1): (a) Forward Outside Oligo, Reverse Mutagenic Oligo and (b) Forward Mutagenic Oligo, Reverse Outside Oligo.

2. Verify success of the PCR reactions by separating a small subsample of the PCR reaction (1–2 µl) using agarose gel electrophoresis (*see* **Note 6**).

3. Separate the remainder of the PCR reaction by agarose gel electrophoresis and extract the PCR product from the agarose gel using a standard gel extraction kit.

4. Measure the purified amplicon yield and dilute isolated amplicons into water to a final concentration of 0.2 nM (*see* **Note 8**).

5. Perform one PCR reaction to pre-assemble the two halves of the inserted sequence into a single-insert sequence. Perform the PCR reaction according to the PCR kit manufacturer's recommended protocol. The template for the PCR reaction will consist of a mix of the isolated amplicons and the primers will be the Forward Outside Oligo and the Reverse Outside Oligo (*see* **Note 9**).

6. Verify the success of the pre-assembly PCR reaction by separating an aliquot of the PCR reaction (1–2 μl) on an agarose gel.

7. Use the PCR cleanup kit to prepare the insert for assembly into the vector (*see* **Note 10**).

8. Measure purified insert amplicon yield and dilute into water to a final concentration of 0.2 nM (*see* **Note 8**).

3.3 Assembly and Verification

1. Set up a Gibson Isothermal Assembly Reaction using the amplicon and the digested template plasmid from **step 4** in Subheading 3.2. Combine equimolar amounts of each DNA fragment with an amount of 2× Gibson isothermal assembly mix equivalent to the total volume of the DNA fragments. Mix the reaction by pipetting up and down several times and keep on ice until transferred directly to a 50 °C water bath or thermocycler. Allow assembly for 30 min to 1 h.

2. Electroporate 1 μl of the assembly reaction into 20 μl high-efficiency electrocompetent cells (*see* **Notes 11** and **12**). Recover the cells from the electroporation cuvette in 1 ml LB, incubate 1 h at 37 °C with shaking, plate 50 μl and 500 μl on an LB plate containing appropriate antibiotics for the plasmid, and incubate overnight.

3. Pick and streak several colonies on fresh LB agar plate containing appropriate antibiotics, and then perform a screening "colony" PCR to identify colonies containing the correctly assembled plasmid construct (*see* **Note 13**).

4. Inoculate positive cell patches into 5–10 ml of LB medium containing the appropriate antibiotic(s). Incubate with vigorous (~300 rpm) shaking at 37 °C overnight.

5. Prepare a frozen stock of the cells by diluting 1:1 in 30 % glycerol (15 % glycerol final concentration) and store at –80 °C.

6. Centrifuge at $12,000 \times g$ the remaining cell culture and purify the plasmid using a plasmid miniprep kit.

7. Digest the plasmid DNA using a restriction enzyme that cuts several times within the sequence to provide a characteristic digestion pattern (*see* **Note 14**).

8. Sequence the newly assembled region of the plasmid to verify that the desired mutagenesis has been performed.

4 Notes

1. We recommend flanking closely (~250 bp) to the mutagenesis sites, so that a minimal number of Sanger sequencing reads are necessary to validate the construction.

2. We recommend known neutral sites, silent mutations in the center of ORFs to minimize phenotypic effects. In the case of Yonemoto et al. [6], due to the paucity of available restriction sites, we moved a unique restriction site by eliminating the site in the center of an ORF and installing the site in the non-translated region between the stop codon of another ORF and its transcriptional terminator. In choosing silent sites, be aware of organismal codon usage and RNA secondary structure considerations when choosing the introduced site.

3. If the insert is sufficiently short, it may be possible to achieve acceptable coverage by sequencing using the outside oligonucleotides themselves, if the forward oligonucleotide reliably sequences the region around the reverse oligonucleotide sequence in the template, and vice versa.

4. 20 bp is a general guideline for Gibson Isothermal DNA Assembly: For AT-rich sequences check to ensure the melting temperature of the overlap sequence is at or above 50 °C; for GC-rich sequences, smaller overlaps may be possible.

5. 2× Gibson isothermal assembly mix can be purchased directly from commercial suppliers (e.g., New England Biolabs or SGI DNA) or it can be made from components. To prepare 2× Gibson isothermal assembly mix, combine the following in a microfuge tube: 320 μl 5× isothermal (5×ISO) reaction buffer (25% PEG-8000, 500 mM Tris–HCl pH 8.0, 50 mM $MgCl_2$, 50 mM DTT, 1 mM each of the four dNTPs, and 5 mM NAD (*see* recipe at the end of this note), 0.64 μl T5 exonuclease (10 U/μl, New England Biolabs), 20 μl Phusion DNA polymerase (2 U/μl, New England Biolabs), 80 μl Taq DNA ligase (40 U/μl, New England Biolabs), 379.8 μl water. Store in aliquots at −20 °C. To prepare 5× isothermal reaction buffer, mix the following reagents: 3 ml of 1 M Tris–HCl, pH 8.0, 300 μl of 1 M $MgCl_2$, 600 μl of 10 mM dNTPs, 300 μl of 1 M Dithiothreitol, DTT, 1.5 g polyethylene glycol, PEG-8000, 300 μl of 100 mM nicotinamide adenine dinucleotide, NAD (Dissolve 0.66 g in 10 ml dH_2O by vortexing and briefly heating at 50 °C), water to 6 ml total volume. Store 5×ISO as aliquots at −20 °C.

6. Use of UV illumination for gel extraction can damage DNA and significantly decrease cloning efficiency. Devices that illuminate the gel with blue light and allow for visualization with orange filters (Such as Safe Imager from Thermo) will improve efficiency of this technique.

7. Run the undigested plasmid on the same gel as the digested plasmid, with at least one lane between the two. Undigested plasmid should have two or three bands, corresponding to supercoiled, nicked and (sometimes) linearized forms of the plasmid. Usually, nicked and linear forms will migrate at a different rate from the digested plasmid, but it is possible for the supercoiled (which typically runs faster than linearized) to co-migrate with the digested plasmid (which is smaller, and thus runs faster than linearized). In this case, run a series of gels with increased or decreased concentrations of buffer (e.g., 0.5× TAE, 0.75× TAE, 1.5× TAE) to achieve resolution between digested and supercoiled plasmids, as the migration of supercoiled DNA is highly sensitive to buffer ionic strength.

8. Occasionally the DNA yield from the gel extraction process will be lower than the reliable detection threshold of the spectrometric instrument, e.g., a Nanodrop-2000 spectrometer. Separating a subsample of several microliters on an agarose gel stained with ethidium bromide can often reveal the presence of the amplicon. We recommend continuing with the procedure using a reasonable quantity of solution, and noting the low quantity, as the succeeding step will often work.

9. For economy of time, a second PCR may be skipped in favor of directly performing Gibson Isothermal DNA Assembly on three components, at the risk of a lowered transformation efficiency. If assembly using the second PCR fails, directly performing the assembly may be an effective alternative strategy, or vice versa. A two-piece Gibson Assembly will often successfully pick up quantities of DNA that are below the reliable detection limit of a Nanodrop instrument.

10. If multiple bands are observed, resecting the DNA band from the gel and using a gel purification kit may be desirable (*see* comment in **Note 6**).

11. We find that chemically competent cells do not successfully transform constructs on the order of 20 kbp, whereas electrocompetent cells can transform constructs in excess of 400 kbp. We also found that DH5-alpha strains (e.g., NEB-5alpha) can have batch-to-batch variation in transformation competence for constructs of this size, while DH10-derivates (e.g., TOP10) seem to consistently transform.

12. For transformation of very large constructs, e.g., >400 kbp, a modified electroporation procedure is suggested: Apply the DNA to the electroporation cuvette using a wide-tip pipette, electroporate as usual. Add recovery media (e.g., SOC) on top, being careful to layer on top of the cells with minimal shear mixing. Gently tap the cuvette to agitate and mix the recovery media and the cells, then transfer to a 37 °C water bath to recover for 1 h, then plate the cells.

13. Colony PCR refers to performing a PCR reaction in which the template comes from live bacterial cells containing the plasmid of interest that are added to the reaction and lysed during the first denaturation step of the PCR reaction. Often it is convenient to use a lower fidelity, less expensive polymerase for this step such as OneTaq from New England Biolabs.

14. Even when the PCR indicates that the insert has been correctly assembled into the plasmid, digestion of the entire plasmid is encouraged to ensure that the entire plasmid is present and still correctly assembled.

References

1. Hutchison CA, Phillips S, Edgell MH, Gillam S, Jahnke P, Smith M (1978) Mutagenesis at a specific position in a DNA sequence. J Biol Chem 253:6551–6560

2. Kunkel TA (1985) Rapid and efficient site-specific mutagenesis without phenotypic selection. PNAS USA 82:488–492

3. Hogrefe HH, Cline J, Youngblood GL, Allen RM (2002) Creating randomized amino acid libraries with the Quikchange® multi site-directed mutagenesis kit. BioTechniques 33:1158–1165

4. Hemsley A, Arnheim N, Toney MD, Cortopassi G, Galas DJ (1989) A simple method for site-directed mutagenesis using the polymerase chain reaction. Nucleic Acids Res 17:8915

5. Gibson DG, Young L, Chuang R-Y, Venter JC, Hutchison CA III, Smith HO (2009) Enzymatic assembly of DNA molecules up to several hundred kilobases. Nat Methods 6:343–345

6. Yonemoto IT, Clarkson BR, Smith HO, Weyman PD (2014) A broad survey reveals substitution tolerance of residues ligating FeS clusters in [NiFe] hydrogenase. BMC Biochem 15:10

Revised Mechanism and Improved Efficiency of the QuikChange Site-Directed Mutagenesis Method

Yongzhen Xia and Luying Xun

Abstract

Site-directed mutagenesis has been widely used for the substitution, addition or deletion of nucleotide residues in a defined DNA sequence. QuikChange™ site-directed mutagenesis and its related protocols have been widely used for this purpose because of convenience and efficiency. We have recently demonstrated that the mechanism of the QuikChange™ site-directed mutagenesis process is different from that being proposed. The new mechanism promotes the use of partially overlapping primers and commercial PCR enzymes for efficient PCR and mutagenesis.

Key words QuikChange, Site-directed mutagenesis, PCR, Overlap PCR

1   Introduction

Site-directed mutagenesis (SDM) is a useful method for the study of relationships between the structure and function of proteins. In the early days of development of the method, site-directed mutagenesis could only be generated by using single-strand DNA as the template [1, 2]; however, this procedure is labor-intensive and time-consuming [3]. Subsequently, PCR-based SDM became popular due to its simplicity and ease of use. Several traditional PCR-based SDM approaches have been reported, for example, Megaprimer, overlap-extension PCR, inverted PCR, and their modified protocols are routinely used [4–6]. The QuikChange™ method, marketed by Stratagene, further simplifies the PCR-based SDM methods, and it has become a widely practiced SDM method. We have recently revised the mechanism of the QuikChange™ method [7]. Instead of producing circular PCR products with staggered nicked ends of the original mechanism, we reported that the PCR product is linear with blunt and homologous ends. After transformation, *Escherichia coli* cells use a RecA-independent pathway to join the homologous

Andrew Reeves (ed.), *In Vitro Mutagenesis: Methods and Protocols*, Methods in Molecular Biology, vol. 1498,
DOI 10.1007/978-1-4939-6472-7_25, © Springer Science+Business Media New York 2017

ends, producing a plasmid with the desired mutations. Thus, we use partially overlapping primers to generate PCR products for SDM, which is more efficient than using completely complementary primers as recommended by the QuikChange™ method [7]. In this chapter, detailed principles of primer design are given to increase the amplification efficiency of the PCR reaction and the SDM procedure. The amount of template DNA is reduced according to the new method, which in turn decreases the formation of false-positive colonies. Phusion DNA polymerase, one of the highest fidelity DNA polymerases reported [8], is employed to increase the accuracy of PCR products. Moreover, greater than 95 % of transformant colonies are positive and contain the desired mutations. Further, the use of partially overlapping primers makes it easier to employ the method for generating deletions, insertions and substitutions of DNA sequences on a plasmid.

2 Materials

2.1 PCR Reagents

1. Phusion DNA polymerase.
2. Phusion 10× Hi-Fidelity buffer or 10× GC (high GC% DNA) buffer (NE Biolabs).
3. 2.5 mM dNTP mix.
4. HPLC-purified primers with desired mutation.
5. Thermocycler.
6. Sterile deionized water.
7. Template plasmid DNA.
8. Fast-digest *Dpn*I endonuclease.
9. 10× Fast Digest buffer.
10. Agarose.
11. 50× TAE solution: 2.0 M Tris-acetate, pH 8.3, 0.05 M EDTA, pH 8.0.

2.2 Competent Cells

1. Luria–Bertani medium (LB): 10 g/L tryptone, 5 g/L yeast extract, 10 g/L NaCl. pH to 7.0 with 5 N NaOH. For preparing agar plates add 20 g/L agar. Sterilize by autoclaving at 121 °C, 15 psig, 20 min.
2. TSB solution: LB-HCL, pH 6.1, 10% PEG-8000, 10% glycerol, 20 mM $MgSO_4$. Autoclave at 121 °C, 20 min. Add DMSO to 5 % before use.
3. 5× KCM solution: 0.5 M KCl, 150 mM $CaCl_2$, 250 mM $MgCl_2$. Autoclave at 121 °C, 20 min.

3   Methods

3.1   SDM PCR

The method presented below outlines the design of oligonucle-otides for site-directed mutagenesis, insertion, deletion, and substitution per reaction, PCR cycling conditions, and *Dpn*I digestion of the PCR product to remove template DNA.

A pair of partially overlapping primers are designed for the SDM PCR (*see* Fig. 1). This method could be used for site-directed mutation, deletion, insertion and substitution on a plasmid by using partially overlapping primers. For generating a site-directed mutation, the mutation site should be in the middle of the overlapping region of the forward (Fr) primer and reverse (Rev) primer (*see* Fig. 1a). For generating a deletion, the overlapping region of these two primers is composed of a 5′ region and a 3′ region adjoining the DNA that will be deleted on the plasmid (*see* Fig. 1b). For generating an insertion, the overlapping region of these two primers is formed from the inserted DNA sequence (*see* Fig. 1c). For generating a substitution, the overlapping region of these two primers is formed from the new DNA sequence and the priming regions of the Fr primer and Rev primer are complementary to the template plasmid adjoining the DNA that will be replaced on the plasmid (*see* Fig. 1d). The partially overlapping sections of the Fr and Rev primers are defined as Fo and Ro, ranging from 14 to 20 bp in length (*see* **Note 1**). The Fc region of the Fr primer and the Rc region of the Rev primer designate the completely complementary sequences of the template plasmid. Under normal conditions, if the GC content of this area is lower than 60%, the overlapping region could be up to 20 nucleotides. However, if the GC content is higher than 60%, the region could be decreased to 14 bp to lower the melting temperature (T_m).

After the Fc, Rc, Fo and Ro region are defined, the main concerns of primer design are the melting temperature, T_m. The T_m calculation is based on a calculator on the Thermo website with the selection of Phusion DNA polymerase (*see* **Note 2**): https://www.thermofisher.com/cn/zh/home/brands/thermo-scientific/molecular-biology/molecular-biology-learning-center/molecular-biology-resource-library/thermo-scientific-web-tools/tm-calculator.html.

Two main concerns should be noted. First, the T_m of the Fc region should be similar to the T_m of the Rc region. The difference in the T_m between these two primers should be no more than 3 °C. Second, the T_m of the Fc region and the Rc region should be at least 5 °C higher than that of the Fo region.

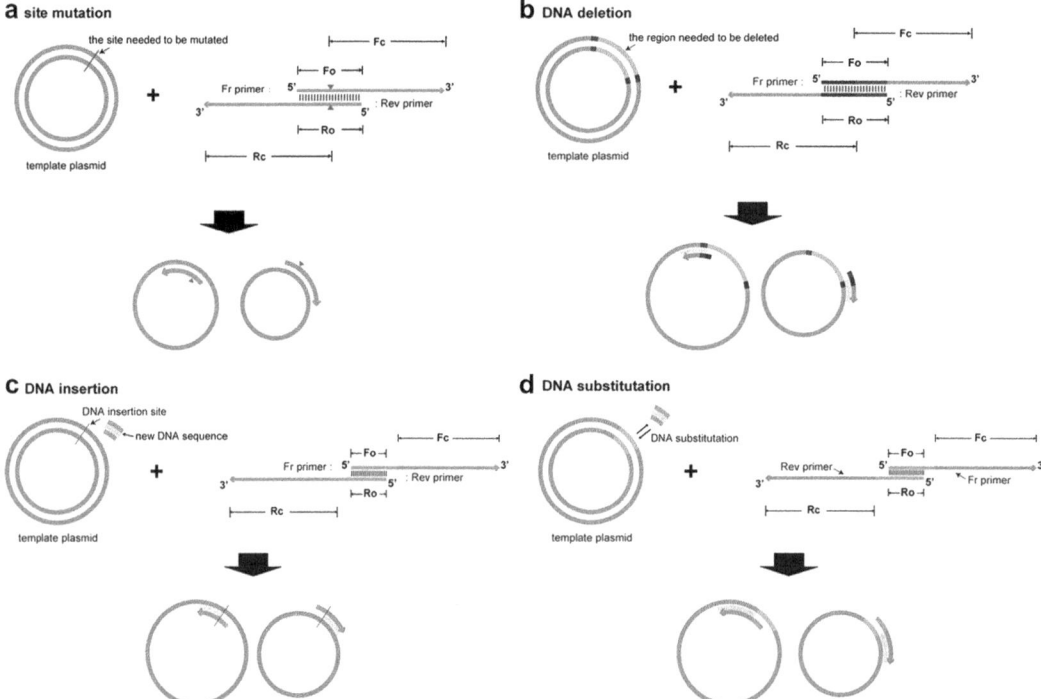

Fig. 1 Illustration showing design of partially overlapping primers. (**a**) Primers for site-directed mutation. Fc, the completely complementary region to the template plasmid of the Fr primer. Rc, the completely complementary region to the template plasmid of the Rev primer. Fo/Ro, the overlapping region between the Fr primer and the Rev primer. The *gray circles* represent the parental plasmid DNA PCR template, while the *red line* marked is the mutation site. *Green arrows* indicate the designed PCR primers. *Red triangles* indicate the location of the mutation site on the primers. (**b**) Primers for DNA deletion. Fc, Rc, Fo, and Ro regions are the same as in **a**, legend. The *gray circles* represent the parental plasmid DNA PCR template. On the plasmid, the *light blue arc* is the region to be deleted. The *dark blue* and *purple arcs* are used as the overlapping regions of the primers, Fo and Ro. (**c**) Primers for DNA insertion. The *gray cycles* represent the parental plasmid DNA PCR template where the *red line* marked is the insertion site. An *orange arc* represents the new DNA to be inserted. The color of the new DNA corresponds with the overlapping region on the primers. (**d**) Primers for DNA substitution. On the plasmid, the *light blue arc* indicates the region to be replaced. The *orange arc* represents the new DNA to be used in the substitution. The color of the new DNA corresponds with the overlapping region on the primers

1. Set up the PCR reaction mixtures as listed in Table 1 (*see* **Notes 3–6**).

2. Set up the PCR cycling reaction conditions as shown in Table 2 (*see* **Notes 7** and **8**).

3. Verify the PCR product using agarose-gel electrophoresis as follows: Run a 0.8 % agarose gel in 1× TAE solution and check the PCR product size, which should be similar to the linear size of the template plasmid. If not, a PCR optimization should be performed.

Table 1
Reaction mixtures (add items in sequence)

Component	20 µL Reaction (µL)	Final concentration
H$_2$O	Add to 20	
5× Phusion HF buffer (*see* **Note 3**)	4	1×
10 mM dNTPs	0.4	200 µM each
20 µM Fr primer	0.5	0.5 µM
20 µM Rev primer	0.5	0.5 µM
Template DNA	X	2–10 ng in total (*see* **Note 4**)
(DMSO, optional)[a]	X	3–10%
Phusion DNA polymerase (2 U/µL) (*see* **Notes 5** and **6**)	0.2	0.02 U/µL

[a]Addition of DMSO is recommended for GC-rich DNA template. DMSO is not recommended for amplicons of low GC% or amplicons of >20 kb

Table 2
PCR cycling conditions

No.	Three-step protocol			Two-step protocol		
	Temperature (°C)	Time		Temperature (°C)	Time	
1	98	1 min		98	1 min	
2	98	20 s		98	20 s	
3	X[a]	10–30 s		–	–	
4	72	30 s/kb (*see* **Note 7**)		72	30 s/kb	
5	Go to step 2	18 cycles (*see* **Note 8**)		Go to step 2	18 cycles	
6	72	10 min		72	10 min	
7	Storage at 15 °C	Forever		Storage at 15 °C	Forever	

[a]The annealing temperature is for 10–30 s at 3 °C above the T_m of the low-T_m primer. If necessary, use a temperature gradient to find the optimal annealing temperature. A two-step protocol is recommended when primer T_m values are at least 69 °C with a combined annealing/extension step at 72 °C

3.2 DpnI Digestion (See Notes 9 and 10)

1. Add 1 µL Fast-Digest *Dpn*I into 20 µL SDM-PCR reaction mix directly without purification. Mix and incubate at 37 °C for 1–3 h to completely remove residual template plasmid.

3.3 Competent Cells (See Notes 11 and 12)

The method of preparing competent cells is adopted from a reported method with minor modifications [9].

1. Streak *E. coli* strain XL-1 Blue MRF' on an LB plate and incubate overnight at 37 °C.

2. Inoculate a single colony in 5 mL of LB and incubate with shaking at 37 °C overnight.

3. Transfer the 5 mL culture to 50 mL of LB and incubate with shaking at 37 °C until the $OD_{600 \, nm}$ is between 0.4–0.6.

4. Centrifuge the cells in two sterile 50-mL tubes at $3000 \times g$ at 4 °C for 15 min.

5. Resuspend the cells in 1 mL of ice-cold TSB.

6. Incubate for 10 min on ice.

7. Aliquot the cells (60 μL/tube) and store at –80 °C.

3.4 DNA Transformation

1. Add 5 μL 5× KCM solution to a 1.5-mL microfuge tube, 5–10 μL *Dpn*I-treated SDM-PCR product, and sterile water to 25 μL.

2. Add 25 μL of thawed (on ice) competent cells to the tube, gently mix by using the pipette tip and incubate on ice for 30 min.

3. Add 250 μL LB and incubate the cells with gentle shaking at 37 °C for 1–1.5 h.

4. Spread an appropriate volume of cells (50–200 μL) on an LB plate with the required antibiotic for selection.

5. Incubate at 37 °C overnight.

3.5 Mutation Confirmation

1. One colony is randomly selected and cultured in 5 mL LB medium with vigorous shaking (~300 rpm) at 37 °C overnight. The plasmid is isolated using a plasmid miniprep kit and sequenced to confirm whether the desired mutation was incorporated. In our experience, the mutation efficiency is greater than 95%. To reduce the required time for finding the right mutation, two or more random colonies could be selected for sequencing.

4 Notes

1. Considering that the longer length of Fo and Ro is inevitable in the case of an insertion or substitution, the Fo and Ro regions are not restricted to 20 nucleotides in such special cases. However, the T_m of Fc/Rc region should be at least 3 °C higher than that of Fo/Ro region.

2. The T_m calculation method using Phusion polymerase is different from the Taq-based polymerase as described in the specifications. Thus, the Thermo website is necessary for T_m calculations when Phusion polymerase is used (https://www.thermofisher.com/cn/zh/home/brands/thermo-scientific/

molecular-biology/molecular-biology-learning-center/
molecular-biology-resource-library/thermo-scientific-web-
tools/tm-calculator.html)

3. If the GC content of the template plasmid is high, a gradient of DMSO concentrations and 10× GC buffer could be used to aid in the denaturation of template DNA. The concentration of DMSO is normally set from 3 to 20% of the reaction volume. If 10% DMSO is used, the annealing temperature should be decreased by 6 °C.

4. The quality and amount of plasmid DNA used in the PCR reaction is an important factor. The amount used in PCR should be carefully controlled. From our experience, 1–5 ng of template DNA per 20 μL PCR reaction is enough in most cases. If that amount is insufficient, an increased amount could be tested from 5 to 50 ng per 20 μL to improve product yield. If a higher amount of the template is used, the amount of *Dpn*I should be increased to ensure complete digestion of the template.

5. Hot-start Phusion DNA polymerase is better in the SDM-PCR amplification reaction.

6. If the SDM product cannot be obtained after optimizations, it is advisable to use another high-fidelity DNA polymerase. However, the PCR procedure should be optimized according to the alternative specifications recommended by the manufacturer.

7. The extension time used in the SDM-PCR reaction is dependent on two factors: the plasmid length and the processivity of the DNA polymerase. The length of the plasmid size is the entire size of the target plasmid including the plasmid backbone and the cloned gene. The processivity of Phusion is about 30 s per kb, but if the size of the plasmid is longer than 5 kb, 40 s/kb is recommended.

8. In most cases, 18 cycles are enough to get the SDM-PCR product. If the yield of PCR product is low, the PCR can be increased to 25 cycles.

9. The plasmid DNA used as the template can be a significant basis of false-positive colonies and it should be thoroughly digested by *Dpn*I. Because *Dpn*I only recognizes a G^{m6}ATC site, the *E. coli* hosts should be *dam*$^+$. Most commonly used *E. coli* strains in the laboratory are *dam*$^+$, such as XL-1 Blue MRF', DH10B, XL-1 Blue, DH5α, JM109, and Top10.

10. The 10x Fast-Digest buffer is also recommended for increasing the digestion rate of Fast-digest *Dpn*I endonuclease.

11. DNA Transformation methods can vary. Alternative chemically competent cell preparation methods or electroporation methods can be adopted.

12. Alternative *E. coli* strains often used in cloning can also be used for DNA transformation.

References

1. Vandeyar MA, Weiner MP, Hutton CJ, Batt CA (1988) A simple and rapid method for the selection of oligodeoxynucleotide-directed mutants. Gene 65(1):129–133, doi:0378-1119(88)90425-8 [pii]
2. Taylor JW, Ott J, Eckstein F (1985) The rapid generation of oligonucleotide-directed mutations at high frequency using phosphorothioate-modified DNA. Nucleic Acids Res 13(24):8765–8785
3. Carrigan PE, Ballar P, Tuzmen S (2011) Site-directed mutagenesis. Methods Mol Biol 700:107–124. doi:10.1007/978-1-61737-954-3_8
4. Vander Kooi CW (2013) Megaprimer method for mutagenesis of DNA. Methods Enzymol 529:259–269. doi:10.1016/B978-0-12-418687-3.00021-5, B978-0-12-418687-3.00021-5 [pii]
5. Simionatto S, Marchioro SB, Galli V, Luerce TD, Hartwig DD, Moreira AN, Dellagostin OA (2009) Efficient site-directed mutagenesis using an overlap extension-PCR method for expressing Mycoplasma hyopneumoniae genes in *Escherichia coli*. J Microbiol Methods 79(1):101–105. doi:10.1016/j.mimet.2009.08.016, S0167-7012(09)00265-6 [pii]
6. Chapnik N, Sherman H, Froy O (2008) A one-tube site-directed mutagenesis method using PCR and primer extension. Anal Biochem 372(2):255–257. doi:10.1016/j.ab.2007.07.020, S0003-2697(07)00476-9 [pii]
7. Xia Y, Chu W, Qi Q, Xun L (2015) New insights into the QuikChange process guide the use of Phusion DNA polymerase for site-directed mutagenesis. Nucleic Acids Res 43(2), e12. doi:10.1093/nar/gku1189, gku1189 [pii]
8. Wang Y, Prosen DE, Mei L, Sullivan JC, Finney M, Vander Horn PB (2004) A novel strategy to engineer DNA polymerases for enhanced processivity and improved performance in vitro. Nucleic Acids Res 32(3):1197–1207, doi:10.1093/nar/gkh271,32/3/1197 [pii]
9. Chung CT, Niemela SL, Miller RH (1989) One-step preparation of competent *Escherichia coli*: transformation and storage of bacterial cells in the same solution. Proc Natl Acad Sci U S A 86(7):2172–2175

Chapter 26

An In Vitro Single-Primer Site-Directed Mutagenesis Method for Use in Biotechnology

Yanchao Huang and Likui Zhang

Abstract

Site-directed mutagenesis is a powerful method to introduce mutation(s) into DNA sequences. A number of methods have been developed over the years with a main goal being to create a high number of mutant genes. The single-mutagenic primer method for site-directed mutagenesis is the most direct method that yields mutant genes in about 25–50 % of transformants in a robust, low-cost reaction. The supercompetent XL10-Gold bacteria used in the Stratagene protocol carry a phage, which may be a problem for some applications; however, in our single-mutagenic primer method the supercompetent bacteria are not needed. A thermostable DNA polymerase with high fidelity and processivity, such as Phusion DNA polymerase, is required for our optimized procedure to avoid extra mutation(s) and enhance mutagenic efficiency.

Key words Site-directed mutagenesis, Single-mutagenic primer method, Competent *Escherichia coli*, *Dpn*I digestion, Thermostable DNA polymerase

1 Introduction

Site-directed mutagenesis (SDM) is an essential method to elucidate structural and functional aspects of proteins and nucleic acids by introducing—in vitro—specific mutation(s) in DNA sequences. SDM with oligonucleotides was first described by Michael Smith [1] and developed to enrich for mutants by Thomas A. Kunkel [2, 3], which greatly extended the application of the method. A number of methods for SDM have been developed over the years with a main goal being to produce a high number of mutant genes. For example, SDM methods such as those commercialized by Stratagene and sold as the QuikChange kit, use a pair of complimentary mutagenic primers [4]. The single-mutagenic primer method is highly efficient (*see* Fig. 1). While the single-mutagenic primer method commercialized by Stratagene [5] has many good features, the supercompetent XL10-Gold bacteria used for the method produce a ϕ80-type bacteriophage at a titer of ~10^8

Andrew Reeves (ed.), *In Vitro Mutagenesis: Methods and Protocols*, Methods in Molecular Biology, vol. 1498,
DOI 10.1007/978-1-4939-6472-7_26, © Springer Science+Business Media New York 2017

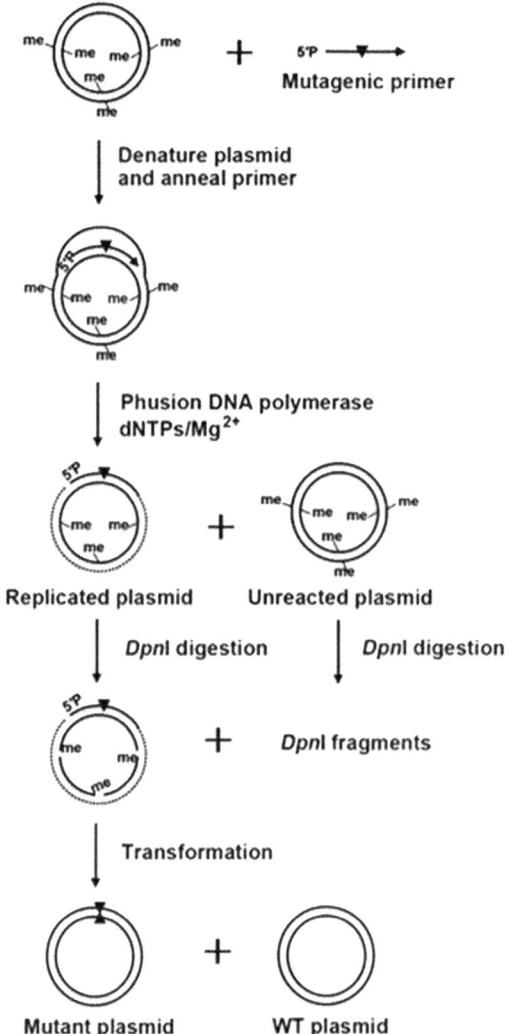

Fig. 1 Schematic of the single primer site-directed mutagenesis method. This figure is adapted from Zhang et al. [9, 10]. The parent plasmid is methylated (Me) and labeled by *solid lines* while the newly synthesized plasmid is labeled in *dashed lines*. The *filled inverted triangle* and *filled triangle* stand for mutation sites. The plasmid DNA is denatured and annealed to the mutagenic primer at 95 and 50 °C, respectively. Phusion DNA polymerase extends the primer by incorporating dNTPs in the presence of Mg^{2+}, which produces the hemi-methylated DNA. *Dpn*I digests both unreacted plasmid DNA and the newly synthesized DNA at different rates. The digested DNA product is transformed into *E. coli* competent cells. The mutant plasmid will be generated by *E. coli* replication polymerases

phage/ml [6], which may be a problem in some applications. However, the supercompetent bacteria are not required in our modified method described herein.

DNA polymerases are a cornerstone of SDM. Fidelity is one of key characteristics of DNA polymerase because high-fidelity DNA polymerases are able to replicate DNA without introducing additional mutations during the SDM process. Furthermore, processivity is another important characteristic of DNA polymerase. Efficient processivity of DNA polymerases allows them to synthesize DNA around a plasmid template during SDM. DNA polymerases with low processivity can be a limitation in some applications such as long PCR. However, processivity can be enhanced by fusing DNA polymerase with other DNA binding domains, or by addition of special "processivity-enhancing" factors such as PCNA (Proliferating Cell Nuclear Antigen) or beta-clamps. One elegant way to increase DNA polymerase processivity is to fuse it with a DNA binding domain at one end of the enzyme. To date, several chimeric archaeal DNA polymerases have been employed commercially in long PCR and site-directed mutagenesis. Our optimized single-primer method for SDM requires a high-fidelity and a highly processive DNA polymerase such as Phusion DNA polymerase.

Phusion DNA polymerase was engineered by fusing the C-terminus of a Pfu-like DNA polymerase with the N-terminus of the Sso7d protein [7]. The Sso7d protein is a highly basic, nonspecific DNA-binding protein from the hyperthermophilic archaeon *Sulfolobus solfataricus*. Since this protein has a strong affinity for DNA, the Sso7d protein was a good candidate for improving DNA polymerase processivity. Biochemical studies revealed that Phusion DNA polymerase was ~10-fold more processive than the wild type Pfu DNA polymerase. In addition, Phusion DNA polymerase showed a sixfold increased in fidelity, as compared to the wild-type Pfu DNA polymerase. As such, Phusion DNA polymerase is useful for both high-fidelity PCR and site-directed mutagenesis due to its high processivity and accurate PCR performance.

A mutagenic primer that is phosphorylated at the 5'-end is annealed to the plasmid carrying the target DNA sequence, which creates a mismatch. Phusion DNA polymerase is then used to extend the 3'-end of the mutagenic primer by incorporating dNTPs; eventually, the entire plasmid is replicated. Because any unreacted plasmid DNA will be the source of non-mutated plasmids, the restriction endonuclease *Dpn*I is used to digest unreplicated plasmid DNA. *Dpn*I has low activity on hemi-methylated DNA and its catalytic activity occurs at a slow rate [8].

The *Dpn*I-digested DNA with the mutation(s) is then transformed into *E. coli* where biological replication yields mutated and non-mutated progeny DNA. The mutant plasmid is distinguished by DNA sequencing. In our optimized procedure, we routinely found that the yield of mutant plasmid ranges from 25 to 50% of the plasmids sequenced [9, 10].

In summary, we optimized a single-mutagenic primer method for SDM that yields mutant genes in ~25–50% of transformants with a robust, low-cost reaction. In this optimized procedure,

Phusion DNA polymerase, a thermostable DNA polymerase with high fidelity and strong processivity, is needed while the supercompetent cells are not required.

2   Materials

Use ultrapure water with a resistance of 18 MΩ at 25 °C to prepare all solutions. Analytical grade reagents and solutions are usually stored at room temperature (unless otherwise specified). All waste materials are disposed of according to proper waste removal procedures. All containers for storing solutions are cleaned with detergent, thoroughly washed and rinsed with ultrapure water three times before using.

2.1 Molecular Biology Reagents

1. Phusion DNA polymerase (Thermo Fisher Scientific) (*see* **Note 1**).

2. 5× HF (High-Fidelity) buffer: this buffer is supplied by Thermo Fisher Scientific and stored at –20 °C (*see* **Note 2**).

3. 2.5 mM dNTPs: Mix 25 μl of 100 mM dATP, dCTP, dGTP, and dTTP in a 1.5 ml eppendorf centrifuge tube. Add 900 μl ultrapure water to the tube to make a final concentration of 2.5 mM dNTPs (*see* **Note 3**).

4. Plasmid DNA purification kit (*see* **Note 4**).

5. Mutagenic primer (*see* **Note 5**).

6. *Dpn*I restriction enzyme (New England Biolabs, USA) (*see* **Note 6**).

2.2 DNA Transformation

1. Competent *E. coli* DH5α cells: The competent cells were prepared as described below (*see* **Note 7**).

2. 0.1 M $CaCl_2$ solution: Weigh 5.549 g $CaCl_2$ and transfer to a 1 l graduated glass beaker. Add ultrapure water up to 500 ml and dissolve completely by using a magnetic stir bar. Autoclave this solution at 121 °C for 20 min and store at room temperature (*see* **Note 8**).

3. 80% glycerol: Mix 20 ml glycerol with 5 ml ultrapure water with the volume. Autoclave 80% Glycerol at 121 °C for 20 min and store at room temperature (*see* **Note 8**).

4. Lysogeny broth (LB) medium: 10 g/l tryptone, 5.0 g/l yeast extract, 10 g/l NaCl and transfer to a 1 l graduated glass beaker respectively. Add 980 ml double-distilled water to a volume of 980 ml and pH to 7.0 with 5 N NaOH. Raise final volume to 1 l with ddH_2O. Autoclave at 121 °C for 20 min and store at room temperature.

5. LB solid medium with appropriate antibiotic: Weigh 2 g tryptone, 1 g yeast extract, 2 g NaCl, 1.5 g agar in a 250 ml conical

flask with 100 ml water. Dissolve by stirring and add water up to a final volume of 200 ml. Seal it with foil and sterilize at 121 °C for 20 min and store at room temperature. Cool down to 50 °C, and then add 200 μl appropriate antibiotic solution and swirl to mix completely (*see* **Note 9**). Pour about 20 ml LB medium into the sterile petri plates. After solidifying, these LB plates were stored at 4 °C (*see* **Note 10**).

3 Methods

All experiments were carried out at room temperature unless otherwise specified.

3.1 PCR

1. DNA amplification components are summarized in Table 1. Thaw all components on ice and add them in a 0.5 ml PCR tube.

2. Insert the PCR tube into a block heater of a thermal cycler (*see* **Note 11**).

3. Run PCR cycling program as described in Table 1. When 95 °C for 3 min is finished, add 0.5 μl Phusion DNA polymerase and mix gently (*see* **Note 12**).

4. When the PCR is complete, vortex and spin down the PCR tube (*see* **Note 13**).

Table 1
Single primer site-directed mutagenesis method [9, 10]

Components	Volume	Comments
H_2O	x μl (volume to produce a final volume of 20 μl)	Mix first five reaction components in a 0.5 ml PCR tube
5× HF buffer[a]	4 μl	
Plasmid DNA	x μl (2 nM)	
Mutagenic primer	x μl (1 μM)	
dNTPs (2.5 mM)	1.5 μl (188 μM)	
Phusion DNA polymerase[b]	0.5 μl (1 U)	Heat reaction for 2 min at 95 °C. Add DNA polymerase. Mix by pipetting up-and-down
Final volume	20 μl	
Thermal cycling program: 95 °C for 1 min, 55 °C for 1 min, 65 °C for 12 min; 30 cycles		
*Dpn*I[c]	1 μl (1 U)	37 °C, 2 h

[a]5× HF (High fidelity) buffer was supplied by the manufacturer (Thermo Fisher Scientific) of Phusion DNA polymerase
[b]Phusion DNA polymerase was purchased from Thermo Fisher Scientific
[c]*Dpn*I was purchased from New England Biolabs, USA

3.2 DpnI Digestion

1. Pipette 1 µl restriction endonuclease *Dpn*I into the DNA replication product and mix gently (*see* **Note 14**). Incubate it in a water bath at 37 °C for 2 h to remove any non-replicated, original plasmid DNA, which would generate non-mutant plasmids.

3.3 DNA Transformation

1. Make competent *E. coli* cells as follows. Transfer *E. coli* DH5α cells from a –80 °C stock into a 20 cm tube containing 5 ml liquid LB medium with a sterile bamboo stick. Culture overnight (~16 h) at 37 °C with shaking at 200 rpm.

2. Transfer 2.5 ml of the overnight culture into a 1-l flask containing 250 ml liquid LB medium. Incubate at 37 °C, 200 rpm overnight. Collect the cells and put flask on ice when an OD_{600} value of 0.4 is reached (about 2 h) (*see* **Note 15**).

3. The cells are spun down using a refrigerated centrifuge at 4000 x *g* at 4 °C for 10 min. The supernatant is discarded and the pellets are resuspended in 30 ml chilled 0.1 M $CaCl_2$.

4. Repeat the centrifugation as described in **step 3** above.

5. Add 8 ml chilled 0.1 M $CaCl_2$ and 2 ml 80% glycerol. Pipette gently to mix.

6. Chill 1.5 ml sterile tubes on ice. Aliquot 100 µl cells/tube and mark the competent cell name and date on the outside of the freezer box. Store them at –80 °C (*see* **Note 16**).

7. Take a tube of competent *E. coli* DH5α cells from –80 °C and put it on ice for 20 min to thaw.

8. Pipette 10 µl of digested DNA PCR products into 100 µl thawed competent *E. coli* cells (*see* **Note 17**). Gently mix and put it on ice for 10 min.

9. Heat this tube in a water bath at 42 °C in a digital block heater for 90 s (*see* **Note 18**). Put cells immediately back on ice for 2 min.

10. Add 150 µl liquid LB medium to the tube and shake gently on an incubator at ~200 rpm at 37 °C for 1 h.

11. Pipette 150 µl of culture onto a LB agar plate containing the appropriate antibiotic. Incubate at 37 °C overnight (*see* **Note 19**).

3.4 Plasmid Sequencing

1. Pick four typical single colonies from the LB plate into 5 ml LB medium containing appropriate antibiotics using a wooden stick or sterile applicator. Incubate the tubes at 37 °C at 200 rpm for 16 h.

2. Pipette 1 ml of the overnight culture into a 1.5 ml Eppendorf tube. Centrifuge at $14,000 \times g$ for 2 min and discard the supernatant. Repeat this step three times (*see* **Note 20**).

3. The plasmid DNA are then purified using a mini-prep plasmid purification kit.

4. Measure the plasmid DNA concentration by Nanodrop reading. Store the plasmid DNA at –20 °C.

5. Sequence the plasmid region where the site-directed mutation was incorporated for confirmation (*see* Fig. 1) (*see* **Note 21**).

---

4   Notes

1. Phusion DNA polymerase is mixed with other DNA replication components by pipetting gently to avoid enzyme inactivation. Phusion DNA polymerase, available from Thermo Fisher Scientific, is recommended because of high fidelity and strong processivity of this enzyme.

2. Mg^{2+} is usually needed for DNA replication catalyzed by DNA polymerase. Mg^{2+} is included in this buffer but the concentration might need to be optimized to yield the highest PCR product concentration.

3. Aliquot dNTPs into small tubes to prevent freezing and thawing repeatedly and store them at –20 °C. Before using, dNTPs must be immediately put on ice to thaw.

4. The covalently closed circular plasmid DNA is best as a template for DNA amplification by Phusion DNA polymerase in our optimized procedure because replication of nicked plasmid by the enzyme would be a source of non-mutant plasmids.

5. The mutation is centralized in the mutagenic primer so that when it is annealed to its target site the mismatch(es) are also centrally located. A mutagenic primer is designed to give a melting temperature (T_m) greater than the annealing temperature of 55 °C. The 3′-end of the mutagenic primer has one or more terminal G or C bases, which will form GC base pairs when the primer is annealed, to enable efficient extension by the polymerase. Even more importantly, the 5′-end of the primer must also have one or more G or C bases because stable annealing is required to prevent the polymerase from dislodging the mutagenic primer after completing one round of synthesis and beginning a second round, which will be a source of non-mutant plasmids. The mutagenic primer must also be phosphorylated at the 5′-end for DNA ligation, which is carried out in vivo by the bacterial DNA ligase. This will significantly increase the yield of mutated plasmids.

6. *Dpn*I selectively cleaves DNA that is methylated on both strands at G^{ME} A↓T C sites [11] and the methylated strand of hemi-methylated DNA, but at a 60-fold slower rate [8].

Thus, DNA synthesized in the reaction is resistant to cleavage. *Dpn*I can convert non-replicated plasmid DNA into several *Dpn*I restriction fragments.

7. The supercompetent XL-10 Gold cells are not needed in our optimized procedure. Competent *E. coli* DH5α cells are enough to produce a large number of transformants.

8. Before autoclaving solutions, the lids of bottles are loosened to prevent excessive pressure build up in the container. After autoclaving, screw the lids on tightly to avoid contamination.

9. Most antibiotic solutions cannot be autoclaved and are sterilized using a 0.22 μM filter. Antibiotic solutions must be added when the temperature of the LB medium reaches about 50 °C.

10. LB solid plates containing antibiotics can be stored for about 4 weeks at 4 °C.

11. Twist the lid of the thermal cycler carefully to avoid distortion of the PCR tubes.

12. A hot-start program is used in our optimized procedure, which may allow Phusion DNA polymerase to replicate DNA with high efficiency.

13. The extension time should be designated as 2 min/kb. Thirty cycles is usually sufficient to produce amplicons for transformation. The length of time for primer extension depends on the size of the plasmid. Twelve minutes is sufficient for a 6–9 kb plasmid.

14. *Dpn*I can be added directly into the DNA replication product because the polymerase buffer is compatible.

15. The transformation efficiency of competent cells will be reduced when using cells harvested at an OD_{600} of more than 0.4.

16. The chemically competent cells should be stored for less than 6 months at −80 °C and typically the transformation frequency is about 1×10^5 μg^{-1} plasmid DNA.

17. 10 μl of digested DNA replication product is enough for generating transformants during transformation. Use 100 μl of competent cells for each transformation of the 10 μl digested DNA replication product. Ideally, about 100 colonies are observed on the LB plate with the appropriate antibiotic(s).

18. Heat shocking is the critical step in chemical transformation. The heat-shock temperature must be carefully controlled in a 42 °C water bath.

19. Incubation time may be prolonged appropriately if colonies are too small.

20. Three milliliter overnight culture is enough to produce enough plasmid DNA for sequencing.

21. Sequences must contain the targeted mutation(s) and have no extra mutation(s).

Acknowledgments

National Natural Science Foundation of China Grant (No. 41306131), Natural Science Foundation for College and University of Jiangsu Province (No. 13KJB180029), and Provincial Natural Science Foundation of Jiangsu, China grant (No. BK20130440) are acknowledged.

References

1. Hutchison CA 3rd, Phillips S, Edgell MH, Gillam S, Jahnke P, Smith M (1978) Mutagenesis at a specific position in a DNA sequence. J Biol Chem 253:6551–6560

2. Kunkel TA (1985) Rapid and efficient site-specific mutagenesis without phenotypic selection. Proc Natl Acad Sci U S A 82:488–492

3. Kunkel TA, Roberts JD, Zakour RA (1987) Rapid and efficient site-specific mutagenesis without phenotypic selection. Methods Enzymol 154:367–382

4. Xia Y, Chu W, Qi Q, Xun L (2014) New insights into the QuikChange™ process guide the use of Phusion DNA polymerase for site-directed mutagenesis. Nucleic Acids Res 43, e12

5. QuikChange® Multi Site-Directed Mutagenesis Kit. Instruction manual. http://www.genomics.agilent.com/files/Manual/200514.pdf

6. Kamal F, Zhang L, Reha-Krantz LJ (2013) Escherichia coli XL10-gold bacteria produce bacteriophage. J Clin Microbiol 51:727

7. Wang Y, Prosen DE, Mei L, Sullivan JC, Finney M, Vander Horn PB (2004) A novel strategy to engineer DNA polymerases for enhanced processivity and improved performance in vitro. Nucleic Acids Res 32:1197–207

8. New England BioLabs. http://www.neb.com/nebecomm/products/faqproductR0176.asp#10

9. Zhang L, Radziwon A, Reha-Krantz LJ (2014) Targeted mutagenesis of a specific gene in yeast. Methods Mol Biol 1163:109–129

10. Zhang L, Kang M, Xu J, Huang Y (2015) Archaeal DNA polymerases in biotechnology. Appl Microbiol Biotechnol 99:6585–6597

11. Smith DW, Garland AM, Herman G, Enns RE, Baker TA, Zyskind JW (1985) Importance of methylation of oriC GATC sites in initiation of DNA replication in Escherichia coli. EMBO J 4:1319–1326

Chapter 27

Use of Megaprimer and Overlapping Extension PCR (OE-PCR) to Mutagenize and Enhance Cyclodextrin Glucosyltransferase (CGTase) Function

Kian Mau Goh, Kok Jun Liew, Kian Piaw Chai, and Rosli Md Illias

Abstract

Protein engineering is a very useful tool for probing structure–function relationships in proteins. Specifically, site-directed mutagenized proteins can provide useful insights into structural, binding and catalytic mechanisms of a protein, particularly when coupled with crystallization. In this chapter, we describe two protocols for performing site-directed mutagenesis of any protein-coding sequence, namely, megaprimer PCR and overlapping extension PCR (OE-PCR). We use as an example how these two SDM methods enhanced the function of a cyclodextrin glucosyltransferase (CGTase) from *Bacillus lehensis* strain G1.

Key words *Bacillus*, CGTase, Megaprimer PCR, Overlapping extension PCR, Protein engineering, Rational design

1 Introduction

Protein engineering is a tool used routinely to generate advances in science, medicine, and the bioprocess industries [1]. The overall aim of modifying proteins is to generate new or improved properties which may include alteration in stability, catalytic activity, receptor binding, and substrate or product specificity [1].

Several protein engineering approaches have been developed and are used extensively to introduce mutation(s) into a protein of interest [2, 3]. Among them, megaprimer and overlapping extension PCR (OE-PCR) are by far the easiest methods to replace codons (*see* **Note 1**), especially for novice researchers [2–5]. The megaprimer approach involves two rounds of PCR, the first PCR is to introduce mutations while the second PCR is to generate the complete gene [4]. Through its simplicity, the length of the megaprimer (often <600 bp) obtained from the first PCR improves the efficiency of the second PCR. In contrast, OE-PCR comprises three rounds of PCR. Mutations are introduced into the gene

Andrew Reeves (ed.), *In Vitro Mutagenesis: Methods and Protocols*, Methods in Molecular Biology, vol. 1498,
DOI 10.1007/978-1-4939-6472-7_27, © Springer Science+Business Media New York 2017

during the first and second PCR, while the third PCR is used to generate the whole gene [5]. With little optimization, OE-PCR can be used to introduce multiple point mutations, deletions, insertions, as well as gene fusions.

CGTase is an industrial enzyme with great commercial potential and its applications have been reviewed extensively [6]. Due to its wide application, numerous mutagenesis studies have been carried out [6], particularly on its product specificity [7] and thermostability [8]. In this chapter, we describe megaprimer and overlapping extension PCR protocols used in the construction of two CGTase mutants.

2 Materials

2.1 Recombinant Plasmid

1. Recombinant plasmid, pET-22b (+) harboring CGTase from *Bacillus lehensis* G1.

2.2 Molecular Biology Reagents

1. High-fidelity Phusion® Flash DNA polymerase Master Mix (New England Biolabs).
2. Hi-fidelity *Bam*HI and *Not*I restriction enzymes.
3. T4 DNA Ligase Master Mix.
4. SYBR Safe DNA stain.
5. PCR purification kit.
6. Gel DNA extraction kit.
7. Plasmid miniprep kit.
8. *E. coli* DH5α and BL21 (DE3) competent cells.

2.3 Equipment

1. Thermal cycler.
2. 1.5 ml microcentrifuge and 0.2 ml PCR tubes.
3. Horizontal agarose gel electrophoresis system.
4. UV transilluminator or gel documentation system.
5. DNA analyser such as a NanoDrop® spectrometer.

3 Methods

3.1 CGTase Sequence

1. Retrieve the nucleotide sequence of *Bacillus lehensis* G1 CGTase (GenBank accession number: AY770576.1) from the NCBI database.
2. Translate the nucleotide sequence into protein sequence (*see* Fig. 1) using an online program such as ExPASy translate (http://web.expasy.org/translate/).

```
gacgtaacaaacaaagtcaattactcaaaagatgtgatttaccaggttgttaccgatcga
 D  V  T  N  K  V  N  Y  S  K  D  V  I  Y  Q  V  V  T  D  R
ttctctgacgggaatcctggcaacaatccttcaggcgctatctttagtcaaaactgtata
 F  S  D  G  N  P  G  N  N  P  S  G  A  I  F  S  Q  N  C  I
gatcttcataagtattgtggtggggactggcaagggattatagacaaaatcaatgacggt
 D  L  H  K  Y  C  G  G  D  W  Q  G  I  I  D  K  I  N  D  G
tacttaactgatttaggcattacggcactatggatttctcagccagtcgaaaacgtttat
 Y  L  T  D  L  G  I  T  A  L  W  I  S  Q  P  V  E  N  V  Y
gccctacacccaagcggctatacctcctaccatggatattgggctcgagattacaaaaag
 A  L  H  P  S  G  Y  T  S  Y  H  G  Y  W  A  R  D  Y  K  K
acaaaccccttactatgggaattttgatgactttgatcgtttaatgagtaccgcacatagc
 T  N  P  Y  Y  G  N  F  D  D  F  D  R  L  M  S  T  A  H  S
aatgggataaaggtaatcatggatttcacgccaaatcattcatcaccggcacttgaaacg
 N  G  I  K  V  I  M  D  F  T  P  N  H  S  S  P  A  L  E  T
aaccctaactatgttgaaaatggggcgatatatgataatggcacattattaggtaactat
 N  P  N  Y  V  E  N  G  A  I  Y  D  N  G  T  L  L  G  N  Y
tcaaatgatcaacaaaacctctttcaccacaatggcggaacagatttctcttcatatgaa
 S  N  D  Q  Q  N  L  F  H  H  N  G  G  T  D  F  S  S  Y  E
gatagcatttacagaaacttatatgatctggcagactatgatttaaacaacacagtcatg
 D  S  I  Y  R  N  L  Y  D  L  A  D  Y  D  L  N  N  T  V  M
gatcaatatttaaaagagtcgattaagttctggttagataaagggattgatggcattcga
 D  Q  Y  L  K  E  S  I  K  F  W  L  D  K  G  I  D  G  I  R
gtagatgccgttaagcatatgtcagaagggtggcaaaccctcttaatgagcgaaatctat
 V  D  A  V  K  H  M  S  E  G  W  Q  T  S  L  M  S  E  I  Y
tcgcataaacctgttttcacatttggagaatggtttttaggatcaggagagttgatccc
 S  H  K  P  V  F  T  F  G  E  W  F  L  G  S  G  E  V  D  P
caaaatcatcacttcgctaatgaaagtggtatgagttttattagatttccaattcggtcaa
 Q  N  H  H  F  A  N  E  S  G  M  S  L  L  D  F  Q  F  G  Q
accattcgtaacgtcttaaaagatcgcacaagcaactggtatgattttaatgaaatgatt
 T  I  R  N  V  L  K  D  R  T  S  N  W  Y  D  F  N  E  M  I
accagtacagaaaaagagtataacgaggtcattgatcaagtaacctttattgataatcac
 T  S  T  E  K  E  Y  N  E  V  I  D  Q  V  T  F  I  D  N  H
gacatgagtcgtttttcggtaggatcatcttcaaaccgtcagacagatatggccctagct
 D  M  S  R  F  S  V  G  S  S  S  N  R  Q  T  D  M  A  L  A
gtcttgcttacttctcgtggtgtaccaacgatttactacgggacagagcagtatgtaaca
 V  L  L  T  S  R  G  V  P  T  I  Y  Y  G  T  E  Q  Y  V  T
ggtggcaacgaccctgaaaatcgcaaaccattgaaaacatttgatcggtctaccaactcc
 G  G  N  D  P  E  N  R  K  P  L  K  T  F  D  R  S  T  N  S
tatcaaatcatcagtaaacttgcttcactacgccaaacaaattccgccttaggctatggc
 Y  Q  I  I  S  K  L  A  S  L  R  Q  T  N  S  A  L  G  Y  G
actacaactgaacgttggctgaacgaagacatttatatttatgaaagaacgtttggcaat
 T  T  T  E  R  W  L  N  E  D  I  Y  I  Y  E  R  T  F  G  N
agtattgtattaactgctgtaaatagcagtaatagtaaccagacgatcactaatttaaac
 S  I  V  L  T  A  V  N  S  S  N  S  N  Q  T  I  T  N  L  N
acctctttacctcaaggaactatacagatgaactacagcaacgtttagatggaaacacg
 T  S  L  P  Q  G  N  Y  T  D  E  L  Q  Q  R  L  D  G  N  T
attactgttaacgccaatggagccgtaaattcctttcaattacgagcaaatagcgtagcg
 I  T  V  N  A  N  G  A  V  N  S  F  Q  L  R  A  N  S  V  A
gtttggcaagtaagcaacccctctacgtctcctctaatcggccaagtgggtcctatgatg
 V  W  Q  V  S  N  P  S  T  S  P  L  I  G  Q  V  G  P  M  M
ggtaagtccgggaataccataacagtaagcggtgaaggatttggtgatgagagaggaagc
 G  K  S  G  N  T  I  T  V  S  G  E  G  F  G  D  E  R  G  S
gttctctttgattcaacctcttctgaaattatttcttggtcaaatacagaaataagcgta
 V  L  F  D  S  T  S  S  E  I  I  S  W  S  N  T  E  I  S  V
aaggtgcctaatgtagcaggcggttattatgatctatccgtcgtaactgcagcaaactta
 K  V  P  N  V  A  G  G  Y  Y  D  L  S  V  V  T  A  A  N  L
aaaagccctacttacaaagagtttgaagtattgtcaggcaatcaagtcagtgtccgctttt
 K  S  P  T  Y  K  E  F  E  V  L  S  G  N  Q  V  S  V  R  F
ggtgttaacaatgccacaacgagcccaggaaccaatttatatatcgttgggaatgtgagc
 G  V  N  N  A  T  T  S  P  G  T  N  L  Y  I  V  G  N  V  S
gagctggggaattgggatgctgataaagcaattggacctatgtttaaccaagtgatgtac
 E  L  G  N  W  D  A  D  K  A  I  G  P  M  F  N  Q  V  M  Y
caataccccaacatggtactatgatattagcgttcctgccggaaaaaaaccttgaatacaaa
 Q  Y  P  T  W  Y  Y  D  I  S  V  P  A  G  K  N  L  E  Y  K
tacattaaaaaagatcagaacggtaacgttgtctggcaaagtggcaacaatcgaacctat
 Y  I  K  K  D  Q  N  G  N  V  V  W  Q  S  G  N  N  R  T  Y
acgtcgcctactaccggaacagatacggttatgattaattggtaa
 T  S  P  T  T  G  T  D  T  V  M  I  N  W  -
```

Fig. 1 Protein sequence of *Bacillus lehensis* G1 CGTase

3. Retrieve the sequences of γ-cyclodextrin producing CGTases that are homologous to *Bacillus lehensis* G1 CGTase by searching against the NCBI database using BlastP.

4. Perform a multiple sequence alignment (*see* Fig. 2) using an online program such as Clustal Omega (http://www.ebi.ac. uk/Tools/msa/clustalo/).

3.2 Primer Design

1. Design and synthesize several forward and reverse primers for mutagenesis grade purity based on the multiple sequence alignment shown in Fig. 2 and in accordance with Table 1 (*see* **Notes 2–7**).

3.3 Megaprimer PCR Protocol

The megaprimer amplification scheme is outlined in Fig. 3.

1. For the first PCR step, transfer 25 ng of DNA template, 25 μl 2× Phusion™ Flash PCR Master Mix, 0.5 μM EFP_H43T primer, and 0.5 μM MIRP_H43T primer into a 0.2 ml thin-wall PCR tube. Add nuclease-free water to a final volume of 50 μl.

2. Perform the PCR according to the conditions stated in Table 2 (*see* **Notes 8–12**).

3. Analyze the amplicon by loading it into a standard 20–26 cm 1% agarose gel and perform gel electrophoresis at 80 V for 50 min.

4. Gel-purify the amplicon from the agarose gel using a DNA-gel extraction kit (*see* **Note 13**).

5. For the second PCR, mix 25 ng DNA template, 25 μl 2× Phusion™ Flash PCR Master Mix, 0.5 μM ERP_H43T primer, and 0.5 μM purified amplicon from the first round of PCR in a 0.2 ml thin-wall PCR tube. Add nuclease-free water to a final volume of 50 μl (*see* **Note 14**).

6. Perform the PCR according to the conditions described in Table 2.

7. Analyze the amplicon by loading it into a 1% agarose gel and perform gel electrophoresis at 80 V for 50 min.

8. Purify the amplicon using a PCR purification kit.

3.4 OE-PCR

The overlapping extension PCR scheme is illustrated in Fig. 4.

1. For the first PCR, mix 25 ng DNA template, 25 μl 2× Phusion™ Flash PCR Master Mix, 0.5 μM EFP_Δ139–144 primer, and 0.5 μM MIRP_Δ139–144 primer in a 0.2 ml thin-wall PCR tube. Add nuclease-free water to a final volume of 50 μl.

2. For the second PCR, mix 25 ng DNA template, 25 μl 2× Phusion™ Flash PCR Master Mix, 0.5 μM ERP_Δ139–144 primer, and 0.5 μM MIFP_Δ139–144 primer in a 0.2 ml

```
G1          ----DVTNKVNYSKDVIYQVVTDRFSDGNPGNNPSGAIFSQNCIDLHKYCGGDWQGIIDK
Clarkii     SNATNDLSNVNYAEEVIYHIVTDRFKDGDPDNNPQGQLFSNGCSDLTKYCGGDWQGIIDE
290-3       ---NENLDNVNYAQEIIYQIVTDRFYDGDPTNNPEGTLFSPGCLDLTKYCGGDWQGVIEK
            :   .:***:::*:*****  **:* ***.* :**   * ** *********:*::

G1          INDGYLTDLGITALWISQPVENVYALHPSGYTSYHGYWARDYKKTNPYYGNFDDFDRLMS
Clarkii     IESGYLPDMGITALWISPPVENVFDLHPEGFSSYHGYWARDFKKTNPFFGDFDDFSRLIE
290-3       IEDGYLPDMGITAIWISPPIENVMELHPGGFASYHGYWGRDFKRTNPAFGSLADFSRLIE
            *:.*** *:****:*** *:***  *** *::****** .**:*:*** :*:.: **.**:.

G1          TAHSNGIKVIMDFTPNHSSPALETNPNYVENGAIYDNGTLLGNYSNDQQNLFHHNGGTDF
Clarkii     TAHAHDIKVVIDFVPNHTSPVD------IEDGALYDNGTLLGHYSTDANNYFYNYGGSDF
290-3       TAHNHDIKVIIDFVPNHTSPVD------IENGALYDNGRLVGHYSNDSEDYFYTNGGSDF
            *** . ***::**.***:**.           :*:**:**** *:*.**.* :: *:  **:**

G1          SSYEDSIYRNLYDLADYDLNNTVMDQYLKESIKFWLDKGIDGIRVDAVKHMSEGWQTSLM
Clarkii     SDYENSIYRNLYDLASLNQQHSFIDKYLKESIQLWLDTGIDGIRVDAVAHMPLGWQKAFI
290-3       SSYEDSIYRNLYDLASLNQQNSFIDRYLKESIQMWLDLGIDGIRVDAVAHMPVGWQKNFV
            *.**:**********.  : :.:.:*:****** :*** ********** **  ***. ::

G1          SEIYSHKPVFTFGEWFLGSGEVDPQNHHFANESGMSLLDFQFGQTIRNVLKDRTSNWYDF
Clarkii     SSVYDYNPVFTFGEWFTGAQGSN-HYHHFVNNSGMSALDFRYAQVAQDVLRNQKGTMHDI
290-3       SSIYDYNPVFTFGEWFTGAGGSD-EYHYFINNSGMSALDFRYAQVVQDVLRNNDGTMYDL
            *.:*.::********** *:    :  *:* *:****  ***::.*.  .:**::..  ..  :*:

G1          NEMITSTEKEYNEVIDQVTFIDNHDMSRFSVGSSSNRQTDMALAVLLTSRGVPTIYYGTE
Clarkii     YDMLASTQLDYERPQDQVTFIDNHDIDRFTVEGRDTRTTDIGLAFLLTSRGVPAIYYGTE
290-3       ETVLRETESVYEKPQDQVTFIDNHDINRFSRNGHSTRTTDLGLAFLLTSRGVPTIYYGTE
            ::  .*:  *:.  **********:.**:      . .. * **:.**.*******:******

G1          QYVTGGNDPENRKPLKTFDRSTNSYQIISKLASLRQTNSALGYGTTTERWLNEDIYIYER
Clarkii     NYMTGKGDPGNRKMMESFDQTTTAYQVIQKLAPLRQENKAVAYGSTKERWINDDVLIYER
290-3       IYMTGDGDPDNRKMMNTFDQSTVAYQIIQQLSSLRQENRAIAYGDTTERWINEDVFIYER
            *:**  ** *** ::::**::* :**:*.:** * *:.** *.:**  *.***:*:*: ****

G1          TFGNSIVLTAVNSS-NSNQTITNLNTSLFQGNYTDELQQRLDCNTITVNANGAVNSFQLR
Clarkii     SFNGDYLLVAINKNVNQAYTISGLLTEMPAQVYHDVLDSLLDGQSLAVKENGTVDSFLLG
290-3       SFNGEYALIAVNRSLNHSYQISSLVTDMPSQLYEDELSGLLDGQSITVDQNGSIQPFLLA
            :*   .  * *:*. .*  *:  .*   *  * *   ***::**. **::: * *

G1          ANSVAVWQVSNP-STSPLIGQVGPMMGKSGNTITVSGEGFGDERGSVLFDSTSSEIISWS
Clarkii     PGEVSVWQHISESGSAPVIGQVGPPMGKPGDAVKISGSGFGSEPGTVYFRDTKIDVLTWD
290-3       PGEVSVWQYSNGQNVAPEIGQIGPPIGKPGDEVRIDGSGFGSSTGDVSFAGSTMNVLSWN
            .*:***  .    :* ***:** :** *: : :.*.***.. * * *  :. :::*.

G1          NTEISVKVPNVAGGYYDLSVVTAANLKSPTYKEFEVLSGNQVSVRFGVNNATTSPGTNLY
Clarkii     DETIVITLPETLGGKAQISVTNSDGVTSNGY-DFQLLTGKQESVRFVVDNAHTNYGENVY
290-3       DDTIIAELPEHNGGKNSVTVTTNSGESSNGY-PFELLTGLQTSVRFVVNQAETSVGENLY
            :  *   :*:  **  .::*:*..   .* * *  *:*:* * ****.*::*  *. * *:*

G1          IVGNVSELGNWDADKAIGPMFNQVMYQYPTWYYDISVPAGKNLEYKYIKKDQNGNVVWQS
Clarkii     LVGNVPELGNWNPADAIGPMFNQVVYSYPTWYYDVSVPADTALEFKFIIVDGNGNVTWES
290-3       VVGDVPELGSWDPDKAIGPMFNQVLYSYPTWYYDVSVPANQDIEYKYIMKDQNGNVSWES
            :**:* ***.*: .*********:**.******** :*:* *   **** *:*

G1          GNNRTYTSPTTGTDTVMINW--
Clarkii     GGNHNYRVTSGSTDTVRVSFRR
290-3       GNNHIYRTPENSTGIVEVNFNQ
            * *:. *     .* * *.:.:
```

Fig. 2 Multiple sequence alignment of *Bacillus lehensis* G1 CGTase (AY770576.1) with other γ-cyclodextrin-producing CGTases including *Bacillus clarkii* CGTase (AB082929.2) and *Bacillus* 290-3 CGTase (A18991.1)

Table 1
Primers for megaprimer and overlap extension PCR

Primer	Primer sequence	Melting temp. (°C)
EFP_H43T	5′-ATC<u>GGATCC</u>GACGTAACAAACAAAGT CAATTACTCAAAAG-3′	65.6
ERP_H43T	5′-AT<u>GCGGCCGC</u>CCAATTAATCATAACC GTATCTGTTCC-3′	63.7
MIRP_H43T	5′-CCACCACAATACTT**CGT**AAGATCTATAC-3′	65.6
EFP_Δ139–144	5′-ATC<u>GGATCC</u>GACGTAACAAACAAAGTCAA TTACTCAAAAGATGTG-3′	69.7
ERP_Δ139–144	5′-AT<u>GCGGCCGC</u>CCAATTAATCATAACCGTAT CTGTTCCGGTAG-3′	70.1
MIFP_Δ139–144	5′-ccggcacttGTTGAAAATGGGGCGATATATG-3′	72.1
MIRP_Δ139–144	5′-CATTTTCAACaagtgccggtgatgaatg-3′	67.2

ªThe overlapping region for OE-PCR is represented in small letters

Table 2
Megaprimer PCR conditions

Conditions	First PCR	Second PCR
Initial denaturation	98 °C, 10 s	98 °C, 10 s
Denaturation	98 °C, 1 s	98 °C, 1 s
Annealing	60 °C, 5 s	60 °C, 5 s
Extension	72 °C, 10 s	72 °C, 25 s
Final extension	72 °C, 60 s	72 °C, 60 s

thin-wall PCR tube. Add nuclease-free water to a final volume of 50 μl.

3. Perform PCR according to the conditions shown in Table 3 (*see* **Note 4**).

4. Analyze the amplicon from the first and second PCR rounds by loading the amplicons into a 3% agarose gel and perform gel electrophoresis at 80 V for 50 min.

5. Gel-purify the amplicons generated from the first and second PCR using a gel extraction kit (*see* **Note 13**).

6. For the third PCR, mix equimolar amplicons from the first and second PCR with 25 μl 2× Phusion™ Flash PCR Master Mix, 0.5 μM EFP_ Δ139–144 primer, and 0.5 μM ERP_ Δ139–144 primer in a 0.2 ml thin-wall PCR tube. Add nuclease-free water to a final volume of 50 μl.

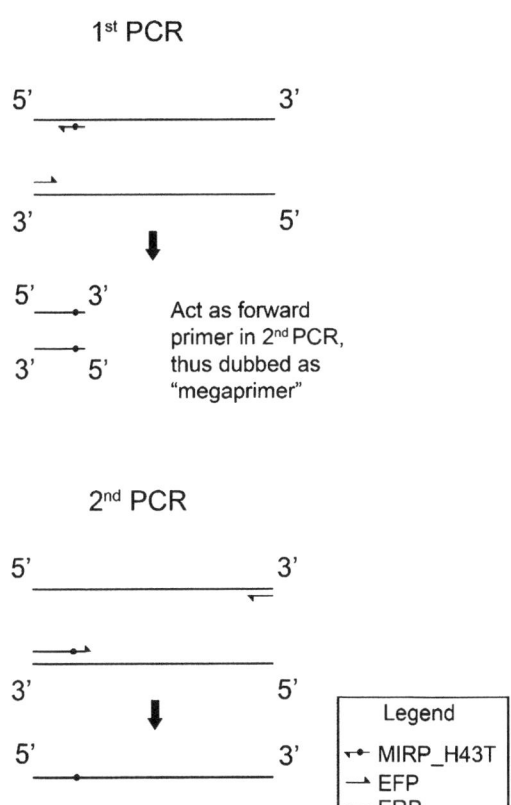

Fig. 3 Simplified flow chart of the megaprimer PCR approach for H43T mutation

7. Perform a PCR according to the reaction conditions shown in Table 3.

8. Analyze the amplicon by loading it into a 1 % agarose gel and perform gel electrophoresis at 80 V for 50 min.

9. Purify the amplicon using a PCR purification kit.

3.5 DNA Transformation

Perform a simultaneous digestion on the purified amplicon (generated from the final round of PCR in megaprimer and overlap extension PCR) and pET-22b (+) using Hi-fidelity *BamH*I and *Not*I.

1. Purify the digested DNA using a DNA clean up or PCR purification kit in accordance with manufacturer's instructions.

2. Perform a ligation reaction by mixing in a 1:3 ratio of purified digested amplicon and pET-22b (+) using a T4 DNA ligase Master Mix. Incubate at 16 °C for 1 h to overnight.

3. Transform the ligated pET-22b (+) into electrocompetent or chemically competent *E. coli* DH5α and select on LB plates containing ampicillin sulfate (100 μg/mL).

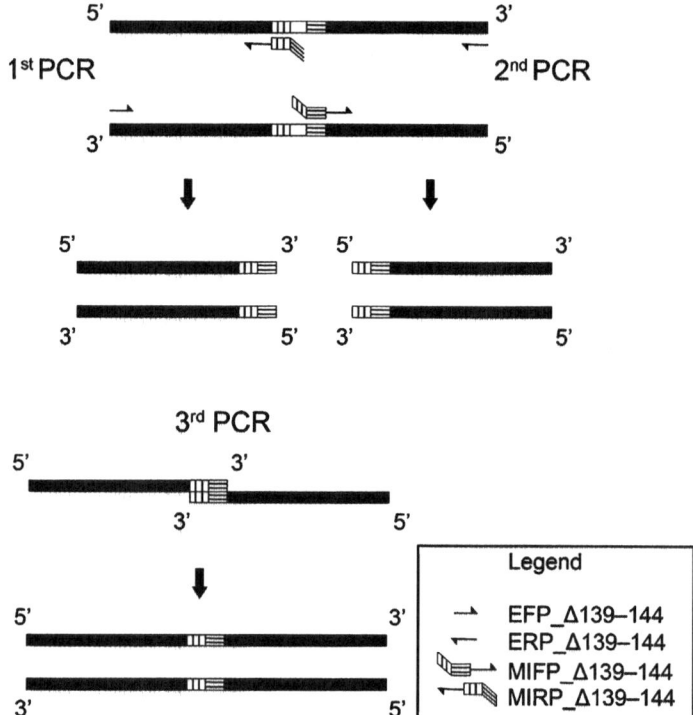

Fig. 4 Simplified flow chart of the OE-PCR approach for generating the loop 139–143 deletion

Table 3
Overlap extension PCR conditions

Conditions	First PCR	Second PCR	Third PCR
Initial denaturation	98 °C, 10 s	98 °C, 10 s	98 °C, 10 s
Denaturation	98 °C, 1 s	98 °C, 1 s	98 °C, 1 s
Annealing	60 °C, 5 s	60 °C, 5 s	60 °C, 5 s
Extension	72 °C, 15 s	72 °C, 25 s	72 °C, 30 s
Final extension	72 °C, 60 s	72 °C, 60 s	72 °C, 60 s

4. Perform colony PCR to screen for recombinant pET-22b (+) and subsequently confirm the mutation by DNA sequencing.

5. Grow single colony isolates of several positive recombinants in LB-amp-(100 μg/mL) medium and purify plasmids for transformation using a plasmid miniprep kit.

6. Transform purified recombinant plasmids into *E. coli* BL21 (DE3) competent cells (via electroporation or chemically competent cells) and plate on an LB-ampicillin (100 μg/mL) plate.

4 Notes

1. A codon table and general properties of amino acids can be obtained from biochemistry textbooks or any reputable internet sources of academic origin.

2. Read the general requirements for good PCR primer design which is freely available on the internet or in genetic engineering textbooks. Good primer design generally enhances the yield and specificity of the amplicons generated from PCR. The best criteria often take into consideration primer length, GC%, GC clamp requirement, melting point (T_m), primer-dimer likelihood, hairpin formation, and primer self-complementarity. There are several online tools that can rapidly determine these criteria for any primer sequence. One useful example is OligoCalc (http://www.basic.northwestern.edu/biotools/oligocalc.html).

3. Readers are encouraged to design primers manually with the help of OligoCalc, instead of using comprehensive primer design programs. Often, these programs will provide the ideal primers, but the primers may not be suitable for mutagenesis work, and sometimes can confuse inexperienced researchers.

4. In certain cases, it is hard to design ideal primers with all the requirements necessary for good amplification. This is particularly true for mutagenic primers in megaprimer and OE-PCR approaches. If this happens, one can choose to ignore this matter, and may proceed with the trials of amplifications using these primers. When needed, optimizing PCR conditions will help improve amplicon yield. If the amplicon yield is low, consider increasing the number of PCR cycles (i.e., 35 cycles, instead of 25 provided a suitable proofreading DNA polymerase is used) or supplement with PCR additives such as glycerol or DMSO into the PCR reaction mixture.

5. For PCR-based mutagenesis, it is common to use longer mutagenic primers (>25 bp) to allow good annealing.

6. Tip: Formula to obtain sequence for forward primer is "Copy." ["Copy" the gene sequence at the 5′ end]. For instance, the forward primer for CGTase without the restriction enzyme recognition site could be 5′-GACGTAACAAACAAAGTCAA TTAC-3′, but this primer (24 bp) could potentially form a hair-pin and it is low in G+C content (33%). Altering the length of the forward primer sequence may help in fulfilling the GC% criteria for a good primer, but may introduce other problems as well. Regions with low GC% content in any typical gene can be a problem. Nevertheless, always try several different designs and determine in silico the quality using OligoCalc before ordering the best possible primer from a commercial supplier.

7. Tip: Formula to obtain the sequence for the reverse primer is "Reverse complement." ("Reverse complement" the gene sequence at the 3′ end). Finding the reverse complement sequence for any reverse primer can be done using an online program such as Reverse Complement: (http://www.bioinformatics.org/sms/rev_comp.html). For instance, the reverse primer for CGTase (*see* Fig. 1), without the restriction enzyme recognition site sequence could be designed as:

   5′-CCAATTAATCATAACCGTATCTGTTCCGG-3′. Please do take note that this primer (29 bp) is longer than the forward primer in order to compensate for the difference in G + C content. The length difference between the forward and reverse primers is not an important concern. It is often suggested that the typical primer length for a conventional PCR is 18–21 bp, but the primer design should not be restricted with this suggestion. Ordinarily, primers for site-directed mutagenesis applications are often longer than normal PCR primers. Shorter primers may be digested by certain proofreading DNA polymerases. However, there is no general rule for the length of a mutagenic primer.

8. Certain regions of a gene are high in GC content. If the mutation site falls in a high GC region, the designed primers will exhibit higher T_m (sometimes greater than 78 °C). Conversely, the T_m might be lower than 40 °C if the mutation site is located in a low GC region. Should this happen, disregard the general theory that suggests the annealing temperature (T_a) should be 5 °C lower than the T_m. In addition, it is difficult to design two primers (forward and reverse) to have identical T_m. Collectively, authors suggest readers to use approximately 55 °C as the starting T_a, and adjust accordingly.

9. In this example, the T_m of both the forward and reverse primers are 50 and 58 °C, respectively. Nevertheless, running the PCR at a T_a of 55 °C generates an excellent product yield (data not shown).

10. The requirement and justification of adding restriction sites when designing primers for PCR cloning will not be explained in this chapter. Readers are expected to check the reading frame upon cloning into a plasmid. Extra care should be taken when designing reverse primers, especially if readers wish to preserve the fusion tags (e.g., poly His-tag) downstream of a DNA insert. The existence of a stop codon (originating from the gene of interest) prior to the fusion tag will result in the expression of a recombinant protein that lacks the fusion tag. Generally, fusion tags are important for protein purification or identification.

11. Use only proofreading DNA polymerase for megaprimer and OE-PCR. Notably, Taq DNA polymerase is not suitable as it might incorporate unwanted mutations due to nucleotide mismatches in each PCR round. Poly-A tails generated by Taq DNA polymerase can introduce unwanted bases (insertions) and cause a reading frame shift which can result in a truncated protein.

12. There are several types of commercially available proofreading DNA polymerases. Examples of these include Q5® DNA polymerase (NEB), *Pfu* DNA polymerase (Promega), KAPA HiFi HotStart DNA polymerase (Kapa Biosystem), and TransStart KD Plus DNA polymerase (TransGenBiotech), Pfx DNA polymerase (Invitrogen and Thermo Fisher Scientific), and KOD DNA polymerase (Merck Millipore). Generally, all these DNA polymerases are suitable for megaprimer PCR and OE-PCR, but they differ in terms of their processivity (polymerization), fidelity, and cost.

13. Readers are advised to gel purify the amplicon generated from each PCR round (i.e., first PCR round in megaprimer; and first and second PCR rounds in OE-PCR), instead of using normal whole PCR cleanup approaches. The chance of obtaining mutant genes will be higher, hence reducing the number of colonies needed to screen.

14. Generally, the longer the length of the megaprimer PCR (generated from the first PCR in the megaprimer approach), the less efficient it is in priming the template during the second PCR round, which reduces the amplicon yield in the later round. Thus, a maximum megaprimer length of 600 bp is suggested. Notably, this length could be calculated from the 5′ or the 3′ end, whichever is closer to the mutation site. However, larger megaprimers (>1000 bp) have been reported in megaprimer PCR experiments [9, 10].

Acknowledgments

This work was financially supported by Universiti Teknologi Malaysia (GUP 09H98 and 06H31).

References

1. Dalby PA (2007) Engineering enzymes for biocatalysis. Recent Pat Biotechnol 1(1):1–9
2. Ling MM, Robinson BH (1997) Approaches to DNA mutagenesis: an overview. Anal Biochem 254(2):157–178
3. Goh KM, Goh PH, Chyi NH, Piaw CK, Rahman RNZRA (2012) Trends and tips in protein engineering, a review. Jurnal Teknologi 59(1):21–31
4. Sarkar G, Sommer SS (1990) The "megaprimer" method of site-directed mutagenesis. BioTechniques 8(4):404–407
5. Ho SN, Hunt HD, Horton RM, Pullen JK, Pease LR (1989) Site-directed mutagenesis by overlap extension using the polymerase chain reaction. Gene 77:51–59
6. Han R, Li J, Shin H-D, Chen RR, Du G, Liu L, Chen J (2014) Recent advances in discov-

ery, heterologous expression, and molecular engineering of cyclodextrin glycosyltransferase for versatile applications. Biotechnol Adv 32: 415–428

7. Goh KM, Mahadi NM, Hassan O, Rahman RNZRA, Illias RM (2009) A predominant β-CGTase G1 engineered to elucidate the relationship between protein structure and product specificity. J Mol Catal B Enzym 57:270–277

8. Goh PH, Illias RM, Goh KM (2012) Rational mutagenesis of cyclodextrin glucanotransferase at the calcium binding regions for enhancement of thermostability. Int J Mol Sci 13:5307–5323

9. Dong YH, Li JF, Hu D, Yin X, Wang CJ, Tang SH, Wu MC (2015) Replacing a piece of loop-structure in the substrate-binding groove of *Aspergillus usamii* β-mannanase, AuMan5A, to improve its enzymatic properties by rational design. Appl Microbiol Biotechnol 100(9):3989–3998. doi:10.1007/s00253-015-7224-7

10. Perez K, Yeam I, Jahn MM, Kang BC (2006) Megaprimer-mediated domain swapping for construction of chimeric viruses. J Virol Methods 135:254–262. doi:10.1016/j.jviromet.2006.03.020

Section VI

In Vitro Mutagenesis for Studies of Protein Structure and Function

Chapter 28

Step-By-Step In Vitro Mutagenesis: Lessons From Fucose-Binding Lectin PA-IIL

Jana Mrázková, Lenka Malinovská, and Michaela Wimmerová

Abstract

Site-directed mutagenesis is a powerful technique which is used to understand the basis of interactions between proteins and their binding partners, as well as to modify these interactions. Methods of rational design that are based on detailed knowledge of the structure of a protein of interest are often used for preliminary investigations of the possible outcomes which can result from the practical application of site-directed mutagenesis. Also, random mutagenesis can be used in tandem with site-directed mutagenesis for an examination of amino acid "hotspots."

Lectins are sugar-binding proteins which, among other functions, mediate the recognition of host cells by a pathogen and its adhesion to the host cell surface. Hence, lectins and their binding properties are studied and engineered using site-directed mutagenesis.

In this chapter, we describe a site-directed mutagenesis method used for investigating the sugar binding pattern of the PA-IIL lectin from the pathogenic bacterium *Pseudomonas aeruginosa*. Moreover, procedures for the production and purification of PA-IIL mutants are described, and several basic methods for characterizing the mutants are discussed.

Key words Site-directed mutagenesis method, Lectins, PA-IIL lectin, Surface plasmon resonance, Isothermal titration calorimetry (ITC), Microscale thermophoresis, Protein crystallography

1 Introduction

Site-directed mutagenesis (also termed "site-specific" mutagenesis) is a practical approach widely used to obtain a deeper understanding of a protein's function is based on specific changes in the protein's structure. It is very often utilized for studies of the functional reaction mechanisms of enzymes [1] and for determining the basis of their interactions with substrates [2]. It is important to investigate which amino acids are essential for the interaction [3] and therefore how they likely play a crucial role in the binding of a substrate [4]. The aim is to determine the residues which are mainly responsible for a particular protein's function [5] and/or substrate specificity [6]. In the protein engineering field, site-directed mutagenesis methods are commonly employed for

Andrew Reeves (ed.), *In Vitro Mutagenesis: Methods and Protocols*, Methods in Molecular Biology, vol. 1498,
DOI 10.1007/978-1-4939-6472-7_28, © Springer Science+Business Media New York 2017

improving or altering various properties of a wide range of proteins [7]. In many cases, the process of site-directed mutagenesis mainly serves to increase the activity of various enzymes, especially those of industrial importance [8]. Moreover, it is often used to positively influence major features of enzymes, such as their pH optimum [9], thermostability [10], and substrate specificity [11].

Experimental procedures which utilize site-directed mutagenesis are often preceded by a rational design approach. For many reasons, it is useful to predict the possible results before starting the hard laboratory work. The domain of rational design is based on knowledge of the protein structure, and it includes a set of computational tools and processes. These enable the prediction of amino acid "hotspots" with a potential functional effect, modeling a protein structure with altered amino acids, and the examination of functional changes in mutant proteins. If the protein structure is not available, sequence alignment procedures [4, 6, 8], sequence conservation analysis [2], and homology modeling [2–4, 6, 10] enable a modified protein structure to be developed. The molecular dynamic simulation method [2, 3, 6, 10] is mainly utilized to refine a structure obtained from homology modeling and to simulate the behavior of the desired biomolecules. This process is often preceded by a substrate-docking model [2–4, 8]. Through the rapid development of these computational methods, the rational design approach has become an indispensable part of site-directed mutagenesis.

Instead of rational design, a random mutagenesis method can also be employed for a preliminary probing into "hotspots" which could be subsequently investigated by site-directed mutagenesis [9, 12]. There are two methods directly derived from site-directed mutagenesis: (1) Site-saturation mutagenesis is used to replace a target amino acid with the remaining 19 amino acids and (2) targeted-random mutagenesis (site-directed random mutagenesis) is utilized to create random substitutions at two or more positions at once.

Lectins are sugar-binding proteins occurring across all kinds of organisms. They are not enzymes because of a lack of catalytic activity. They are not produced by B cells or plasmocytes of the immune system, therefore they are not carbohydrate-specific antibodies. They are neither protein sensors nor part of the class of proteins transporting free monosaccharides and oligosaccharides. The fundamental characteristics of lectins include the ability to bind diverse carbohydrate moieties non-covalently, reversibly and with high specificity [13]. Due to these facts, lectins play an important role mainly in the recognition of glycoproteins and carbohydrate structures displayed on the cell surface. They are often involved in an interaction between a pathogen and its host organism because they mediate adhesion to available glycoconjugates on the surface of the host cells. These processes are related to bacterial pathogenicity and lectins often are considered virulence factors. Hence, lectins from pathogenic bacteria and their adhesion modes are the focus of intense study [14]. Lectins and their binding properties are also intensively studied and have been engineered via site-directed mutagenesis in order to understand the nature of their interactions and to potentially enhance their carbohydrate-binding properties [15]. The outcomes of this process are often the production of proteins with novel affinities [16], fine-tuned specificities [17], or changed anomer preferences [18]. The improved lectins can then be applied, for example, as probes for diagnostic assays.

Site-directed mutagenesis was successfully employed for an investigation of the binding properties of the PA-IIL lectin from *Pseudomonas aeruginosa*. *P. aeruginosa* is an opportunistic human pathogen that is often involved in nosocomial infections and it invades immunocompromised patients, e.g., those suffering from cystic fibrosis. Three amino acids responsible for the sugar binding specificity of PA-IIL were replaced with those present in the RS-IIL from the bacterium *Ralstonia solanacearum*, which is an important plant pathogen [19]. Both of these lectins are homologs belonging to the PA-IIL superfamily with significant sequential and structural similarities. They are able to bind saccharides with similar stereochemistry, but they differ in their specificity. PA-IIL prefers l-fucose and its derivatives, whereas RS-IIL prefers d-mannose and its derivatives. The key to this phenomenon lies in the architecture of their binding pockets, in the difference of three consecutive amino acids belonging to a so-called "specificity binding loop" that is responsible for the sugar preference and affinity. However, the influence of each amino acid in the loop on specificity is still unclear. Therefore, three point mutants with amino acid changes at positions 22, 23 and 24 were prepared and subjected to a detailed examination. It was discovered that replacing the serine at position 22 leads to PA-IIL preferring mannose instead of fucose and therefore it is crucial for determining sugar binding specificity. On the other hand, substitutions of the amino acids at positions 23 and 24 only led to an increased affinity towards fucose [20]. Subsequently, a work was published that focused on a comparison of two docking programs, which were used to study the interaction of the PA-IIL lectin by means of molecular modeling. The three aforementioned one-point mutants were prepared in silico and subjected to the docking process with a set of monosaccharide ligands [21].

Following these results, a study that utilized an in silico mutagenesis and docking model with a subsequent in vitro analysis of promising mutants was performed. The saturation in silico mutagenesis was used for the preparation of three sets of nineteen single-point mutants with substituted amino acids at positions 22, 23 and 24 to obtain a deeper understanding of the importance of each amino acid belonging to the "specificity binding loop".

Subsequent docking of two benchmark ligands was performed (methyl α-l-fucopyranoside and methyl α-d-mannopyranoside), followed by calculation of the binding free energy. The results showed that almost all mutations of serine at position 22 led to a decrease or loss of binding ability. It was found that mutations of the serine at position 23 had no negative influence on sugar binding in general. On the other hand, it seemed that any substitution of glycine at position 24 increased the affinity towards the methyl α-d-mannopyranoside. Four mutants with a change at position 24 were prepared in vitro and their binding properties were examined by isothermal titration calorimetry. Mutants G24W, G24N, and G24R exhibited a higher affinity towards both benchmark ligands in comparison to the G24D mutant, which had a lower affinity than the wild-type PA-IIL lectin [22].

A combination of in silico and in vitro mutagenesis was also utilized for investigating the effect of a terminal glycine114 deletion on PA-IIL binding properties. The terminal glycine of an adjacent monomer is a part of the binding site. It was found that it contributes to the interactions with saccharides, because its deletion resulted in a decreased affinity towards fucose. Despite being unable to crystallize this mutant protein and experimentally solve its structure, the combination of isothermal titration calorimetry and molecular dynamic simulation was able to provide sufficient structural information about its binding properties [23].

In the following protocols we describe the procedures used in our laboratory for the site-directed mutagenesis of our proteins of interest via amplification of the whole plasmid using a pair of complementary primers carrying the desired base pair substitutions (*see* Fig. 1) and the production of mutant proteins by an *E. coli* expression system. In addition, we describe the purification of the proteins by one-step affinity chromatography and we discuss several methods used to investigate their binding properties.

2   Materials

2.1   Molecular Biology Reagents

1. 0.8 % agarose solution: Mix agarose for electrophoresis with 1× TBE buffer (Tris–borate–EDTA) in an Erlenmeyer flask, place it in a microwave oven and boil it three times shaking each time to dissolve the agarose (*see* **Note 1**).

2. Agarose and agarose gel electrophoresis apparatus (*see* **Note 2**).

3. DNA staining dye (e.g., ethidium bromide or Biotium GelRed nucleic acid stain).

4. Appropriate antibiotics.

5. Plasmid preparation kit.

6. Enzymes for PCR (e.g., PfuUltra HF polymerase).

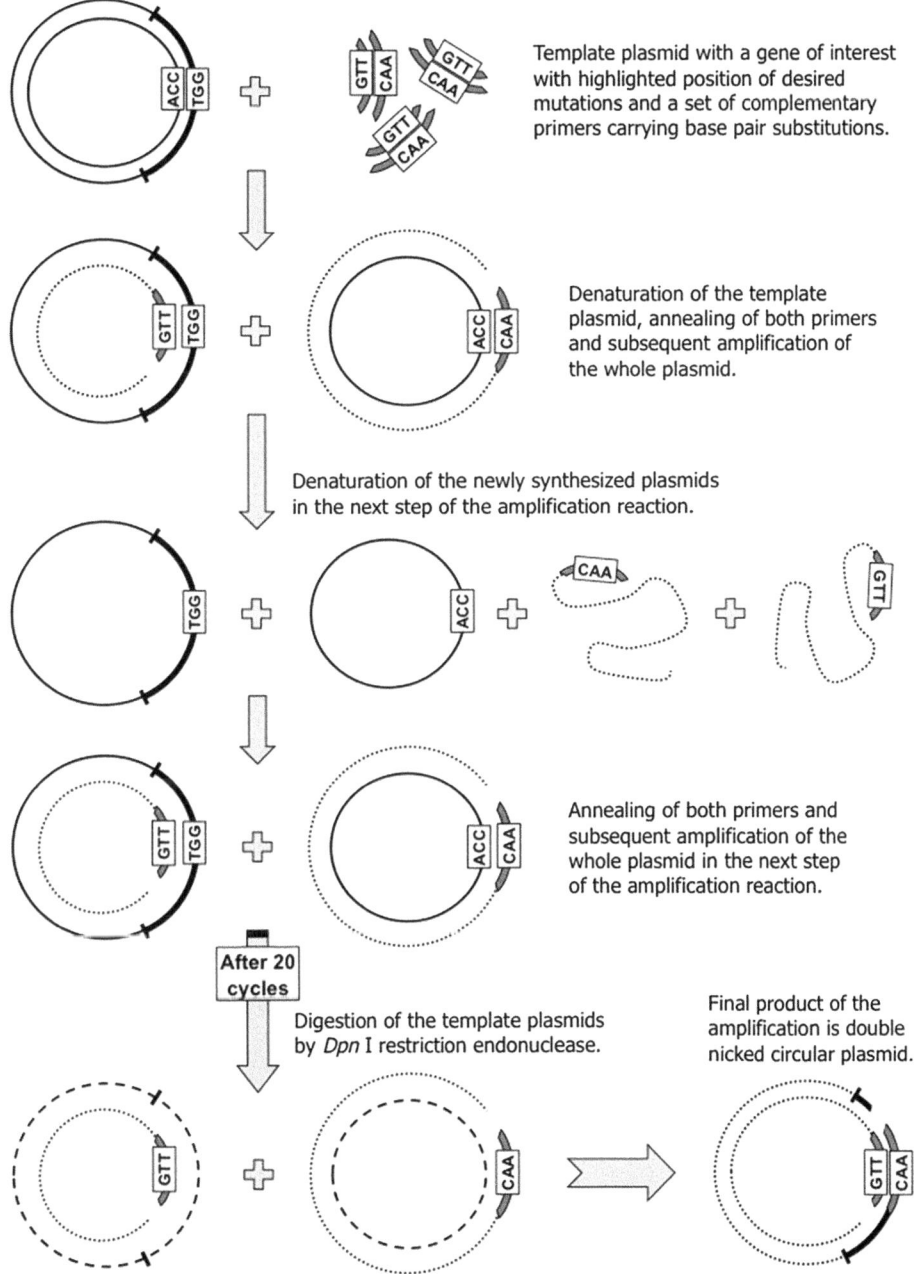

Fig. 1 Schematic representation of the site-directed mutagenesis principle. Amplification of a whole plasmid using a pair of complementary primers carrying the desired base pair substitutions

7. Mutagenic primers for PCR.

8. Thin-walled PCR tubes (0.5 μL).

9. Falcon tubes (50 mL).

10. *Dpn*I restriction enzyme.

11. Chemically competent *E. coli* cells (e.g., DH10B).

12. Isopropyl-β-d-1-thiogalactopyranoside (IPTG) stock solution (50 mM).

2.2 Media and Buffers

Perform all steps of medium production using aseptic technique. Use deionized water for preparation of buffers. Store the prepared buffers at 4 °C.

1. Lysogeny broth (LB) medium: 10 g/L tryptone, 5 g/L yeast extract, 10 g/L NaCl. pH to 7.0 with 5 N NaOH. Autoclave for 15 min at 121 °C.

2. 10× TBE buffer: 89 mM Tris–HCl, 89 mM boric acid, 2 mM EDTA, pH 8.0. Dilute the buffer with distilled water to 1× concentration.

3. Binding buffer: 20 mM Tris–HCl, 100 mM NaCl, 100 μM $CaCl_2$ (*see* **Note 3**), pH 7.5. Prepare 1 L of the buffer, filter it through a 0.45 μm filter and degas it subsequently.

4. Elution buffer: 20 mM Tris–HCl, 100 mM NaCl, 100 mM d-mannose, 100 μM $CaCl_2$, pH 7.5. Make 0.5 L of the buffer, filter it through a 0.45 μm filter and degas it subsequently.

2.3 Equipment

1. Thermal cycler.

2. Gel documentation system or transilluminator.

3. Microcentrifuge.

4. SDS-PAGE apparatus.

5. Affinity chromatography materials.

6. Dialysis membranes.

7. Lyophilizer.

8. Sonicator and probe.

3   Methods

3.1 Mutagenic Primer Design

The commercial on-line program QuikChange Primer Design provided by Agilent Technologies can be used for design of primers carrying specific base substitutions. The website address to the program is: http://www.genomics.agilent.com/primerDesignPro gram.jsp.

The on-line program OligoAnalyzer 3.1 provided by Integrated DNA Technologies may be used for analyzing primer properties (hairpins and self-dimers mainly). The website address to the program is https://eu.idtdna.com/calc/analyzer (*see* **Note 4**).

1. Open the website with the program QuikChange Primer Design.

2. Leave the "QuikChange II" choice in the form.

3. Paste your DNA sequence and click "Upload Translated."

4. After uploading the sequence, click on the circular button at the end of the sentence "Select up to seven amino acids that you want to change."

5. In the numbered list check amino acids that you want to change.

6. Change selected amino acids to those that you want (*see* **Note 5**).

7. Click the "Design Primers" button to obtain sequences of complementary primers.

8. Open the website with the program OligoAnalyzer 3.1.

9. Analyze both sequences for the presence of hairpins and self-dimers.

10. Modify the primers by adding or removing bases at the ends if it is necessary (*see* **Note 6**).

11. When ordering synthesized primers for a targeted-random mutagenesis experiment, it is necessary to replace the chosen triplet by "NNN" (all four bases incorporated) in the ordered sequence. For example:

    5′-GGT GGT ACA CCG GT**N NN**T TCA ACG AGC CCG GCG-3′ (*see* **Note 7**).

3.2 Electrophoresis

Always prepare the agarose gel fresh and precool it before use.

1. Dilute the appropriate volume of 10× TBE buffer with distilled water to a 1× concentration.

2. Mix 0.8 % agarose powder for electrophoresis with 1× TBE buffer in an Erlenmeyer flask, put it in a microwave oven and boil it three times as described in Subheading 2.

3. Set up an apparatus for DNA electrophoresis with an appropriate comb.

4. Cool down the solution to approximately 60 °C in a water bath and add an appropriate nucleic acid stain (*see* **Note 8**).

5. Pour the agarose solution into a casting tray with the well comb in place, remove air bubbles with a pipette tip and let it solidify for 15 min or more.

6. Cover gel with a thin layer of 1× TBE buffer before use.

3.3 Competent E. coli Cells

Perform all steps of the procedure in a sterile environment with sterile solutions.

1. Add 2 mL of LB medium without any antibiotic into a sterile 50 mL conical centrifuge tube. Scrape a piece of frozen stock

cells and resuspend in the medium. Incubate the cell suspension at 37 °C and 180 rpm overnight.

2. Add 2×50 mL of LB medium in 250 mL shake flasks without any antibiotic and inoculate it with the overnight cell culture. Incubate the cultures at 37 °C and 220 rpm until they reach an OD_{600} of between 0.3 and 0.5.

3. Chill four sterile 50 mL conical centrifuge tubes on ice. Pour the cell cultures into the tubes and cool them on ice for 10 min. Afterwards, centrifuge them for 7 min at $2700 \times g$ at 4 °C.

4. Discard the supernatant and gently but thoroughly resuspend each cell pellet in 10 mL of sterile cold 50 mM $CaCl_2$.

5. Cool the cell suspensions on ice for 15 min. Centrifuge them for 7 min at $2700 \times g$ at 4 °C.

6. Discard the supernatant and gently but thoroughly resuspend each cell pellet in 500 µL of sterile cold 72 mM $CaCl_2$.

7. Add 115 µL of 80 % sterile glycerol to each cell suspension and mix it gently but thoroughly. Transfer 100 µL aliquots of the suspension into prechilled 1.5 mL tubes.

8. Immediately cool the tubes down in liquid nitrogen. Store the cells at −80 °C.

9. Always test transformation efficiency of newly prepared competent cells.

3.4 Site-Directed Mutagenesis

It is not necessary to keep the reaction mixture on ice during the preparation procedure.

1. Dilute the primers (*see* Subheading 2.1.) with the desired mutation(s) to a final concentration 10 µM.

2. Purify a template plasmid carrying a gene of interest (*see* **Note 9**).

3. Reaction mixture: In a thin-walled PCR tube mix 5 µL of 10× *PfuUltra* HF reaction buffer, 1 µL of 40 mM of dNTPs (10 mM each), 50–100 ng of template plasmid (*see* **Note 10**), 2.5 µL of primer 1, 2.5 µL of primer 2 (*see* **Note 11**), 1 µL of *PfuUltra* HF polymerase (2.5 U/mL). Add ddH_2O to a final volume 50 µL. Mix the reaction mixture thoroughly by gently pipetting the solution up and down. Spin the mixture down before putting it into a thermal cycler.

4. Use the PCR following PCR cycling conditions: (a) 95 °C 2 min, (b) 20 cycles (*see* **Note 12**): 95 °C 50 s, 58 °C 50 s, 72 °C 1 min per kilobase of template plasmid length plus 1 extra minute (*see* **Note 13**), and (c) 72 °C, 10 min (*see* **Note 14**).

5. Cool the reaction mixture and spin it down shortly (to run all traces of liquid condensed on the cap and the walls to the bottom of the tube).

6. Take 5 μL of the undigested reaction mixture as a control for an agarose gel DNA electrophoresis.

7. Add 1 μL of *Dpn* I restriction endonuclease (20,000 U/mL) (*see* **Note 15**) directly into the 45 μL reaction mixture and mix it thoroughly by gently pipetting the solution up and down. Spin the digestion mixture down and incubate it at 37 °C for 3 h (*see* **Note 16**).

8. Spin the digested mixture down shortly (long enough to run all traces of liquid condensed on the cap and the walls to the bottom of the tube).

9. Take 5 μL of the digested mixture as a control for agarose gel DNA electrophoresis.

10. Use the digested mixture for transformation of competent cells.

11. Store the digested mixture at −20 °C.

3.5 Analysis of PCR Products

Perform the electrophoresis in a cold environment (~4 °C).

1. Pour a 0.8 % agarose gel with a nucleic acid stain (*see* Subheading 2.2).

2. Electrophoresis of samples: 5 μL of undigested amplification reaction mixture, 5 μL of digested mixture and approximately 100 ng of template plasmid. Mix each sample with an appropriate DNA loading dye.

3. Load all samples into the gel and then load an appropriate DNA ladder for size determination.

4. Run the electrophoresis at 70 V for at least 1 h or until all forms of the plasmids are visibly separated in each sample (check on UV transilluminator regularly).

5. Stop the electrophoresis and place the gel on a UV transilluminator or gel documentation system.

6. Check it if there is sufficient yield of the required PCR product in the undigested mixture and whether all of the template DNA has been digested in the *Dpn*I reaction (*see* Fig. 2).

3.6 DNA Transformation

Use chemically competent *E. coli* cells.

1. Gently thaw 50 μL of competent *E. coli* cells on ice (*see* **Note 17**).

2. Add 4 μL of the digestion mixture into the chilled competent cells and gently flick the transformation mixture to disperse the DNA (*see* **Note 18**).

3. Incubate the transformation reaction mixture on ice for 30 min.

4. Heat-shock the mixture at exactly 42 °C for 45 s and immediately cool it on ice for 2 min.

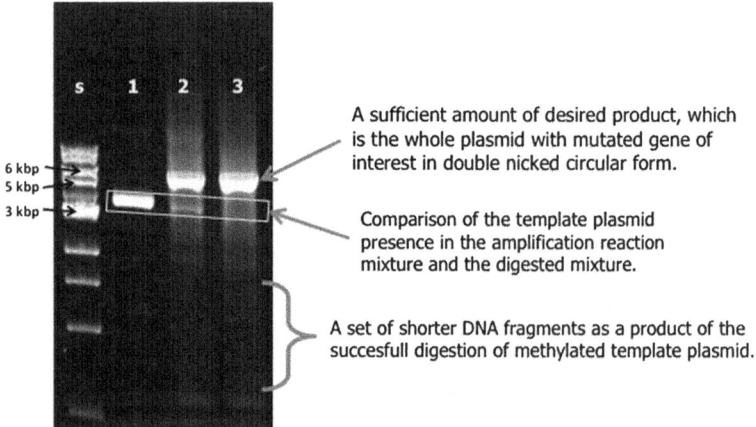

Fig. 2 Example of an agarose gel after DNA electrophoresis. Lane 1, contains approximately 100 ng of the template plasmid pET25b*pa2l* bearing the gene for the PA-IIL lectin. Lane 2, contains 5 µL of the undigested amplification reaction mixture. The thick band of the desired product and the thin band of the template plasmid are highly visible. *Lane 3*, contains 5 µL of the *Dpn*I digested mixture. The thick band is still highly visible. The thin band of the template plasmid disappeared (highlighted with a *white square*), while a set of shorter DNA fragments appeared after digestion. This indicates that the digested mixture could be used for the transformation of competent cells. *Lane S*, 1 kb DNA marker ladder (New England Biolabs)

5. Add 1 mL of SOC medium into the transformation reaction mixture and incubate it at 37 °C and 600 rpm for 1 h in a thermomixer or in a shaker incubator at 200 rpm (1-in. orbital rotation).

6. In a sterile, laminar flow hood place a closed LB agar plate containing the appropriate selective antibiotic.

7. Pellet the transformed cells by centrifugation for 6 min at 5,200 × *g* in a benchtop microcentrifuge (*see* **Note 19**).

8. Remove about 800 µL of the supernatant and resuspend the cell pellet in the remaining volume (~50–100 µL).

9. Spread the cell suspension on the LB-antibiotic agar plate, allow a few minutes to absorb, and incubate the plate bottom side up at 37 °C overnight (at least 16 h).

10. Isolate plasmid DNA using a miniprep plasmid isolation kit from one cell colony and send it for sequence analysis to check if the gene of interest carries the desired mutations.

3.7 Mutant Protein Expression

If sequence analysis confirms the presence of the desired mutations in your gene of interest, use the isolated plasmid for transformation of competent *E. coli* expression cells (*see* **Note 20**).

If chemically competent protein expression strains are needed, simply repeat the method of transformation of competent cells mentioned above (*see* Subheading 3.3.). In **step 2** in Subheading 3.3 add 200 ng/µL of the pure plasmid into a cell suspension instead of 4 µL of the digested mixture and then continue with the procedure as described in the protocol.

The protocol described below represents a typical procedure of production and purification of wild-type PA-IIL lectin and its mutants.

It is necessary to keep the cells in a sterile environment until protein production is finished.

1. Transfer 10 mL of sterile LB medium with an appropriate antibiotic into a sterile 50 mL falcon tube. Pick one colony from the agar plate with newly transformed expression cells and resuspend it in the medium (*see* **Note 21**). Incubate the cell suspension at 37 °C and 180 rpm overnight.

2. Centrifuge the overnight culture for 15 min at $3200 \times g$ at room temperature and discard the supernatant. Resuspend the cell pellet in 2 mL of fresh LB medium containing the appropriate antibiotic and transfer it to a 1-L Erlenmeyer flask containing 250 mL of fresh LB medium plus the appropriate antibiotic. Incubate the cell culture at 37 °C and 180 rpm overnight.

3. Centrifuge the overnight culture for 15 min at $3200 \times g$ at room temperature and discard the supernatant (*see* **Note 22**). Resuspend the cell pellet in fresh LB medium (a few mL) containing the appropriate antibiotic and inoculate 8×500 mL of the LB medium with antibiotic in 8×2-L Erlenmeyer cultivation flasks. Incubate the cell cultures at 37 °C at 220 rpm in a shaker incubator until they reach an OD_{600} of between 0.5 and 0.6.

4. Take 1 mL of the cell culture as a sample for analysis of protein production (but before induction) by SDS-PAGE.

5. Add isopropyl β-d-1-thiogalactopyranoside to a final concentration 0.5 mM to all cell cultures (*see* **Note 23**). Incubate the cell cultures for 3 h at 37 °C and 220 rpm.

6. At the end of cultivation, take 1 mL of the cell culture, dilute it with LB medium to the same density as the sample before induction and prepare a sample for analysis of protein production after induction by SDS-PAGE.

7. Harvest the cells by centrifugation for 10 min at $8100 \times g$ and 4 °C. Resuspend the cell pellet in a few mL of binding buffer (*see* **item 3** in Subheading 2.2).

8. Disrupt the cells by sonication (*see* **Note 24**) and use 100 µL of the cell lysate as a sample for analysis of protein production in soluble or insoluble form. Centrifuge the cell lysate at $21,000 \times g$ at 4 °C for at least 1 h.

9. Filter the supernatant through a 0.45 μm filter (*see* **Note 25**) and keep it at 4 °C before loading it onto an affinity chromatography column.

3.8 Protein Purification

We use a one-step purification procedure by affinity chromatography for purification of PA-IIL lectin and its mutants (*see* **Note 26**).

1. Wash the column with 3 column volumes of the elution buffer (*see* **step 4** in Subheading 2.2) and equilibrate it with at least 3 column volumes of the binding buffer (*see* **Note 27**).

2. Reduce the flow of binding buffer and load the filtered supernatant onto the column. When the supernatant is loaded, wash out ballast proteins by increasing the flow of the binding buffer.

3. When all unbound ballast proteins are washed out from the column, elute the protein of interest with the elution buffer.

4. Wash the column and store it for later use according to standard laboratory procedures.

5. Prepare a set of samples from all purification fractions for analysis by SDS-PAGE.

6. Dialyze the purified protein against distilled water for 1 week (change water once per day) (*see* **Note 28**).

7. Lyophilize the protein and keep it at –20 °C for long-term storage.

3.9 Measuring Lectin Binding

The methods described below are routinely used to characterize the binding properties of the PA-IIL lectin and its mutants. These are some of the methods that can be utilized to investigate changes in the properties of mutated proteins.

3.9.1 Surface Plasmon Resonance (SPR)

SPR is an optical technique that is used to characterize the interaction of two binding partners (antibody–antigen, ligand–receptor, enzyme–substrate, protein–saccharide, etc.) and to analyze the specificity, affinity, and kinetic parameters of this interaction. One of the partners is immobilized on a chip surface, while the second floats over it in a continuously flowing solution. The presence of the binding interaction leads to a change in surface plasmon resonance, which is determined optically [24].

An SPR measurement enables the direct determination of the association rate constant (k_a) and the dissociation rate constant (k_d), therefore the equilibrium dissociation constant (K_D) can be calculated. Hence, it provides insight into the biomolecular interaction kinetics, which is typically determined by using a range of sample concentrations, each run in a separate cycle. In addition to determining the binding affinity (from mM to pM), the stoichiometry of the interaction can be calculated from the measurement. Also, some SPR instruments enable an analysis of thermodynamic parameters, but this analysis is not as straightforward as with ITC measurement.

The advantages of this method include high sensitivity (detection limit on the order of 10 pg/mL) and specificity, a label-free environment, real-time and continuous measurement that is very fast (in minutes), and small sample volume consumption. It is necessary to take into account some of the difficulties of this method, such as nonspecific interactions of an analyte with the chip surface, possible steric hindrances during the binding of the analyte to the immobilized ligand, and negative effects of the immobilization procedure on the ligand. Also, the initial costs of an SPR machine are very high [25].

When characterizing mutated lectins, this method is usually used for the rapid acquisition of preliminary information about mutant activity and specificity. In the four-channel arranged chip, each channel contains a different immobilized sugar. Hence, a qualitative comparison of the wild-type lectin and its mutant can be performed.

3.9.2 Isothermal Titration Calorimetry (ITC)

ITC is the most convenient method for the analysis of changes in the sugar-binding specificity of mutated lectins, because it enables a complex and detailed insight into the binding mechanism from a functional point of view. This method is based on directly measuring the heat evolved or absorbed during interactions between the ligand and macromolecule in the solution. The greatest advantage of this technique is that it enables the direct determination of enthalpy changes, stoichiometry and association constants. In addition, Gibbs free energy changes and entropy changes can be calculated. Therefore, a complete thermodynamic profile is obtained in one measurement.

An ITC measurement is usually performed at constant temperature. In addition, a change in heat capacity can be directly determined from a set of titration experiments performed at different temperatures, which provide information about the temperature dependence of the interaction.

ITC is also an appropriate method for studying the kinetics and thermodynamics of enzyme catalysis. It is able to measure native proteins in buffer solutions mimicking their natural environment. The same buffer needs to be used for the ligands and macromolecules, therefore a problem may occur with the solubility of the binding partners or with the buffer ionization enthalpies. Nonetheless, this method is label-free and it does not require any immobilization or chemical modification of the ligand or macromolecule. It has no temperature, pH or molecular weight limitation. It does not depend on the transparency of the measured solutions because ITC is not an optical method, but it is very sensitive to air bubbles or aggregates formation. In addition, only a very low concentration of detergents can be used, where necessary.

The most important disadvantage of this method is the requirement of a high concentration of sample (in milligrams) and relatively high sample volume consumption (about 1.7 mL for VP-ITC and 300 μL for ITC200) for one measurement. In addition, a stirring of the sample may lead to protein precipitation during the measurement. Another important disadvantage may be lower sensitivity (K_D in the nM to mM range). Problems may occur when high affinity reactions are measured [26, 27]. Currently, the problem with low-throughput has been circumvented by automatic ITC machines (e.g., AutoITC). However, the sample volume consumption is higher, due to the robotic pipetting (450 μL).

3.9.3 Microscale Thermophoresis (MST)

MST is a relatively new method for the analysis of biomolecular interactions. This optical method is based on analyzing the fluorescence intensity of molecules which move through the temperature gradients induced by an infrared laser. Hence, covalently attached fluorophores or fluorescent fusion proteins are needed. Moreover, it is possible to measure the fluorescence of intrinsic tryptophans.

This method is fast and very sensitive (K_D in the pM/nM to mM range) because it effectively combines the principle of thermophoresis and the detection of fluorescence. Therefore, a change in the size, conformation, charge and hydration shell of the labeled molecule can be detected. In addition, this technique provides the possibility of carrying out the measurement in complex buffers or directly in cell lysates. For example, a membrane protein can be measured in its native environment with this method. Various interactions of diverse binding partners can be studied as well by MST.

MST is an appropriate method for the determination of binding constants (regular mode of measurement) and stoichiometries (saturation mode of measurement). The experiment is performed as a set of measurements with a constant concentration of the labeled molecule and increasing concentration of the unlabelled binding partner. The samples for the analysis are prepared by serial dilution, and therefore, inaccuracy during the dilution procedure may affect the results of the measurement. However, this inaccuracy can be recognized through the "cap scan" measurement, which is usually performed before each set of regular measurements. Also, it is possible to perform measurements of K_D at different temperatures, which can be used for the subsequent calculation of thermodynamic parameters. Moreover, this method is suitable for the analysis of cooperative and competitive binding modes and also protein folding properties.

In contrast with ITC, MST is able to work with a low concentration of analyzed macromolecules (nM) and it is able to quantify high-affinity interactions. Due to its capillary format, a small sample volume is used (usually 16×4 μL). However, problems with proteins sticking to the walls of capillaries may occur. Hence, it is necessary to optimize the composition of the buffer used in the experiment. A disadvantage of this method may also be higher costs due to the consumption of capillaries and dyes for fluorescent labeling. In addition, the ITC arrangement enables the condition of the sample to be visually checked after each measurement, in contrast to MST [28, 29].

3.10 Protein Structure Determination

The determination of protein structure is an important process, because it helps with rationalizing possible functional changes in the protein of interest carrying the desired mutations.

Usually, X-ray protein crystallography is utilized to determine the structures of proteins with their binding partners, because it enables deep insight into the molecular basis of the interaction. This is due to its ability to provide three-dimensional protein structure with resolution at the atomic level. However, this approach involves several challenges, such as obtaining sufficient data from X-ray diffraction measurements or successful solution of the phase problem [30].

In addition, protein crystallography is often limited by the ability to obtain appropriate protein crystals. In contrast to small molecules, the process of crystallizing proteins is much more difficult. A huge amount of protein is needed for crystallization, because protein crystals initiate their development at only a very high level of supersaturation. Hence, the preparation of a solution with a protein concentration of tens of milligrams per milliliter is sometimes necessary. During the crystallization process, the protein could form amorphous precipitates instead of crystals. These precipitates are unsuitable for X-ray diffraction measurements. The process of obtaining appropriate crystals can also take a long time, such as weeks, months or even years [31].

Alternatively, nuclear magnetic resonance and cryo-electron microscopy are methods which also enable the determination of protein structure. Which type of method will prove to be the best for solving the structure of the mutated protein of interest depends on many factors that are beyond the scope of this chapter [32, 33].

4 Notes

1. Always keep an eye on the gel solution in an Erlenmeyer flask during warming in a microwave oven. When bubbles first start to appear, stop the heating, pull the Erlenmeyer flask from the microwave oven and gently swirl the solution by rotating of the flask. Repeat this procedure three times.

2. We utilize a Scie-Plas Easigel H1-Set with H1-Cl-16 comb. Usually, we use 30 mL of 0.8% agarose in 1× TBE buffer for preparation of a thin gel.

3. All lectins belonging to a PA-IIL superfamily bind sugars through two calcium cations. Therefore, presence of the calcium cations in the buffers used for purification of these lectins is necessary.

4. Both oligonucleotide primer programs are freely available.

5. If you want to perform a targeted random mutagenesis, select a type of amino acid which will result in a triple-base substitution to obtain primers designed with three mismatches. If you want to make amino acid changes near the start or end of your protein of interest, it is necessary to add the preceding or following part of a sequence of plasmid carrying the gene of interest. Otherwise, the program is not able to design the required primers.

6. Usually, we design oligonucleotide primers with lengths from 30 to 40 bases with the desired substitutions situated exactly in the middle of the sequence. The melting temperature should be higher than 78 °C and the GC content should be at least 40%. It is recommended to have one or more G or C bases at the 3′-end of the primer sequences. It is necessary to analyze both primers for the occurrence of hairpins and homodimer formation (mainly at the 3′ end of the primer sequences).

7. During the synthesis of the primers, each base is coupled to the previous one by a step-by-step synthetic cycle according to the requested sequence. When comes to coupling of the base(s) which are marked "N", all four types of bases are added to the chemical reaction during the synthetic cycle. The final output of this process is a set of primers having random base substitution at "N" marked positions.

8. We add Biotium GelRed Nucleic Acid Gel Stain into the warm agarose solution for visualization of DNA (in gel) after electrophoresis.

9. It is necessary to isolate the template plasmid from a *dam+ E. coli* strain (majority of the commonly used *E. coli* strains). Only plasmids with *dam* methylated adenines in the sequence 5′-Gm⁶ATC-3′ are suitable templates for this site-directed mutagenesis technique. We recommend using a miniprep DNA kit for isolation of plasmid DNA. We recommend using sterile ddH$_2$O for elution in the last step of the miniprep kit protocol to obtain a pure plasmid in water instead of in buffer.

10. If the yield of the newly synthesized plasmid is low it is possible to slightly increase the input amount of template DNA. The newly synthesized complementary strands are nicked therefore they cannot serve as templates in the next cycle of amplification because they linearize during the denaturation step (*see* Fig. 1).

11. If a high number of nonspecific products occurred, it is possible to decrease the input volume of both primers to 1 μL.

12. It is not recommended to increase the number of cycles in the case of a low yield of newly synthesized plasmid. The higher number of cycles may increase susceptibility towards random mutations anywhere in the newly synthesized plasmids.

13. We often use pET25b plasmids containing our genes of interest. The length of the constructs is often around 6 kbp, therefore, duration of the elongation step is usually 7 min.

14. These reaction conditions work very well in our hands. If a problem with no yield/low yield of newly synthesized plasmid occurs, duration of the initialization step can be extended to 2 min per kilobase of the template plasmid length or duration of the initialization step together with duration of the denaturation step can be extended as well. In addition, the primer annealing temperature can be decreased and/or duration of the annealing step can be extended. If a problem with a higher number of nonspecific products occurs, primer annealing temperature can be increased and/or duration of the annealing step can be shortened.

15. *Dpn* I restriction endonuclease recognizes the sequence 5'-Gm^{6}ATC-3', therefore it is used for the digestion of template DNA in this method because new plasmids, as products of this reaction, are unmethylated. Plasmids that were isolated from *dam+ E. coli* strains carry this pattern of methylation or hemi-methylation, and thus can serve as templates in this method.

16. Usually, we use a thermal cycler or a thermoblock for PCR tubes with a top lid for incubation of the digestion mixture.

17. We always use *E. coli* DH5-α cells with good results. We use the competent cells prepared according to the procedure described in the previous section (*see* Subheading 2.3). It is possible to use any type of commercially available competent cells that are suitable for transformation by unmethylated DNA (with *hsd*R mutation in chromosomal genotype) and high transformation efficiency (e.g., XL10-Gold Ultracompetent cells).

18. If the transformation procedure is unsuccessful with no colonies appearing on the agar plate, it is possible to decrease the volume of the digested mixture added to the competent cells to 1 μL.

19. Alternatively, it is possible to centrifuge the transformed cells for 2 min at 10,000 rpm in the benchtop microcentrifuge.

20. Usually, we use *E. coli* expression strains BL21 (DE3) or Tuner (DE3) for production of mutant lectins with good results. We

prepare the competent cells according to the procedure described in the previous section (*see* Subheading 2.3). Usually, the *E. coli* expression cells represent the first choice for production of recombinant proteins [34, 35]. However, some problems may occur during production of the proteins in the *E. coli* cells. For example, desired mutations can lead to reduced protein solubility. This issue can be overcome by the addition of a solubility-enhancing fusion tag [36] or by using another type of expression system [37].

21. It is recommended to make a fresh agar plate with the newly transformed expression cells before each production and purification of the protein of interest.

22. Usually, we use sterile 50 mL conical centrifugation tubes for cell centrifugation in this step.

23. Use an appropriate inducer based on the plasmid used containing the gene of interest.

24. Usually, we use Sonics Vibra-Cell ultrasonic processor for cell disruption. It is necessary to adjust the conditions of the sonication procedure according to the cell density and to check the amount of disrupted cells regularly. It is possible to check it visually (a lysate with well-broken cells is transparent). Also, it is possible to check the quantity of intact cells under the microscope or by centrifugation of a small sample of the cell lysate.

25. It is necessary to filter the sample before loading it onto the column to remove all traces of the pellet.

26. Usually, each protein is purified by a procedure which is optimal for obtaining it in a sufficient amount and purity. An optimization of the procedure can be required, when it is used for purification of a mutated variant of the protein. Utilization of affinity fusion tags could significantly facilitate protein purification. There are plenty of publications dealing with optimization of protein purification methods [38–40]. In addition, many manuals and protocols for protein purification are available on websites of commercial manufacturers.

27. We routinely use GE Healthcare ÄKTApurifier for purification of PA-IIL lectin and its mutants.

28. In the case of PA-IIL lectin and its mutants that are purified by affinity chromatography and eluted by d-mannose, it is necessary to perform the dialysis procedure at least 6 days to remove saccharides from all binding sites properly. Usually, it is possible to perform dialysis of PA-IIL lectin and its mutants against water due to their high stability. It depends on the properties of each protein and on which dialysis conditions are required [41].

Acknowledgments

The work was supported by the Czech Science Foundation (13-25401S) and Student Project Grant at MU (MUNI/A/1265/2015).

References

1. Vashishtha AK, West AH, Cook PF (2015) Probing the chemical mechanism of saccharopine reductase from *Saccharomyces cerevisiae* using site-directed mutagenesis. Arch Biochem Biophys 584:98–106. doi:10.1016/j.abb.2015.08.023

2. Lee J, Daniels V, Sands ZA, Lebon F, Shi J, Biggin PC (2015) Exploring the interaction of SV2A with racetams using homology modelling, molecular dynamics and site-directed mutagenesis. PLoS One 10(2), e0116589. doi:10.1371/journal.pone.0116589

3. Nicole P, Couvineau P, Jamin N, Voisin T, Couvineau A (2015) Crucial role of the orexin-B C-terminus in the induction of OX1 receptor-mediated apoptosis: analysis by alanine scanning, molecular modelling and site-directed mutagenesis. Br J Pharmacol 172(21):5211–5223. doi:10.1111/bph.13287

4. Jung IP, Cho JH, Koo BS, Yoon MY (2015) Functional evaluation of residues in the herbicide-binding site of *Mycobacterium tuberculosis* acetohydroxyacid synthase by site-directed mutagenesis. Enzyme Microb Technol 78:18–26. doi:10.1016/j.enzmictec.2015.06.009

5. Hiromori Y, Aoki A, Nishikawa J, Nagase H, Nakanishi T (2015) Transactivation of the human retinoid X receptor by organotins: use of site-directed mutagenesis to identify critical amino acid residues for organotin-induced transactivation. Metallomics 7(7):1180–1188. doi:10.1039/C5MT00086F

6. Fang Z, Zhang J, Liu B, Du G, Chen J (2015) Insight into the substrate specificity of keratinase KerSMD from *Stenotrophomonas maltophilia* by site-directed mutagenesis studies in the S1 pocket. RSC Adv 5(91):74953–74960. doi:10.1039/c5ra12598g

7. Lu Y, Li L, Chen W, Wu M (2015) Enhanced anti-tumor (anti-proliferation) activity of recombinant human interleukin-29 (IL-29) mutants using site-directed mutagenesis method. Appl Biochem Biotechnol 177(5):1164–1175. doi:10.1007/s12010-015-1804-y

8. Chen Z, Li Y, Feng Y, Cheng L, Yuan Q (2015) Enzyme activity enhancement of chondroitinase ABC I from *Proteus vulgaris* by site-directed mutagenesis. RSC Adv 5(93):76040–76047. doi:10.1039/c5ra15220h

9. Fan C, Xu W, Zhang T, Zhou L, Jiang B, Mu W (2015) Engineering of *Alicyclobacillus hesperidum* L-Arabinose isomerase for improved catalytic activity and reduced pH optimum using random and site-directed mutagenesis. Appl Biochem Biotechnol 177(7):1480–1492. doi:10.1007/s12010-015-1828-3

10. Hesampour A, Siadat SER, Malboobi MA, Mohandesi N, Arab SS, Ghahremanpour MM (2015) Enhancement of thermostability and kinetic efficiency of *Aspergillus niger* PhyA phytase by site-directed mutagenesis. Appl Biochem Biotechnol 175(5):2528–2541. doi:10.1007/s12010-014-1440-y

11. Furuya T, Shitashima Y, Kino K (2015) Alteration of the substrate specificity of cytochrome P450 CYP199A2 by site-directed mutagenesis. J Biosci Bioeng 119(1):47–51. doi:10.1016/j.jbiosc.2014.05.028

12. Hu D, Tateno H, Kuno A, Yabe R, Hirabayashi J (2012) Directed evolution of lectins with sugar-binding specificity for 6-sulfo-galactose. J Biol Chem 287(24):20313–20320. doi:10.1074/jbc.M112.351965

13. André S, Kaltner H, Manning JC, Murphy PV, Gabius HJ (2015) Lectins: getting familiar with translators of the sugar code. Molecules 20(2):1788–1823. doi:10.3390/molecules20021788

14. Audfray A, Varrot A, Imberty A (2013) Bacteria love our sugars: interaction between soluble lectins and human fucosylated glycans, structures, thermodynamics and design of competing glycocompounds. C R Chim 16(5):482–490. doi:10.1016/j.crci.2012.11.021

15. Hu D, Tateno H, Hirabayashi J (2015) Lectin engineering, a molecular evolutionary approach to expanding the lectin utilities. Molecules 20(5):7637–7656. doi:10.3390/molecules20057637

16. Moriuchi H, Unno H, Goda S, Tateno H, Hirabayashi J, Hatakeyama T (2015) Mannose-recognition mutant of the galactose/N-acetylgalactosamine-specific C-type lectin CEL-I engineered by site-directed mutagenesis. Biochim Biophys Acta 1850(7):1457–1465. doi:10.1016/j.bbagen.2015.04.004

17. Romano PR, Mackay A, Vong M, DeSa J, Lamontagne A, Comunale MA, Hafner J, Block T, Lec R, Mehta A (2011) Development of recombinant *Aleuria aurantia* lectins with altered binding specificities to fucosylated glycans. Biochem Biophys Res Commun 414(1):84–89. doi:10.1016/j.bbrc.2011.09.027

18. Keogh D, Thompson R, Larragy R, McMahon K, O'Connell M, O'Connor B, Clarke P (2014) Generating novel recombinant prokaryotic lectins with altered carbohydrate binding properties through mutagenesis of the PA-IL protein from Pseudomonas aeruginosa. Biochim Biophys Acta 1840(6):2091–2104. doi:10.1016/j.bbagen.2014.01.020

19. Sudakevitz D, Kostlánová N, Blatman-Jan G, Mitchell EP, Lerrer B, Wimmerová M, Katcoff DJ, Imberty A, Gilboa-Garber N (2004) A new *Ralstonia solanacearum* high-affinity mannosebinding lectin RS-IIL structurally resembling the *Pseudomonas aeruginosa* fucose-specific lectin PA-IIL. Mol Microbiol 52(3):691–700. doi:10.1111/j.1365-2958.2004.04020.x

20. Adam J, Pokorná M, Sabin C, Mitchell EP, Imberty A, Wimmerová M (2007) Engineering of PA-IIL lectin from *Pseudomonas aeruginosa* – unravelling the role of the specificity loop for sugar preference. BMC Struct Biol 7:36. doi:10.1186/1472-6807-7-36

21. Adam J, Kříž Z, Prokop M, Wimmerová M, Koča J (2008) *In silico* mutagenesis and docking studies of *Pseudomonas aeruginosa* PA-IIL lectin – Predicting binding modes and energies. J Chem Inf Model 48(11):2234–2242. doi:10.1021/ci8002107

22. Kříž Z, Adam J, Mrázková J, Zotos P, Chatzipavlou T, Wimmerová M, Koča J (2014) Engineering the *Pseudomonas aeruginosa* II lectin: designing mutants with changed affinity and specificity. J Comput Aided Mol Des 28(9):951–960. doi:10.1007/s10822-014-9774-7

23. Wimmerová M, Mishra NK, Pokorná M, Koča J (2009) Importance of oligomerisation on *Pseudomonas aeruginosa* Lectin-II binding affinity. *In silico* and *in vitro* mutagenesis. J Mol Model 15(6):673–679. doi:10.1007/s00894-009-0464-7

24. Nguyen HH, Park J, Kang S, Kim M (2015) Surface plasmon resonance: a versatile technique for biosensor applications. Sensors 15(5):10481–10510. doi:10.3390/s150510481

25. Helmerhorst E, Chandler DJ, Nussio M, Mamotte CD (2012) Real-time and label-free bio-sensing of molecular interactions by surface plasmon resonance: a laboratory medicine perspective. Clin Biochem Rev 33(4):161–173

26. Freyer MW, Lewis EA (2008) Isothermal titration calorimetry: experimental design, data analysis, and probing macromolecule/ligand binding and kinetic interactions. Methods Cell Biol 84:79–113. doi:10.1016/S0091-679X(07)84004-0

27. Duff MR Jr, Grubbs J, Howell EE (2011) Isothermal titration calorimetry for measuring macromolecule-ligand affinity. J Vis Exp 55, e2796. doi:10.3791/2796

28. Jerabek-Willemsen M, André T, Wanner R, Roth HM, Duhr S, Baaske P, Breitsprecher D (2014) MicroScale thermophoresis: interaction analysis and beyond. J Mol Struct 1077:101–113. doi:10.1016/j.molstruc.2014.03.009

29. Seidel SAI, Dijkman PM, Lea WA, van den Bogaart G, Jerabek-Willemsen M, Lazic A, Joseph JS, Srinivasan P, Baaske P, Simeonov A, Katritch I, Melo FA, Ladbury JE, Schreiber G, Watts A, Braun D, Duhr S (2013) Microscale thermophoresis quantifies biomolecular interactions under previously challenging conditions. Methods 59(3):301–315. doi:10.1016/j.ymeth.2012.12.005

30. Krauss IR, Merlino A, Vergara A, Sica F (2013) An overview of biological macromolecule crystallization. Int J Mol Sci 14(6):11643–11691. doi:10.3390/ijms140611643

31. McPherson A, Gavira JA (2014) Introduction to protein crystallization. Acta Crystallogr F Struct Biol Commun 70(Pt 1):2–20. doi:10.1107/S2053230X13033141

32. Krishnan VV, Rupp B (2012) Macromolecular structure determination: comparison of X-ray crystallography and NMR spectroscopy. John Wiley & Sons Ltd, Chichester, http://www.els.net. doi:10.1002/9780470015902.a0002716.pub2

33. Wang H (2015) Cryo-electron microscopy for structural biology: current status and future perspectives. Sci China Life Sci 58(8):750–756. doi:10.1007/s11427-015-4851-2

34. Rosano GL, Ceccarelli EA (2014) Recombinant protein expression in *Escherichia coli*: advances and challenges. Front Microbiol 5:172. doi:10.3389/fmicb.2014.00172

35. Papaneophytou CP, Kontopidis G (2014) Statistical approaches to maximize recombinant protein expression in *Escherichia coli*: a general review. Protein Expr Purif 94:22–32. doi:10.1016/j.pep.2013.10.016

36. Young CL, Britton ZT, Robinson AS (2012) Recombinant protein expression and purification: a comprehensive review of affinity tags and microbial applications. Biotechnol J 7(5):620–634. doi:10.1002/biot.201100155

37. Fernández FJ, Vega MC (2013) Technologies to keep an eye on: alternative hosts for protein production in structural biology. Curr Opin Struct Biol 23(3):365–373. doi:10.1016/j.sbi.2013.02.002

38. Saraswat M, Musante L, Ravidá A, Shortt B, Byrne B, Holthofer H (2013) Preparative purification of recombinant proteins: current status and future trends. Biomed Res Int 2013:312709. doi:10.1155/2013/312709

39. Guan D, Chen Z (2014) Challenges and recent advances in affinity purification of tag-free proteins. Biotechnol Lett 36(7):1391–1406. doi:10.1007/s10529-014-1509-2

40. Pina AS, Lowe CR, Roque ACA (2013) Challenges and opportunities in the purification of recombinant tagged proteins. Biotechnol Adv 32(2):366–381. doi:10.1016/j.biotechadv.2013.12.001

41. Bansal P, Ajay D (2012) Laboratory dialysis—past, present and future. Recent Pat Biotechnol 6(1):32–44. doi:10.2174/187220812799789217

Chapter 29

Analytical Methods for Assessing the Effects of Site-Directed Mutagenesis on Protein–Cofactor and Protein–Protein Functional Relationships

Esha Sehanobish, Brian A. Dow, and Victor L. Davidson

Abstract

To completely understand the role of an amino acid residue that is targeted for site-directed mutagenesis a thorough analysis of the impact that the mutation has on the function of the protein is required. General methods for performing site-directed mutagenesis and expressing the recombinant protein variant are described. Protein–cofactor interactions are important because cofactors are often directly involved in facilitating catalysis by enzymes and in electron transfer by redox proteins. Many cofactors also have characteristic spectroscopic properties. As such, general methods are described to analyze the spectroscopic, redox and catalytic properties of protein-bound cofactors. Methods for assessing the effects of a mutation on protein–protein interactions are also described. Lastly, methods for assessing the overall structural integrity of the protein are described, as this is important to ensure that the mutation has not caused a global disruption of protein structure, rather than a specific effect on function.

Key words Binding constant, Circular dichroism, Coenzymes, Metalloprotein, Polymerase chain reaction, Protein expression, Redox potential, Steady-state kinetics, Tryptophan fluorescence

1 Introduction

This chapter describes methods that can be used to assess the effects of site-directed mutagenesis on cofactor–protein and protein–protein interactions. Cofactors are molecules which are associated with proteins and are required for the activity of the protein [1]. There are four general classifications of cofactors. Organic cofactors are often referred to as coenzymes as they are typically required for the enzymatic activity of the protein. Many are derived from essential vitamins. They can be either covalently bound to the protein or tightly or reversibly non-covalently bonded to the protein. An example of a dissociable organic cofactor is nicotinamide adenine dinucleotide (NAD+). Examples of tightly bound organic cofactors which do not dissociate unless the protein is denatured include flavin adenine mononucleotide

Andrew Reeves (ed.), *In Vitro Mutagenesis: Methods and Protocols*, Methods in Molecular Biology, vol. 1498,
DOI 10.1007/978-1-4939-6472-7_29, © Springer Science+Business Media New York 2017

(FMN) and pyridoxal phosphate (PLP). Organometallic cofactors are organic molecules which contain a metal ion. These are typically tightly bound to the protein and do not dissociate. They are involved in catalytic and redox reactions. Two examples are heme, which is a porphyrin ring that contains iron, and cobalamin, which is a corrin ring that contains cobalt. Inorganic metal ions can be bound to proteins and act as cofactors. These include Cu, Mg, Fe, Ni, Mn, and Zn. Amino acid residues of the protein provide ligands for the metals that stabilize binding and influence the properties of the metals. Protein-derived cofactors are not exogenous cofactors that associate with proteins but ones that have been generated by posttranslational modification of amino acids within a protein [2]. The modifications endow the protein with a catalytic or redox function, or both. In this case, amino acid residues in the protein not only interact with the cofactor to perform the biological function, but also may be involved in the biogenesis of the cofactor. Two examples are tryptophan tryptophylquinone (TTQ) which is derived from two tryptophan residues, and lysine tyrosylquinone (LTQ) which is derived from a lysine and a tryptophan residue.

Many cofactors are chromophores or fluorophores that have characteristic spectral features. These features are usually influenced by interactions with the protein. As such, various spectroscopic techniques such as absorbance and fluorescence spectroscopy may be used to probe protein–cofactor interactions. For cofactors that are redox-active, interactions with the host protein will influence the oxidation–reduction midpoint potential (E_m) value. When cofactors are present in the active site of an enzyme, interactions with the protein will influence their catalytic properties. Steady-state kinetic analysis of the enzyme-catalyzed reaction is therefore a useful approach to assess the effect of altered protein–cofactor interactions on the steady-state kinetic parameters, k_{cat} and K_m.

Protein–protein interactions do not necessarily give rise to a spectral change that can be used to monitor and quantitate the interaction. However, in cases in which there is a spectral change associated with a reaction between two proteins, which is often the case with cofactor-bearing proteins, it may be possible to determine the dissociation constant (K_d) for the protein–protein complex using spectrophotometric or kinetic techniques.

Sometimes a protein will lose its function as a consequence of the introduction of a site-directed mutation. In this case it is necessary to assess the structural integrity of the protein to ensure that the mutation has not caused a global disruption of protein structure, rather than it being a specific effect on function. Useful techniques to assess this property include circular dichroism (CD) and intrinsic fluorescence spectroscopy.

2 Materials

2.1 Spectroscopy equipment

1. Quartz cuvettes.
2. 0.4 cm × 0.4 cm rectangular quartz cuvette.
3. Scanning UV–visible spectrophotometer.
4. Fluorimeter, equipped with a temperature-controlled cuvette holder, if necessary.
5. pH meter.
6. Spectropolarimeter equipped with a Peltier temperature controller (e.g., Jasco J-810) or comparable equipment.
7. Spectrofluorimeter.

2.2 Site-Directed Mutagenesis

1. Plasmid containing the target gene of interest.
2. Forward and reverse DNA primers.
3. Thermal cycler (e.g., Bio-Rad T100) or comparable instrument.
4. QuikChange Lightning Site-Directed Mutagenesis Kit (Agilent) or comparable kit.
5. XL10-Gold Ultracompetent Cells (Agilent) or a comparable strain of competent cells.
6. Luria Bertani (LB) medium: 10 g/L tryptone, 5 g/L yeast extract, and 5 g/L NaCl. pH to 7.0 with 5 N NaOH. Add agar to 1.5 % for making agar plates.

2.3 Protein Expression

1. *E. coli* BL21 cells or other cells optimized for protein expression.
2. LB medium as described above.
3. Isopropyl-β-d-1-thiogalactopyranoside (IPTG) solution: Add from a 1 M stock to 1.0 mM final concentration.
4. Sonicator or French Press.

2.4 Spectroscopic Analyses

1. UV-Visible Spectrophotometer.
2. Orion 370 PerpHecT benchtop pH/ORP/ISE/T meter or comparable equipment.
3. MI-800 ORP electrode (Microelectrodes, Inc.) or comparable redox electrode.
4. Hamilton 600 series gastight syringe equipped with PB600 repeating dispenser.
5. Quartz cuvettes.
6. Quinhydrone solution (a saturated solution, approximately 100 μM quinhydrone in pH 7.0 buffer).
7. 100 μM sodium dithionite dissolved in water.
8. 50 μM potassium ferricyanide dissolved in water.

3 Methods

3.1 Site-Directed Mutagenesis

There are several commercially available kits that can be used for performing site-directed mutagenesis. It is simpler and more efficient to use a kit rather than purchasing each of the components separately or making them oneself. Each of these kits is based on essentially the same principle; they use the Polymerase Chain Reaction (PCR) to incorporate mutations into the gene sequence. However, the exact methods for doing this vary from kit to kit. The protocol described here uses the QuikChange Lightning Site-Directed Mutagenesis Kit from Agilent technologies (*see* **Note 1**).

1. Design the forward primer. The forward primer will anneal to the template DNA strand in the 5′–3′ direction and will consist of a nucleotide sequence of 25–45 bases. Change the base(s) in the codon for the target amino acid to give the desired mutation (*see* **Note 2**).

2. Design the reverse primer. This is simply the sequence complementary to the forward primer in the reverse 5′–3′ direction. The forward and reverse primers need to anneal for the PCR reaction to take place.

3. Determine the melting temperature (T_m) and GC content of the primers once they are designed. This can be done using online programs such as www.basic.northwestern.edu/biotools which will analyze the sequence. These programs can also predict whether the sequence is likely to form hairpin loops or undergoes self-complementarity, both of which are undesirable. If all looks well then have the primers synthesized either in-house or commercially (*see* **Note 3**).

4. Perform the PCR reaction to synthesize the mutated DNA strand according to the instructions provided in the QuikChange Lightning Kit using the desired host plasmid and the designed forward and reverse primers and the reagents provided in the kit. The reaction will proceed as follows: thermal denaturation of the DNA, primer annealing and primer extension. These steps will be repeated 25–30 times and then the mixture will be incubated at 37 °C with the restriction enzyme *Dpn*I to remove the methylated parent DNA strands. After this step the mixture can be stored at 4 °C (*see* **Note 4**).

5. Transform the plasmid containing the mutation which was generated in the previous step into competent cells which have been optimized for this process. This involves incubating the plasmid DNA with the cells under appropriate conditions for transforming as an example XL10-Gold Ultracompetent Cells (*see* **Note 5**).

6. Incubate the transformed cells in a shaker for 1 h at 37 °C in growth medium and then spread them on an LB agar plate which includes the antibiotic to which the plasmid confers resistance. The plate is incubated for about 16 h (overnight) at 37 °C to visualize colonies.

7. Select a single colony after overnight incubation and grow it in 4 ml LB medium containing the appropriate antibiotic. Next, extract the plasmids from the cells using a miniprep DNA extraction kit. Many such kits are commercially available.

8. Sequence the region of the plasmid containing the desired mutated gene region to ensure that the mutation has indeed been correctly incorporated. The sequencing may be done in-house or outsourced. Several companies will perform sequencing at reasonable prices.

4 Recombinant Variant Proteins

1. Transform the plasmid encoding the mutated gene into an *E. coli* strain that has been optimized for protein expression, such as BL21 cells, using the method described above in step 4 in Section 3.1.

2. Plate the transformed cells on LB agar plates containing the appropriate antibiotic (typically at a concentration of 25–100 μg/ml) and a colony is picked and grown in liquid culture. Depending on the level of protein expression and amount of protein needed, the culture sizes may vary from a few milliliters to several liters.

3. Induce the culture with IPTG once the culture reaches an optical density (OD 600 nm) of 0.6. IPTG is added to a final concentration of 1 mM to induce selective expression of the protein under the control of the lac operator, which is commonly designed into commercial expression plasmids (*see* **Note 6**). The time and temperature of induction varies but typically is around 4 h and 30 °C, respectively (*see* **Note 7**).

4. Cells are harvested and disrupted by sonication or French press or an alternative method. After centrifugation, the soluble contents of the cells are subjected to chromatography using a resin that is designed to bind the tag which has been engineered at the end of the protein. Resins are commercially available with instructions for their use (*see* **Note 8**).

4.1 General Mutational Effects

Many protein cofactors have characteristic visible absorbance. Some examples are shown in Fig. 1. Mutation of an amino acid residue in the proximity of the cofactor may affect its absorbance properties. It may shift the absorbance wavelength maximum of

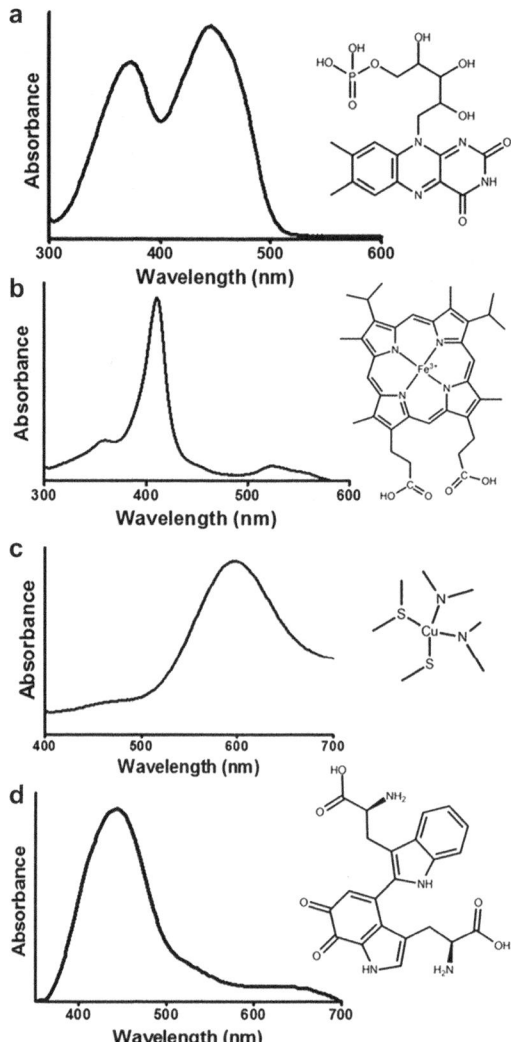

Fig. 1 Examples of protein-bound cofactors and their absorbance spectra. (**a**) Flavin mononucleotide. (**b**) Heme. (**c**) Type-1 copper site with ligands provided by two histidine nitrogens and sulfurs from a cysteine and methionine residue. (**d**) The protein-derived TTQ cofactor which is derived from two tryptophan residues

the spectral feature or it may affect the intensity of the absorbance feature (*see* **Note 9**).

1. Record the UV-visible absorbance spectrum of a cuvette containing only buffer and no protein to obtain a baseline spectrum.

2. Prepare solutions containing the native protein and the protein harboring the mutation. They should contain the same concentrations of protein.

3. Record the spectrum from ~250 nm to a sufficiently high wavelength to visualize the spectral features of the cofactor.

4. The absorbance maximum near 280 nm (A_{280}) corresponds to concentration of protein. The exact wavelength maximum may vary depending on the amino acid content of the protein. Both protein samples should have the same A_{280} in order to accurately determine any effects of the mutation on the cofactor spectrum. It is important to be certain that differences in the intensity of the cofactor absorbance are not due to differences in protein concentration. The ratio of absorbance of the peak maxima corresponding to the cofactor absorbance (A_{cof})/A_{280} is the best way to compare the intensities of the absorbance features attributable to the cofactor.

5. Shifts in wavelength maxima of the cofactor absorbance may be subtle. The best way to compare the spectra of the native and variant proteins is to construct a difference spectrum. To do this one must download the two data files of absorbance versus wavelength and then subtract one from the other. An example of a difference spectrum is shown in Fig. 2.

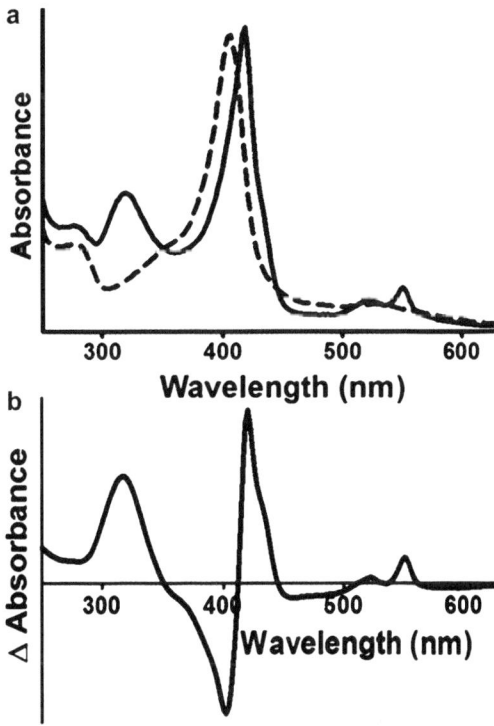

Fig. 2 Spectral changes associated with the redox state of a protein-bound cofactor, the diheme cofactor of MauG. (**a**) The *dashed line* indicates the fully oxidized state of the heme and the *solid line* indicates the fully reduced state of the cofactor. (**b**) An oxidized minus reduced difference spectrum constructed from the spectra in panel **a**. Features above the x-axis reflect increases in absorbance and those below reflect decreases in absorbance

4.2 Redox Properties of Cofactors

Many cofactors are redox-active and exhibit different spectroscopic properties depending upon their redox state. The properties provide a convenient handle for monitoring the redox reactions of protein-bound cofactors. An important property of redox-active cofactors is the oxidation-reduction midpoint potential (E_m) value associated with the interconversion between redox states. This value is very sensitive to the protein environment of the cofactor and protein–cofactor interactions. There are different methods for determining E_m values of proteins. The protocol described below is for a spectrochemical titration to determine the E_m value. This technique requires only a spectrophotometer and pH meter which can serve as a potentiometer. The technique is particularly well-suited for proteins with cofactors with visible spectra that change with redox state (see **Note 10**). A general protocol is described that would be applicable to any protein and to illustrate the method data is shown in Fig. 3 for a spectrochemical titration of amicyanin which possesses a copper cofactor [3].

1. Determine the reference spectra for the oxidized and reduced forms of the proteins. Chemical oxidants and reductants can be used to generate these forms. In this protocol sodium dithionite is used as a reductant and potassium ferricyanide is used as an oxidant.

2. The potentiometer is calibrated using a quinhydrone solution. Place the electrode in the solution and zero the meter so the measured potential of the hydroquinone solution reads as 0 mV. Since the known potential of the quinhydrone solution is

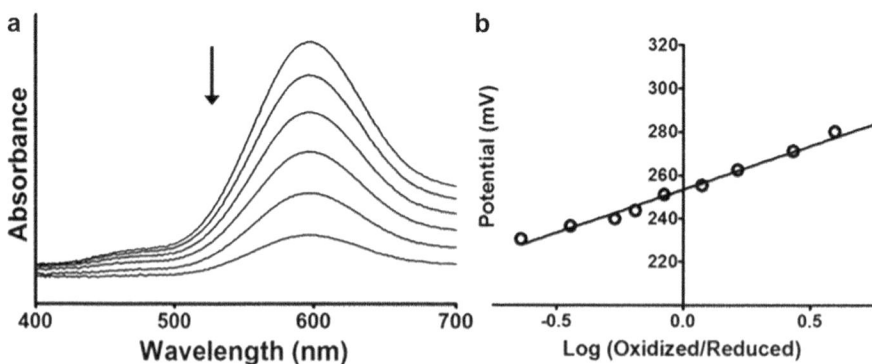

Fig. 3 Spectrochemical redox titration of a protein-bound cofactor, the type I copper on amicyanin. (**b**) Spectra of amicyanin recorded during the reductive titration after additions of sodium dithionite. Amicyanin contains a type-1 copper center, which has a strong absorbance feature centered near 600 nm when oxidized and no visible absorbance when reduced. Thus, the absorbance is lost as the cofactor is reduced. The direction of the changes in absorbance is indicated by the *arrow*. (**b**) Plot of measured ambient potential of the solution (*E*) versus the log of the ratio fraction of amicyanin which is oxidized over the fraction which is reduced at each point during the titration. The line was determined using linear regression analysis. The *y*-interecept represents the redox potential (E_m value) at which 50 % of the protein is reduced and 50 % is oxidized

+286 mV, the recorded potentials obtained during the procedure are adjusted by +286 mV to account for the calibration.

3. A cuvette containing a solution of the oxidized protein is placed in the spectrophotometer and its absorbance spectrum is recorded. The solution may also include mediators, small redox-active molecules, such as ferricyanide, dichlorophenolindophenol, or p-benzoquinone, that facilitate equilibration of the electrode and the protein [4]. The electrode is also placed in the solution and the potential of the solution is recorded (*see* **Note 11**).

4. Using a gastight syringe with a repeating dispenser, add a small amount of an anaerobic solution of sodium dithionite and mix. Record the absorbance spectrum and the potential.

5. Repeat step 4 several times until no further change in the spectrum is observed upon addition of dithionite. This completes the reductive titration.

6. While not absolutely necessary, it is also desirable to perform a titration in the reverse direction. Initially record the absorbance spectrum and the potential of the solution of the fully reduced spectrum.

7. Using a gastight syringe with a repeating dispenser, add a small amount of an anaerobic solution of potassium ferricyanide and mix well. Record the absorbance spectrum and the potential.

8. Repeat step 7 several times until no further change in the spectrum is observed on addition of dithionite. This completes the oxidative titration (*see* **Note 12**).

9. To determine the E_m value plot the recorded potential (E) versus the log of the fraction of oxidized/reduced protein, which is determined by comparison of the recorded spectrum to those of the reference spectra for the fully oxidized and reduced proteins. The data are fit to the equation:

$$E = Em + (2.3RT / nF) \log([oxidized] / [reduced])$$

where R is the gas constant, T is temperature, n is the number of electrons transferred and F is the Faraday constant. For a one-electron redox couple at 25 °C the term $(2.3RT/nF)$ is equal to 59 V. The point where the line crosses the Y-axis on the linear plot represents the E_m value of the titration.

4.3 Catalytic Properties of Cofactors

If the protein of interest is an enzyme then the effects of a mutation on enzymatic activity may be determined by performing a steady-state analysis of the native and mutated enzymes. This will allow comparison of the maximum reaction velocity (V_{max}) and the K_m for the substrate. K_m is defined as the concentration of the substrate at which the initial velocity of the enzyme is 50 % of V_{max}, and as such reflects the affinity of the enzyme for the substrate. V_{max}/K_m is considered the catalytic efficiency. Since different enzymes have different substrates and products, the protocol for assaying each enzyme is different. If the substrate or product has characteristic absorbance or fluorescence properties then the reaction can be directly monitored by the rate of change in absorbance or fluorescence. If this is not the case, then it may be possible to design a coupled assay in which the product undergoes a subsequent reaction which gives rise to a change in absorbance or fluorescence. Alternatively, it may be necessary to remove aliquots of the reaction mixture at different time points and physically analyze the aliquot to determine the concentration of product. A general protocol is described below for performing a steady-state enzyme assay which may be monitored spectrophotometrically.

1. An appropriate buffer containing the enzyme is placed in a cuvette in the spectrophotometer and allowed to equilibrate to the desired temperature.

2. The reaction is initiated by the addition of substrate. The concentration of the enzyme should be several fold less than the amount of substrate as the enzyme is a catalyst. The substrate should be added from a concentrated stock solution so that the change in the volume of the sample is minimal.

3. The absorbance or fluorescence of the sample is monitored at the wavelength maximum that corresponds to the change in absorbance that is associated with formation of product or consumption of substrate.

4. The initial rate of the reaction (v_i) is determined. This is the linear change in absorbance or fluorescence with time at the very beginning of the reaction.

5. The same experiment is repeated for a range of different concentrations of the substrate.

6. The data is analyzed to determine V_{max} and K_m (*see* **Note 13**). The initial velocity (v_i) is plotted against substrate concentration [S] and fit to the Michaelis–Menten equation:

   $$v_i = V_{max}[S]/(K_m + [S])$$ (*see* **Notes 14** and **15**). Examples of plots of v versus [S] and a double reciprocal plot of the data, using simulated data are shown in Fig. 4.

7. It is also possible that a mutation may affect the dependence of the kinetic parameters on temperature or pH or both. To determine this the steps above may be repeated over a range of temperatures and a range of pH values.

4.4 Site-Directed Mutagenesis and Protein Interactions

Several methods may be used to study protein–protein interactions. In some cases the techniques will provide confirmation that the proteins interact but do not provide quantitative information

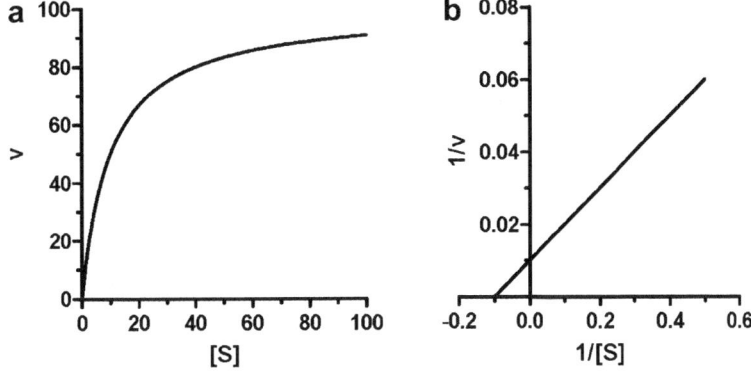

Fig. 4 Michaelis–Menten plots of simulated steady-state kinetic data. (**a**) A hyperbolic curve which describes the dependence of the initial rate (v) of an enzyme reaction on substrate concentration ($[S]$). (**b**) A double reciprocal plot of the same data used in panel **a**. In this linear transformation of the data, the y-intercept is equal to $1/V_{max}$ and the x-intercept is equal to $-1/K_m$

about the interaction. Such techniques include chemical crosslinking, co-immunoprecipitation, GST pull-down assays, and size exclusion chromatography. A single site-directed mutation will not necessarily abolish a protein–protein interaction but could affect the affinity for the protein–protein interaction. Ideally, one would like to obtain the K_d for the protein–protein interaction and then determine whether and to what extent the mutation impacted this parameter. For interactions involving one or more cofactor-bearing proteins the protein–protein interaction may be accompanied by a change in the absorbance or fluorescence of the chromophore. If so, then the protein–protein interaction may be monitored spectrophotometrically. Two general approaches, an equilibrium binding technique and a kinetic technique are described:

4.4.1 Equilibrium Binding Approach to Determine K_d

1. Place a solution of the protein of known concentration into a cuvette in a spectrophotometer and record the spectrum.

2. Prepare a more concentrated stock of the second protein.

3. Sequentially add small volumes of the second protein to the solution and record the final concentration of the second protein in the cuvette and record the spectrum.

4. Continue additions of the second protein until the change in absorbance or fluorescence is complete.

5. Plot the intensity of absorbance or fluorescence at the appropriate wavelength versus the concentration of the added protein to determine the K_d for complex formation between the proteins. Those data can be fit to the following equation for a straight line:

$Y = Y_{max}[P]/(K_d + [P])$, where Y is the intensity of absorbance or fluorescence and $[P]$ is the concentration of the added protein.

4.4.2 Single-Turnover Kinetic Approach to Determine K_d

1. Place a solution of the protein with the chromophore that will be monitored into a cuvette in a spectrophotometer and set the spectrophotometer to record absorbance or fluorescence at the wavelength of interest.

2. Prepare a concentrated stock of the second protein.

3. Add a small volume of the second protein to the solution with rapid mixing and monitor the rate of change in absorbance or fluorescence during the entire reaction. In most cases the rate of change can be fit to the equation for a single exponential transition ($\Upsilon = A_o \exp(-k_{obs}t)$), where A_o is the initial absorbance and t is time. The observed rate constant k_{obs} is determined from the fit of the data (*see* **Note 16**).

4. Repeat the experiment with increasing concentrations of the second protein until an increase in the rate constant is no longer observed.

5. Plot the observed rate constant for the reaction versus concentration of the protein whose concentration was varied to determine the K_d for complex formation between the proteins as well as the limiting first order rate constant (k_{max}) for the reaction. Those data can be fit to the following equation: $k_{obs} = k_{max}[P]/(K_d + [P])$ (*see* **Note 17**).

4.4.3 Site-Directed Mutagenesis and Protein Structure

Site-directed mutagenesis may cause unanticipated changes in the overall protein structure and stability of the protein. If the effect of site-directed mutagenesis is loss of protein function then it is important to confirm that this is a consequence of the mutation altering function and not a nonspecific effect on the structure or stability of the protein. Circular dichroism (CD) spectroscopy can be used to determine and compare the secondary structures of the native and variant proteins. Alpha-helices, beta sheets and unordered structures give rise to different CD signals. Fluorescence spectroscopy can be used to ascertain whether the protein is folded into its native structure. Aromatic residues, particularly tryptophan, will have a different fluorescence intensity depending if they are in a hydrophobic environment (i.e., the interior of the protein) or an aqueous environment (i.e., exposed at the protein surface). To ascertain whether a mutation has affected the stability of the protein one may monitor the effect of temperature on the CD or fluorescence spectrum and determine a T_m that corresponds to the thermal stability of the protein [5].

4.4.4 Circular Dichroism (CD)

1. Make a solution of the native and variant proteins of sufficient concentration that the spectroscopic feature and subsequent changes to it may be monitored and measured (*see* **Note 18**). Place the protein solution in a cuvette in the spectropolarimeter.

2. Record the spectra at the temperature at which the protein function is typically studied.

3. Convert the measured ellipticity (θ_{meas}) to mean residue molar ellipticity (θ) by the formula $\theta = \theta_{meas}/nlc$, where n is the number of residues of the protein, l is the optical pathlength in mm, and c is the molar concentration of the protein. Plot (θ) versus wavelength to obtain the CD spectra of the native and variant proteins for comparison to determine whether or not there have been any alterations in the secondary structure of the protein.

4. To assess the thermal stability of the proteins record the CD spectra over a range of temperatures.

5. After recording the spectrum as described above, increase the temperature by 1–2 °C increments and after incubation of the protein for a few minutes at each temperature record the CD spectrum.

6. Continue recording spectra at increasing temperatures until the protein is denatured as indicated by no further changes in the spectrum with increasing temperature. An intense negative signal with a mean residue molar ellipticity at 200 nm will appear with increasing temperature, which corresponds to an increasing amount of unordered structure. The temperature at which 50 % of this transition occurs is the T_m value, which reflects the thermal stability of the protein (*see* **Note 19**). An example of the results of such an experiment showing the thermal unfolding of a protein is shown in Fig. 5.

4.4.5 Intrinsic Fluorescence Spectroscopy

1. Prepare solutions of the native and variant proteins. Place each protein solution in a cuvette in the spectrofluorometer.

2. Excite the sample at 280 nm, which is the absorbance maximum for tryptophan residues and record the fluorescence emission spectrum which will have a wavelength maximum between 300 and 350 nm depending on the polarity of the tryptophan environment.

3. To assess the thermal stability of the proteins record the fluorescence spectra over a range of temperatures as described above for the CD protocol.

4. Continue recording spectra until the protein is denatured at which point the change in fluorescence intensity will no longer change with temperature. The temperature at which 50 % of this transition occurs is the T_m value which reflects the thermal stability of the protein (*see* **Note 20**).

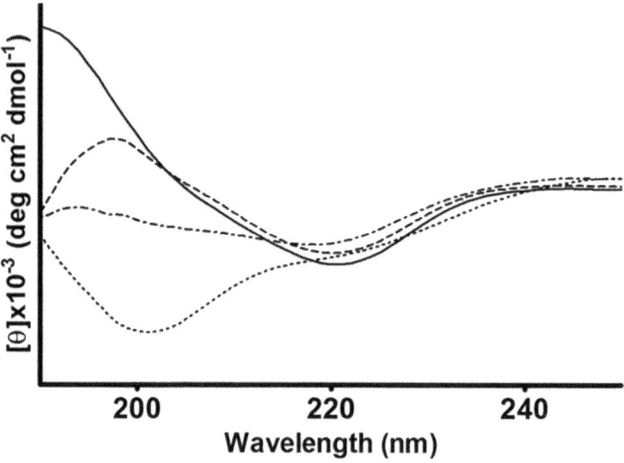

Fig. 5 Temperature-dependent changes in the CD spectrum of a protein, amicyanin. The *solid, dashed, dashed/dotted,* and *dotted lines* represent the CD spectra of amicyanin recorded at 25, 40, 58, and 80 °C, respectively. At 25 °C, the spectrum of amicyanin displays a maximum around 195–198 nm which is indicative of a beta-sheet secondary structure and is predominant in amicyanin in its native, folded state. As the temperature increases, this feature is lost and a minimum near 200 nm appears. This new feature is characteristic of unordered structure, which predominates at the highest temperature when the protein is unfolded as a consequence of thermal denaturation

5   Notes

1. With this kit one can perform mutagenesis on any double stranded circular plasmid DNA. One doesn't need to linearize it for amplification or ligate it afterwards. An example of an alternative approach is the Phusion site directed mutagenesis kit (New England Biolabs). It utilizes the "Round the Horn" method, which linearizes the plasmid first, then amplifies the linearized plasmid, and then ligates the DNA strands.

2. Try using the least number of changes in the bases as necessary. The more bases that are changed the less efficient the annealing will be due to the greater number of mis-matches. The mutation should be in the middle of the primer sequence to maximize the efficiency of annealing of the flanking sequences.

3. It should be noted that very high GC content of the primer will increase the T_m. The melting temperature (T_m) should be less than 78 °C.

4. It may be desirable at this point to subject the PCR product to agarose gel electrophoresis to check its size to be sure that the reactions worked.

5. (a) Prechill two 14-ml BD Falcon polypropylene round-bottom tubes on ice. (One tube is for the experimental transformation and one tube is for the pUC18 control.) Preheat NZY+ broth to 42 °C. (b) Thaw the cells on ice. When thawed, gently mix and aliquot 100 μl of cells into each of the two prechilled tubes. (c) Add 4 μl of the β-ME mix provided with this kit to each aliquot of cells. (d) Swirl the tubes gently. Incubate the cells on ice for 10 min, swirling gently every 2 min. (e) Add 0.1–50 ng of the experimental DNA (or 2 μl of a ligation mixture) to one aliquot of cells. Dilute the pUC18 control DNA 1:10 with sterile dH₂O, then add 1 μl of the diluted pUC18 DNA to the other aliquot of cells. (f) Swirl the tubes gently, then incubate the tubes on ice for 30 min. (g) Heat-pulse the tubes in a 42 °C water bath for 30 s. The duration of the heat pulse is critical. (h) Incubate the tubes on ice for 2 min. (i) Add 0.9 ml of preheated (42 °C) NZY+ broth and incubate the tubes at 37 °C for 1 h with shaking at 225–250 rpm. (j) Plate ≤200 μl of the transformation mixture on LB agar plates containing the appropriate antibiotic (and containing IPTG and X-gal if color screening is desired). For the pUC18 control transformation, plate 5 μl of the transformation on LB-ampicillin agar plates. (k) Incubate the plates at 37 °C overnight. If performing blue-white color screening, incubate the plates at 37 °C for at least 17 h to allow color development (color can be enhanced by subsequent incubation of the plates for 2 h at 4 °C). For the pUC18 control, expect 250 colonies (≥5 × 10⁹ cfu/μg pUC18 DNA). For the experimental DNA, the number of colonies will vary according to the size and form of the transforming DNA, with larger and non-supercoiled DNA producing fewer colonies.

6. Commercial plasmid vectors should have an origin of replication and a multiple cloning site at which genes may be inserted using restriction enzymes. An inducible promoter, such as the lac operon, is located before the gene. This vector should also possess a drug-resistance gene for a particular antibiotic such as ampicillin or kanamycin. Many vectors also have gene sequences at the N- or C-terminus of the inserted gene that encode polyhistidine, maltose-binding protein or glutathione S-transferase tags to facilitate purification of the expressed protein.

7. If expression levels are less than expected, then the induction conditions can be altered. Sometimes it is necessary to use lower temperature during the induction or decrease or increase the time. Some trial and error may be necessary to optimize conditions.

8. Common tags for protein purification include a polyhistidine tag, a maltose-binding protein tag and a glutathione S-transferase tag. Resins containing ligands for these tags are commercially available that can be used to purify the tagged proteins by affinity chromatography.

9. Absorbance spectroscopy may not be appropriate in all cases. Fluorescence spectroscopy may be better for some cofactors. More sophisticated spectroscopic techniques such as electron paramagnetic resonance spectroscopy may be useful for some metallo-cofactors.

10. E_m values may also be obtained using a technique called cyclic voltammetry [6] which does not require a spectroscopic change to monitor. However, this technique requires specialized equipment and the protein must be adsorbed to a surface during the measurement, rather than be present in solution.

11. It is usually necessary to perform this experiment under anaerobic conditions if the redox protein or reagents used are sensitive to oxygen. One way for this to be done is to perform the experiments in an anaerobic glove box. Alternatively, all stock solutions can be prepared anaerobically and the cuvette may be fitted with a septum through which the electrode is inserted and made anaerobic. One method for doing the latter is to seal containers and the cuvette with a septum and insert a needle which is attached to tubing. Using a vacuum pump and argon tank the sample is made anaerobic by subjecting it to several cycles of alternating vacuum followed by bubbling with argon.

12. The final spectrum should be identical to the starting spectrum of the oxidized protein. This means that the titration was reversible and the protein was not altered during the process. If the spectrum is altered, then this means that the protein may have been damaged or denatured during the process.

13. If an enzyme uses two substrates then the experiment should be done with one fixed substrate present in large excess and the other substrate varied. Then the fixed and varied substrates can be switched and the experiment repeated. V_{max} should be the same in both cases. The K_m values of each substrate would be considered apparent K_m values but still are useful for assessing the effects of mutation. More sophisticated experiments would be required to obtain the true K_m values of each substrate.

14. If one does not have a program that can fit the data using nonlinear curve fitting the data can be plotted as a double reciprocal plot of $1/v_i$ versus $1/[S]$ and fit by the following equation of a straight line: $1/v_i = \left(K_m / V_{max}\right) / \left(1/[S]\right) + 1/V_{max}$.

15. Most enzymes obey Michaelis–Menten kinetics but some do not. The most common exception are allosteric enzymes which exhibit cooperativity. In this case the plot of v_i versus $[S]$ would yield a sigmoidal plot rather than a hyperbolic plot, and an alternative method of data analysis would be required.

16. This kinetic experiment is different from steady-state kinetic experiments in that it is not the initial rate that is measured but the overall rate constant for the entire reaction. One is not monitoring product formation of a reaction but the change in the properties of the individual protein during a single reaction.

17. The reaction may be too fast to observe in a conventional spectrophotometer using manual mixing. More specialized instrumentation would then be required. For reactions occurring in the millisecond range a stopped-flow spectrophotometer can be used. In this configuration the two reactants are rapidly transferred into a flow cell and mixed within a dead-time of a few milliseconds. Analysis of the data would be the same as described in the protocol.

18. The amount of protein is usually very small, less than 10 μM depending on the size of the protein and the buffer composition. Small amounts of protein are necessary due to the sensitivity of the signal at lower wavelengths, which result from excess light scattering in the far UV region as protein concentration increases. High concentrations of salt in the buffer can cause similar problems.

19. The signal at 200 nm, which corresponds to the amount of unordered structure, should be used to determine Tm, as opposed to signals which may reflect other secondary structures, such as alpha helices or beta sheets. This is because these structures may not fully denature due to their high numbers of hydrogen bonds. However, simpler structures such as loops and turns will denature easily and allow for the protein to distort, collapse, and/or aggregate and also allow for water to enter the hydrophobic core. Thus, even though some structures may not become fully denatured themselves, the protein as a whole can be considered denatured, and this can be seen by the negative signal around 200 nm, which corresponds to unordered structure.

20. Depending upon the polarity of the environment of the tryptophan, the wavelength maximum will be between 300 and 350 nm. As the protein unfolds, not only will the intensity of the tryptophan fluorescence emission spectra change, but it is likely that the peak will shift as the tryptophan is further exposed to water or is further sequestered in the hydrophobic interior of aggregates.

References

1. Broderick JB (2001) Coenzymes and cofactors. In: Encyclopedia of life sciences (eLS). John Wiley & Sons Ltd, Chichester. http://www.els.net. doi:10.1038/npg.els.0000631.

2. Davidson VL (2007) Protein-derived cofactors. Expanding the scope of post-translational modifications. Biochemistry 46:5283–5292

3. Zhu Z, Cunane LM, Chen Z, Durley RC, Mathews FS, Davidson VL (1998) Molecular basis for interprotein complex-dependent effects on the redox properties of amicyanin. Biochemistry 37:17128–17136

4. Durst RA, Fultz ML (1982) Mediator compounds for the electrochemical study of biological redox systems: a compilation. Anal Chim Acta 140:1–18

5. Sreerama N, Woody RW (2000) Circular dichroism of peptides and proteins. In: Berova N, Nakanishi K, Woody RW (eds) Circular dichroism: principles and applications, 2nd edn. John Wiley & Sons, New York, NY, pp 601–620

6. Mabbott GA (1983) An introduction to cyclic voltammetry. J Chem Ed 60:697–702

Chapter 30

Biochemical and Biophysical Methods to Examine the Effects of Site-Directed Mutagenesis on Enzymatic Activities and Interprotein Interactions

Misaki Kinoshita, Ju Yaen Kim, Yuxi Lin, Natalia Markova, Toshiharu Hase, and Young-Ho Lee

Abstract

Mutations in proteins often affect interactions with partner molecules, sequentially changing their activities and functions. In order to examine mutagenic effects, we herein describe practical and detailed protocols for enzymatic activity assays using ferredoxin (Fd)-NADP$^+$ reductase (FNR) and sulfite reductase (SiR), which are electron-transferring enzymes for the Calvin cycle and sulfur assimilation in various organisms, respectively. Methods for isothermal titration calorimetry and nuclear magnetic resonance spectroscopy, which are very useful thermodynamically and mechanically for investigating the effects of mutations on intermolecular interactions, are also described with practical examples of the Fd–FNR binding system.

Key words Electron transfer, Enzymatic activity, Ferredoxin, Ferredoxin-NADP$^+$ reductase, Isothermal titration calorimetry, Nuclear magnetic resonance spectroscopy, Redox protein, Sulfite reductase

1 Introduction

The interfaces of protein–protein complexes were previously reported to statistically consist of 57 % apolar residues, 19 % charged residues, and 24 % polar residues [1]. This finding indicates that the molecular recognition of proteins depends on the physico-chemical properties of protein surfaces and also that non-covalent interactions, both electrostatic and non-electrostatic forces (*see* **Note 1**), are indispensable in protein–protein and protein–ligand interactions. Thus, the intermolecular non-covalent interactions between enzymes and binding partners including substrates are critical for enzymes to exert their effects [2–5]. Any disruption of the forces on intermolecular interfaces by mutations or environmental stresses such as changes in pH, salt concentrations, and temperature often decreases enzymatic activity.

Andrew Reeves (ed.), *In Vitro Mutagenesis: Methods and Protocols*, Methods in Molecular Biology, vol. 1498,
DOI 10.1007/978-1-4939-6472-7_30, © Springer Science+Business Media New York 2017

We herein focus on the mutagenic effects on intermolecular interactions between oppositely charged proteins and how mutations affect enzyme activity using the combined methods of biophysics and biochemistry. We show and explain how to measure and interpret changes in the enzyme activities and interactions between ferredoxin (Fd) and either ferredoxin-NADP($^+$) reductase (FNR) or sulfite reductase (SiR) with mutations on the basis of a detailed protocol and useful suggestions.

Fd is a small acidic protein (*see* Fig. 1) that transfers electrons to various enzymes including FNR [6–9] and SiR [2, 5, 10] in many organisms. FNR receives two electrons from Fd and reduces NADP$^+$ to NADPH (*see* Fig. 2a). Six electrons flow from Fd to SiR by forming the Fd–SiR complex with the reduction of SO_3^{2-} to S_2^- (*see* Fig. 2b). This interprotein electron transfer is performed through the formation of a complex (*see* Fig. 1). Thus, the formation of an electron transfer complex is essential for the activities of FNR and SiR.

We previously demonstrated that basic patches of FNR and SiR formed a complementary electrostatic network with acidic patches of Fd [2, 5, 6, 9] (*see* also the crystal structures of the Fd–SiR complex: PDB ID, 5H8V, 5H8Y, and 5H92), suggesting

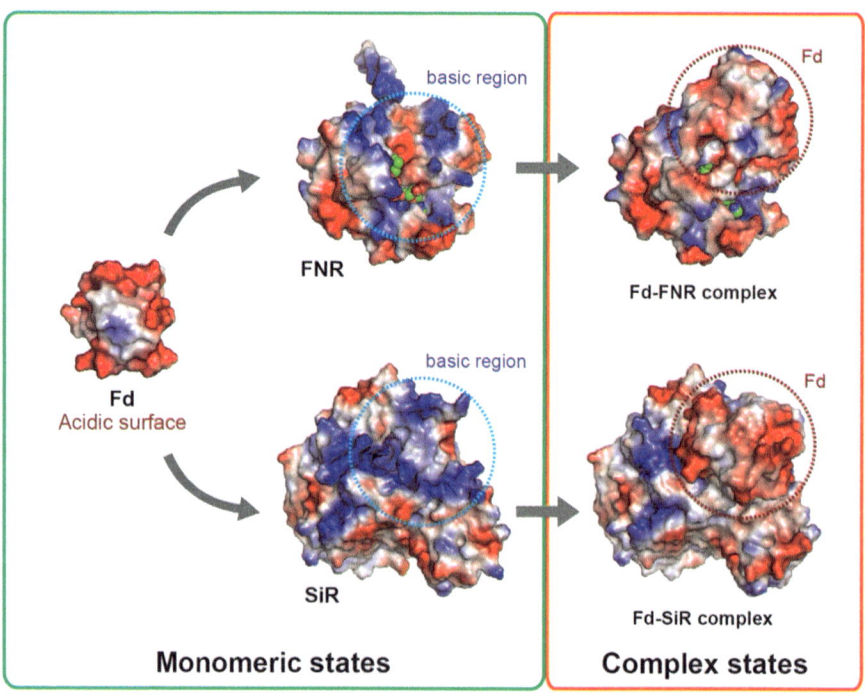

Fig. 1 Three dimensional structures of free Fd, FNR, and SiR and their complexes. X-Ray crystal structures of Fd (PDB ID: 1GAQ), FNR (PDB ID: 1GAQ), and SiR (PDB ID: 5H92) in their free forms (*left*) and the complex of Fd with FNR (*right upper*) (PDB ID: 1GAQ) and SiR (*right lower*) (PDB ID: 5H92) are shown. Electrostatic surface potential on the interaction side of SiR with Fd is shown. The positive potential is colored in *blue* and negative potential in *red*

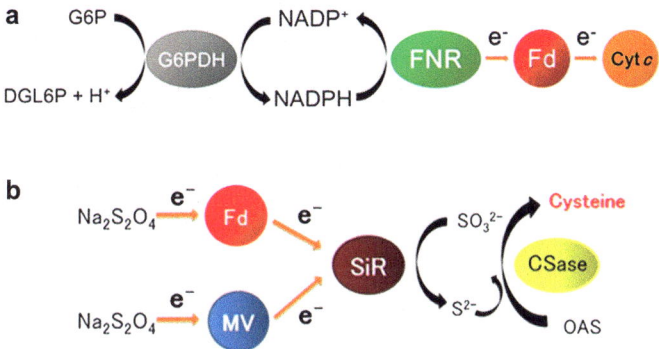

Fig. 2 Assay schemes of FNR and SiR activities. (**a**, **b**) The three assays for the activities of FNR (**a**) and SiR (**b**) are illustrated. Proteins used for activity assays are schematically represented by ellipses: Fd, FNR, cytochrome c (cyt c), SiR, and cysteine synthase (CSase). Electrons are shown with an e^-. (**a**) Glucose-6-phosphate dehydrogenase (G6PDH), reduced by glucose-6-phosphate (G6P), provides NADP$^+$ with electrons. FNR is sequentially reduced by NADPH and an electron flows to Fd. The reduction of cyt c is monitored by an increase in absorption at 550 nm. DGL6P indicates d-glucono-1, 5-lactone-6-phosphate. (**b**) Fd and methyl viologen (MV) are reduced by sodium dithionite (Na$_2$S$_2$O$_4$). Reduced Fd (the Fd-dependent SiR activity assay, upper pathway) or MV (the MV-dependent SiR activity assay, lower pathway) transfers an electron to SiR. Sulfide (S^{2-}) produced by SiR is converted to cysteine in the presence of an excess amount of O-acetyl serine and CSase. The production of cysteine is monitored by an increase in the absorption intensity at 546 nm resulting from the reaction of cysteine with acid-ninhydrin reagent

that attractive electrostatic interactions on interfaces stabilize electron transfer complexes. Based on our X-ray crystal structures of the Fd–FNR complex [6] and Fd–SiR complex (PDB ID, 5H8V, 5H8Y, and 5H92) (*see* Fig. 1), we prepared a series of mutant proteins: Fd mutants, the interfacial physicochemical properties of which were altered in order to perturb FNR activity and interactions with FNR [4], and a SiR mutant, the interfacial positive charge of which was neutralized in order to affect SiR activity and interactions with Fd [5].

We herein describe three types of activity assays using these mutants: the NADPH-dependent activity assay for overall FNR activity, the Fd-dependent SiR activity assay, which provides information on interprotein interactions between Fd and SiR, and the methyl-viologen (MV)-dependent SiR activity assay for evaluating the intramolecular activity of SiR. We provide a detailed protocol for isothermal titration calorimetry (ITC) measurements as well as practical suggestions. An interpretation of thermodynamic parameters is also given. We also describe solution-state NMR spectroscopy of ^{15}N-labeled FNR in the absence and presence of Fds, which provides useful information on the interacting site and mode at atomic resolution in solution.

2   Materials

Prepare all solutions using doubly deionized water (DDW) and analytical grade reagents. DDW may be further filtrated using a 0.2- or 0.5-μm filter. Perform all procedures at room temperature unless otherwise noted.

2.1   Chemicals, Media, Reagents

1. Lysogeny Broth (LB) medium: dissolve 80 g of tryptone, 50 g of yeast extract, and 80 g of NaCl into 8 L of DDW. pH to 7.0 using 5 N NaOH. Sterilize by autoclaving at 121 °C, 15 psig for 1 h.

2. M9 minimal medium: To make M9 salts stock solution, aliquot 800 mL H_2O and add 64 g Na_2HPO_4-$7H_2O$, 15 g KH_2PO_4, 2.5 g NaCl, 5.0 g NH_4Cl. Stir until dissolved. Adjust to 1000 mL with distilled H_2O. Sterilize by autoclaving. To make M9 medium measure ~700 mL of distilled H_2O (sterile), add 200 mL of M9 salts, 2 mL of 1 M $MgSO_4$ (sterile), 20 mL of 20% glucose (or other carbon source) and 100 μL of 1 M $CaCl_2$ (sterile). Adjust to 1000 mL with distilled H_2O. Add $^{15}NH_4Cl$ as the sole nitrogen source for tracer studies.

3. 2% IPTG (isopropyl-β-d-thiogalactopyranoside) stock solution.

4. IPTG-inducible *E. coli* strains expressing selected proteins.

5. *E. coli* strain JM109.

6. Buffer A: 50 mM Tris–HCl, 150 mM NaCl, 1 mM $MgCl_2$, 1 mM EDTA, and 0.1% β-mercaptoethanol. Final pH of 7.5.

7. Ammonium sulfate.

8. 50 mM Tris–HCl buffer (pH 7.5) containing 0.5 M NaCl.

9. Solution #1: 5 M NaCl, 1 M Tris–HCl buffer (pH 7.5), 0.5 M glucose-6-phosphate (G6P), 4 mM horse heart cytochrome *c* (cyt *c*), G6P dehydrogenase (1 U/μL), 10 μM FNR, 1 mM Fd (wild-type, S43W, and S43D), and 2.5 mM NADPH.

10. Solution #2: 100 mM NaCl, 50 mM Tris–HCl (pH 7.5), 3 mM G6P, 200 μM cyt *c*, G6P dehydrogenase (3 U/μL), and 20 nM FNR.

11. Solution #3: 10 μM Fd (wild-type, S43W, and S43D) and 50 μM NADPH.

12. Solution #4 (Fd) and 5 (MV): consists of Fd (0–40 μM) in 50 mM Tris–HCl buffer (pH 7.5), for Fd-dependent activity (solution 4) (*see* Fig. 2b, upper pathway) or methyl viologen (MV) (1 mM) (solution 5) (*see* Fig. 2b, lower pathway) in DDW, for MV-dependent activity, and wild-type or mutant SiR (200 nM) and cysteine synthase (0.4 U) in 50 mM Tris–HCl buffer (pH 7.5), sodium sulfite (2 mM) and O-acetyl

serine (10 mM) in DDW, and 1 M Tris–HCl buffer (pH 7.5) (50 mM) at 30 °C. The concentrations in parentheses indicate the final concentrations in the reaction mixtures. Confirm a final pH value of 7.5.

13. DE52, anion-exchange resin (GE Healthcare).

14. Size-exclusion chromatography (Superdex-75G).

15. PMSF protease inhibitor.

16. Trichloroacetic acid.

17. Ninhydrin (2,2-dihydroxyindane-1,3-dione) solution: dissolve 250 mg of ninhydrin powder in 6 mL of acetic acid and 4 mL of hydrochloric acid.

2.2   Equipment

1. AVANCE-III H/D 800 spectrometer equipped with a cryogenic probe (e.g., Bruker BioSpin, Germany) or similar spectrometer.

2. Isothermal titration calorimeter (e.g., VP-ITC (MicroCal™, Malvern Instruments, UK).

3. Thermo-Vac unit (e.g., MicroCal™, Malvern Instruments, UK).

4. Sonicator.

5. DEAE-Toyopearl-650 anion-exchange resin (Tosoh Biosciences).

6. Toyopearl Phenyl-650 resin (Tosoh Biosciences).

7. Phenylsepharose column.

8. Shigemi and standard NMR tubes (5 mm in diameter).

3   Methods

Prepare all solutions using ultrapure water and analytical grade reagents. Ultrapure water may be further filtered using a 0.2- or 0.5-μm filter. Perform all procedures at room temperature unless otherwise noted.

3.1   Purification of Ferredoxin (Fd)

1. Transform *E. coli* strain *JM109* cells with the expression vector carrying wild-type or mutant Fd cDNA (for S43W and S43D) and the ampicillin resistance gene and incubate them overnight in 50 mL of LB medium containing 50 μg/mL ampicillin at 37 °C. This will act as the seed culture.

2. Inoculate the seed culture into 8 L of the same medium. After 3 h incubation at 37 °C, add IPTG (isopropyl-β-d-thiogalactopyranoside) to reach a final concentration of 0.5 mM, and continue incubating the culture overnight.

3. Harvest all the cells by centrifugation at 6,800×*g* at 4 °C for 10 min and measure and record the weight of the harvested cells.

4. Resuspend the pelleted cells in 200 mL of 50 mM Tris–HCl buffer (pH 7.5) containing 150 mM NaCl, 1 mM MgCl$_2$, 1 mM EDTA, and 0.1 % β-mercaptoethanol.

5. Add lysozyme to the above (**step 4**) solution at 0.1 % of the cell weight.

6. Disrupt the cells using an ultrasound disintegrator (e.g., Sonifier 450) with a power output of 80 % for 2 min. Repeat this process four times.

7. Centrifuge the disrupted solution at 25,000×g at 4 °C for 10 min and collect the supernatant containing the Fd.

8. Add the same weight of the anion exchange resin (cellulose DE52) to the supernatant as that of the harvested cells.

9. Fill an open column with Fd-adsorbed resins and wash the column using 0.5 L of buffer A.

10. Elute the Fd from the resins using 50 mM Tris–HCl buffer (pH 7.5) containing 0.5 M NaCl.

11. Add ammonium sulfate to the eluted Fd solution in order for the saturation rate of ammonium sulfate to reach 70 %.

12. Repeat the processes described in **steps 7–9** in this section.

13. Dialyze the solution with 50 mM Tris–HCl buffer (pH 7.5) in order to remove NaCl.

14. Apply the dialyzed protein solution to the DEAE-Toyopearl column, and elute Fd-containing fractions with a linear gradient of NaCl from 0 to 0.5 M.

15. Add ammonium sulfate to reach a final concentration of 70 % saturated.

16. Apply the protein solution to the Phenyl-Toyopearl column. Elute and separate the Fd solution with a linear gradient of ammonium sulfate from 70 to 0 %.

17. Concentrate the Fd solution to a volume of 5 mL or less.

18. Apply the Fd solution to size-exclusion chromatography resin (Superdex-75G). Separate the Fd from other compounds using a constant flow of 50 mM Tris–HCl buffer (pH 7.5).

19. The two Fd mutants (S43W and S43D) are expressed and purified in an identical manner to that of wild-type Fd.

3.2 Ferredoxin-NADP+ Reductase

1. Transform *E. coli* strain JM109 cells with the expression vector carrying wild-type ferredoxin-NADP$^+$ reductase FNR cDNA and the ampicillin resistance gene. Incubate the cells at 37 °C in LB medium overnight. This acts as the seed culture.

2. Inoculate the seed cells into 8 L of the same medium and incubate at 37 °C for 3 h. Add IPTG to reach a final concentration of 0.5 mM and incubate for 20 h.

3. Harvest whole cells by centrifugation at $6,800 \times g$ at 4 °C for 10 min, and measure and record the weight of the harvested cells.

4. Suspend the collected cells in 50 mM Tris–HCl buffer (pH 7.5) including 150 mM NaCl, 1 mM MgCl$_2$, 1 mM EDTA, and 0.1 % β-mercaptoethanol, and add lysozyme to the solution at 0.1 % of the cell weight.

5. Disrupt the cells using an ultrasound disintegrator with a power output of 80 % for 2 min. Repeat this process four times.

6. Centrifuge the cell lysate at $25,000 \times g$ at 4 °C for 10 min and collect the supernatant including FNR.

7. Add the same weight of the anion exchange resin (cellulose DE52) to the supernatant as that of the harvested cells.

8. Elute FNR using solution #4.

9. Add ammonium sulfate to reach a final concentration of 35 %.

10. Centrifuge the solution at $25,000 \times g$ at 4 °C for 10 min and collect the supernatant.

11. Add ammonium sulfate to reach a final concentration of 70 %.

12. Centrifuge the solution at $25,000 \times g$ at 4 °C for 10 min and collect the pellet.

13. Suspend the pellet containing FNR in 50 mM Tris–HCl buffer (pH 7.5).

14. Dialyze the solution against 25 mM Tris–HCl buffer (pH 7.5).

15. Apply the dialyzed protein solution to a DEAE-Toyopearl column, and elute the FNR fractions with a linear gradient of NaCl from 0 to 0.5 M in 50 mM Tris–HCl buffer (pH 7.5).

16. Add ammonium sulfate to reach a final concentration of 35 %.

17. Apply the solution to a Phenyl-Toyopearl column. Separate and elute FNR with a linear gradient of ammonium sulfate from 35 to 0 %.

18. Perform the same procedure as described in **steps 11–14** in this section above.

19. Apply the solution to a Fd-affinity column and elute the FNR with a linear gradient of NaCl from 0 to 0.5 M in Tris–HCl buffer (pH 7.5).

20. Collect fractions including FNR and dilute them by adding a four-fold volume of 50 mM Tris–HCl buffer (pH 7.5). FNR uniformly labeled with ^{15}N for NMR measurements is obtained by culturing bacterial cells in M9 minimal medium containing ^{15}NH$_4$Cl as the sole nitrogen source. The same procedure for the purification of non-labeled FNR is applicable to isotope-labeled FNR.

3.3 Sulfite Reductase (SiR)

1. Transform *E. coli* strain JM109 cells with the expression vector carrying wild-type and mutant SiR cDNA and incubate 25 mL cells at 37 °C overnight in LB medium containing 50 μg/mL ampicillin. This acts as the seed culture.

2. Inoculate the seed culture into 8 L of the same medium and incubate with vigorous aeration at 37 °C for 3 h.

3. Reduce the culture temperature to 27 °C and add IPTG to reach a final concentration of 0.5 mM. After further cultivation overnight, collect the cells by centrifugation at $6,800 \times g$ for 10 min.

4. Suspend the collected cells in Buffer A.

5. Add PMSF at a final concentration of 0.5 mM and perform a sonication treatment (on ice) until complete cell lysis of the cell suspension.

6. Centrifuge the solution containing broken cells at $25,000 \times g$ for 15 min.

7. Pass the supernatant through a DE52 resin solution pre-equilibrated with Solution #4.

8. Recover the solution containing SiR from the flow-through fraction and fractionate the solution by salting out with ammonium sulfate concentrations between 35 and 60%.

9. Solubilize the precipitates with Solution #4 and dialyze them against 25 mM Tris–HCl buffer (pH 7.5) at 4 °C overnight.

10. Apply the protein solution dialyzed on the DEAE-Toyopearl column and chromatograph with a linear gradient of NaCl from 0 to 200 mM in 50 mM Tris–HCl buffer (pH 7.5).

11. Add ammonium sulfate to the SiR fraction at a final concentration of 30%.

12. Apply the SiR solution to a phenylsepharose column and elute SiR with a reverse linear gradient of ammonium sulfate from 40 to 0% in 50 mM Tris–HCl buffer (pH 7.5). The SiR mutant (R111Q) is expressed and purified in the same manner as that described for the wild-type SiR.

3.4 NADPH-Dependent FNR Assay

1. Make a sterile stock of Solution #1 consisting of 5 M NaCl, 1 M Tris–HCl buffer (pH 7.5), 0.5 M glucose-6-phosphate (G6P), 4 mM horse heart cytochrome *c* (cyt *c*), G6P dehydrogenase (1 U/μL), 10 μM FNR, 1 mM Fd (wild-type, S43W, and S43D), and 2.5 mM NADPH.

2. Adjust the volume of Solution #1 to 440 μL containing the following final concentrations: 100 mM NaCl, 50 mM Tris–HCl (pH 7.5), 3 mM G6P, 200 μM cyt *c*, G6P dehydrogenase (3 U), and 20 nM FNR.

3. Make Solution #2 to a volume of 60 μL with the following final concentrations: 10 μM Fd (wild-type, S43W, and S43D) and 50 μM NADPH.

4. Select the kinetic mode of the UV–Vis spectrometer in order to observe reductions in cyt c based on changes in absorbance at 550 nm (*see* Fig. 2a) and set the temperature to 25 °C.

5. Add solution #1 to a cuvette and incubate it in the spectrometer at 25 °C for several minutes. Use clean 2-mm glass cuvettes for measurements.

6. Add solution #2, maintained at 25 °C, to a cuvette containing Solution #1 and then quickly mix by pipetting several times.

7. Immediately start the measurement, monitor absorbance until it saturates (it finishes within 1 min), and calculate the initial velocity.

8. Repeat the same measurements and calculations by increasing the concentration of Fds (wild-type, S43W, and S43D). The concentrations of wild-type Fd are 1.25, 2.5, 5, 10, 20, and 40 μM while those of S43W Fd are 0.25, 0.5, 1, 2.5, 5, and 10 μM. S43D Fd is used at concentrations of 5, 10, 20, and 40 μM.

9. Plot each initial velocity against the concentration of each Fd (*see* Fig. 3a).

10. By fitting the initial velocity to the Michaelis–Menten equation (*see* Fig. 3a), the Michaelis–Menten constant (K_m) and turnover number (k_{cat}) are obtained (*see* Figs. 3b, 3d) (*see* **Notes 2** and **3**).

11. Perform the same measurements and analyses at specific temperatures (288, 293, and 303 K) and calculate the activation energy (E_a) from temperature-dependent k_{cat} (*see* Fig. 3c) (*see* **Note 3**).

3.5 SiR Activity

Perform all procedures at 30 °C unless otherwise noted.

1. Generate Solution#4 reaction mixtures by using (1) Fd (0–40 μM) in 50 mM Tris–HCl buffer (pH 7.5), for Fd-dependent activity (*see* Fig. 2b, upper pathway), or (2) methyl viologen (MV) (1 mM) (*see* Fig. 2b, lower pathway) in doubly deionized water (DDW), for measuring MV-dependent activity, containing wild-type or mutant SiR (200 nM) and cysteine synthase (0.4 U, in 50 mM Tris–HCl buffer, pH 7.5) containing sodium sulfite (2 mM) and O-acetyl serine (10 mM) in DDW, and 1 M Tris–HCl buffer (pH 7.5) (50 mM) at 30 °C. The concentrations in parentheses indicate the final concentrations in the reaction mixtures. Confirm a final pH value of 7.5.

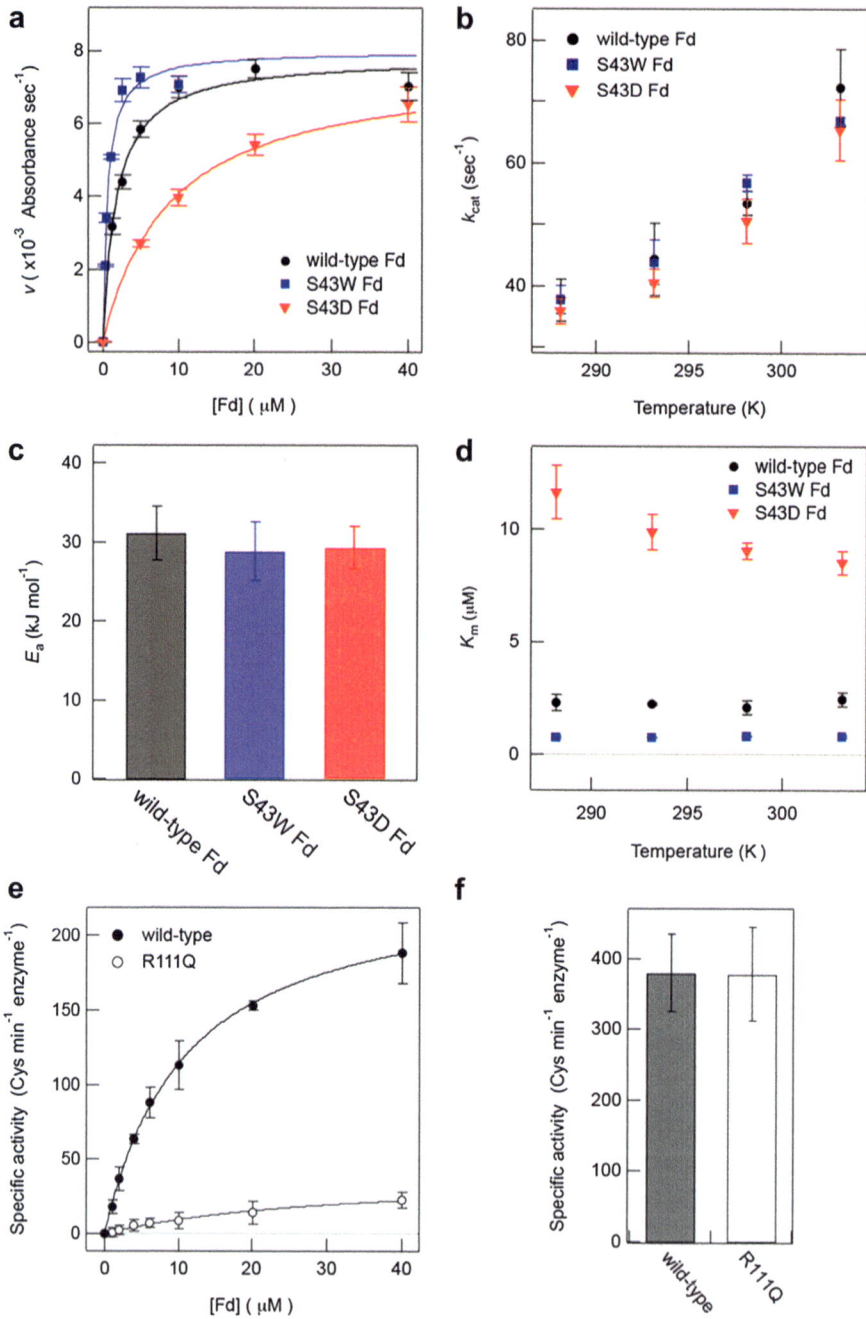

Fig. 3 FNR activity dependent on mutations and temperature. (**a**) The steady-state kinetics of FNR based on the reduction of cytochrome *c* at 298 K are shown. (**b**) k_{cat} values are plotted against temperature. (**c**) Schematic bar presentations of activation energies (E_a). (**d**) Temperature-dependent K_m values. Values corresponding to wild-type, S43W, and S43D Fds are shown in *black* or *gray*, *blue*, and *red*, respectively. Figures **a**–**d** were modified with permission [4]. (**e**) The steady-state kinetics of wild-type SiR (*closed circle*) and R111Q SiR (*open circle*) at various Fd concentrations. (**f**) Comparison of SiR activities obtained by Fd- (*closed bar*) and MV-dependent assays (*open bar*). The *continuous lines* in **a** and **e** indicate the fitting curves based on the Michaelis–Menten equation

2. Perform the Fd-dependent SiR activity assay. Initiate sulfite reduction by adding 20 mM sodium dithionite ($Na_2S_2O_4$) in 50 mM sodium hydrogen carbonate to the reaction mixture in step 1 above (*see* Fig. 2b).

3. Stop the reaction by adding trichloroacetic acid at a final concentration of 20% (vol/vol) after a 3-min incubation (*see* Fig. 3e).

4. Centrifuge the reacted solution at 25,000 × *g* for 3 min to remove insoluble materials.

5. Add 150 μL of acetic acid and 150 μL of acidic 2,2-dihydroxyindane-1,3-dione (ninhydrin) solution to 150 μL of the supernatant. Ninhydrin solution is prepared by dissolving 250 mg of ninhydrin powder in 6 mL of glacial acetic acid and 4 mL of hydrochloric acid.

6. Heat 450 μL of the acidic solution to 95 °C for 10 min and then add 450 μL of ethanol to the ninhydrin-containing solution.

7. Confirm the production of cysteine by observing increases in the intensity of absorption at 546 nm.

8. Perform the same measurements by increasing the concentration of Fd (e.g., 1, 2, 3, 4, 6, 10, 20, and 40 μM) and then perform the MV-dependent SiR activity assay (*see* **Note 4**).

9. Plot each activity obtained from the Fd-dependent assay as a function of the concentration of Fd and fit the plotted data to the Michaelis–Menten equation (*see* Figs. 3e, f) (*see* **Note 2**). Summarize the values of K_m and k_{cat}.

3.6 Isothermal Titration Calorimetry

1. Perform a DDW-to-DDW titration in order to confirm the cleanliness of the ITC sample cell and to check the overall performance of the ITC instrument. For a well-functioning ITC instrument with a clean cell, the resulting peaks should be small and invariant with injection number. If (relatively) large exothermic or endothermic peaks are observed, rinse and wash the ITC cell and syringe again as carefully as possible (for more details, visit www.malvern.com) (*see* **Note 5**).

2. Prepare FNR at a concentration of 50 μM and each Fd (titrant) at a concentration of 1 mM in 50 mM Tris–HCl buffer (pH 7.5). Care is needed to prevent a mismatch of solution compositions, except for Fd and FNR (*see* **Note 6**).

3. Open the ITC Controls window, set the experimental parameters for the ITC measurement (*see* **Note 7**) (more information is available at www.malvern.com), and then stabilize the ITC instrument at a desired temperature before the degassing procedure described below. Detailed setting parameters are provided in parentheses: initial delay (1800 s), spacing time (600 s), stirring speed (264 rpm), number of injections (40), duration (4 s

for the first injection and 14 s for the other injections), injection volume (2 µL for the first injection and 7 µL for other injections), reference power (10 µcal/s), feedback (Low), ITC Equilibrium (fast equilibrium and auto), and the filter period (4 s).

4. Place ~2.5 mL of the FNR solution and ~0.6–0.7 mL of the Fd solution in plastic vials with magnetic stirring bars.

5. Degas the FNR solution for 5 min using the Thermo-Vac unit (*see* **Note 8**).

6. Gently fill the ITC sample cell with the degassed FNR sample solution using a long-needled 2.5-mL glass syringe and leave the ITC instrument for further stabilization (*see* **Note 9**).

7. Degas the Fd solution for 3 min using a Thermo-Vac unit.

8. Transfer the degassed Fd solution to an Eppendorf tube (1.5 mL) and place this tube so that the tip of the ITC pipette is soaked.

9. Fill the ITC syringe (or pipette) with the Fd solution by using the plastic filling syringe and the commands in the Pipette Controls window (*see* **Note 10**).

10. Carefully insert the ITC pipette down into the ITC sample cell while avoiding bending the needle. Monitor the value of differential power (DP).

11. DP values should settle at the value set for the Reference Power within approximately (±) 1 µcal/s. Monitor the ITC thermogram (Fig. 4a, upper panels) (*see* **Note 11**).

12. Rinse and wash the ITC cell and syringe as thoroughly as possible.

13. Perform the following three sets of titration measurements as a reference (50 mM Tris–HCl buffer (pH 7.5) is used): (a) 1 mM Fd (syringe) to buffer (cell), (b) buffer (syringe) to 50 µM FNR solution (cell), and (c) buffer (syringe) to buffer (syringe) (*see* **Note 12**).

14. Perform the baseline correction for four ITC measurements using the MicroCal Origin 7.0 software and calculate the net heat of binding as denoted (*see* **Note 12**).

15. Fit a binding isotherm to a theoretical equation with a single binding-site model (one set of the site model in MicroCal Origin 7.0) (*see* Fig. 4a, lower panels) (*see* **Note 13**).

16. Summarize the thermodynamic parameters of the wild-type Fd–FNR binding reaction (*see* **Note 14**).

17. Repeat ITC measurements using mutant Fds (S43W and S43D) at different temperatures (*see* Figs. 4a, b) (*see* **Note 15**).

18. Plot the thermodynamic parameters against the temperature in order to characterize the Fd–FNR interactions thermodynamically (*see* Fig. 4b).

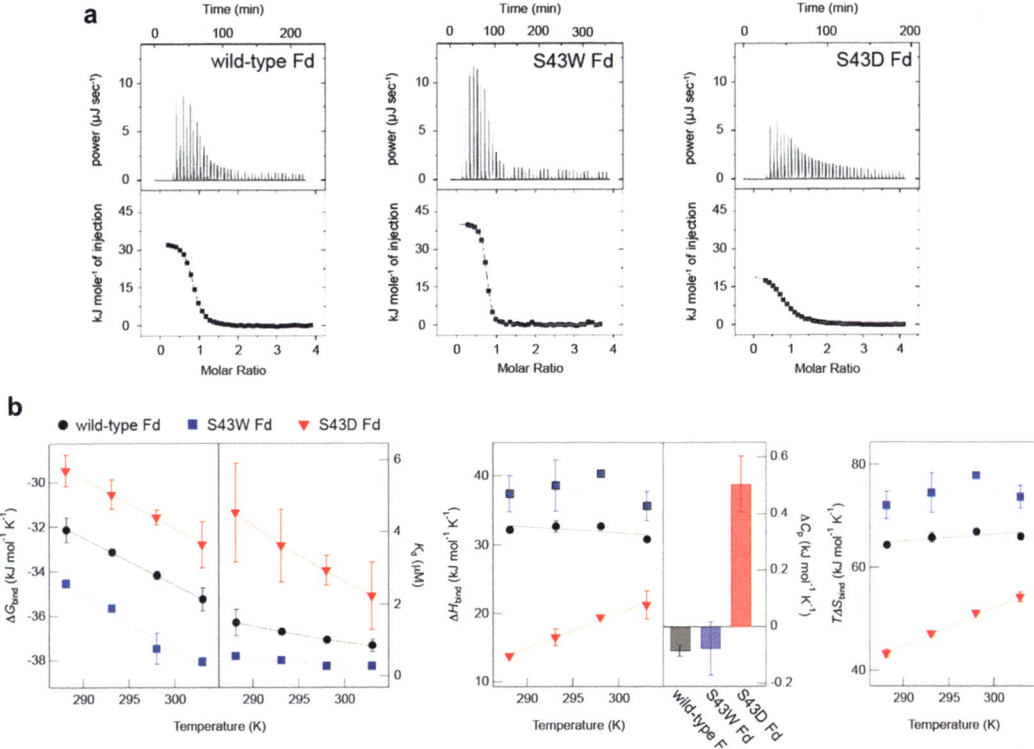

Fig. 4 ITC-based thermodynamic characterization of interprotein interactions between Fd and FNR. (**a**) ITC thermograms of the titration of wild-type Fd (*left*), S43W Fd (*middle*), and S43D Fd (*right*) to FNR at 298 K are shown in the upper panel. Normalized heat values are plotted against the molar ratio ([Fd]/[FNR]) in the lower panel. Fitted curves are exhibited using continuous lines. (**b**) Temperature-dependent ΔG_{bind} (left *panel*) and K_d (*right panel*) (*left thermogram*), ΔH_{bind} (*left panel*) and ΔCp (*right panel*) (*middle*), and $-T\Delta S_{bind}$ (*right*) are plotted. Reproduced from ref. [4] with permission

3.7 NMR Spectroscopy

1. Use ^{15}N-labeled FNR at a concentration of 0.1 mM with and without each Fd at a concentration of 0.2 mM (wild-type, S43W, and S43D) dissolved in 50 mM Tris–HCl buffer (pH 7.5) containing 10% D_2O for solution-state NMR measurements. The sample volumes are generally 280 μL for the Shigemi NMR tube (5 mm in diameter) and 550 μL for the normal NMR tube (5 mm in diameter).

2. Fill the NMR tubes with free ^{15}N-labeled FNR solution and insert the NMR tube into the spinner.

3. Set the temperature to 25 °C and create a new dataset for the two-dimensional heteronuclear single quantum correlation (HSQC) measurement (*see* **Note 16**).

4. Insert the NMR sample tube into the AVANCE-III H/D 800 spectrometer equipped with a cryogenic probe and perform the HSQC measurement (*see* **Note 17**).

5. Perform the same measurements on ^{15}N-labeled FNR in the presence of wild-type Fd, S43W Fd, or S43D Fd.

6. Process all NMR data using the NMRPipe software at: (http://spin.niddk.nih.gov/NMRPipe/) [11] and show the ^{1}H-^{15}N HSQC spectrum (*see* Fig. 5a).

7. Assign each peak (*see* Figs. 5a, b) and analyze the data using Sparky software at: (http://www.cgl.ucsf.edu/home/sparky/)

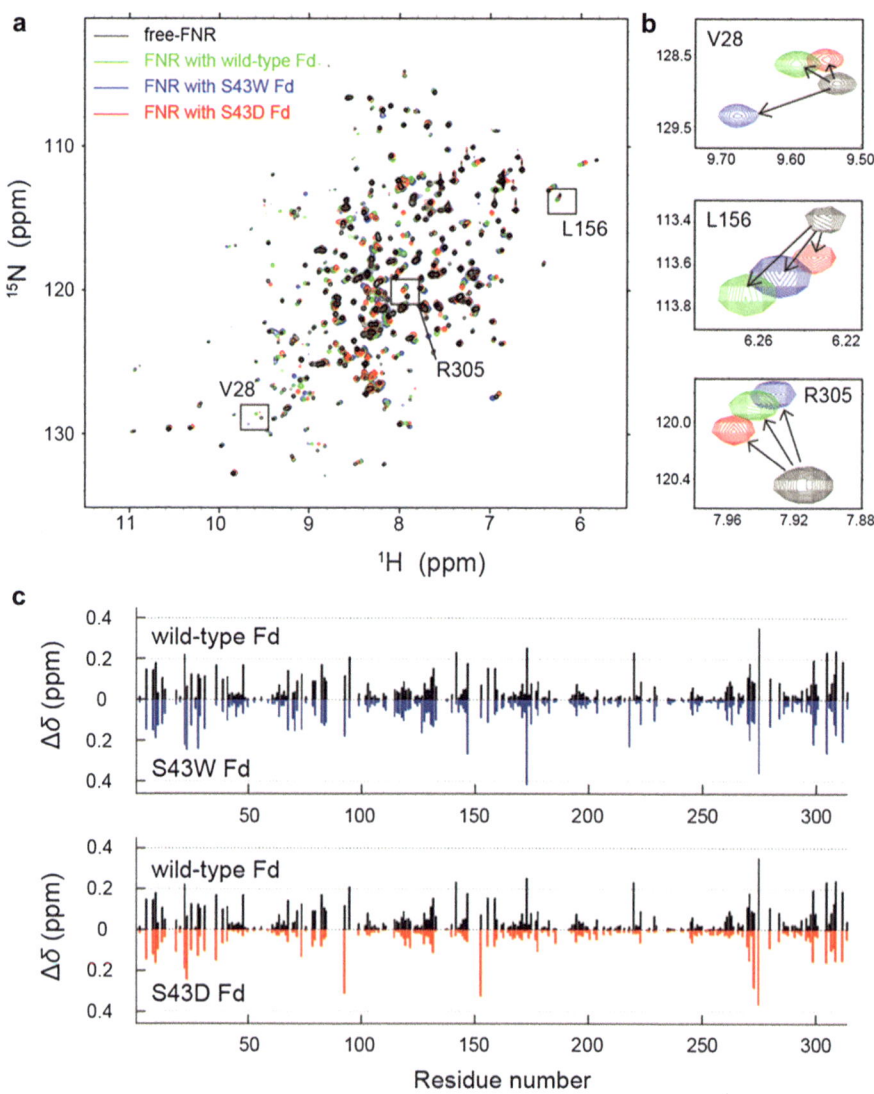

Fig. 5 Multidimensional NMR spectroscopy of ^{15}N-labeled FNR with and without Fd and chemical shift perturbation analyses. (**a**) The superposition of ^{1}H-^{15}N HSQC spectra of FNR without Fd (*black*) and with wild-type (*green*), S43W (*blue*), or S43D Fds (*red*). (**b**) Shifts in the peak (V28, L156, and R305) are representatively magnified. (**c**) Chemical shift perturbations with the addition of each type of Fd were plotted against the residue number of FNR. Reproduced from ref. [4] with permission

[12] and information on assignments previously reported [9, 13, 14].

8. Calculate the chemical shift perturbation (CSP) and plot CSP against the residue number of FNR (*see* Fig. 5c) (*see* **Note 18**).

9. Map CSP onto the crystal structure of FNR (1GAQ) [6] depending on the degree of CSP (*see* Figs. 6a–c, left) (*see* **Note 19**). We use PyMOL software at (http://pymol.org/) to view and handle three-dimensional structures.

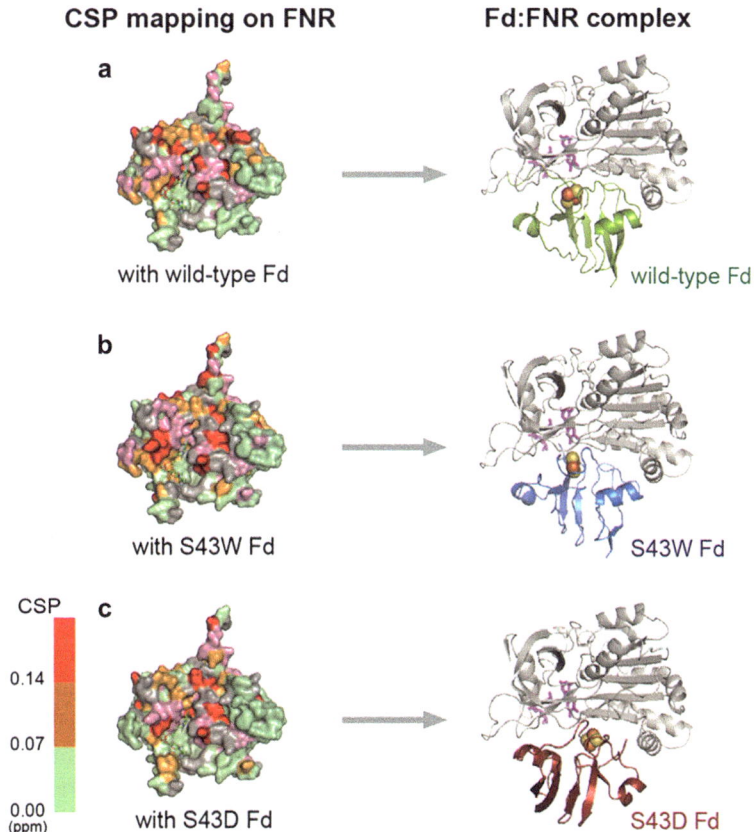

Fig. 6 Mapping of NMR chemical shifts in FNR and modeled Fd–FNR complexes. (**a–c**) Chemical shift perturbations in ^{15}N-labeled FNR in the presence of wild-type Fd (**a**, *left*), S43W Fd (**b**, *middle*), and S43D Fd (**c**, *right*) were mapped onto the X-ray crystal structure of FNR (PDB ID: 1GAQ). The color code represents the following: *red*, CSP > 0.14; *orange*, 0.14 > CSP > 0.07; *light green*, CSP < 0.07. The unassigned residues due to the ambiguity of assignments are shown in *pink*. The three Fd–FNR complex structures obtained by the docking simulation of HADDOCK in combination with NMR data are shown. FNR is colored in *gray*. Wild-type (**a**, *right*), S43W (**b**, *right*), and S43D Fds (**c**, *right*) are shown in *green*, *blue*, and *red*, respectively. *Magenta sticks* in FNR indicate FAD. *Red* and *yellow spheres* in Fd represent iron and sulfur, respectively. Reproduced, with minor modifications, from ref. [4] with permission

10. In order to visualize each complex, perform the HADDOCK simulation at: http://www.nmr.chem.uu.nl/haddock2.1/ [15] by incorporating NMR CSP information (*see* Fig. 6a–c, right) (*see* **Notes 20** and **21**).

4 Notes

1. Intermolecular forces may be largely classified into two types: electrostatic interactions involving charge interactions and non-electrostatic interactions such as hydrophobic interactions and van der Waals contacts. The grouping of polar interactions among partial charges is confusing. Some text books classify polar interactions into electrostatic interactions whereas others do not.

2. The Michaelis–Menten equation is as follows: $v = (k_{cat}[\text{FNR}]_0[\text{Fd}])/([\text{Fd}] + K_m)$, where v indicates the initial velocity of the catalytic reaction of FNR, and $[\text{FNR}]_0$ and $[\text{Fd}]$ indicate the concentrations of total FNR and free Fd in the reaction mixture, respectively.

3. Although no appreciable changes are observed in the chemical reaction (k_{cat}) (*see* Fig. 3b) or activation energy (E_a) for S43W or S43D (*see* Fig. 3c), physical changes are identified in K_m (*see* Fig. 3d), revealing that the mutation in Fd disrupts the interprotein interaction. These results suggest that the interprotein interaction between Fd and FNR is a dominant factor controlling the overall activity of FNR, and changing the physicochemical properties at interfaces by mutations affects the formation of an electron transfer complex between Fd and FNR.

4. Fd and MV are physiological and artificial electron donors, respectively, for the reduction of SiR. SiR activity obtained by the Fd-dependent assay (*see* Fig. 2b, upper) includes the effects of Fd–SiR interactions. Meanwhile, SiR activity obtained by the MV-dependent assay (*see* Fig. 2b, lower) only reflects the intramolecular activity of SiR due to excess amounts of MV in the reaction mixture. By comparing the SiR activities of the two assays, the effects of the SiR mutation on intermolecular interactions and intramolecular activity may be distinguished. The significant decrease observed in Fd-dependent SiR activity (*see* Fig. 3e) and no change in MV-dependent SiR activity by mutations (*see* Fig. 3f) suggest that the neutralization of the positive charge of SiR on interfaces for Fd markedly decreases SiR activity due to weakened affinity stemming from the disruption of the attractive electrostatic interaction between the oppositely charged residues of two proteins.

5. If the ITC sample cell is dirty, rinse it and wash it again and check the magnitude of the ITC peaks of the DDW-to-DDW titration. If aggregates strongly adhere to the cell walls, incubate a cell filled with surfactant solution at a higher speed (e.g., >600 rpm) and temperature (~50 °C). Clean the cell reservoir thoroughly because proteins are frequently adsorbed to the surfaces.

6. The two sample solutions for the ITC sample cell and syringe need to be matched with regard to composition otherwise the heat of mixing (or dilution) signals will occur, which may hide the heat of binding reactions by producing a large amount of heat. In general, two sample solutions are dialyzed together against a large volume of working buffer for an ITC measurement. The buffer exchange of a sample solution with a desired buffer using a disposable desalting column (e.g., PD-10, NAP-10, or NAP-5) is also effective.

7. The C value, defined by $C = [M]t\,K_a\,n$, is useful for determining the appropriate concentration of macromolecules ($[M]t$) in the cell with known or expected values for the association constant (K_a) and binding stoichiometry (n). C values in the range $1 \leq c \leq 1000$, which falls into an intermediate binding affinity, are practically appropriate for ITC measurements [16]. Set the first injection to a volume of 2 μL in order to expel small bubbles. Data obtained for this first injection cannot be used for curve fitting. If proteins are prone to aggregation, use a short initial delay and spacing time as well as a low stirring speed in order to minimize aggregation.

8. Long degassing changes the concentration of sample solutions, which hampers accurate and precise ITC analyses. If possible, it is advisable to confirm the protein concentration using the remaining sample solution after filling the cell and syringe. Set the temperature to less than that for the ITC experiment to achieve the stabilization of the ITC instrument more rapidly. Degassing at a temperature more than 30 °C often causes sudden boiling. Thus, careful monitoring is needed in order to prevent overflow by adjusting the strength of the vacuum with the bleeder screw. Proper stirring is effective for shortening the time for degassing. However, care is needed regarding stirring-induced aggregation.

9. In order to remove bubbles, gently stir the solution with the syringe needle and spurt the solution several times.

10. The detailed procedures are to be followed in this order: Click the "Open Fill Port" button → Withdraw the plunger of the plastic syringe (ensure a bubble-free state) → Click the "Close Fill Port" button → Remove the plastic syringe → Click the "Purge → ReFill" button to remove bubbles (perform this pro-

cedure 2–3 times) → Wash the tip of the ITC pipette using the working buffer → Click the "Dn" (i.e., down) button after setting 2 μL, clean the tip again using wet Kimwipes, and wipe very carefully with dry Kimwipes.

11. Pay careful attention to the completion of the first (2 μL) and second titrations (7 μL). If the initial spacing time is not sufficient to completely trace the heat generated or is too long between titrations, reset the spacing time and click the "Update Run Parameter" button prior to the initiation of the next titration.

12. In principle, the net binding heat is obtained as follows: net binding heat = (heat from the titration of Fd to FNR) – (heat from the titration of Fd to buffer) – (heat from the titration of buffer to FNR) + (heat from the titration of buffer to buffer). However, as a rule, the subtraction of the heat of dilution of the titrant (Fd) is sufficient to obtain the net binding heat, that is, net binding heat = (heat from the titration of Fd to FNR) – (heat from the titration of Fd to buffer). Although a more detailed description is not given here, the contribution of each term of ΔH to the apparent ΔH (ΔH_{app}) obtained by ITC is worthy of consideration.

$$H_{app} = H_{bind} + \Delta H_{str} + H_{hyd} + H_{pro} + H_{ion}$$

where ΔH_{bind} is the true value for the binding reaction and ΔH_{str} reflects heat accompanying structural changes with binding. ΔH_{hyd} indicates the contribution of (de)hydration from proteins. ΔH_{pro} and ΔH_{ion} are involved in (de)protonation of the ionizable group and the ion–protein interaction (attachment/detachment of ions), respectively. In most cases, ΔH_{app} has been used to indicate ΔH_{bind} without careful inspection.

13. The observed enthalpy changes (ΔH_{bind}) for binding and the dissociation constant (K_d) are directly calculated from the integrated heat using the one-set independent binding sites model supplied by MicroCal Origin 7.0 software. The equation for this binding model is:

$$Q = \frac{n[P]_t \; H_{bind} V_0}{2}\left[1 + \frac{L_R}{n} + \frac{K_d}{n[P]_t} - \sqrt{\left(1 + \frac{L_R}{n} + \frac{K_d}{n[P]_t}\right)^2 - \frac{4L_R}{n}} \right]$$

where Q is the change in heat in the system, V_0 is the effective volume of the calorimeter cell (1.43 mL), L_R is the ratio of the total Fd concentration to total FNR concentration ($[P]t$) at any given point during the titration, and n is the binding stoichiometry of Fd per FNR. Using ΔH_{bind} and K_d, the observed Gibbs free energy change for binding (ΔG_{bind}) and observed entropy change for binding (ΔS_{bind}) are obtained by the following thermodynamic relationships: $\Delta G_{bind} = RT \ln K_d$ and $\Delta G_{bind} = \Delta H_{bind} - T\Delta S_{bind}$, where R is the universal gas constant and T is the temperature in Kelvin.

14. We provide a very general way to interpret thermodynamic parameters for a binding reaction, ΔG_{bind}, ΔH_{bind}, ΔS_{bind}, and K_d. ΔG_{bind} is balanced by the two driving forces, ΔH_{bind} and ΔS_{bind}, and is also adjusted by K_d. Thus, to decrease ΔG_{bind}, i.e., increase binding affinity, the favorable contributions of decreasing ΔH_{bind} and increasing ΔS_{bind} are required. Reductions in K_d lower ΔG_{bind}. In most cases, the formation of salt bridges and hydrogen bonds, van der Waals contacts, and hydrophobic interactions decrease ΔH_{bind}. The detachment of hydrated water around charged residues and/or structural changes is accompanied by the unfavorable absorption of heat ($\Delta H > 0$). Binding-induced folding often decreases ΔH ($\Delta H < 0$). Increases in the amount of dehydrated water from surfaces and/or the interior of proteins and in conformational dynamics favorably increase ΔS. Here are practical examples for the thermodynamic description of mutation-induced changes in binding thermodynamics. Complexation-interrupting positive ΔH_{bind} is detected in all complexes (*see* Fig. 4b, middle) formed as a result of energy costs in the dehydration of charged residues, which drives complex formation solely with entropy (*see* Fig. 4b, right). The higher affinity of S43W for FNR than the wild-type (*see* Fig. 4b, left) is attributed to the gain of entropy (increases in ΔS_{bind}), which overwhelms the unfavorable enthalpic loss (increases in ΔH_{bind}). The dehydration of water from hydrophobic regions around the tryptophan residue, i.e., the hydrophobic effect, and electrostatic/polar regions may largely stabilize the Fd–FNR complex. The weakened affinity of S43D for FNR with the loss of $\Delta\Delta G = 2.63$ kJ/mol (*see* Fig. 4b, left) is explained by the penalty of entropy (*see* Fig. 4b, right). The formation of an electrostatic and hydrogen bond network around the negative charge of aspartic acid at 43 may engender favorable contributions to the formation of the S43D:FNR complex by reducing ΔH_{bind}.

15. One more important thermodynamic parameter, the change in heat capacity (ΔCp), is obtained from the temperature (T)-dependence of ΔH, $\Delta Cp = (\partial \Delta H / \partial T)p$ where 'p' is pressure (*see* Fig. 4b, middle). ΔCp is a thermodynamic indicator of the physicochemical properties for the buried surface area on complexation or protein folding [9, 17–19]. ΔCp shows small negative values for wild-type Fd–FNR binding and S43W–FNR binding (*see* Fig. 4b, middle), which demonstrates that the thermodynamic contributions of buried hydrophobic surface areas are slightly more dominant than those of electrostatic/

polar surface areas. These results indicate the thermodynamic importance of hydrophobic forces in stabilizing the Fd–FNR complex. A positive ΔCp value is obtained for S43D–FNR binding, suggestive of the large contribution of the burial of electrostatic/polar surface areas. Distinct, individual thermodynamic parameters for complexation depending on the type of Fds suggest different binding modes and/or orientations of individual Fds to FNR, which are fit toward maximizing energetic benefits for physical interactions between proteins under a given set of conditions.

16. HSQC is one of the most popular multidimensional NMR measurements in protein NMR. It is easily measured and provides useful information on the stability, dynamics, and intermolecular interactions of proteins at the atomic level in solution. ^{1}H-^{15}N HSQC measurements mainly reflect the correlation between backbone amides and amide protons. Thus, one peak in the ^{1}H-^{15}N HSQC spectrum corresponds to one amino acid residue.

17. Perform the following procedures with TopSpin software. Locking (select $H_2O + D_2O$) → Tuning and matching the probe (^{1}H → ^{15}N) → Shimming with topshim (1D → 3D) → Calibration of the 90 degree pulse for a proton (p1) → Setting various parameters using the p1 value determined and the command of "getprosol ^{1}H pulse length power" → Adjusting the receiver gain with the command of "rg" → Start the measurement.

18. An NMR chemical shift is susceptible to the electromagnetic environments induced by binding with partner molecules or structural changes. Differences in CSP in the NMR peaks of ^{15}N-labeled FNR in the absence and presence of Fd are calculated using the following equation:

$$\mathrm{CSP}\left(\Delta\delta_{\mathrm{ave}}\right) = \left(\left(\Delta\delta_{\mathrm{HN}}\right)^2 + \left(\Delta\delta_{\mathrm{N}} \times 0.158\right)^2\right)^{0.5}$$

where $\Delta\delta_{\mathrm{HN}}$ and $\Delta\delta_{\mathrm{N}}$ are changes in ^{1}H and ^{15}N chemical shifts in ppm, respectively. The weighting factor of 0.158 was used to adjust the relative magnitudes of the amide nitrogen chemical shift range and the amide proton chemical shift range. By plotting CSP against the residue number, the interaction region on FNR for Fd is revealed.

19. The spatial representation of CSP on protein molecules is more understandable than the simple plot of CSP against the residue number when several regions, which show high CSP distance from each other in the primary sequence, are spatially very close in the proteins.

20. The docking simulation between Fd (wild-type, S43W, or S43D) and FNR is performed using the HADDOCK easy

interface server with semiflexible refinements. Active residues for interaction restraints are based on NMR CSP data and are filtered for the accessible surface area calculated by VARDAR (version 1.8). Passive residues are defined automatically around the active residues. In the simulation of the complex of FNR with mutant Fds, homology modeling of S43W- and S43D-substituted Fd is performed using the Modeller program v9.11 and the crystal structure of wild-type Fd for a template structure. The qualities of the modeled structures of the Fd mutants are confirmed using the VERIFY3D program. The three Fd–FNR complexes share similar binding interfaces and the orientation of each complex is different.

21. Changes in the physicochemical nature on the interfacial residues of Fd by mutations (i.e., increasing hydrophobicity in S43W or an attractive electrostatic interaction in S43D) affect the interprotein affinity and binding mode. Solution-state NMR may detect small changes in the topological orientation of proteins in a complex in terms of NMR chemical shifts. Furthermore, it is important to note that the direction of peak shifts directly provides key information on subtle differences in binding modes between proteins in solution (*see* Fig. 5b). The distinct direction of a peak shift is detected among three Fd–FNR complexes, indicating that the H-bond network and dihedral angle at interfaces are not the same.

Acknowledgments

We thank Prof. Takahisa Ikegami (Yokohama City University, Japan), Prof. Chojiro Kojima (Osaka University, Japan), and Prof. Toshihiko Sugiki (Osaka University, Japan) for their valuable comments on NMR spectroscopy, Dr. Satoshi Kume for the HADDOCK simulation (RIKEN, Japan), and Prof. Yuji Goto for ITC (Osaka University, Japan). Y.-H.L. is supported by a Grant-in-Aid for Young Scientists (B) (15K18518 and 25870407). T.S., J.Y.K., and Y.L. received financial support from a Grant-in-Aid for Scientific Research (B) (24370021) and the Japan Society for the Promotion of Science (13J03956) and (14J04433), respectively. M.K. and J.Y.K. contributed equally to this work.

References

1. Wodak SJ, Janin J (2002) Protein modules and protein-protein interaction. Adv Protein Chem 61:21–23

2. Saitoh T, Ikegami T, Nakayama M, Teshima K, Akutsu H, Hase T (2006) NMR study of the electron transfer complex of plant ferredoxin and sulfite reductase: mapping the interaction sites of ferredoxin. J Biol Chem 281:10482–10488

3. Kume S, Lee YH, Nakatsuji M, Teraoka Y, Yamaguchi K, Goto Y, Inui T (2014) Fine-tuned broad binding capability of human lipocalin-type prostaglandin D synthase for various small lipophilic ligands. FEBS Lett 588:962–969

4. Kinoshita M, Kim JY, Kume S, Sakakibara Y, Sugiki T, Kojima C, Kurisu G, Ikegami T, Hase T, Kimata-Ariga Y, Lee YH (2015) Physicochemical nature of interfaces controlling ferredoxin NADP(+) reductase activity through its interprotein interactions with ferredoxin. Biochim Biophys Acta 1847:1200–1211

5. Kim JY, Lee YH (2015) Sulfur. Springer 61:21–23

6. Kurisu G, Kusunoki M, Katoh E, Yamazaki T, Teshima K, Onda Y, Kimata-Ariga Y, Hase T (2001) Structure of the electron transfer complex between ferredoxin and ferredoxin-NADP(+) reductase. Nat Struct Biol 8:117–121

7. Nogues I, Martinez-Julvez M, Navarro JA, Hervas M, Armenteros L, de la Rosa MA, Brodie TB, Hurley JK, Tollin G, Gomez-Moreno C, Medina M (2003) Role of hydrophobic interactions in the flavodoxin mediated electron transfer from photosystem I to ferredoxin-NADP+ reductase in Anabaena PCC 7119. Biochemistry 42:2036–2045

8. Nogues I, Perez-Dorado I, Frago S, Bittel C, Mayhew SG, Gomez-Moreno C, Hermoso JA, Medina M, Cortez N, Carrillo N (2005) The ferredoxin-NADP(H) reductase from *Rhodobacter capsulatus*: molecular structure and catalytic mechanism. Biochemistry 44:11730–11740

9. Lee YH, Ikegami T, Standley DM, Sakurai K, Hase T, Goto Y (2011) Binding energetics of ferredoxin-NADP+ reductase with ferredoxin and its relation to function. Chembiochem 12:2062–2070

10. Nakayama M, Akashi T, Hase T (2000) Plant sulfite reductase: molecular structure, catalytic function and interaction with ferredoxin. J Inorg Biochem 82:27–32

11. Delaglio F, Grzesiek S, Vuister GW, Zhu G, Pfeifer J, Bax A (1995) NMRPipe: a multidimensional spectral processing system based on UNIX pipes. J Biomol NMR 6:277–293

12. Lee W, Tonelli M, Markley JL (2015) NMRFAM-SPARKY: enhanced software for biomolecular NMR spectroscopy. Bioinformatics 31:1325–1327

13. Maeda M, Lee YH, Ikegami T, Tamura K, Hoshino M, Yamazaki T, Nakayama M, Hase T, Goto Y (2005) Identification of the N- and C-terminal substrate binding segments of ferredoxin-NADP+ reductase by NMR. Biochemistry 44:10644–10653

14. Lee YH, Tamura K, Maeda M, Hoshino M, Sakurai K, Takahashi S, Ikegami T, Hase T, Goto Y (2007) Cores and pH-dependent dynamics of ferredoxin-NADP+ reductase revealed by hydrogen/deuterium exchange. J Biol Chem 282:5959–5967

15. Dominguez C, Boelens R, Bonvin AM (2003) HADDOCK: a protein-protein docking approach based on biochemical or biophysical information. J Am Chem Soc 125:1731–1737

16. Markova N, Hallen D (2004) The development of a continuous isothermal titration calorimetric method for equilibrium studies. Anal Biochem 331:77–88

17. Loladze VV, Ermolenko DN, Makhatadze GI (2001) Heat capacity changes upon burial of polar and nonpolar groups in proteins. Protein Sci 10:1343–1352

18. Aoki M, Ishimori K, Fukada H, Takahashi K, Morishima I (1998) Isothermal titration calorimetric studies on the associations of putidaredoxin to NADH-putidaredoxin reductase and P450cam. Biochim Biophys Acta 1384:180–188

19. Bergqvist S, Williams MA, O'Brien R, Ladbury JE (2004) Heat capacity effects of water molecules and ions at a protein-DNA interface. J Mol Biol 336:829–842

Chapter 31

Use of Random and Site-Directed Mutagenesis to Probe Protein Structure–Function Relationships: Applied Techniques in the Study of *Helicobacter pylori*

Jeannette M. Whitmire and D. Scott Merrell

Abstract

Mutagenesis is a valuable tool to examine the structure–function relationships of bacterial proteins. As such, a wide variety of mutagenesis techniques and strategies have been developed. This chapter details a selection of random mutagenesis methods and site-directed mutagenesis procedures that can be applied to an array of bacterial species. Additionally, the direct application of the techniques to study the *Helicobacter pylori* Ferric Uptake Regulator (Fur) protein is described. The varied approaches illustrated herein allow the robust investigation of the structural–functional relationships within a protein of interest.

Key words Bacterial mutagenesis, Random mutagenesis, Site-directed mutagenesis, Structure–function, Protein analysis

1 Introduction

Mutagenesis can be a useful tool to investigate the structure–function relationship of proteins within numerous bacterial species. The alteration of amino acid residues within a protein followed by observation of a subsequent phenotype in comparison to the wild-type protein can provide insight into the functionality of individual structural elements of the protein. The mutagenesis approach can be accomplished via a variety of strategies. Site-directed mutagenesis is particularly helpful in targeting specific, often conserved, sites within the protein when some information about structure and function is known or can be inferred from homologous proteins [1]. In addition, in the absence of specific target sites, random mutagenesis techniques can be employed to generate unbiased mutations that span the protein of interest [2]. Together, these approaches can provide a more complete picture of the structure–function relationship of a protein of interest within a bacterium. Herein, we present several

Andrew Reeves (ed.), *In Vitro Mutagenesis: Methods and Protocols*, Methods in Molecular Biology, vol. 1498,
DOI 10.1007/978-1-4939-6472-7_31, © Springer Science+Business Media New York 2017

practical random and site-directed mutagenesis techniques that can be utilized for virtually any bacterial gene for which an observable phenotype exists. In addition, we provide information on the direct application of the techniques to the Fur protein within *H. pylori* [3].

2　Materials

2.1　Molecular Biology Reagents

1. Computer with access to the Internet.
2. GeneMorph II Random Mutagenesis Kit (Agilent Technologies).
3. Nuclease-Free Water.
4. QiaQuick Gel Extraction Kit (Qiagen).
5. Phusion Hot Start II High-Fidelity DNA Polymerase Kit (ThermoFisher Scientific).
6. MinElute PCR Purification Kit (Qiagen).
7. Ethyl methanesulfonate (EMS) (Sigma).
8. 2% sodium thiosulfate.
9. TE Buffer.
10. Isopropanol.
11. 1× PBS, pH 7.2.
12. RNA isolation kit (Qiagen).
13. QuantiTect Reverse Transcription Kit (Qiagen).
14. Quantitect SYBR Green PCR Kit (Qiagen).
15. Brucella Broth (Neogen Corp.): dissolve 28 g in 1 L ddH$_2$O and autoclave. Store at room temperature.
16. *H. pylori* freezing media: 10% fetal bovine serum and 20% glycerol in Brain Heart Infusion medium. Filter sterilize and store at +4 °C.
17. Columbia agar-based plates: 5% horse blood, 0.2% β-cyclo-dextrin, 10 μg/mL vancomycin hydrochloride, 8 μg/mL amphotericin B, 5 μg/mL trimethoprim, 5 μg/mL cefsulodin sodium Salt, and 0.33 μg/mL polymyxin B sulfate. Store at +4 °C.

3　Methods

3.1　Selecting Sites for Mutagenesis in silico

Perform basic protein alignments to observe conserved regions using CLUSTAL [4] for site selection.

1. Use the NCBI database to obtain the sequence of the protein from the organism of interest along with the sequences of

orthologous proteins from other organisms (http://www.ncbi. nlm.nih.gov/protein). (If the protein is found in other organisms, it is useful to compare the amino acid sequence among the various organisms to identify regions of conservation.)

2. Proceed to the CLUSTAL Omega website: http://www.ebi. ac.uk/Tools/msa/clustalo/.

3. Enter the sequences for comparison using FASTA format (*see* **Note 1**).

4. Select the "Clustal w/ Numbers" option.

5. Click Submit.

6. The output file will include markings that will annotate completely conserved residues, conserved substitutions, and semi-conserved substitutions. These conserved residues and substitutions provide useful information when determining residues to target using site-directed mutagenesis; conserved regions are predicted to play important roles in the structure–function of orthologous proteins.

3.2 Identification of Functional Domains

1. Even in the absence of true orthologs, identification of conserved protein domains can be useful. The BLAST search database can identify conserved functional domains within the protein of interest and can compare these domains to homologous proteins in other organisms [5]. Proceed to the BLASTP website: http:// blast.ncbi.nlm.nih.gov/Blast.cgi?PAGE=Proteins.

2. Enter the protein sequence of interest using FASTA format (*see* **Note 1**).

3. Choose the nonredundant protein database (nr).

4. If interested, specific organisms can be included or excluded from the BLAST search.

5. Click on BLAST to submit the sequence.

6. The output file will identify any putative conserved domains, along with potential metal binding sites, DNA-binding sites, catalytic sites, etc., depending on the sequence that was entered. The conserved amino acid residues identified within these sites can be selected for site-directed mutagenesis as a means to clearly elucidate their functional role within the protein of interest.

3.3 SWISS-MODEL Comparisons

This protocol describes using the SWISS-MODEL server for generating comparisons of protein structures [6].

1. Proceed to the SWISS-MODEL website: http://swissmodel. expasy.org/.

2. Click on Start Modeling.

3. Enter the protein sequence of interest using FASTA format (*see* **Note 1**).

4. Click on Search for Templates (*see* **Note 2**).

5. Select any templates that are of interest.

6. Click on sequence similarity to view a comparison of the protein of interest with the protein template identified by SWISS-MODEL.

7. Click on Templates again and select any other templates of interest and view the sequence similarity as above.

8. Proceed back to Templates, and select any of the templates that should be incorporated into the protein model.

9. Click on Build Models.

10. The generated models will provide insight into the potential structure of the protein in relation to other proteins of similar sequence and function. The modeled protein can then be used to select residues to target for mutagenesis.

3.4 Random Mutagenesis

This is a PCR-based random mutagenesis technique (*see* **Note 3**) (*see* Fig. 1).

1. Prepare a 50 μL reaction using the GeneMorph II kit: 41 μL of Nuclease-free water, 5 μL 10× Mutazyme II reaction buffer, 1 μL 40 μM dNTP mix, 0.5 μL of each primer (Primers #5 and #6; (*see* Fig. 1), 1 μL Mutazyme II DNA polymerase, and 1 μL template DNA (*see* **Notes 4–6**).

2. Amplify the gene of interest using the following cycling conditions: Initial denaturation at 95 °C for 2 min; 30 cycles with a denaturation at 95 °C for 30 s, annealing 55 °C for 30 s, and extension at 72 °C for 60 s; and a final extension at 72 °C for 10 min (*see* **Notes 7–9**).

3. Gel purify the PCR reaction product using a QiaQuick Gel Extraction Kit.

4. Determine the concentration of DNA from the gel purified PCR reaction. Store at +4 °C.

5. In separate reactions, amplify the 500-bp regions immediately upstream and downstream from the coding sequence using genomic DNA as a template. Prepare the reactions with the Phusion Enzyme kit as follows: 50 ng of genomic DNA, 1 μM each primer (Primers #1 and #2 for the upstream region and primers #3 and #4 for the downstream region (*see* Fig. 1), 1 μM of dNTPs, 1–2 units of Phusion enzyme, and balance with the required amount of buffer and nuclease-free water (*see* **Notes 10 and 11**).

6. Place the reactions in a thermal cycler and amplify using the following conditions: Initial Denaturation at 98 °C for 2 min; 30 cycles with a denaturation at 98 °C for 5 s, annealing at the primer T_m for 20 s, and an extension at 72 °C for 20 s; and a final extension at 72 °C for 5 min (*see* **Notes 8 and 9**).

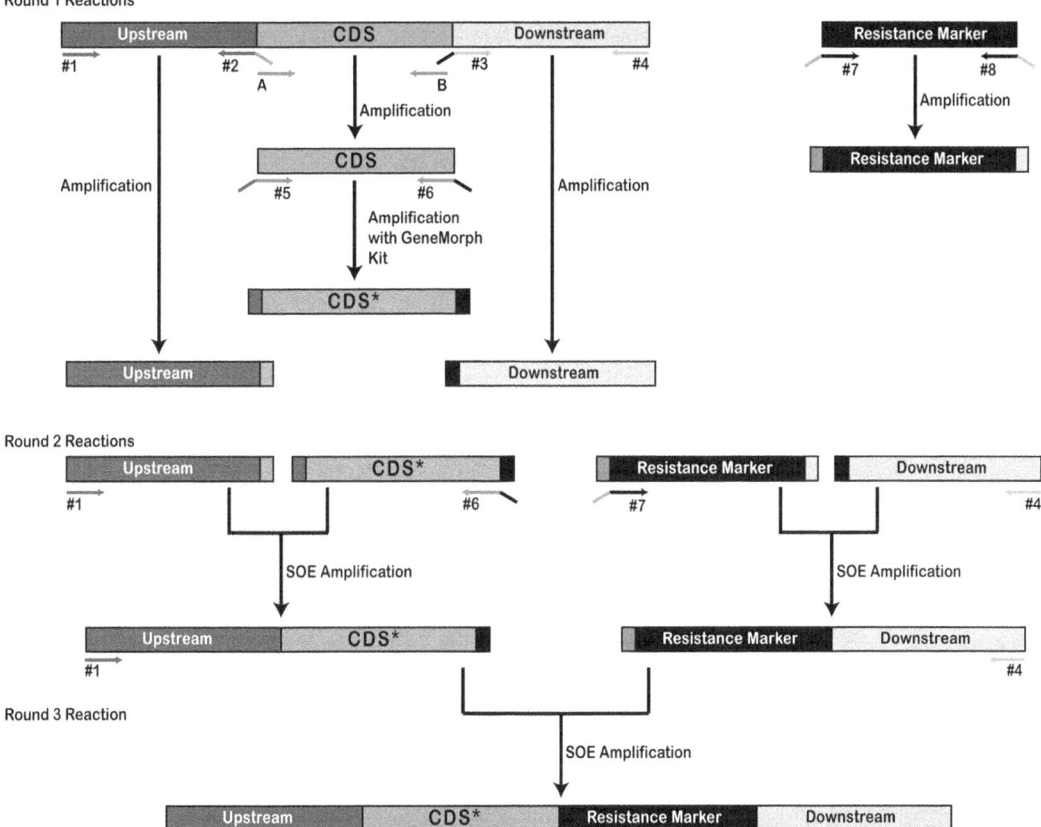

Fig. 1 Schematic representation of the random mutagenesis procedure. Round 1 reactions include amplifying the *coding DNA sequence* (CDS) of the gene of interest with a high fidelity enzyme (*see* **Note 5**) using primers *A* and *B* followed by amplification of the resulting product with the GeneMorph Kit using primers #5 and #6. Additionally, the regions upstream and downstream of the CDS along with the resistance marker need to be amplified with a high fidelity polymerase. Primer pairs #1 and #2 will amplify the upstream regions, while primers #3 and #4 amplify the downstream region. The resistance marker amplicon is generated with primers #7 and #8. As depicted above, each primer that is located at a splice site must incorporate a portion of the sequence from the fragment to which it will be spliced. Round 2 reactions include two separate SOE amplifications: (1) upstream region combined with the mutated CDS using primers #1 and #6 and (2) downstream region combined with the resistance marker using primers #4 and #7. The round 3 reaction includes the final SOE amplification of the upstream region/CDS amplicon combined with the resistance marker/downstream region amplicon using primers #1 and #4. Note that primers designed to incorporate homologous regions that are used in the downstream SOE fusion reactions are indicated by bent "tails." The *shading* of these tailed regions match the corresponding regions in the amplified products and indicate the sites of fusion

7. Purify the PCR reaction products using a MinElute PCR Purification Kit.

8. Determine the concentration of DNA from the purified PCR reaction. Store at +4 °C.

9. PCR amplify a resistance marker that is useful in the bacterium of interest. Prepare the reactions with the Phusion Enzyme kit

as follows: 50 ng of template DNA, 1 μM of each primer (Primers #7 and #8) (*see* Fig. 1), 1 μM of dNTPs, 1–2 units of Phusion enzyme, and balance with the required amount of buffer and nuclease-free water (*see* **Note 12**).

10. Place the reaction in a thermal cycler using the following conditions: Initial denaturation at 98 °C for 2 min; 30 cycles with a denaturation at 98 °C for 5 s, annealing at the primer T_m for 20 s, and an extension at 72 °C for 20 s; and final extension at 72 °C for 5 min (*see* **Notes 8** and **9**).

11. Purify the PCR reaction products using a MinElute PCR Purification Kit.

12. Determine the concentration of DNA from the purified PCR reaction. Store at +4 °C.

13. Initialize the first set of splicing by overlap extension (SOE) reactions using the following templates and primer sets: (1) amplicons from the upstream amplification reaction (generated in **steps 5–8**) and the random mutagenesis reaction (generated in **steps 1–4**) along with the primers #1 and #6, (2) amplicons from the resistance marker amplification reaction (generated in **steps 9–12**) and the downstream amplification reaction (generated in **steps 5–8**) along with primers #4 and #7 (*see* Fig. 1).

14. Set up the initial step of the SOE reaction (50 μL) as follows: 50–100 ng of each amplicon, 1 μM of dNTPs, 1–2 units of Phusion enzyme, and balance with the required amount of buffer and nuclease-free water. *Note*: This step excludes primers.

15. Place reaction in a thermal cycler with the following conditions: Initial denaturation at 98 °C for 30 s; 20 cycles with a denaturation at 98 °C for 5 s, annealing at 55 °C for 20 s, and an extension at 72 °C for 20 s; and final extension at 72 °C for 5 min (*see* **Note 9**).

16. Remove the reaction tubes from the thermal cycler and add an additional 1 unit of Phusion enzyme along with 1 μM of each primer noted in **step 13** in Subheading 3.4.

17. Return the reaction tubes to the thermal cycler with the following cycling conditions: Initial denaturation at 98 °C for 30 s; 29 cycles with a denaturation at 98 °C for 5 s, annealing at 55 °C for 20 s, and an extension at 72 °C for 35 s; and final extension at 72 °C for 10 min (*see* **Note 9**).

18. Gel purify the PCR reaction products using a QiaQuick Gel Extraction Kit.

19. Initialize the final set of SOE reactions using the amplicons from the upstream/random mutagenesis SOE reaction and the resistance marker/downstream SOE reaction along with primers #1 and #4 (*see* Fig. 1).

20. Set up the 50 µL reaction tube as follows: 50–100 ng of each SOE amplicon, 1 µM of dNTPs, 1–2 units of Phusion enzyme, and balance with the required amount of buffer and nuclease-free water. *Note*: this step excludes primers.

21. Place the reaction tube in a thermal cycler with the following conditions: Initial denaturation at 98 °C for 30 s; 20 cycles with a denaturation at 98 °C for 5 s, annealing at 55 °C for 20 s, and an extension at 72 °C for 20 s; and a final extension at 72 °C for 5 min (*see* **Note 9**).

22. Remove the reaction tubes from the thermal cycler and add an additional 1 unit of Phusion enzyme along with 1 µM of each primer noted in **step 19** of Subheading 3.4.

23. Return the reaction tubes to the thermal cycler with the following conditions: Initial denaturation at 98 °C for 30 s; 29 cycles with a denaturation at 98 °C for 5 s, annealing at 55 °C for 20 s, and an extension at 72 °C for 35 s; and a final extension at 72 °C for 10 min (*see* **Note 9**).

24. The resulting spliced amplicon can be further PCR amplified for direct transformation into the bacterium of interest, or it can be ligated into a suicide vector for transformation (*see* **Note 13**).

25. The linear PCR product or random-mutation bearing suicide plasmid can be transformed into the bacterial strain using natural transformation, electroporation, or chemically induced transformation. The method chosen will depend on the natural competency of the organism.

26. After transformation, the mutants can be selected via resistance to the antibiotic marker chosen in **step 9** of Subheading 3.4.

27. Multiple transformants should be expanded from the selection plates to ensure that a variety of random mutations is represented in the protein of interest.

28. Following expansion, the gene of interest should be sequenced in the mutant strains to identify the location of the random mutations and to ensure the proper insertion of the mutated gene (*see* **Note 14**).

29. Multiple random mutants can be combined to create a pooled random mutant library for downstream selection and screening assays. Alternatively, single colony isolates can be banked in 96-well plates such that each well contains a single mutant.

3.5 Chemical-Based Mutagenesis

This protocol describes a random mutagenesis method adapted from Lai et al. [7].

1. Amplify the gene of interest in a 50 µL reaction using the Phusion Enzyme kit as follows: 50 ng of genomic DNA, 1 µM each primer, 1 µM of dNTPs, 1–2 units of Phusion enzyme,

and balance with the required amount of buffer and nuclease-free water.

2. Gel purify the PCR reaction product using a QiaQuick Gel Extraction Kit.

3. Set up a 20 μL reaction with the purified amplicon and a final concentration of 10 mM EMS.

4. Incubate at 37 °C for 1 h.

5. Add 70 μL of 1× PBS, pH 7.2.

6. Add 10 μL 2 % sodium thiosulfate to terminate the reaction.

7. Precipitate DNA by adding 100 μL isopropanol.

8. Centrifuge the reaction mix and remove the supernatant.

9. Resuspend the DNA in TE buffer.

10. The resulting mutated DNA can be ligated into a plasmid for direct insertion into the bacterium of interest, or it can be used in SOE reactions as above to create a marked mutant strain bearing random mutations induced by EMS (*see* **Note 15**).

3.6 Site-Directed Mutagenesis

Generating defined mutations through Splicing by Overlap Extension (SOE) PCR (*see* Fig. 2).

1. Design a forward and reverse primer that both incorporate the codon change required to generate the desired amino acid at the specific site (*see* **Note 16**).

2. Amplify the coding DNA sequence (CDS) of the gene in two segments: the region upstream and including the codon change and the region downstream and including the codon change.

3. Use the Phusion Enzyme kit for the 2–50 μL reactions as follows: 50 ng of genomic DNA, 1 μM each primer (primers #1 and #2 for the upstream CDS and primers #3 and #4 for the downstream CDS) (*see* Fig. 2), 1 μM of dNTPs, 1–2 units of Phusion enzyme, and balance with the required amount of buffer and nuclease-free water.

4. Place the reaction in a thermal cycler using the following conditions: Initial denaturation at 98 °C for 2 min; 30 cycles with a denaturation at 98 °C for 5 s, annealing at primer T_m for 20 s, and an extension at 72 °C for 20 s; and a final extension at 72 °C for 5 min (*see* **Notes 8** and **9**).

5. Gel purify the PCR reaction products using a QiaQuick Gel Extraction Kit.

6. Initialize the splicing by overlap extension (SOE) reactions using the templates from the amplification reactions of both the upstream and downstream regions that incorporated the codon change.

7. Set up the initial step of the SOE reaction (50 μL) as follows: 50–100 ng of each amplicon, 1 μM of dNTPs, 1–2 units of

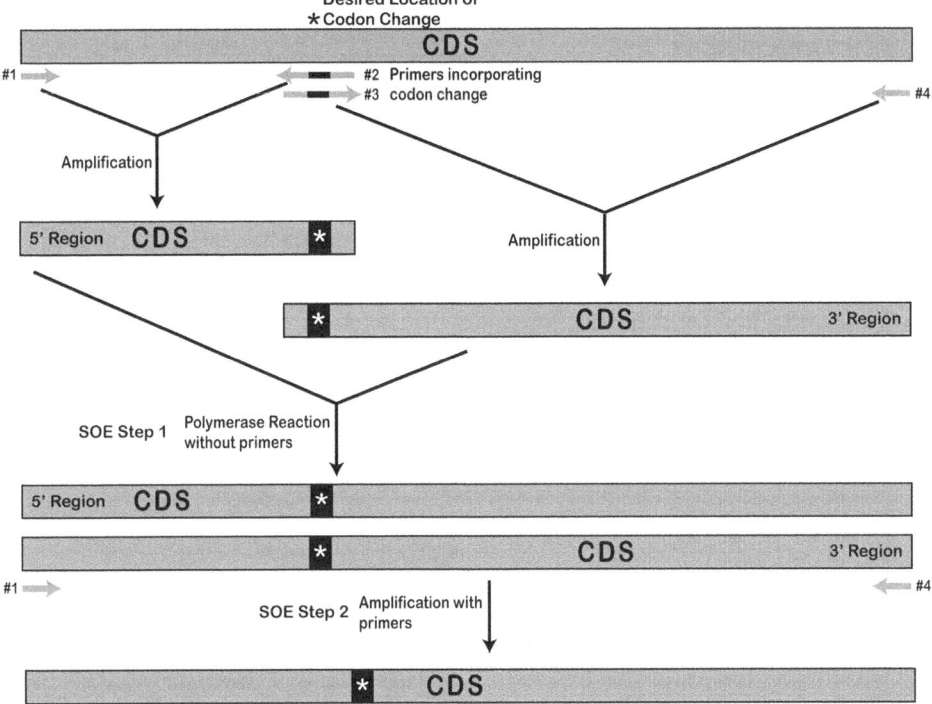

Fig. 2 Diagram of site-directed mutagenesis using SOE. Primers must be designed that incorporate the desired codon change (primers #2 and #3) to allow the amplification of the coding DNA sequence (CDS) of the gene of interest in two reactions: the 5′ region upstream and including the codon change with primers #1 and #2 and the 3′ region downstream and including the codon change using primers #3 and #4. The resulting two amplicons are combined with polymerase but without primers to complete the first round of SOE; this step fills in the missing 5′ sequence on the 3′ region amplicon and the missing 3′ sequence on the 5′ amplicon. The second and final SOE step includes the original forward and reverse primers (#1 and #4) to complete the amplification of the entire CDS incorporating the codon change

Phusion enzyme, and balance with the required amount of buffer and nuclease-free water. *Note*: This step excludes primers.

8. Place the reaction mix in a thermal cycler with the following conditions: Initial denaturation at 98 °C for 30 s; 20 cycles with a denaturation at 98 °C for 5 s, annealing at 55 °C for 20 s, and an extension at 72 °C for 20 s; and a final extension at 72 °C for 5 min (*see* **Note 9**).

9. Remove the reaction tubes from the thermal cycler and add an additional 1 unit of Phusion enzyme and 1 μM of primers #1 and #4 (*see* Fig. 2).

10. Return the reaction tubes to the thermal cycler with the following conditions: Initial Denaturation at 98 °C for 30 s; 29 cycles with a Denaturation at 98 °C for 5 s, Annealing at 55 °C for 20 s, and Extension at 72 °C for 20 s; and Final Extension at 72 °C for 10 min (*see* **Note 9**).

11. The resulting spliced product should be sequenced to verify the incorporation of the desired codon change.

12. The spliced product can then be ligated into a vector for transformation, or it can be spliced together with a marker as in the random mutagenesis procedure described in **steps 9–24** of Subheading 3.4.

13. The spliced PCR product or plasmid can be transformed into the host organism as described in **steps 25–26** of Subheading 3.4.

14. Any resulting transformants should be sequenced to confirm the proper insertion of the mutated coding sequence.

3.7 Mutations Using Synthetic DNA

1. To simplify the generation of site-directed mutations, synthetic DNA can be ordered from companies, including Integrated DNA Technologies and ThermoFisher. These oligonucleotides will be double-stranded and can be designed to incorporate multiple codon changes.

2. Design the DNA sequence for the gene incorporating all desired codon alterations.

3. Upon receipt, resuspend the synthetic oligonucleotide in TE buffer.

4. The synthetic DNA can be ligated into a vector for transformation, or it can be spliced together with a marker as in the random mutagenesis procedure described in **steps 9–24** of Subheading 3.4.

5. The spliced PCR product or plasmid can be transformed into the host organism as described in **steps 25–26** of Subheading 3.4.

6. Any resulting transformants should be sequenced to confirm the proper insertion of the mutated coding sequence.

3.8 Analysis of Constructed Mutant Proteins

This protocol involves the transcriptional analysis of constructed mutant proteins (*see* **Note 17**) using qPCR.

1. When investigating regulatory proteins, qPCR is an efficient method to investigate the effect that the mutation has on the transcription of known, downstream effectors.

2. Isolate RNA from the mutant strain and the wild-type strain (*see* **Note 18**).

3. Using the QuantiTect Reverse Transcription Kit to generate cDNA, combine 1 μg of RNA in a volume of 12 μL of nuclease-free water with 2 μL of gDNA Wipeout buffer.

4. Incubate at 42 °C for 2 min and immediately transfer to ice.

5. Add 4 μL 5× RT Buffer, 1 μL RT primer mix, and 1 μL reverse transcriptase (*see* **Notes 19** and **20**).

6. Incubate the tubes at 42 °C for 15 min.

7. Inactivate the reverse transcriptase enzyme at 95 °C for 3 min.

8. Immediately place tubes on ice.

9. Dilute the cDNA 1:1 with nuclease-free water. Store at –20 °C until ready to use.

10. Set up the qPCR reaction tubes using Quantitect SYBR Green PCR Kit: 1 μL diluted cDNA, 8 μL nuclease-free water, 10 μL SYBR green master mix, 0.5 μL (1.5 pmol) forward primer, and 0.5 μL (1.5 pmol) reverse primer (*see* **Notes 21** and **22**).

11. Run the qPCR reactions with the following cycling conditions: 95 °C for 15 min; 35 cycles with denaturation at 94 °C 15 s, annealing at 50 °C for 20 s, and an extension at 72 °C for 30 s (*see* **Note 23**).

12. Analyze the cycle number (C_t) values from the reactions to identify any gene expression differences between the mutant bacterial strain and the wild-type strain (*see* **Note 24**).

3.9 Additional Methods to Analyze Mutant Proteins

This section describes alternative methods for analyzing mutant proteins.

1. Perform Western blotting to analyze levels of the protein of interest or downstream effector proteins, comparing the wild-type strain to the mutant strain.

2. Conduct electrophoretic mobility shift assays (EMSA's) using purified wild-type and mutant proteins to identify any alterations in DNA–protein binding efficiency.

3. Run fluorescence anisotropy (FA) experiments using purified wild-type and mutant proteins to test DNA–protein binding interactions.

4. Use atomic absorption spectrophotometry to identify any differences in metal binding efficiencies between the purified proteins.

5. Employ circular dichroism (CD) spectroscopy to characterize the relative secondary structures of the purified proteins.

6. Perform oligomerization assays to determine differences in oligomer formation between the purified mutant and wild-type proteins.

3.10 Mutagenesis Techniques in H. pylori

The mutagenesis techniques described above have been successfully applied to study the ferric uptake regulator (Fur) protein from *H. pylori* [3, 8]. Fur is a regulatory protein with an iron (Fe^{2+}) cofactor. Interestingly, Fur regulates genes not only in its iron-bound form, but also in its iron-free, *apo* form. Additionally, Fur can regulate genes via repression or activation [9, 10]. To investigate the role of various amino acid residues in the structure and function of *H. pylori* Fur, random and site-directed mutagenesis procedures were employed [3, 8]. The resulting random mutants were selected as described below. Following selection and screening, the mutants were analyzed using the qPCR, Western blot, FA, CD spectroscopy, atomic absorption spectrophotometry, and oligomerization assays described above.

3.11 Generation of a Random Mutant Library

1. Pellet 1 mL of log-phase *H. pylori* cells growing in liquid media and remove the supernatant (*see* **Notes 25** and **26**).

2. Resuspend the pellet in 100 μL of Brucella broth.

3. Pipet 50 μL of the cell suspension on a blood agar plate (*see* **Note 27**).

4. Allow the spotted cells to dry for approximately 10 min.

5. Place the plate in a microaerophilic environment for 4–5 h at 37 °C, allowing the cells to outgrow (*see* **Note 28**).

6. Pipet 5 μL of the linear SOE product amplified as described in Subheading 3.4. Use approximately 500 ng of PCR product (*see* **Note 29**).

7. Allow the spotted DNA to dry for 2 min.

8. Place the plate in a microaerophilic environment overnight at 37 °C, allowing uptake of the DNA (*see* **Note 28**).

9. Remove the cells from the plate using a sterile cotton swab and resuspend the cells in 1 mL Brucella broth.

10. Pellet the cell suspension and remove the supernatant.

11. Resuspend the cells in 100 μL Brucella broth.

12. Plate 25 μL aliquots of the cell suspension on blood agar plates supplemented with the antibiotic corresponding to the resistance marker integrated during the SOE procedure (*see* **Note 30**).

13. Incubate the plates in a microaerophilic environment for 4–6 days at 37 °C (*see* **Note 28**).

14. Independently repeat the transformations to obtain at least 10,000 independent colonies.

15. Pool the colonies by using a sterile swab to collect the colonies from an entire plate and resuspend in 1 mL Brucella broth.

16. Pellet the resuspended cells and remove the supernatant.

17. Resuspend in *H. pylori* freezing media (*see* **Note 31**).

18. Store the random mutant library at −80 °C (*see* **Note 32**).

3.12 Selection of Fur Mutant Strains

This protocol describes a method for selecting Fur mutants using manganese (*see* **Notes 33** and **34**).

1. Pellet a 5 mL liquid culture of the random mutant library and remove the supernatant.

2. Resuspend the pelleted cells in 1 mL of Brucella broth.

3. Plate 100 μL aliquots on individual Mn-supplemented plates (*see* **Note 35**).

4. Incubate the plates in a microaerophilic environment for 4–6 days at 37 °C (*see* **Note 28**).

5. Pick individual colonies with a toothpick and streak them on a blood agar plate (*see* **Note 27**).

6. Incubate the plates containing the streaks in a microaerophilic environment for 1–2 days at 37 °C (*see* **Note 28**).

7. Remove the streaked cells with a sterile cotton swab dampened with Brucella broth and lawn the cells onto a fresh blood agar plate (*see* **Note 27**).

8. Incubate the plates containing the lawned cells in a microaerophilic environment for 1 day at 37 °C (*see* **Note 28**).

9. Remove the lawned cells with a sterile cotton swab and resuspend in *H. pylori* freezing media (*see* **Notes 31** and **36**).

10. Store the frozen cells at –80 °C.

3.13 Site-Specific Fur Mutants

1. Pellet 1 mL of log-phase *H. pylori* cells growing in liquid media and remove the supernatant (*see* **Note 25**).

2. Resuspend pellet in 100 μL of Brucella broth.

3. Pipet 50 μL of the cell suspension on a blood agar plate (*see* **Note 27**).

4. Allow the spotted cells to dry for approximately 10 min.

5. Place the plate in a microaerophilic environment for 4–5 h at 37 °C, allowing the cells to outgrow (*see* **Note 28**).

6. Pipet 5 μL of the linear site-specific SOE product amplified as described in Subheading 3.6. Use approximately 500 ng of PCR product (*see* **Notes 29** and **37**).

7. Allow the spotted DNA to dry for 2 min.

8. Place the plate in a microaerophilic environment overnight at 37 °C, allowing uptake of the DNA (*see* **Note 28**).

9. Remove the cells from the plate using a sterile cotton swab and resuspend the cells in 1 mL Brucella broth.

10. Pellet the cell suspension and remove the supernatant.

11. Resuspend the cells in 100 μL Brucella broth.

12. Plate 10 and 90 μL aliquots of the cell suspension on blood agar plates supplemented with the antibiotic corresponding to the resistance marker integrated during the SOE procedure (*see* **Note 30**).

13. Incubate the plates in a microaerophilic environment for 4–6 days at 37 °C (*see* **Note 28**).

14. Pick individual colonies with a toothpick and streak them on a blood agar plate (*see* **Note 27**).

15. Incubate the plates containing the streaks in a microaerophilic environment for 1–2 days at 37 °C (*see* **Note 28**).

16. Remove the streaked cells with a sterile cotton swab dampened with Brucella broth and lawn the cells onto a fresh blood agar plate (*see* **Note 27**).

17. Incubate the plates containing the lawned cells in a microaerophilic environment for 1 day at 37 °C (*see* **Note 28**).

18. Remove the lawned cells with a sterile cotton swab and resuspend in *H. pylori* freezing media (*see* **Notes 31** and **38**).

19. Store the frozen cells at –80 °C.

4 Notes

1. FASTA format is generated using a greater-than symbol ">" followed by the sequence name or description. The descriptor line is ended with a "hard return" and followed by the sequence with no spaces or "hard returns".

2. "Templates" are files containing known structures of proteins, and the SWISS-MODEL program will identify template sequences homologous to the protein of interest. The algorithm of the program then uses these known structures to generate the structural model of the protein of interest.

3. The strategy described in Subheading 3.4 involves the integration of the randomly mutated gene directly into the chromosome of the organism being studied. This same random mutagenesis strategy can also be used to generate the mutant gene for use in a heterologous protein expression system in *Escherichia coli*. The mutagenesis primers would simply need to be adapted as needed for insertion into the desired expression plasmid.

4. The number of mutations induced by the GeneMorph II Kit can be optimized by varying the amount of input DNA and the number of amplification cycles. Please refer to the recommendations provided by the manufacturer to obtain 1–2 mutations in the target DNA sequence.

5. To improve the efficiency of the random mutagenesis, it is critical to use either PCR product or plasmid DNA as the template for the mutagenesis reaction and to avoid the use of genomic DNA. Thus, the coding sequence of the gene of interest should first be amplified using a high fidelity polymerase and primers "A" and "B" (*see* Fig. 1). The resulting amplicon should then be gel purified and can be used for the mutagenesis reaction. If

desired, the purified amplicon can be ligated to a high copy number plasmid and transformed into *E. coli*. The resulting transformants should be screened and the plasmid insert sequenced to select a strain bearing a plasmid with the correct coding sequence. The purified plasmid can then be used for the random mutagenesis reaction.

6. The primers for the mutagenesis reaction should include any sequences necessary for subsequent SOE reactions.

7. The GeneMorph manufacturer recommends an annealing temperature 5 °C below the Primer T_m.

8. When determining primer T_m, use only the portion of the primer sequence that will anneal during the reaction.

9. The extension time for the amplification reaction will depend on the length of the region being amplified. Please check the instructions provided by the manufacturer of the polymerase to determine the appropriate amount of time.

10. The length of the amplified regions may need to be altered from the suggested 500 bp length. The ultimate length of these regions will be determined by the crossover efficiency of the organism being studied.

11. Each piece of amplified DNA will be joined together using splicing by overlap extension (SOE). To effectively perform SOE, the primers must be designed with overlapping sequences that will allow joining of the regions of interest. For example, as depicted in Fig. 1, the coding DNA sequence (CDS) of the gene of interest will be spliced to both the amplified upstream fragment and the amplified resistance marker. To accomplish this, the amplification reaction of the CDS must include a forward primer (#5) containing a sequence to match the upstream region and a reverse primer (#6) with a sequence to match the resistance marker sequence (*see* Fig. 1).

12. Since the resistance marker will be placed internally on the final DNA segment, ensure the primers used to amplify the marker contain the appropriate sequences for the subsequent SOE reaction (*see* Fig. 1).

13. If using a vector for transformation, multiple single colonies from the plasmid-bearing strain must be expanded to ensure multiple random mutations are represented. Plasmid DNA should be isolated from each expanded strain and combined prior to the transformation procedure. Additionally, the plasmids can be sequenced to determine the location of the random mutations.

14. It is important to sequence a multitude of transformants to ensure random mutations are represented throughout the protein sequence. Continue sequencing the mutant strains until

saturation is reached, such that additional sequenced strains have duplicate mutations to strains already sequenced in previous rounds of sequencing.

15. EMS can induce multiple mutations within the exposed DNA fragment. The concentration of EMS and the exposure time will need to be optimized to obtain the desired number of mutations.

16. When selecting an amino acid to replace the existing residue, it is important to consider the polarity, charge, and size of the amino acid side chain.

17. Methods for analyzing the mutated protein will vary greatly depending on its type and function. All downstream analyses will depend on the protein under investigation and the availability of assays to determine any potential phenotypic changes caused by the mutant protein. For this reason, only a select group of specific techniques that we have employed in our laboratory are mentioned.

18. Depending on the method of regulation, experiments can be conducted to compare an array of growth conditions and their effect on the transcription of the downstream effector. For example, if the regulator is sensitive to metal concentration, chelation or metal addition experiments can be performed. If the regulator is responsive to pH stress, the pH of the growth media can be modified to determine any transcriptional modulation. Comparing the response of the mutant strains to the wild-type strain will reveal if the modified residue or residues affect the regulatory ability of the protein in response to the various stresses.

19. When performing RT-qPCR on samples recovered from bacterial cells, it is important to run "no RT" control reactions, where the RT enzyme is left out of the reaction and 1 µL of RNAse-free water is substituted. This step is useful to verify that no amplification of contaminating gDNA is occurring during the qPCR step.

20. Keep on ice until ready for the next step.

21. Several online tools exist to aid in proper primer design for SYBR Green qPCR reactions. Proper primer design is crucial to successful interpretation of the qPCR data for the gene of interest [11].

22. It is also necessary to run qPCR reactions for a housekeeping or reference gene, such as 16S rRNA, at the same time as the qPCR reactions for the gene of interest. Data obtained from the reference gene can then be used to normalize for any variation in starting material [12, 13].

23. The annealing temperature will vary based on the primer design, and some real-time PCR kits are designed to allow a combined annealing and extension step during which the fluorescence signal should be collected. Depending on the features of the real-time PCR instrument, it is also useful to run a dissociation/melting curve analysis after the final extension step when possible. The melting curve analysis will identify the production of a single amplicon, ensuring that the designed primers are specific to one region [14, 15]. Additionally, the products from the PCR reactions can be electrophoresed on an agarose gel and visualized to verify the presence of a single amplicon.

24. Typically, the C_t from the reference gene is subtracted from the C_t of the gene of interest for all the samples. The resulting values, known as ΔC_t, are compared among the various samples. To calculate differences between samples, the ΔC_t values are subtracted to create the $\Delta\Delta C_t$ value, which becomes the negative exponent with a base of 2 ($2^{-\Delta\Delta C_t}$) [16].

25. G27 is a frequently used lab strain that can be naturally transformed [17, 18].

26. Our lab uses *H. pylori* liquid growth media comprised of Brucella broth supplemented with 10% fetal bovine serum (FBS) and 10 μg/mL vancomycin hydrochloride [19].

27. Our lab successfully uses Columbia agar-based plates containing 5% horse blood, 0.2% β-cyclodextrin, 10 μg/mL vancomycin hydrochloride, 8 μg/mL amphotericin B, 5 μg/mL trimethoprim, 5 μg/mL cefsulodin sodium salt, and 0.33 μg/mL polymyxin B sulfate [19].

28. Our lab employs an Anoxomat (Advanced Instruments, Inc., Norwood, MA) evacuation and replacement system to create a microaerophilic environment of 5% O_2 and 10% CO_2 [19].

29. To obtain sufficient quantities of DNA, further amplify the SOE products generated in Subheadings 3.4 and 3.6 in 50 μL reactions with the Phusion Enzyme kit as follows: 1 μL of the final SOE product, 1 μM of each primer (#1 and #4; *see* Figs. 1 and 2), 1 μM of dNTPs, 1–2 units of Phusion enzyme, and balance with the required amount of buffer and nuclease-free water. Amplify the SOE product using the following cycling conditions: Initial denaturation at 98 °C for 30 s; 30 cycles with a denaturation at 98 °C for 15 s, annealing 50 °C for 20 s, and an extension at 72 °C for 45 s; and a final extension at 72 °C for 5 min.

30. Our lab integrated a chloramphenicol resistance marker into the final SOE product, so the blood agar plates were supplemented with 8 μg/mL chloramphenicol.

31. Our lab uses an *H. pylori* freezing media comprising 10% fetal bovine serum and 20% glycerol in Brain Heart Infusion medium [19].

32. Sequence the gene in question from a selection of the colonies to determine the frequency of mutations within the gene of interest to ensure the mutation rate is as expected. If optimized as recommended by the manufacturer, the GeneMorph II kit produces a rate of 1–2 mutations per 1 kb.

33. Manganese selection has been used for the identification of amino acid residues important for Fur function in a variety of bacterial species [20–23], including *H. pylori* [3]. Mn acts as a substitute for Fe; thus, Fur mutants deficient in Fe-binding will grow on Mn-supplemented plates while cells containing a normal Fe-binding Fur cannot grow in the presence of the excess Mn.

34. In addition to Mn selection to identify mutants lacking Fe-bound Fur regulation, our lab also sought to identify mutants deficient in *apo*-Fur regulation. To this end, we screened the random mutant library for mutants expressing altered levels of the Pfr protein, a ferritin, upon exposure to an iron-depleted environment [3]. The *pfr* gene is repressed by *apo*-Fur; thus, the absence of *apo*-Fur regulation results in high levels of the Pfr protein when Fe is absent from the culture [24]. Using basic Coomassie staining of protein gels, the increased levels of Pfr can be readily identified [3, 24].

35. Brucella broth-based agar plates (Brucella broth with 1.3% agar (w/v)) work well for *H. pylori* strain G27 when supplemented with 4 mM $MnCl_2$ in addition to 10% fetal bovine serum, 10 μg/mL vancomycin, 8 μg/mL amphotericin B, 5 μg/mL trimethoprim, and 2.5 units/mL polymyxin B.

36. The *fur* gene can be sequenced at this point to identify the mutations that resulted in the Mn-resistant phenotype. Once the altered codons are identified, it is best to create the same mutation again following the site-directed mutagenesis protocol delineated in Subheading 3.6 and transforming the mutated gene into the bacterium. If the phenotype is observed in the mutant created in a clean genetic background, then the mutation can be confirmed to cause the observed phenotype.

37. To select the specific sites for mutagenesis, our lab took into account the location of the random mutations along the *fur* coding sequence and chose sites primarily in the N-terminus, which was not well represented by the colonies identified through *pfr* screening and Mn selection experiments. We primarily chose to mutate the amino acid residues to alanine residues, since alanine has a small, nonpolar side chain. Additionally, a negatively charged amino acid (glutamic acid) located in a

metal coordinating site was selected to be mutated to a positively charged amino acid (arginine) to investigate the effects of the change in charge on metallation capability. Thus, the size and charge of amino acid side chains are important considerations when selecting sites to mutate. It is also important to include a "mutant" that contains the wild-type sequence attached to the resistance marker sequence generated through the SOE process; transforming the bacteria with this sequence will ensure any phenotypic observations are due to the site-specific mutations and not the bacterial manipulation process.

38. The *fur* gene can be sequenced at this point to ensure that the site-specific mutation has been accurately integrated.

References

1. Glass JI, Heinz BA (2003) Site-Directed Mutagenesis. In: Khudyakov YE, Fields HA (eds) Artificial DNA: methods and applications. CRC Press, Boca Raton, FL, pp 199–224

2. Li ITS, Pham E, Truong K (2007) Current approaches for engineering proteins with diverse biological properties. In: Chan WCW (ed) Bio-applications of nanoparticles, vol 620, Advances in experimental medicine and biology. Springer and Landes Bioscience, New York, NY, pp 18–33

3. Gilbreath JJ, Pich OQ, Benoit SL, Besold AN, Cha JH, Maier RJ, Michel SL, Maynard EL, Merrell DS (2013) Random and site-specific mutagenesis of the *Helicobacter pylori* ferric uptake regulator provides insight into Fur structure-function relationships. Mol Microbiol 89(2):304–323. doi:10.1111/mmi.12278

4. Li W, Cowley A, Uludag M, Gur T, McWilliam H, Squizzato S, Park YM, Buso N, Lopez R (2015) The EMBL-EBI bioinformatics web and programmatic tools framework. Nucleic Acids Res 43(W1):W580–W584. doi:10.1093/nar/gkv279

5. Altschul SF, Madden TL, Schaffer AA, Zhang J, Zhang Z, Miller W, Lipman DJ (1997) Gapped BLAST and PSI-BLAST: a new generation of protein database search programs. Nucleic Acids Res 25(17):3389–3402

6. Schwede T, Kopp J, Guex N, Peitsch MC (2003) SWISS-MODEL: an automated protein homology-modeling server. Nucleic Acids Res 31(13):3381–3385

7. Lai YP, Huang J, Wang LF, Li J, Wu ZR (2004) A new approach to random mutagenesis *in vitro*. Biotechnol Bioeng 86(6):622–627. doi:10.1002/bit.20066

8. Carpenter BM, Gancz H, Benoit SL, Evans S, Olsen CH, Michel SL, Maier RJ, Merrell DS (2010) Mutagenesis of conserved amino acids of *Helicobacter pylori* Fur reveals residues important for function. J Bacteriol 192(19):5037–5052. doi:10.1128/jb.00198-10

9. Carpenter BM, Whitmire JM, Merrell DS (2009) This is not your mother's repressor: the complex role of *fur* in pathogenesis. Infect Immun 77(7):2590–2601. doi:10.1128/iai.00116-09

10. Pich OQ, Merrell DS (2013) The ferric uptake regulator of *Helicobacter pylori*: a critical player in the battle for iron and colonization of the stomach. Future Microbiol 8(6):725–738. doi:10.2217/fmb.13.43

11. Thornton B, Basu C (2011) Real-time PCR (qPCR) primer design using free online software. Biochem Mol Biol Educ 39(2):145–154. doi:10.1002/bmb.20461

12. Edmunds RC, McIntyre JK, Luckenbach JA, Baldwin DH, Incardona JP (2014) Toward enhanced MIQE compliance: reference residual normalization of qPCR gene expression data. J Biomol Tech 25(2):54–60. doi:10.7171/jbt.14-2502-003

13. Rocha DJ, Santos CS, Pacheco LG (2015) Bacterial reference genes for gene expression studies by RT-qPCR: survey and analysis. Antonie Van Leeuwenhoek 108(3):685–693. doi:10.1007/s10482-015-0524-1

14. Bustin SA, Benes V, Garson JA, Hellemans J, Huggett J, Kubista M, Mueller R, Nolan T, Pfaffl MW, Shipley GL, Vandesompele J, Wittwer CT (2009) The MIQE guidelines: minimum information for publication of quantitative real-time PCR experiments. Clin Chem 55(4):611–622. doi:10.1373/clinchem.2008.112797

15. Ririe KM, Rasmussen RP, Wittwer CT (1997) Product differentiation by analysis of DNA melting curves during the polymerase chain

reaction. Anal Biochem 245(2):154–160. doi:10.1006/abio.1996.9916

16. Livak KJ, Schmittgen TD (2001) Analysis of relative gene expression data using real-time quantitative PCR and the 2(-delta delta C(T)) method. Methods 25(4):402–408. doi:10.1006/meth.2001.1262

17. Baltrus DA, Amieva MR, Covacci A, Lowe TM, Merrell DS, Ottemann KM, Stein M, Salama NR, Guillemin K (2009) The complete genome sequence of *Helicobacter pylori* strain G27. J Bacteriol 191(1):447–448. doi:10.1128/jb.01416-08

18. Covacci A, Censini S, Bugnoli M, Petracca R, Burroni D, Macchia G, Massone A, Papini E, Xiang Z, Figura N et al (1993) Molecular characterization of the 128-kDa immunodominant antigen of *Helicobacter pylori* associated with cytotoxicity and duodenal ulcer. Proc Natl Acad Sci U S A 90(12):5791–5795

19. Whitmire JM, Merrell DS (2012) Successful culture techniques for *Helicobacter* species: general culture techniques for *Helicobacter pylori*. Methods Mol Biol 921:17–27. doi:10.1007/978-1-62703-005-2_4

20. Funahashi T, Fujiwara C, Okada M, Miyoshi S, Shinoda S, Narimatsu S, Yamamoto S (2000) Characterization of *Vibrio parahaemolyticus* manganese-resistant mutants in reference to the function of the ferric uptake regulatory protein. Microbiol Immunol 44(12):963–970

21. Hantke K (1987) Selection procedure for deregulated iron transport mutants (*fur*) in *Escherichia coli* K 12: *fur* not only affects iron metabolism. Mol Gen Genet 210(1):135–139

22. Lam MS, Litwin CM, Carroll PA, Calderwood SB (1994) *Vibrio cholerae fur* mutations associated with loss of repressor activity: implications for the structural-functional relationships of *fur*. J Bacteriol 176(16):5108–5115

23. Loprasert S, Sallabhan R, Whangsuk W, Mongkolsuk S (2000) Characterization and mutagenesis of *fur* gene from *Burkholderia pseudomallei*. Gene 254(1-2):129–137

24. Bereswill S, Greiner S, van Vliet AH, Waidner B, Fassbinder F, Schiltz E, Kusters JG, Kist M (2000) Regulation of ferritin-mediated cytoplasmic iron storage by the ferric uptake regulator homolog (Fur) of *Helicobacter pylori*. J Bacteriol 182(21):5948–5953

Section VII

Random Mutagenesis: Novel PCR-Based Methods

Chapter 32

Novel Random Mutagenesis Method for Directed Evolution

Hong Feng, Hai-Yan Wang, and Hong-Yan Zhao

Abstract

Directed evolution is a powerful strategy for gene mutagenesis, and has been used for protein engineering both in scientific research and in the biotechnology industry. The routine method for directed evolution was developed by Stemmer in 1994 (Stemmer, Proc Natl Acad Sci USA 91, 10747–10751, 1994; Stemmer, Nature 370, 389–391, 1994). Since then, various methods have been introduced, each of which has advantages and limitations depending upon the targeted genes and procedure. In this chapter, a novel alternative directed evolution method which combines mutagenesis PCR with dITP and fragmentation by endonuclease V is described. The kanamycin resistance gene is used as a reporter gene to verify the novel method for directed evolution. This method for directed evolution has been demonstrated to be efficient, reproducible, and easy to manipulate in practice.

Key words DNA shuffling, Molecular evolution, dITP, PCR, Mutagenesis, Endonuclease V

1 Introduction

In vitro directed evolution is a powerful tool for gene mutagenesis and is widely used for improving the catalytic properties of enzymes and for investigating the relationship between their structure and function [3–5]. The method of routine directed evolution was originally developed by Stemmer [1, 2], which included primarily three steps: (1) introduction of random mutations by error-prone PCR; (2) fragmentation with DNase I; and (3) assembly and reamplification by PCR. In the first step, error-prone PCR is a primary method to introduce mutations, which can be achieved by adjusting concentrations of Mg/Mn and dNTPs, or by using error-prone Taq DNA polymerase. However, error-prone PCR usually induces sequence biases during the in vitro recombination steps [6]. Thus, other methods have also been developed to introduce random mutations. For example, dITP can be used for introducing mutations into DNA strands by PCR [7], which has been successfully used in random mutagenesis of certain genes [8, 9]. For fragmentation of DNA, DNase I is usually used for this purpose, but it

Andrew Reeves (ed.), *In Vitro Mutagenesis: Methods and Protocols*, Methods in Molecular Biology, vol. 1498,
DOI 10.1007/978-1-4939-6472-7_32, © Springer Science+Business Media New York 2017

requires careful optimization of the digestion conditions [1, 2]. Instead, several alternative methods for DNA fragmentation have been developed [10–12]. For example, a variety of restriction enzymes can be used to generate the DNA fragments [13, 14].

Endonuclease V (endo V) is a DNA modification enzyme involved in repairing DNA deamination by hydrolyzing the second phosphodiester bond 3′ from the base lesion [15]. Several characterized endo V from bacteria have been demonstrated to efficiently cleave the damaged DNA strands which contain dITP, dUTP, etc. [16]. As a result, endo V has been used for cleavage U-contained DNA strands to prepare DNA fragments in DNA shuffling [17]. In this chapter, we introduce an alternative method to combine mutagenic PCR with dITP and fragmentation by endo V for directed evolution using a kanamycin resistance gene (*kan*) as the target gene [18]. This method is demonstrated to be efficient, reproducible, and easy to manipulate in practice.

2 Materials

2.1 Chemical Reagents and Enzymes

1. *Taq* DNA polymerase.
2. *Tma* endonuclease V.
3. *Dpn* I.
4. Appropriate restriction enzymes to fragment DNA.
5. T4 DNA ligase.
6. dITP (100 mM).
7. dNTP (2.5 mM each).
8. Primers (K1302F: 5′-ACAGGCAGCCCATCAGTC-3′; K1302F: 5′-CGGTGCTACAGAGTTCTTG-3′) for amplification of kanamycin resistance gene (*kan*) from the plasmid pCAMBIA1302 (*see* **Note 1**).
9. DNA markers, such as pUC19 DNA/*Msp*I (*Hpa*II).
10. Ethidium bromide: 5 mg/mL stock solution in distilled water.
11. TAE-agarose gels (0.7 and 2%).

2.2 Plasmids and Strains

1. Template DNA (pCAMBIA1302 containing a kanamycin gene) (*see* **Note 2**).
2. pMD-T cloning vector (*see* **Note 3**).
3. *Escherichia coli* DH5α or any other appropriate *E. coli* plasmid maintenance strains.

2.3 Reagents and Equipment

1. PCR Purification Kit.
2. Gel DNA Recovery Kit (Qiagen).
3. Eppendorf Mastercycler thermocycler capable of gradient PCR.

4. Visible/UV spectrometer.

5. Microcentrifuge.

6. DNA electrophoresis equipment.

3 Methods

3.1 Mutagenic PCR

1. Set up a 50 µL PCR mixture in a 200-µL thin-wall Eppendorf tube by the addition of 1× PCR buffer containing 1.5 mM MgCl, 50 µM dNTPs, 150 µM dITP, 400 nM of each primer, 10 ng template DNA (pCAMBIA1302), and 5 U Taq DNA polymerase.

2. Place the tube in the PCR thermocycler. Set up the PCR parameters as follows: (a) an initial denaturing at 95 °C for 2 min; (b) 30 cycles at 94 °C for 0.5 min, 54 °C for 0.5 min, and 72 °C for 1 min; and (c) a final extension at 72 °C for 10 min (*see* **Note 4**).

3. Load 5 µL of PCR product into a 0.7 % agarose gel for electrophoresis analysis (*see* **Note 5**) (*see* Fig. 1).

3.2 Endo V Digestion of Mutagenized DNA

1. Add 20 U *Dpn*I directly into the 50 µL PCR mixture and incubate at 37 °C for 2–4 h to remove the template DNA.

2. Purify the mutagenesis PCR products using the PCR Purification Kit following the manufacturer's protocol, and calculate the DNA concentration by OD_{260} absorbance with a spectrometer.

3. In a 500 µL Eppendorf tube, add 500 ng purified DNA, 5 µL Endo V buffer, and 10 U endo V. Mix gently.

4. Incubate at 60 °C for 2 h or more to fragment the DNA.

5. Load all the fragmented DNA sample on a 2 % agarose gel with the DNA marker (pUC19 DNA/*Msp*I) to resolve the DNA fragments by electrophoresis (*see* Fig. 1).

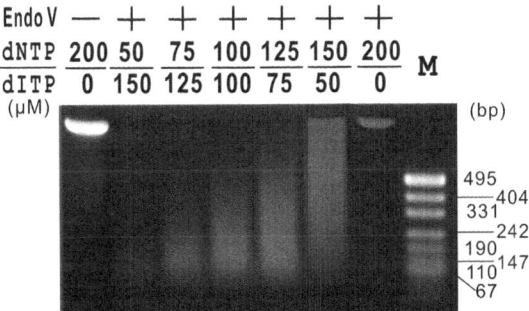

Fig. 1 Endonuclease V cleavage of the mutagenesis PCR products with dITP incorporated. About 500 ng of PCR products was digested by 10 U *Tma* endo V at 60 °C for 2 h. Reproduced from ref. [18] with permission

6. After electrophoresis, immerse the gel into a 1 % ethidium bromide solution for staining for 10 min.

7. Rinse the gel with ddH$_2$O several times.

8. Place the gel under ultraviolet light and quickly but carefully excise the gel region containing the DNA fragments with a size in the range of 50–100 bp.

9. Recover the DNA from the agarose gel using a gel DNA cleanup kit and elute with 30 μL elution buffer.

10. Load 5 μL of the recovered, resuspended DNA in a 2 % agarose gel for electrophoresis analysis, and calculate the DNA concentration by reading the OD$_{260}$ value on a spectrometer.

3.3 Assembly and Reamplification

1. Set 100 μL of the PCR mixture in a 200 μL thin-wall tube by addition of 100 ng fragmented DNA, 10 μL PCR buffer, 200 μM dNTP, and 5 U Taq DNA polymerase.

2. Set up the PCR cycling parameters to: 95 °C for 2 min followed by 45 cycles of 94 °C for 2 min, 45 °C for 30 s, and 72 °C for 0.5 min and 2 s per cycle (32 s) and a final extension reaction for 10 min (see **Note 6**).

3. Load 5 μL PCR products on a 0.7 % agarose gel for electrophoresis analysis. The PCR products should occur as a smear in the desired size range.

4. For reamplification of the target DNA, set up a 100 μL PCR mixture containing 1 μL of the assembled products, 200 μM dNTPs, 400 nM each primer, and 2.5 U Taq DNA polymerase. The PCR parameters are set up with an initial denaturation step at 95 °C for 2 min followed by 25 cycles of 94 °C for 0.5 min, 52 °C for 0.5 min, and 72 °C for 1.5 min and a final extension cycle at 72 °C for 10 min (see **Note 7** and Figs. 2 and 3).

5. Load 5 μL of the PCR products on a 0.7 % agarose gel for electrophoresis. Normally, the target DNA band should be readily observed in the gel.

6. Recover the target DNA from the agarose gel by using a Gel DNA Recovery Kit.

3.4 Cloning and Sequencing (See Note 8)

1. Set up a 10 μL ligation mixture by adding 1 μL T4 DNA ligase buffer, 25 ng target DNA, 5 ng pMD19-T, 0.5 U T4 DNA ligase.

2. Place the ligation mixture at 16 °C overnight.

3. Transform 200 μL of E. coli chemically competent cells with 2 μL ligation mixture, and then dispense all the cells onto several LB agar plates containing 100 μg/mL ampicillin (see **Note 9**).

$$\frac{\text{dNTP}}{\text{dITP}} \quad \frac{75}{125} \quad \frac{50}{150} \quad \text{M}$$
(μM)

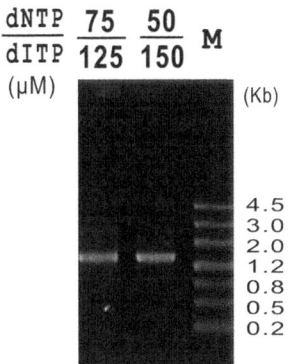

Fig. 2 Electrophoresis analysis of reamplified target DNA fragments on 0.7 % agarose gel. The PCR mixture (100 μL) contained 1 μL assembly PCR products, 2.5 U *Taq* DNA polymerase, 200 μM dNTPs, and 400 nM of each primer. The reaction was performed with an initial denaturation at 95 °C for 2 min followed by 25 cycles of 94 °C for 0.5 min, 52 °C for 0.5 min, and 72 °C for 1.5 min and a final extension at 72 °C for 10 min. Reproduced from ref. [18] with permission

$$\frac{75}{125} \quad \frac{50}{150} \quad \frac{75}{125} \quad \frac{50}{150} \quad \frac{\text{dNTP}}{\text{dITP}}$$
(μM)

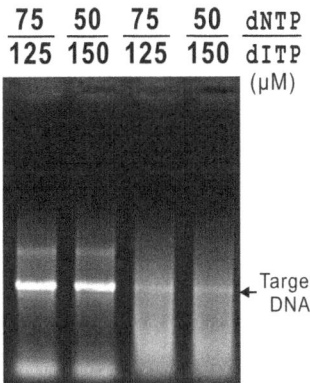

Fig. 3 Electrophoresis analysis of reamplified target DNA fragments on 0.7 % agarose gel. The PCR reaction was set up as described in Fig. 2, except the cycling number was set up as 32 for the *left two lanes* and the 5 μL assembly PCR products were used as template for the *right two lanes*

4. Incubate the LB agar plates at 37 °C overnight.

5. Randomly pick 30 distinct colonies for DNA sequencing to evaluate the mutation frequency (*see* **Note 10**).

6. For construction of the mutagenesis library, collect the transformant cells on each plate by washing with 5 mL sterile ddH₂O. Pool all the cells collected from the transformation plates to isolate the plasmid DNA by the standard mini-plasmid preparation techniques. In our experience, the plasmid DNA library is suitable for storage at −20 °C.

4 Notes

1. The primers should be designed based on the target gene sequence. The restriction enzyme sites may be engineered into the primers at the 5′ end and appropriate 3–6 bp-sized clamps at their 5′ ends to allow for efficient digestion of the PCR products and subsequent cloning.

2. The plasmid DNA containing the target gene should be prepared and serve as the template for mutagenic PCR. The concentration of the plasmid DNA should be determined, such as by reading the OD_{260}.

3. An appropriate expression vector should be prepared for any other target genes.

4. The PCR parameters reported here may not be suitable for the other target genes, which should be set up based on the primer sequences used for any other target gene.

5. The yield of PCR products tends to be reduced with increasing dITP concentration including in mutagenic PCR. The ratio of dITP vs. dNTP may be changed to modulate the mutation frequency. The frequency of dITP incorporation into the newly synthesized DNA strands could be directly evaluated by endonuclease V cleavage. Predictably, more dITP should be incorporated with an increased ratio of dITP vs. dNTP because the size range of DNA fragments after endonuclease V cleavage becomes smaller with an increasing amount of dITP (*see* Fig. 1).

6. Assembly PCR may require more cycles than standard PCR, resulting in a smear of DNA products when observed in an agarose gel. Too many cycles is not suggested, which may lead to issues in the reamplification PCR.

7. The reamplification PCR is important to obtain a distinct target DNA (*see* Fig. 2). Too much template DNA from the assembly PCR and too many cycles may lead to a smear of DNA products or a reduced yield (*see* Fig. 3).

8. The PCR products obtained from the reamplification PCR should be digested with the appropriate restriction enzymes prior to cloning. In order to achieve appropriate amounts of digested DNA for generating the mutant library, highly competent cells must be used or a highly efficient electroporation transformation will need to be performed.

9. DNA transformation procedure: (a) Gently thaw 200 μL of competent *E. coli* cells on ice. (b) Add 2 μL of the ligation mixture into the chilled competent cells and gently flick the transformation mixture to disperse the DNA. (c) Incubate the transformation reaction mixture on ice for 30 min. (d) Heat-shock the mixture at exactly 42 °C for 60 s and immedi-

ately cool it on ice for 2 min. (e) Add 1 mL of SOC medium into the transformation reaction mixture and incubate it at 37 °C and 600 rpm for 1 h in a thermomixer or in a shaker incubator at 200 rpm (1-in. orbital rotation). (f) In a sterile, laminar flow hood place a closed LB agar plate containing the appropriate selective antibiotic. (g) Pellet the transformed cells by centrifugation for 6 min at $3{,}500 \times g$ in a benchtop microcentrifuge. (h) Remove about 1,000 µL of the supernatant and resuspend the cell pellet in the remaining volume (~200 µL). (i) Spread the cell suspension on the LB-antibiotic agar plate, allow a few minutes to absorb, and incubate the plate bottom side up at 37 °C overnight (at least 16 h). (j) Isolate plasmid DNA using a miniprep plasmid isolation kit from one cell colony and send it for sequence analysis to check if the gene of interest carries the desired mutations.

10. Evaluation of the mutation frequency is necessary for the mutagenized DNA library. In doing so, 30 or more clones should be randomly picked for DNA sequencing. For example, a mutation frequency of about 3 per kb of DNA is easily achievable. In our experience, the mutations will be distributed randomly over the full PCR fragment and the occurrence of deletions and insertions does not seem to be an issue using this protocol. However, the mutagenesis pattern is not randomly distributed over the DNA segments, but has an obvious bias for transitions (A/G–G/A or T/C–C/T). Statistically, the transitions (A/T–G/C and G/C–A/T) account for 78.15 % of substitutions, while transversions (A/T–T/A, G/C–C/G, A/T–C/G and G/C–T/A) account for 21.85 % of the total mutations observed.

Acknowledgments

This study was supported by the High Technology Research and Development Program of China (2006AA02Z221) and the National Natural Science Foundation of China (31171204) to H.F.

References

1. Stemmer WPC (1994) DNA shuffling by random fragmentation and reassembly: in vitro recombination for molecular evolution. Proc Natl Acad Sci USA 91:10747–10751

2. Stemmer WPC (1994) Rapid evolution of a protein in vitro by DNA shuffling. Nature 370:389–391

3. Tee KL, Wong TS (2013) Polishing the craft of genetic diversity creation in directed evolution. Biotechnol Adv 31:1707–1721

4. Yuan L, Kurek I, English J, Keenan R (2005) Laboratory directed protein involution. Microbiol Mol Biol Rev 69:373–392

5. Ruff AJ, Dennig A, Schwaneberg U (2013) To get what we aim for—progress in diversity generation methods. FEBS J 280:2961–2978

6. Joern JM, Meinhold P, Arnold FH (2002) Analysis of shuffled gene libraries. J Mol Biol 316:6443–6656

7. Dierick H, Stul M, De Kelver W, Marynen P, Cassiman J (1993) Incorporation of dITP or 7-deaza dGTP during PCR improves sequencing of the product. Nucleic Acids Res 21:4427–4428

8. Kuipers OP (1996) Random mutagenesis by using mixtures of dNTP and dITP. Methods Mol Biol 57:351–356

9. Spee JH, de Vos WM, Kuipers O (1993) Efficient random mutagenesis method with adjustable mutation frequency by use of PCR with dITP. Nucleic Acids Res 21:777–778

10. Zhao H, Giver L, Shao Z, Affholter JA, Arnold FH (1998) Molecular evolution by staggered extension process (StEP) in vitro recombination. Nat Biotechnol 16:258–261

11. Müller KM, Stebel SC, Knall S, Zip G, Bernauer HS, Arndt KM (2005) Nucleotide exchange and excision technology (NExT) DNA shuffling: A robust method for DNA fragmentation and directed evolution. Nucleic Acids Res 33, e117

12. Wong TS, Tee KL, Hauer B, Schwaneberg U (2004) Sequence saturation mutagenesis (SeSaM): a novel method for directed evolution. Nucleic Acids Res 32, e26

13. Engler C, Gruetzner R, Kandzia R, Marillonnet S (2009) Golden gate shuffling: a one-pot DNA shuffling method based on type IIs restriction enzymes. PLoS One 4, e5553

14. Rosic NN, Huang W, Johnston WA, DeVoss JJ, Gillam WMJ (2007) Extending the diversity of cytochrome P450 enzymes by DNA family shuffling. Gene 395:40–48

15. Cao W (2013) Endonuclease V: an unusual enzyme for repair of DNA deamination. Cell Mol Life Sci 70:3145–3156

16. Feng H, Klutz AM, Cao W (2005) Active site plasticity of endonuclease V from *Salmonella typhimurium*. Biochemistry 44:675–683

17. Miyazaki K (2002) Random DNA fragmentation with endonuclease V: application to DNA shuffling. Nucleic Acids Res 30, e139

18. Wang Z, Wang H-Y, Feng H (2013) A simple and reproducible method for directed evolution: combination of random mutation with dITP and DNA fragmentation with endonuclease V. Mol Biotechnol 53:49–54

Chapter 33

Random Mutagenesis by Error-Prone Polymerase Chain Reaction Using a Heavy Water Solvent

Toshifumi Minamoto

Abstract

Heavy water is a form of water that contains a heavier isotope of hydrogen (^{2}H, also known as deuterium, D) or oxygen (^{17}O or ^{18}O). When using heavy water as a solvent, error-prone polymerase chain reaction (epPCR) can induce random mutations independent of the polymerase used or the composition of the PCR reaction mixture. This relatively new method can easily be combined with the existing epPCR methods to increase the rate of mutations.

Key words Random mutagenesis, Error-prone polymerase chain reaction (epPCR), Heavy water, D$_2$O, H$_2$^{18}O, Stable isotope

1 Introduction

Random mutagenesis is used to generate DNA changes in the protein coding regions of genes in order to screen for proteins with altered or optimized functions. Traditionally, mutations were generated by UV irradiation or chemical mutagenesis [1, 2]. More recently, a variety of error-prone PCR (epPCR) techniques have been developed and are being utilized [3]. The DNA polymerases used in epPCR, such as *Thermus aquaticus* (*Taq*) DNA polymerase or *Thermus thermophilus* (*Tth*) DNA polymerase, lack $3' \rightarrow 5'$ exonuclease activity (also known as proofreading activity). Error-prone conditions, such as an increased concentration of Mn^{2+}, biased nucleotide concentration, or the use of nucleotide analogues [3, 4], are generally used to enhance the rate of mutations. However, epPCR utilizing *Taq* DNA polymerase shows biased patterns of mutation. When heavy water is used as a solvent in epPCR, it induces mutations by a completely different principle from the techniques listed above. Thus, protocols for using heavy water in epPCR have been developed [5].

Heavy water is a form of water that contains the hydrogen isotope deuterium (depicted as D or ^{2}H), or an oxygen isotope

Andrew Reeves (ed.), *In Vitro Mutagenesis: Methods and Protocols*, Methods in Molecular Biology, vol. 1498,
DOI 10.1007/978-1-4939-6472-7_33, © Springer Science+Business Media New York 2017

(^{17}O or ^{18}O). In this chapter, methods in which D_2O and $H_2^{18}O$ are used as a solvent in PCR reactions are described. The chemical characteristics of the solvent not only affect enzymatic reactions, but the use of heavy water may affect the substrate specificity of the DNA polymerase as well. As a result, more mutations are induced when a heavy water solvent is used [5]. The preparation of the PCR mixture is similar to standard PCR methods. However, in order to increase the concentration of the heavy water in the solvent, regular water contained in the reagent is first evaporated, and then the dried reagents are redissolved in the heavy water. After the PCR reaction is completed, the amplicons are subsequently cloned and sequenced. The reported maximum error rate obtained from this method is 1.8×10^{-3} errors/bp [5].

One of the advantages of using heavy water in epPCR is that it can easily be applied to all related random mutagenesis PCR protocols. Thus, the method described here is only a representative example.

2   Materials

2.1   epPCR with Heavy Water

1. >99.9 atom % of D_2O that is commercially available.

2. DNA polymerase which does not have proofreading activity ($3' \rightarrow 5'$ exonuclease activity) (*see* **Note 1**), such as *Taq* DNA polymerase or *Tth* DNA polymerase.

3. PCR buffers and deoxynucleotide (dNTP) solution mixtures.

4. 25 mM $MnCl_2$ should be prepared using ultrapure water (PCR grade water) and autoclaved.

2.2   Subcloning of Amplicons

1. Luria Bertani (LB) broth with ampicillin: tryptone 10 g, yeast extract 5 g, NaCl 10 g. Add deionized water to a total volume of 1000 mL. Autoclave at 121 °C, 15 psig for 20 min. After cooling, add 50 mg of sterile ampicillin sodium salt and mix well.

2. LB agar plates with ampicillin/IPTG/X-Gal: tryptone 1 g, yeast extract 0.5 g, NaCl 1.0 g, agar 1.5 g, and pure water to a total volume of 100 mL. Autoclave at 121 °C, 15 psig for 20 min. After cooling (~50 °C), add 5 mg of sterile ampicillin sodium salt, mix well and pour into culture dishes before solidification.

3. 2% X-gal (5-bromo-4-chloro-3-indolyl β-D-galactopyranoside).

4. 20 μl of 0.1 M IPTG (isopropyl β-D-1-thiogalactopyranoside).

5. Plasmid DNA miniprep kit.

6. PCR DNA cleanup kit.

7. Agarose.

8. Agarose electrophoresis equipment.

9. PCR product cloning vector (e.g., TA cloning vector).

3 Methods

3.1 PCR with Heavy Water

1. Mix all reagents except *Taq* DNA polymerase in 0.5 ml PCR tubes. For a typical 50 µl PCR reaction add the following: 5 µl of 10× buffer or 10 µl of 5× buffer (final concentration of buffer is 1×); 1 µl of 10 mM solution of each dNTP (final concentration of each dNTP is 200 µM); 3–4.2 µl of 25 mM MnCl$_2$ (range for final concentration is 1.5–2.1 mM) (*see* **Note 2**); 20 pmol of each primer (volume dependent on stock concentration); and template DNA.

2. Dry the PCR reaction mixture using a centrifugal evaporator (*see* **Note 3**). The pellet might not be visible at this step.

3. Resuspend the pellet in 49.5 µl of D$_2$O (*see* **Note 4**). Place the reaction tube on ice.

4. Add 0.5 µl of *Taq* DNA polymerase (*Taq* DNA polymerase is typically provided at a concentration of 5 units/µl) and gently mix the reaction solution.

5. Amplify the DNA using a standard three-step PCR protocol. Typical PCR conditions are as follows: 30 cycles at 94 °C for 30 s (denaturation), 50–60 °C (depending on Tm value of the primers) for 30 s (annealing), and 72 °C for 30–60 s (extension; varies depending on the target length).

6. Visualize the products of the PCR amplification in a 1–2 % agarose gel by electrophoresis (*see* Fig. 1).

3.2 Amplicon Cloning

1. The PCR products can be used directly in a TA cloning vector ligation reaction without further processing (*see* **Note 5**).

2. Transform the high-efficiency competent cells with TA cloning vector ligation reaction product (*see* **Note 6**).

3. Culture the cells on LB/ampicillin/IPTG/X-Gal plates, and incubate at 37 °C overnight.

4. Pick individual white colonies and inoculate into separate tubes each containing 5 mL of LB/ampicillin medium. Grow for 12–16 h at 37 °C with gentle shaking.

5. Purify the recombinant plasmid using a standard minipreparation procedure (*see* **Note 7**).

6. Sequence the purified plasmids to identify and characterize mutations.

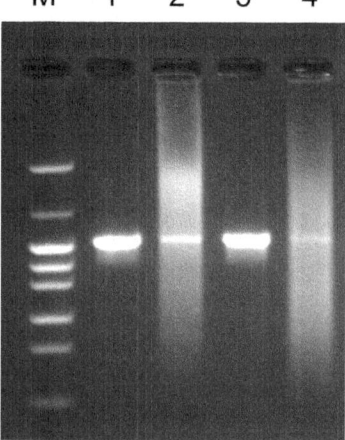

Fig. 1 epPCR reaction products resolved by agarose gel electrophoresis. The PCR products were generated using the following conditions: *Lane 1*, 100 % H_2O solvent with 1.5 mM Mn^{2+}. *Lane 2*, 100 % H_2O solvent with 2.1 mM Mn^{2+}. *Lane 3*, 99 % D_2O solvent with 1.5 mM Mn^{2+}. *Lane 4*, 99 % D_2O solvent with 2.1 mM Mn^{2+}. The sample volume applied in *Lanes 1* and *3* was 1 μl, and that of *Lanes 2* and *4* was 10 μl. *Lane M* indicates a molecular weight marker lane

4   Notes

1. While a number of factors contribute to the induced mutation rate in the PCR reaction, the type of polymerase used is the most influential variable. Polymerases with proofreading activity, such as *Pfu* DNA polymerase (Promega) and KOD DNA polymerase (EMD Millipore), drastically decrease the number of mutations.

2. Higher concentrations of Mn^{2+} increase mutations, but simultaneously decrease the PCR efficiency, as shown in Fig. 1. The highest concentration of Mn^{2+} that may be used depends on the type of DNA polymerase, the PCR buffers, and the target genes and primers. Note that the PCR buffer provided with the commercially purchased DNA polymerase may contain Mn^{2+}.

3. Drying time should be carefully determined beforehand, as over-drying may result in a pellet that may not readily resuspend in buffer.

4. The use of $H_2^{18}O$ instead of D_2O as solvent results in a lower level of mutations, but causes more transversions. Thus, it may be possible to control the mutation spectrum by mixing D_2O and $H_2^{18}O$ [5].

5. Most TA cloning vectors utilize a gene that confers ampicillin resistance and a multiple cloning site within the *lacZα* sequence. This construction allows selection for ampicillin resistance and blue/white colony selection (where white colonies contain a cloned PCR amplicon).

6. To utilize blue-white colony screening by disruption of the *lacZα* in the vector, competent cells must have an appropriate *lacZ* deletion genotype (typically *lacZΔM15*), to allow for *α*—complementation. Many types of appropriate competent cells are commercially available. Follow the manufacturer's instructions provided.

7. Many commercial kits are available to perform small-scale plasmid preparations. Follow the manufacturer's instructions provided.

References

1. Kadonaga JT, Knowles JR (1985) A simple and efficient method for chemical mutagenesis of DNA. Nucleic Acid Res 13:1733–1745

2. Witkin EM (1976) Ultraviolet mutagenesis and inducible DNA repair in *Escherichia coli*. Microbiol Mol Biol Rev 40:869–907

3. Cadwell RC, Joyce GF (1994) Mutagenic PCR. Genome Res 3:S136–S140

4. Biles BD, Connolly BA (2004) Low-fidelity *Pyrococcus furiosus* DNA polymerase mutants useful in error-prone PCR. Nucleic Acid Res 32:e176–e176

5. Minamoto T, Wada T, Shimizu I (2012) A new method for random mutagenesis by error-prone polymerase chain reaction using heavy water. J Biotechnol 157:71–74

Chapter 34

Development and Use of a Novel Random Mutagenesis Method: In Situ Error-Prone PCR (*is*-epPCR)

Weilan Shao, Kesen Ma, Yilin Le, Hongcheng Wang, and Chong Sha

Abstract

Directed evolution methods are increasingly needed to improve gene and protein properties. Error-prone PCR is the most efficient method to introduce random mutations by reducing the fidelity of the DNA polymerase. However, a highly efficient process is required for constructing and screening a diverse mutagenesis library since a large pool of transformants is needed to generate a desired mutant. We developed a method called in situ error-prone PCR (*is*-epPCR) to improve the efficiency of constructing a mutation library for directed evolution. This method offers the following advantages: (1) closed-circular PCR products can be directly transformed into competent *E. coli* cells and easily selected by using an alternative antibiotic; (2) a mutant library can be created and screened by one-step error-prone amplification of a variable DNA region in an expression plasmid; and (3) accumulation of desired mutations in one sequence can be obtained by multiple rounds of *is*-epPCR. *Is*-epPCR offers a novel, convenient, and efficient approach for improving genes and proteins through directed evolution.

Key words Amplification of circular plasmids, Directed evolution, In situ error-prone PCR, Random mutagenesis, Selection marker swapping, Thermostable DNA ligase

1 Introduction

Directed evolution methods are increasingly needed to improve protein properties, including enzyme activity, stability, substrate specificity, and reaction temperatures [1–4]. Success of directed protein evolution hinges on the efficiency of creating mutagenesis libraries and screening the libraries for the desired protein properties [5]. Lutz and Patrick reviewed in vitro mutagenesis approaches developed for directed evolution of enzymes [6]. The most useful method is error-prone mutagenic PCR (epPCR), which introduces random mutations by reducing the fidelity of the DNA polymerase [7]. Early techniques of epPCR are similar to standard PCR cloning of a target sequence, except that the target gene is amplified under error-prone conditions. However, a large pool of

Andrew Reeves (ed.), *In Vitro Mutagenesis: Methods and Protocols*, Methods in Molecular Biology, vol. 1498,
DOI 10.1007/978-1-4939-6472-7_34, © Springer Science+Business Media New York 2017

transformants is needed to find a desired mutant, and therefore, a highly efficient process is required for constructing and screening a diverse mutagenesis library.

Megaprimer PCR of whole plasmid (MegaWHOP) is a method developed for improving the efficiency of mutation library construction by using epPCR. This method amplifies a target gene by epPCR, then uses the epPCR products as the PCR primers (Megaprimers), and the expression vector harboring the wild-type gene as the template to produce full-length plasmid [8]. In these procedures, the newly extended, linear ssDNA (single-strand DNA) cannot serve as a template to achieve exponential amplification, and some of these ssDNA strands can ultimately anneal to form nicked circular double-strand DNA (dsDNA) to be directly transformed into *Escherichia coli*. The template DNA needs to be methylated and digested before the plasmids carrying epPCR products are transformed.

To improve the efficiency of directed evolution *via* random mutagenesis, we developed a novel method designated as "in situ error-prone PCR" (*is*-epPCR) on the bases of the technique that generates circular PCR products by a four-step cycled PCR of "denaturation–annealing–elongation–ligation" [9]. The procedure of *is*-epPCR is schematically illustrated in Fig. 1. In this method, a long complementary primer pair is designated as vector-primers which contains the entire vector sequence but missing the target gene in the expression region, and can be employed for different target genes cloned into the expression vector. The *is*-epPCR products are directly coupled to a selection and screening procedure, facilitating directed evolution of the target gene or protein in a simple process of "*is*-epPCR → transformation → screening," This method possesses the following novelties:

(1) Vector-primers are used to extend the target gene over a circular template under error-prone conditions, followed by DNA ligation of the resultant nick between the 3′-end of the newly extended strand and the 5′-end of the primer; (2) When the vector-primers are designed to contain a second selection marker, the resulting plasmid product will harbor this new marker, and the template plasmid carrying the original marker is eliminated when transformed host cells are cultured under the second selection pressure; (3) The ability to swap the selection marker facilitates multiple rounds of mutagenesis and selection resulting in additional desired (positive) mutations. A second pair of primers containing a different selection marker can be similarly prepared so that multiple rounds of directed evolution can be conveniently performed over a target gene; (4) Besides a generic vector-primer pair to be used for different target genes, *is*-epPCR can be tailored for different applications with proper design of the primer pair.

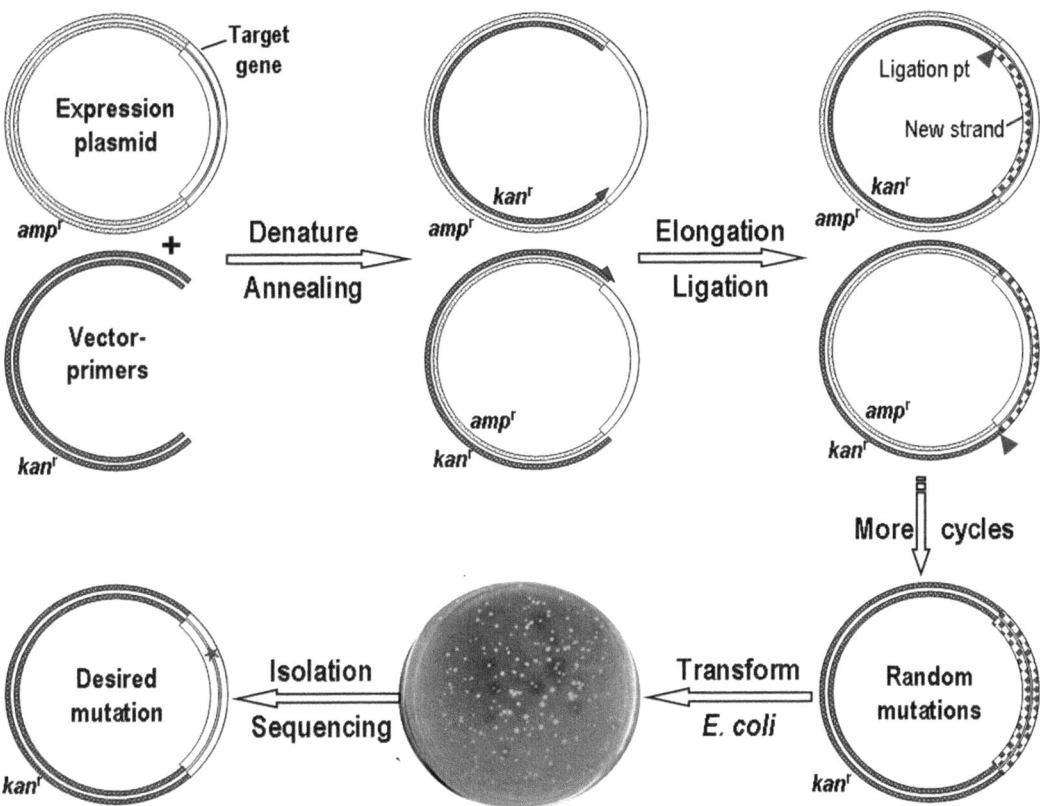

Fig. 1 Schematic procedure of directed evolution by in situ error-prone PCR (*is*-epPCR) coupled to transformation and mutant screening. In the *is*-epPCR reaction system, an expression plasmid and DNA fragment amplified from the vector with a swapped selection marker are used as a template and a primer pair. A directed evolution is performed in a simple process of "*is*-epPCR → transformation → screening": (*1*) Being denatured, the single-strand primer anneals to a template strand, and covers the entire template plasmid sequence minus that of the target sequence. (*2*) The new strand is elongated over the target gene under error-prone conditions, and the nick is ligated between the 3′-end of the newly extended strand and the 5′-end of the primer. (*3*) Carrying mismatched bases in the target gene, the circular PCR products are ready for transforming into the host cells, and transformants are plated and screened on indicator medium plates

Three examples of these potential applications include the random mutations of (a) only one domain of a multi-domain gene, (b) the regulatory region for a reporter gene, and (c) the second replication origin (replicon) of a shuttle vector by transforming the second host.

Using the directed evolution method on the xylanase A gene from *Thermomyces lanuginosus* as an example, we describe in this chapter the procedures to perform *is*-epPCR, and demonstrate how the mutant gene expression plasmids are produced and identified from original gene expression plasmids in a single step.

2   Materials

2.1   Strains, Plasmids, and Media

1. Competent cells: *Escherichia coli* DH10B, BL21, or BL21-CodonPlus (DE3)-RIL is used as the host strain for gene cloning and expression.

2. Vectors (*see* **Note 1**): pHsh-amp and pHsh-kan (GenBank accession FJ571619 and FJ571621), a pair of identical expression vectors comprising different selection markers [10].

3. Expression plasmid: pHsh-ex-*xynA2*, expresses a *T. lanuginosus* gene coding for a secreted xylanase in pHsh-amp [11].

4. Luria–Bertani (LB) medium: LB broth containing (%, w/v) 1 % tryptone, 0.5 % yeast extract, 1.0 % NaCl, pH 7. LB agar: add 2 % agar for making plates.

5. Selective LB media: LB medium supplemented with 100 μg/ml ampicillin (amp) or 50 μg/ml kanamycin (kan) for selecting transformants harboring plasmids with an amp- or kan-resistance gene marker (*amp^r* or *kan^r*), respectively.

6. SOC medium: 2 % (w/v) tryptone, 0.5 % (w/v) yeast extract, 10 mM NaCl, 2.5 mM KCl, 1 ml of filter-sterilized 2 M $MgSO_4$, and 1 ml of filter-sterilized 2 M glucose.

7. Indicator plates: Selective LB media supplemented with 2 % xylan (w/v), a reaction substrate, for screening xylanase activity.

2.2   In Situ epPCR

1. Oligo-primers: G-rev, 5′-pCCTCCATGGGTATATCTCCTT-3′ and G-fwd, 5′-pAAGCTTGAAGGCCGCTTCCGA-3′, a pair of primers designed to amplify the entire pHsh expression vector by regular PCR for preparing the general vector-primer pairs from pHsh-amp and pHsh-kan.

2. High-fidelity DNA polymerase: PrimeSTAR HS DNA polymerase, PrimeSTAR buffer and dNTP mix.

3. *Tma* DNA ligase (20×): DNA ligase from *Thermotoga maritima*, 0.5 μg/μl.

4. TE buffer: 10 mM Tris–HCl buffer, pH 8.0, containing 1 mM EDTA.

5. Stock mix (5×): PCR buffer containing 10 mM $MgCl_2$, 1.25 mM $MnCl_2$, 1 mM dNTPs, and 2.5 mM NAD^+.

6. Zyme mix (20×): 0.5 U/μl *Taq* DNA polymerase, 0.5 μg/μl *Tma* DNA ligase in 10 mM Tris–HCl buffer (pH 8.0) containing 20 % glycerol and 0.01 % NaN_3.

2.3   Kits, Equipment, Services

1. DNA manipulation kits: QIAquick Gel Extraction Kit; QIAGEN Plasmid Mini Kit (Qiagen).

2. Thermal cycler (e.g., PE Applied Biosystems 9700).

3. Electroporator (e.g., GenePulser Xcell, Bio-Rad).

4. Electrophoresis: Mini-PROTEAN II Dual Slab Cell.

5. Ultracentrifuge.

6. Sequencing and primer synthesis services.

3   Methods

3.1   Vector-Primers (See Note 2)

1. Prepare a 50 μl-reaction mix: 10 ng of pHsh-kan as template, 15 pmol each of oligonucleotide primers, 10 μl of PrimeSTAR buffer, 4 μl of dNTP mix, and H_2O to 50 μl.

2. Start the PCR by adding 0.5 μl (1.25 U) of PrimeSTAR HS DNA polymerase after 95 °C for 5 min followed by 30 cycles at 94 °C for 30 s, 54 °C for 30 s and 72 °C for 2 min and end the cycling with a single cycle of 10-min at 72 °C.

3. Separate the PCR fragments in an agarose gel and excise the target band with a clean, sharp scalpel.

4. Extract and purify the DNA from an agarose gel using a Qiagen PCR Purification Kit. Resuspend the purified DNA in 100 μl TE buffer.

5. Precipitate the PCR product by adding 10 μl of 3 M Na-acetate (pH 5.0) and 100 μl of isopropanol, centrifuge and discard supernatant. Wash the pellet 2× by adding 70% ethanol and microcentrifuging. Discard the supernatant.

6. Reconstitute the DNA in TE buffer to a concentration of 0.2 μg/μl and label as Kan^r-vector primers.

7. Prepare Amp^r-vector primers from pHsh-amp (as template) using the same oligonucleotide primers and the same procedure described in **steps 1–5** in this section.

3.2   Is-epPCR Mutagenic Library

1. Prepare an *is*-epPCR mix for a 20 μl reaction in a PCR tube: add 30 ng pHsh-ex-*xynA2* (*see* **Note 3**), 0.2 μg Kan^r-vector primers, 4 μl stock mix, and supplement with distilled water (dH_2O) to 19 μl of total volume.

2. Begin the cycling reaction by adding 1 μl Zyme mix after incubating the *is*-epPCR mixture at 95 °C for 5 min, then perform 15 cycles consisting of 94 °C for 1 min, 72 °C for 2 min, and 60 °C for 2 min and finish cycling with a single cycle at 72 °C for 10 min (*see* **Note 4**).

3. Transfer the PCR tube containing the *is*-epPCR product to an ice bath. In the meantime, thaw competent cells of *E. coli* BL21 on ice.

4. Mix 5 μl of the *is*-epPCR product with 100 μl competent cells (*see* **Note 5**) and transfer the DNA and cells to a prechilled electroporation cuvette (0.2 cm electrode gap). Wipe any ice or water from the sides of the cuvette using a Kimwipes.

5. Place the cuvette into the sample chamber, set the electroporator (e.g., GenePulser) to 2.5 kV, 25 μF, and the pulse controller to 200 Ω. Deliver the pulse by pushing in both charging buttons simultaneously and holding until a short beep is heard.

6. Remove the cuvette from the sample chamber. Immediately add 1 ml SOC medium and transfer the cells to a sterile culture tube using a Pasteur pipette.

7. Incubate the cultures for 60 min at 30 °C in an incubator with moderate shaking (~200 rpm) to allow for selection marker gene expression (*see* **Note 6**). Plate aliquots of the electroporation mixture onto LB indicator plates supplemented with xylan (0.2–0.5 % w/v) and kan (50 μg/ml). Incubate the plates at 30 °C overnight.

8. Transfer the plates to a 42 °C incubator to induce target gene expression for 4–6 h and detect xylanase activity by observing a clear zone surrounding individual colonies (*see* Fig. 2a). Mark the colonies having the largest clear zones, pick each into an 8-ml LB broth containing kan (50 μg/ml) and aerobically incubate overnight at 30 °C.

9. When the OD_{600} reaches 2.5–3.0, take 4 ml of each culture for isolating plasmid DNA using a Qiagen plasmid mini-prep kit and transfer the tube with the remaining culture to a shaking water bath incubator to induce gene expression at 42 °C for 2 h (*see* **Note 7**).

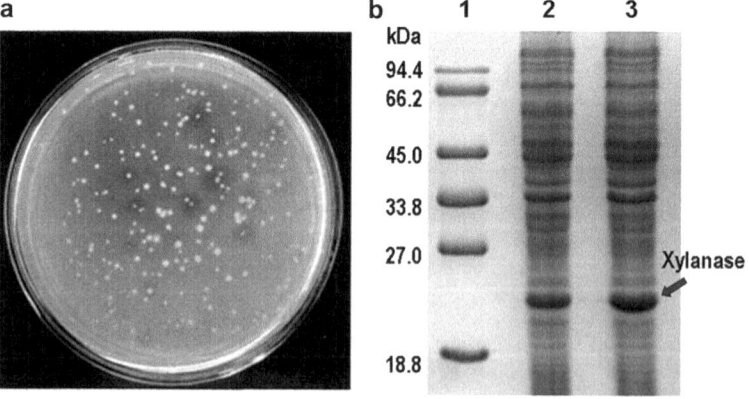

Fig. 2 Directed evolution of *Thermomyces lanuginosus* xylanase by an in situ error-prone PCR over the expression plasmid pHsh-ex-*xynA2*. (**a**) Functional screening of *E. coli* transformants for xylanase activity on an indicator plate. Positive clones were identified by clear zones surrounding xylanase-expressing colonies. (**b**) Comparison of the expression levels from starting gene *xynA2* and mutant gene *xyn-m1*. Lanes: *1*, Protein markers for molecular mass; *2* and *3*, soluble protein of *E. coli* cells harboring plasmids pHsh-ex-*xynA2* [2] and pHsh-ex-*xyn-m1*, respectively [3]

10. Pellet cells at $6000 \times g$ for 5 min and resuspend them in 0.4 ml of 20 mM Tris–HCl buffer (pH 7.2).

11. Disrupt the cells by sonication, centrifuge at $10,000 \times g$ for 30 min at 4 °C.

12. Extract the supernatant for protein and xylanase activity analysis (*see* Fig. 2b).

13. Determine the properties of mutants in comparison with the parent gene *xynA2*, and sequence the mutants to identify the mutation sites. From pHsh-ex-*xynA2*, a desired mutant was produced, and designated as *xyn-m1* in expression plasmid pHsh-ex-*xyn-m1* containing *kan*^{r} [9] (*see* **Note 8**).

3.3 Successive is-epPCR

1. Perform multiple rounds of mutagenesis.

2. Make a new *is*-epPCR mix using 0.2 μg Amp^r-vector primers with 30 ng pHsh-ex-*xyn-m1* as template, 4 μl stock mix and dH$_2$O to 19 μl.

3. Repeat **steps 2–11** in Subheading 3.2 except that the selection medium contains ampicillin-sodium salt at 100 μg/ml.

4. Perform further rounds of *is*-epPCR by swapping the selection markers to achieve additional desired mutations as shown in the schematic protocol (*see* Fig. 3).

5. Facilitate *is*-epPCR over a particular region of a target gene.

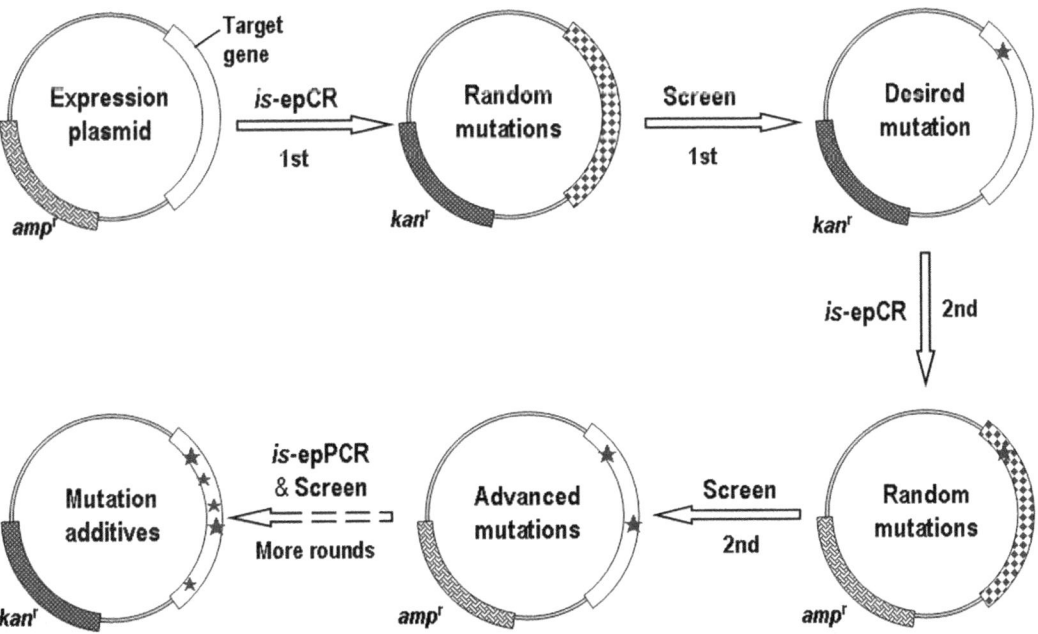

Fig. 3 Protocol for using multiple rounds of "*is*-epPCR–transformation–screening" to obtain successive positive mutations. By using two pairs of vector-primers containing different selection markers, *kan*^{r} or *amp*^{r}, multiple rounds of directed evolution can be conveniently performed over a target gene

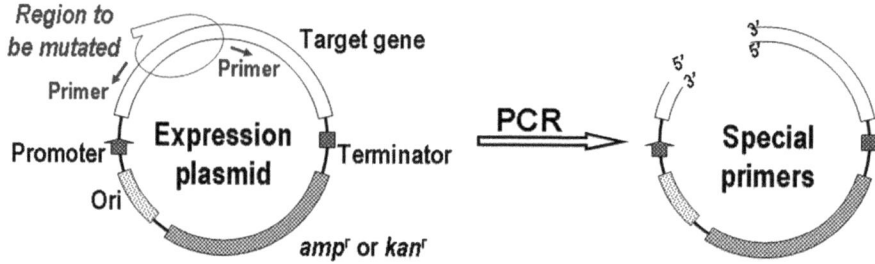

Fig. 4 Schematic illustration of design and amplification of special primer pairs used for *is*-epPCR in a special region of a target gene by a conventional PCR. The 5′-ends of these primers have to be phosphorylated for the ligase reaction in *is*-epPCR. High-fidelity DNA polymerase is employed to amplify these primer pairs with blunt-ends

6. Swap the selection marker from pHsh-amp-*target* to pHsh-kan-*target* by using Kan^r-vector primers, and PrimeSTAR HS DNA polymerase system with pHsh-amp-*target* as target and the addition of *Tma* DNA ligase (*see* **Note 2**).

7. Transform *E. coli* DH10B and select for the *E. coli* DH10B/ pHsh-kan-*target* on LB plates containing kan (50 μg/ml).

8. Isolate pHsh-kan-*target* from *E. coli* cells, design and synthesize a pair of Special oligo-primers flanking the region to be mutated (*see* Fig. 4).

9. Use the special oligonucleotide primers to amplify Kan^r-special and Amp^r-special primers from pHsh-kan-*target* and pHsh-amp-*target* by using the same method as for the preparation of Kan^r-vector and Amp^r-vector primers (*see* Subheading 3.1), respectively.

10. Use Kan^r-special primers and Amp^r-special primers instead of Kan^r-vector primers, Amp^r-vector primers, to perform *is*-epPCR on an expression plasmid containing the original target gene as described in Subheading 3.2.

4   Notes

1. The Hsh (heat shock) system of expression vectors such as pHsh-*amp* and pHsh-*kan* is recommended for *is*-epPCR, which contains an Hsh promoter recognizable by an alternative sigma factor (σ^{32}) of *E. coli*, and the expression of foreign genes is induced by heat shock (a rapid temperature up-shift) [10]. Heat shock induction makes it easy to control gene expression on plates when screening library by protein activity. Besides, the small size of these vectors allows large fragment cloning and efficient transformation.

2. The Kan^r- and Amp^r-vector primers are amplified from expression vectors by using a pair of conventional PCR primers. The

5′-ends of these primers have to be phosphorylated for the ligase reaction in *is*-epPCR. To prevent mutation of the vector, the primer pairs with blunt-ends are amplified by a high-fidelity polymerase such as PrimeSTAR, *Pyrobest* or *Pfu* DNA polymerase. The generic primer pair contains the entire vector sequence but missing the target gene in the cloning site, and can be employed for different target genes cloned into the multi-cloning sites of an expression plasmid vector. Kan[r]- and Amp[r]-vector primers are now commercially available in *is*-epPCR Kit (Shine E Biotech).

3. pHsh-ex-*xynA2* is an expression plasmid ready to be used for an *is*-epPCR experiment [11]. Any target gene can be cloned into pHsh-*amp* and pHsh-*kan* (GenBank accession FJ571619 and FJ571621) at multiple cloning sites by using any standard gene cloning method [10]. Expression T-vectors have been constructed as pHsh-amp-T and pHsh-kan-T for one-step TA cloning and expression without redundant sequence in the expression region (Shine E Biotech).

4. The number of thermal cycles and Mn^{2+} concentration are key factors affecting the mutation frequency [9]. Taking a gene of about 1 kb as an example, we obtained a mutation frequency of about 75 % when performing 15 thermal cycles at 0.25 mM $MnCl_2$. Setting Mn^{2+} concentration at 0.25 mM, we can now change the mutation frequency as desired by the number of thermal cycles. For a target gene larger than 2 kb, we lower the mutation frequency by reducing the number of thermal cycles.

5. A gene expression in pHsh system vectors does not require a particular *E. coli* strain genotype. However, there are differences among *E. coli* strains when they are used as cloning and expression hosts. We use *E. coli* DH10B for construction and preparation of expression plasmids when expression should be limited. But *E. coli* BL21 is a better strain when expression is to be induced, and *E. coli* BL21-CodonPlus (DE3)-RIL will be employed when a gene is not significantly expressed for activity detection.

6. Most genes can be overexpressed in pHsh system vectors. When the expression levels are high, even a low ratio of leak-through expression before induction will overload the bacterial synthesis and processing systems, which would result in the modification of genes or growth suppression of *E. coli* cells. The gene expression in pHsh plasmids is induced by a shift in temperature from ≤30 °C to 40–42 °C. Before inducing gene expression, the higher the cell growth temperature, the more leakthrough gene expression will occur. Therefore, the cell growth temperature should not be higher than 30 °C before the time of induction of gene expression.

7. A rapid rise of temperature is important for an efficient induction of foreign gene expression. It has been found that gene expression levels are about 50% higher when raising the temperature in a water bath than in an gravity incubator. Therefore, we induce gene expression by transferring the test tubes or flasks into a shaking water bath incubator preheated to 42 °C, and continuing cultivation for 2–6 h as desired. Alternatively, when a shaking water bath incubator is not available, we balance the temperature of culture tubes or flasks by manually shaking in a 42 °C water bath for 10–20 min before transferring them into a gravity incubator for further growth and gene expression.

8. Properties of mutants should be confirmed by using purified protein. In this experiment, xynA-m1 is sequenced, and a single mutation site at 180 is changed from GGT to AGT (G180S), resulting in an increase of expression level from 31,967 to 40,963 U/mg of total protein. Both XynA2 and XynA-m1 were purified for a comparison of their specific activities, and no significant difference was found between them, indicating that the mutation largely improved the expression level rather than an enzyme property [9].

References

1. Chirumamilla RR, Muralidhar R, Marchant R, Nigam P (2001) Improving the quality of industrially important enzymes by directed evolution. Mol Cell Biochem 224:159–68

2. Cherry JR, Fidantsef AL (2003) Directed evolution of industrial enzymes: an update. Curr Opin Biotechnol 14:438–43

3. Eijsink VGH, Gaseidnes S, Borchert TV, van den Burg B (2005) Directed evolution of enzyme stability. Biomol Eng 22:21–30

4. Yuan L, Kurek I, English J, Keenan R (2005) Laboratory-directed protein evolution. Microbiol Mol Biol Rev 69:373–392

5. Otten LG, Quax WJ (2005) Directed evolution: selecting today's biocatalysts. Biomol Eng 22:1–9

6. Lutz S, Patrick WM (2004) Novel methods for directed evolution of enzymes: quality, not quantity. Curr Opin Biotechnol 15:291–7

7. Cadwell RC, Joyce GF (1992) Randomization of genes by PCR mutagenesis. PCR Methods Appl 2:28–33

8. Miyazaki K, Takenouchi M (2002) Creating random mutagenesis libraries using megaprimer PCR of whole plasmid. Biotechniques 33:1033–8

9. Le Y, Chen H, Zagursky R, Wu JHD, Shao W (2013) Thermostable DNA ligase-mediated PCR production of circular plasmid (PPCP) and its application in directed evolution via *in situ* error-prone PCR. DNA Res 20:375–382

10. Wu H, Pei J, Jiang Y, Song X, Shao W (2010) pHsh vectors, a novel expression system of *Escherichia coli* for the large-scale production of recombinant enzymes. Biotechnol Lett 32:795–801

11. Le Y, Peng J, Wu H, Sun J, Shao W (2011) A technique for production of soluble protein from a fungal gene encoding an aggregation-prone xylanase in *Escherichia coli*. PLoS One 6(4), e18489

INDEX

Andrew Reeves (ed.), *In Vitro Mutagenesis: Methods and Protocols*, Methods in Molecular Biology, vol. 1498,
DOI 10.1007/978-1-4939-6472-7, © Springer Science+Business Media New York 2017

Printed by Printforce, the Netherlands